S0-BPR-637

Estuarine Research

Academic Press Rapid Manuscript Reproduction

produced by the
Estuarine Research Federation
with support from the
Office of the Coastal Zone Management
National Oceanic and Atmospheric Administration
United States Department of Commerce

Estuarine Research

VOLUME II

Geology and Engineering

Edited by

L. Eugene Cronin

Estuarine Research Federation

Academic Press, Inc. NEW YORK SAN FRANCISCO LONDON 1975

A Subsidiary of Harcourt Brace Jovanovich, Publishers

ACADEMIC PRESS, INC.
111 Fifth Avenue, New York, New York 10003

United Kingdom Edition published by
ACADEMIC PRESS, INC. (LONDON) LTD.
24/28 Oval Road, London NW1

Library of Congress Cataloging in Publication Data

International Estuarine Research Conference, 2d,
Myrtle Beach, S. C., 1973.
Estuarine research.

Papers presented at a conference held by the
Estuarine Research Federation and cosponsored by the
American Society of Limnology and Oceanography and the
Estuarine and Brackish Water Sciences Association.
Bibliography: p.
Includes index.
CONTENTS: v. 1. Chemistry and biology.—
v. 2. Geology and engineering.
1. Estuaries—Congresses. 2. Estuarine ocean-
ography—Congresses. 3. Estuarine biology—Congresses.
I. Cronin, Lewis Eugene, (date) II. Estuarine
Research Federation. III. American Society of
Limnology and Oceanography. IV. Estuarine and
Brackish-water Sciences Association. V. Title.
GC96.I57 1973 551.4'609 75-29370
ISBN 0-12-197502-9 (v. 2)

PRINTED IN THE UNITED STATES OF AMERICA

Contents

Part I
Geology: Coarse Grained Sediment Transport and Accumulation in Estuaries

Part II
Engineering: 1) Use of Vegetation in Coastal Engineering

Preface

These publications are the first of a biennial series planned by the Estuarine Research Federation to present new information and concepts relating to the estuaries of the world. Volumes I and II contain the papers presented in the Second International Estuarine Research Conference, held by the Federation at Myrtle Beach, South Carolina in October of 1973. The Conference was cosponsored by the American Society of Limnology and Oceanography and by the Estuarine and Brackish Water Sciences Association.

There has been a rapid and recent increase in research on estuaries, their components and processes, and their responses to human activities. The increase has followed recognition of the exceptional value of these coastal systems, awareness of the abuse many of them have received, and expanding scientific interest in these complex and highly dynamic bodies of water which link the fresh water and the seas. As the number of persons engaged in estuarine research, and of those who wish to use the product of such research increased, so, too, did the need for improved communications among and from investigators. A small Atlantic Estuarine Research Society was organized in 1947 to provide frequent, informal exchange. In later years, the New England Estuarine Society, the South Atlantic Estuarine Research Society, and the Gulf Estuarine Research Society have emerged to serve their respective regions. All of these have joined to form the Estuarine Research Federation, an umbrella organization for the constituent societies and their 1200 members, with potential for adding additional, interested organizations. The Federation conducts and publishes biennial symposia on "Recent Advances in Estuarine Research," implements estuarine research, and provides assistance on national and international policies and practices related to estuaries.

A valuable symposium on estuaries was held under multiple sponsorship in 1964 at Jekyll Island, Georgia, and produced the classic volume Estuaries edited by George Lauff and published by AAAS. That volume was comprehensive. The Federation held its First International Conference on Long Island in 1971 but publication of papers was not feasible. The Federation recognizes that total coverage is no longer feasible at any one point in time because of the expanding production of new results of research. The Executive Board has therefore decided to select, for each biennial meeting, those topics in which major recent advances have indeed been achieved, design a symposium for their presentation and discussion, and arrange for publication. These are the first products. Volume I contains papers on *Chemistry*, focused on the Cycling of Elements and Estuaries; *Biology*, including sessions on the Dynamics of Food Webs, Nutrient Cycling, Zooplankton, Nekton, and Benthos; and *The Estuarine System*. Volume II provides publications on *Geology*, with collections on Estuaries with Small Tidal Ranges, Intermediate Tidal Ranges, and Large Tidal Ranges, and an additional section on Wide-Mouthed Estuaries. It also includes new materials on *Engineering*, with emphasis on Use of Vegetation in Coastal Engineering and on Estuarine Dredging Problems and Effects. The Third International Conference will be held by the Federation in October of 1975 at Galveston, Texas. The present publications are somewhat delayed in production, but rapid completion of future volumes is a foremost goal and commitment.

We wish to express exceptional appreciation to the conveners, chairman, and contributors, identified elsewhere, for the innovative and dedicated efforts they put into the creation and conduct of the Conference. Dr. Robert J. Reimold of the University of Georgia gave excellent supervision to the preparation and arrangement of all materials for camera-ready copy.

Quite special acknowledgment is given to the Office of Coastal Zone Management of the U.S. National Oceanic and Atmospheric Administration and its Director, Dr. Robert Knecht, for considerately administered financial support which made possible participation by scientists from distant laboratories and the preparation of final materials for publication.

L. Eugene Cronin
Chairman

Austin B. Williams

Jerome Williams

For the Editorial Committee

CONTENTS OF VOLUME I

Part I
Chemistry: Cycling of Elements in Estuaries

Part II
Biology: Dynamics of Food Webs in Estuaries

Part III
The Estuarine System: Estuarine Modeling

Estuarine Research

PART I

GEOLOGY: COARSE GRAINED SEDIMENT TRANSPORT AND ACCUMULATION IN ESTUARIES

Convened By:
Miles O. Hayes
Department of Geology
University of South Carolina
Columbia, South Carolina 29208

MORPHOLOGY OF SAND ACCUMULATION IN ESTUARIES:

AN INTRODUCTION TO THE SYMPOSIUM

Miles O. Hayes[1]

ABSTRACT

The morphology of sand deposits in estuaries is determined by the interaction of a number of process variables, including: (a) tidal range, (b) tidal currents, (c) wave conditions, and (d) storm action. Of these, variations in tidal range have the broadest effect in determining large-scale differences in the morphology of sand accumulation. The papers in this symposium have, therefore, been arranged according to differences in tidal range of the areas discussed, following the classification scheme proposed by Davies (4):

I. **Coarse-grained sediment accumulation in estuaries with small tidal ranges (microtidal estuaries: tidal range (T.R.) = 0 - 2 m).**

Wave action and storm deposition are more important in this class than in any other. Galveston Bay, Texas, is an example of this type of estuary.

II. **Coarse-grained sediment accumulation in estuaries with intermediate tidal ranges (mesotidal: T.R. = 2 - 4 m).**

Tidal deltas and tidal-current-formed sand bodies increase noticeably in this class. The estuaries of New England, South Carolina, and Georgia are prototypes.

III. **Coarse-grained sediment accumulation in estuaries with large tidal ranges (macrotidal: T.R. > 4 m).**

Funnel-shaped, wide-mouthed estuaries that contain linear sand bodies are the most common types occuring in this category. Prototypes are Bristol Bay, Alaska, and the Ord River estuary, Australia.

1. Coastal Research Division, Department of Geology, University of South Carolina, Columbia, South Carolina 29208.

IV. Wide-mouthed estuaries.

This category was created in order to include in the symposium papers covering the large entrances into such major bodies of water as the Baltic Sea and Chesapeake Bay.

Much of the emphasis in these papers has been placed on estuaries in the mesotidal category, principally because these are the ones that have been studied most. Despite the fact that mesotidal estuaries show a wide range in morphological and hydrographic characteristics, the sand shoals affiliated with them are remarkably similar from place to place. For example, flood-tidal deltas usually contain the same major components, including a flood ramp, flood channels, ebb shields, ebb spits, and spillover lobes, regardless of the variations in current and wave conditions under which they occur. Similarly, the ebb-tidal deltas, although they are exposed to great variations in open-ocean-wave intensity, are strikingly consistent in morphology.

INTRODUCTION

At first view, sand deposits occurring in estuaries are extremely complicated. The morphology of these sand bodies is controlled by the interaction of numerous process parameters, including tidal-range conditions, tidal currents, wave conditions, and coastal storms. After several years of studying tidal deltas under different wave and tidal regimes, as well as studying the coastal charts of the world, I have concluded that tidal range has the principal control over the distribution and form of sand deposits affiliated with estuaries. That is, estuaries occurring in areas with small tidal ranges have a suite of sand shoals associated with them that is distinctly different from sand shoals occurring in estuaries with large tidal ranges.

Davies (4) recognized how important tidal range is to coastal morphology, and proposed the following classification of tides:

Microtidal – tidal range 0 - 2 m [2]
Mesotidal – tidal range 2 - 4 m
Macrotidal – tidal range > 4 m

The papers of the symposium have been grouped according to this classification scheme.

The importance of tidal range in controlling coastal geomorphology was first called to my attention by W. Armstrong Price, who feels that coastal-plain shorelines can be defined on the basis of whether they are wave-dominated or tide-dominated. In compiling information on shorelines of the world for the

2. In actuality, Davies' boundaries were 0-6 ft., 6-12 ft., and 12 ft. I have rounded off these numbers to the nearest whole metric unit. On the basis of study of details of coastal morphology on the coast of North America, I feel there is much justification for considering changing the mesotidal boundaries or perhaps splitting the mesotidal class into two categories; however, the boundaries proposed by Davies will be maintained in this paper.

Defense Research Laboratory at the University of Texas, I came to a similar conclusion. Figure 1 shows the relative abundance of different morphological features of coastal-plain shorelines that occur under microtidal, mesotidal, and macrotidal conditions. Note that river deltas and barrier islands are best developed in microtidal regions, whereas offshore linear sand ridges (built by tidal currents), tidal flats, and salt marshes are most abundant under macrotidal conditions. Tidal deltas and tidal inlets are most common in mesotidal areas because of the increasing abundance of tidal inlets on barrier island shorelines under these tidal conditions.

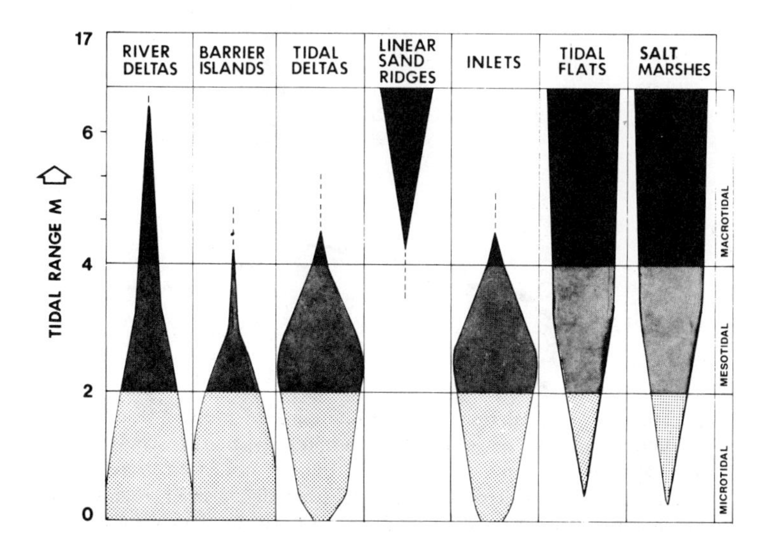

Figure 1. Variation of morphology of coastal-plain shorelines with respect to differences in tidal range. River deltas and barrier islands are most common in microtidal areas, tidal deltas and tidal inlets are most common in mesotidal areas, and linear sand ridges (sand ridges deposited on shallow continental shelf by tidal currents), tidal flats, and salt marshes are most common in macrotidal areas.

The coast of Western Europe (Figs. 2 and 3) clearly demonstrates the relationship of coastal morphology to changes in tidal range. Stunted barrier islands and abundant tidal inlets and tidal deltas occur on the mesotidal shoreline of The Netherlands. Near the mouths of the Elbe and Weser rivers in West Germany, the shoreline becomes more tide-dominated (approaching macrotidal), and the orientation of the shoreline sand bodies changes from a parallel to the coast orientation in the mesotidal area to a perpendicular orientation in the macrotidal area. To the north, on the west coast of northern Denmark and on the shore of the Baltic Sea, where tides are extremely small, the coastal morphology is dominated by long linear barrier islands with few tidal

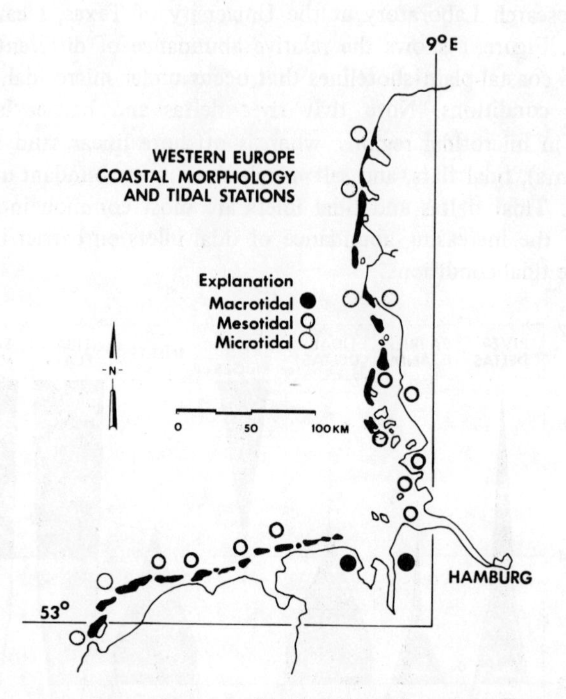

Figure 2. Location of tidal stations on the shoreline of western Europe. Note the relationship of tidal range to coastal morphology. Barrier islands of the coast of The Netherlands grow progressively shorter from west to east until they disappear in the macrotidal area off Germany. Barrier islands and barrier beaches become better developed to the north until they become dominant features on the microtidal coast of Denmark.

inlets. This general correlation between morphological types and tidal range can be seen on the map in Figure 2.

The kind of strict control of coastal geomorphology by tidal range demonstrated by the example of the shoreline of Western Europe also applies to the morphology of sand deposits in estuaries. Therefore, it is possible to construct general depositional models for estuaries occurring in microtidal, mesotidal, and macrotidal areas. These models will now be discussed individually.

MICROTIDAL ESTUARIES

A simplified microtidal estuary model is given in Figure 4. Characteristic sand deposits occurring in microtidal estuaries include: (a) washover fans deposited during storm-surge floods of major storms (Fig. 5); (b) wave-built features such as aligned beaches, recurved spits, cuspate spits, and bay mouth bars (Figs. 6 and

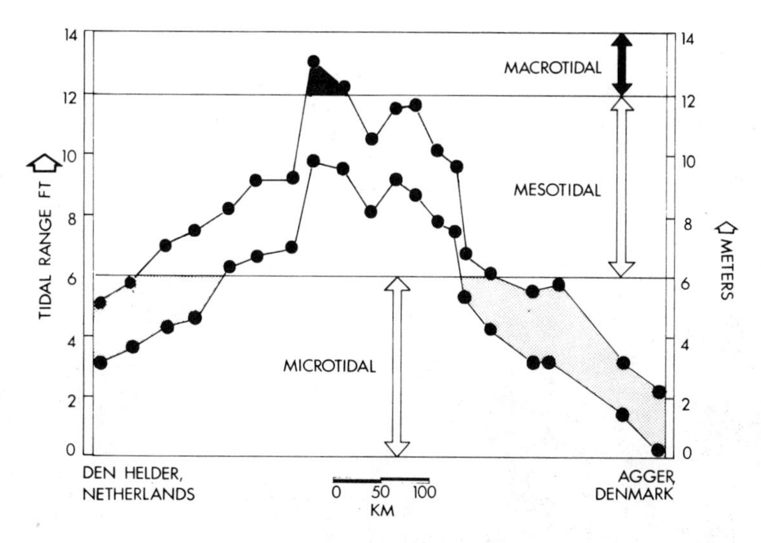

Figure 3. Variation of tidal range along the shoreline of Western Europe (tide stations are located on Figure 2). Lower line represents mean neap tide, and upper line represents mean spring tide at each station.

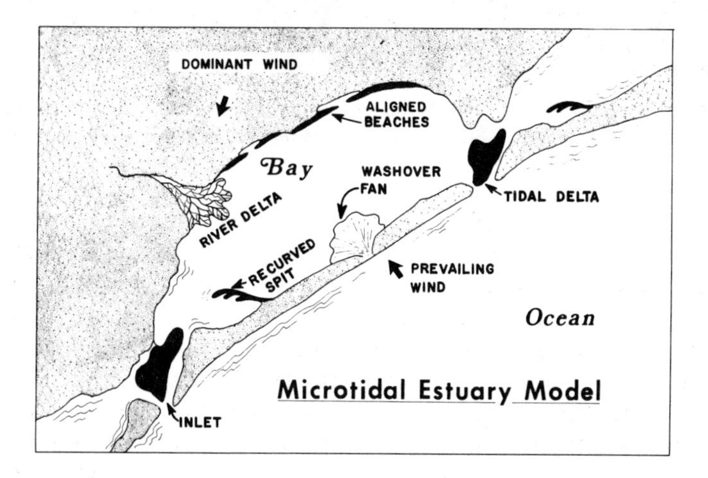

Figure 4. Microtidal estuary model. Sand bodies in this type of estuary are mostly storm- or river-generated (e.g., washover fans; river deltas) or wave-generated (e.g., aligned beaches; recurved spits). Tidal deltas are usually small. Most of the estuaries of the Gulf of Mexico are of this type.

Figure 5. Washover fan deposits building into a microtidal estuary in the Magdalen Islands, Gulf of St. Lawrence. These fans are built by the storm surges of winter northeasterly storms.

7); (c) river deltas; and (d) small flood-tidal deltas. Several of these features are discussed in other papers in this volume. Storm deposits are discussed at length in the paper by McGowan and Scott (8), and the occurrence of cuspate spits on estuarine shorelines is discussed by Rosen (11). A typical cuspate spit on the shoreline of a microtidal estuary is shown in Figure 6.

The processes that dominate in microtidal estuaries are created by wind and wave effects. Wind tides are commonly generated and extensive wind-tidal flats may develop. The wave-formed features include aligned bay beaches, recurved spits, and cuspate spits. Tidal currents generated by the astronomical tide are important only at the inlet mouth. In some instances, however, important flood-tidal delta deposits can develop, as at Destin Inlet, Florida (12).

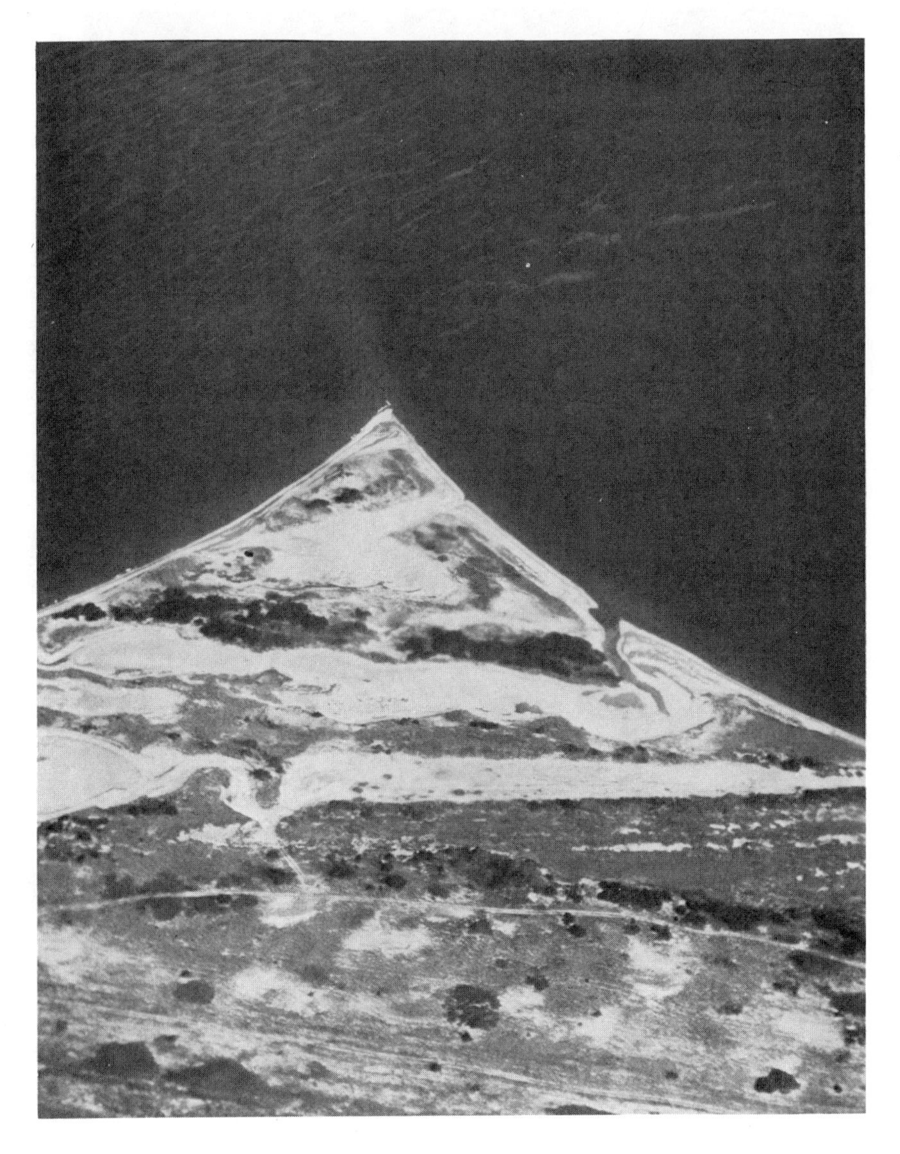

Figure 6. Cuspate spit on a microtidal estuary on Nantucket Island, Massachusetts. This feature is built by waves approaching from opposite directions, i.e., from right and left (11).

Figure 7. Aligned bay-beach and bay-mouth bars on microtidal estuary in northwest Florida. These features, as well as the ones shown in Figures 5 and 6, are typical of microtidal estuaries.

MESOTIDAL ESTUARIES

Most of the papers in ths symposium are devoted to the topic of mesotidal estuaries. Several papers are directed to the topic of the morphology and processes of tidal inlets, whereas others delve into the subject of the morphology, sediments, and bedforms occurring on the affiliated tidal deltas. Figure 8 gives the mesotidal estuary model, and Figure 9 is an aerial photograph of a typical mesotidal estuary. These estuaries differ from the microtidal estuaries in that sediments deposited by tidal currents begin to predominate. The barrier islands themselves are short and stubby, and the tidal deltas are large and conspicuous. Meandering tidal channels occur behind the barriers; point-bar deposits containing bedforms generated by tidal currents usually predominate in these channels. the principal sand deposits in mesotidal estuaries are the tidal deltas. The Coastal Research Group (3) proposed the following terminology: (a) *ebb-tidal delta* - sediment accumulation seaward of a tidal inlet, deposited by ebb-tidal currents, and (b) *flood-tidal delta* - sediment accumulation formed on the landward side of an inlet by flood-tidal currents. Other terminology was proposed by Oertel (9). Engineers often refer to these sand deposits as the interior shoal (flood-tidal delta) and the outer shoal (ebb-tidal delta) (5).

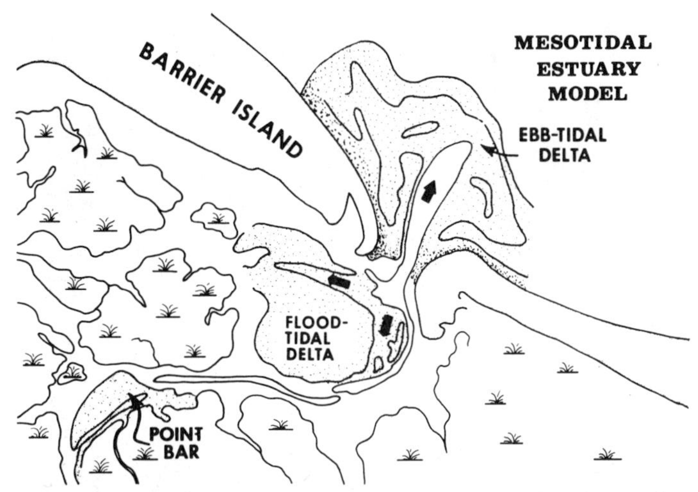

Figure 8. Mesotidal estuary model. Tidal deltas and point-bar deposits are the principal sand bodies occurring in this type of estuary. Examples of this type include most of the estuaries on the shoreline of New England and those of the Wadden Sea, The Netherlands.

Ebb-tidal Deltas

Studies of numerous ebb-tidal deltas on the east coast of the United States, as

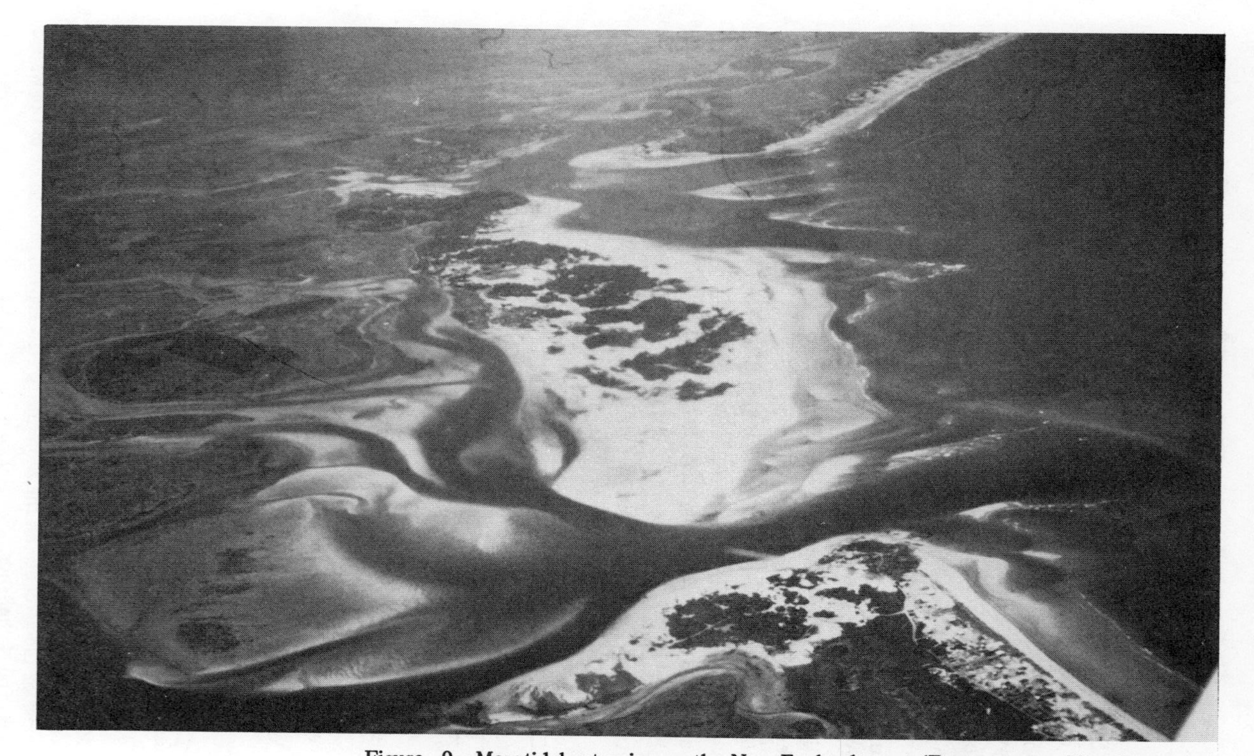

Figure 9. Mesotidal estuaries on the New England coast (Essex estuary, Massahcusetts, in foreground, and Parker River estuary, Massachusetts, in background). Note large tidal deltas. Photograph taken at low tide (T.R. = 3m). (Photograph by Jon C. Boothroyd)

well as reconnaissance studies on the coasts of Alaska, Baja California, and the Gulf of St. Lawrence, by the Coastal Research Division, University of South Carolina, indicate that the morphology of these sand bodies is similar from place to place. Therefore, we are able to construct what we consider to be a standard model of ebb-tidal delta morphology (Figs. 10 and 11).

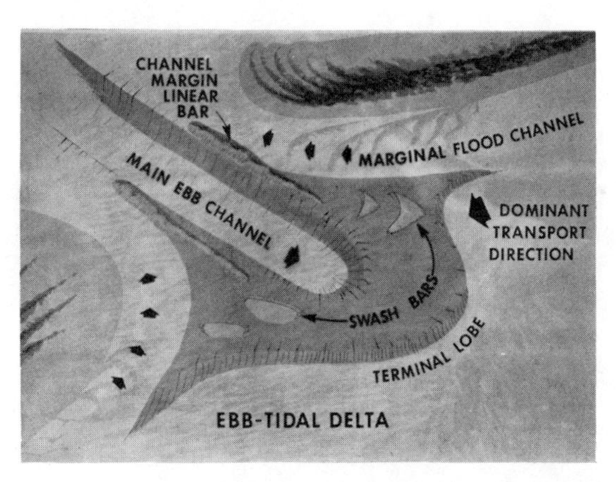

Fiture 10. Model of morphology of ebb-tidal deltas. Arrows indicate dominant direction of tidal currents. Model is based on studies of tidal inlets on the coasts of New England, South Carolina, Alaska, and Baja California by the Coastal Research Division, University of South Carolina.

The components of a typical ebb-tidal delta include a *main ebb channel,* which usually shows a slight-to-strong dominance of ebb-tidal currents over flood-tidal currents. The main ebb channel is flanked on either side by *channel-margin linear bars,* which are levee-like deposits built by the interaction of ebb- and flood-tidal currents with wave-generated currents. At the end of the main channel is a relatively steep, seaward-sloping lobe of sand called the *terminal lobe.* Broad Sheets of sand, called *swash platforms,* flank both sides of the main channel. Usually, isolated *swash bars,* built by the swash action of waves (7) occur on the swash platforms. Marginal tidal channels dominated by flood-tidal currents, called *marginal flood channels,* usually occur between a swash platform and the adjacent updrift and downdrift beaches.

The overall morphology of the ebb-tidal delta is a function of the interaction of tidal currents and waves. Of prime importance is a phenomenon called *time-velocity asymmetry of tidal currents.* As described by Postma (10) this means that maximum ebb- and flood-tidal currents do not occur at mid-tide. Of crucial importance in most of the estuaries our group has studied is the fact that maximum ebb currents occur late in the tidal cycle, near low water. This means that at low water, as the tide turns, strong currents are still flowing seaward out

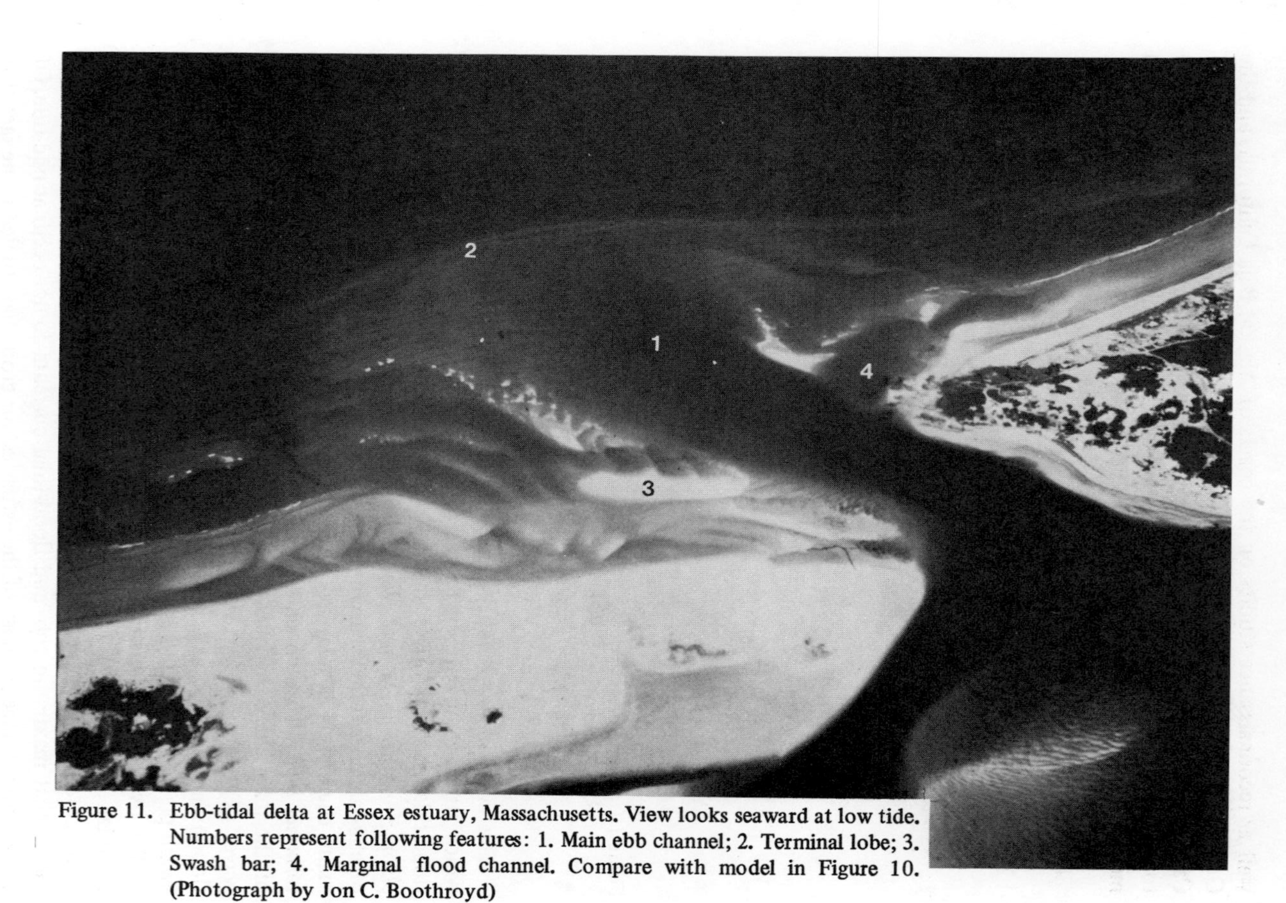

Figure 11. Ebb-tidal delta at Essex estuary, Massachusetts. View looks seaward at low tide. Numbers represent following features: 1. Main ebb channel; 2. Terminal lobe; 3. Swash bar; 4. Marginal flood channel. Compare with model in Figure 10. (Photograph by Jon C. Boothroyd)

of the main ebb channel. As the water level rises, the flood currents seek the paths of least resistance around the margin of the delta. This creates the horizontal segregation of flood and ebb currents in the tidal channels that is depicted by the model (Fig. 8). For details on the ebb-tidal delta at Chatham Harbor estuary, Massachusetts, **see** Hine (6).

Flood-tidal Deltas

The morphology and bedforms of flood-tidal deltas have been described in several papers by our research group (2, 3, 6). An example of a flood-tidal delta is shown in Figure 12. On the basis of these studies, we can describe a typical model for flood-tidal delta morphology (Fig. 13). We feel this model is even more universal than the ebb-tidal delta model. A typical flood-tidal delta consists of:

(a) *Flood ramp* – Seaward-facing slope on the sand body over which the main force of the flood current is directed. Always covered with flood-oriented sand waves.[3]

(b) *Flood channels* – Channels dominated by flood currents that bifurcate off the flood ramp.

(c) *Ebb shields* – Topographically high rims or margins around the tidal delta that protect portions of it from modification by ebb currents.

(d) *Ebb spits* – Spits formed by ebb-tidal currents.

(e) *Spillover lobes* – Following the definition of Ball (1), lobate bodies of sediment formed by unidirectional currents.

Flood-tidal deltas are remarkably similar in morphology from place to place. Compare the one at Fire Island Inlet, New York (Fig. 14), with the model shown in Figure 13.

Point Bars

Many mesotidal estuaries have meandering tidal creeks with well-developed point-bar deposits. The example of the Parker River estuary, Massachusetts is given in Figure 15. The basic morphology of tidal point bars is similar to that of fluvial point bars; however, bedform orientations are more complex because of the presence of reversing tidal curents (Fig. 15).

3. As defined by Coastal Research Group (3): **Sand wave** - asymmetrical bedform with spacing (or wave length) 6m. **Megaripple** - asymmetrical bedform with spacing (or wave length) 60 cm 6 m. **Ripple** - asymmetrical bedform with spacing (or wave length) 60 cm.

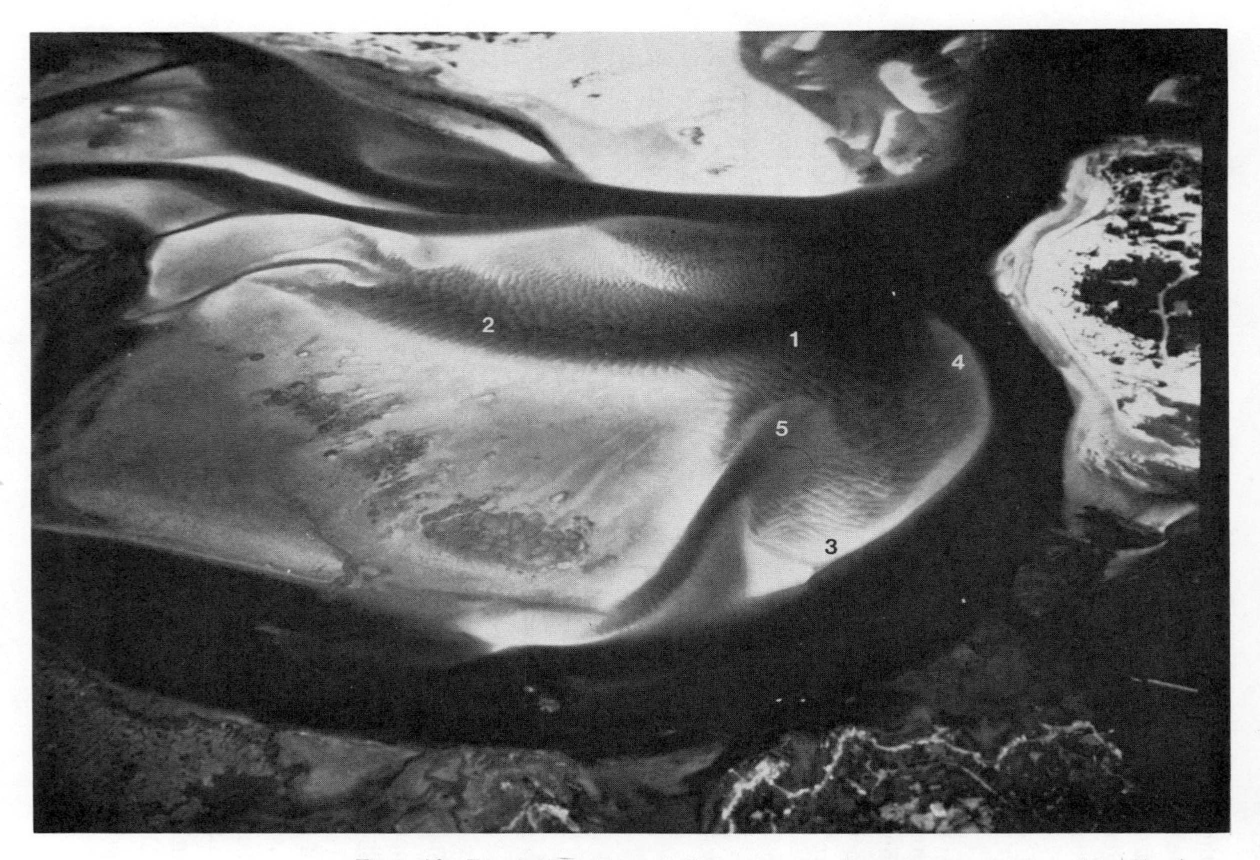

Figure 12. Flood-tidal delta at Essex estuary, Massachusetts, at low tide. Numbers represent following features: 1. Flood ramp; 2. Flood channel; 3. Ebb shield; 4. Ebb spit; 5. Spillover lobe. (Photograph by Jon C. Boothroyd)

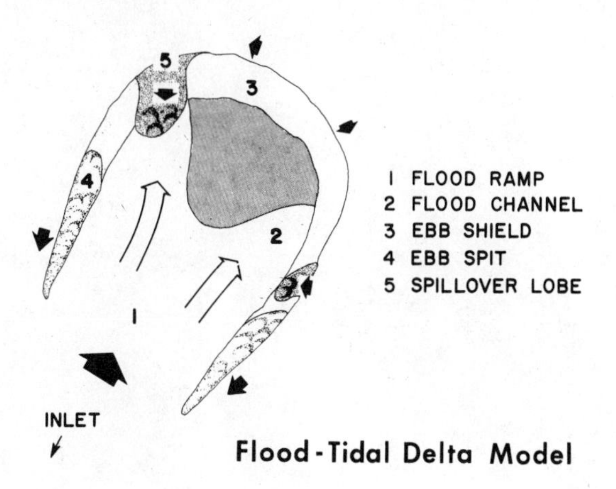

Figure 13. Model of morphology of flood-tidal deltas. Arrows indicate dominant direction of tidal currents. Based on studies of tidal inlets on the coasts of New England, South Carolina, Alaska, and Baja California by the Coastal Research Division, University of South Carolina.

MACROTIDAL ESTUARIES

This group is the least studied of the three principal classes. The papers by Wright and Coleman, Dalrymple, Knight and Middleton, and Green in this volume describe the sediments of the Ord River estuary, Australia, the Bay of Fundy, and the Tay estuary, Scotland. Other examples of macrotidal estuaries include Bristol Bay, Alaska (Fig. 16), the Gulf of Cambay, and the upper reaches of the Persian Gulf and the Bay of Benegal. The most prominent feature of this type of estuary is the overwhelming dominance of tidal currents. Such estuaries are usually broadmouthed and funnel-shaped. Sand deposition is normally concentrated in the center of the estuary, away from the shore, which is usually dominated by broad, muddy tidal flats. The sand bodies are long linear features oriented parallel with the tidal currents. A morphological model is given in Figure 17. There are few real field data on this type of estuary; the papers in this symposium are a welcome addition to the literature.

Figure 14. Vertical areial view of flood-tidal delta at Fire Island Inlet, New York. The inlet is located ot the left. Numbers represent the following features: 1. Flood ramp; 2. Spillover lobe; 3. Ebb spit; 4. Ebb shield. Bedforms visible in photograph are mostly flood-oriented sand waves.

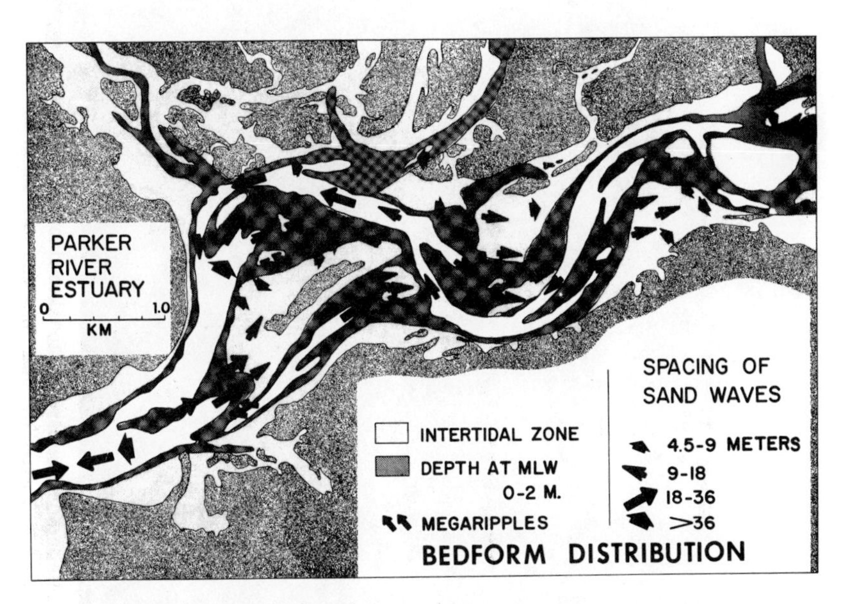

Figure 15. Morphology and bedform distribution and orientations in Parker River estuary, Massachusetts. Based on data compiled by Boothroyd and Hubbard (2).

Figure 16. Bristol Bay, Alaska, a typical macrotidal estuary. Photo taken at low tide. Note large sand body in foreground and extensive tidal flats and salt marsh in background.

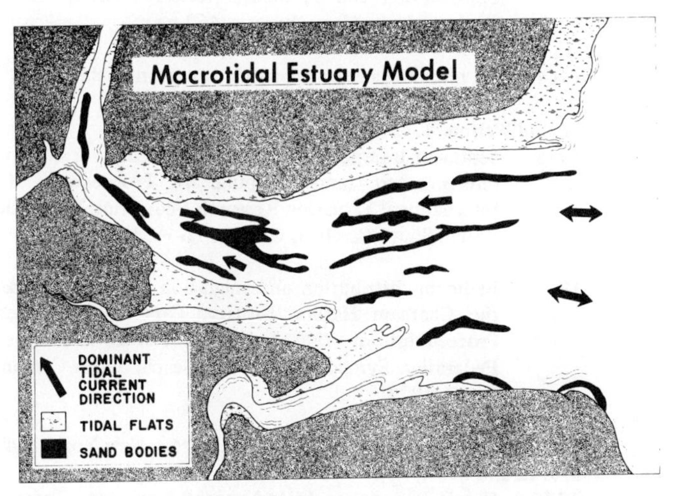

Figure 17. Macrotidal estuary model. The principal sand bodies are linear sand bars located in central portions of the estuary that are built by tidal currents.

CONCLUSION

Sand deposition in estuaries is a function of the interaction of a number of dynamic processes. The symposium papers have been organized so as to emphasize the importance of tidal range, and in this paper three basic models of estuarine sedimentation are proposed:

(a) *Microtidal* model in which waves and wind tides dominate as the major processes;

(b) *Mesotidal* model in which tidal delta deposits predominate; and

(c) *Macrotidal* model which is dominated by tidal-current deposition.

Details on sedimentation patterns and processes that occur in each type of estuary are presented in the papers that follow.

REFERENCES

1. Ball, M. M.
1967 Carbonate sand bodies of Florida and the Bahamas. **Jour. Sed. Pet.,** 37: 556-591.

2. Boothroyd, J. C. and Hubbard, D. K.
1975 Genesis of bedforms in mesotidal estuaries. In **Proceedings of the 2nd International Estuarine Research Federation Symposium,** Myrtle Beach, South Carolina, October 16-18, 1973.

3. Coastal Research Group
1969 Coastal Environments: NE Mass. and N.H. (ed. Hayes, M. O.), Contribution No. 1, Coastal Research Group, Department of Geology, Univ. Massachusetts, 462 p.

4. Davies, J. L.
1964 A morphogenetic approach to world shorelines. **Zeits fur Geomorph.,** 8: 127-142.

5. Dean, R. G. and Walton, T. C.
1975 Sediment transport processes in the vicinity of inlets with special reference to sand trapping. In **Proceedings of the 2nd International Estuarine Research Federation Symposium,** Myrtle Beach, South Carolina, October 16-18, 1973.

6. Hine, A. C., III
1975 Bedform distribution and migration patterns on tidal deltas in the Chatham Harbor Estuary, Cape Cod, Massachusetts. In **Proceedings of the 2nd International Estuarine Research Federation Symposium,** Myrtle Beach, South Carolina, October 16-18, 1973.

7. King, C. A. M.
1972 **Beaches and coasts.** St. Martin's Press, New York, 2nd ed., 570 p.

8. McGowan, J. H. and Scott, A. S.
1975 Hurricanes as geological agents on the Texas Coast. In **Proceedings of the 2nd International Estuarine Research Federation Symposium,** Myrtle Beach, South Carolina, October 16-18, 1973.

9. Oertel, G. F.
1972 Sediment transport of estuary entrance shoals and the formation of swash platforms. **Jour. Sed. Pet.,** 42: 857-863.

10. Postma, H.
1967 Sediment transport and sedimentation in the estuarine environment. In **Estuaries,** (ed. Lauff, G. A.), AAAS, Washington, D.C., 158-179.

11. Rosen, P. S.
1975 Origin and processes of cuspate spit shorelines. In **Proceedings of the 2nd International Estuarine Research Federation Symposium,** Myrtle Beach, South Carolina, October 16-18, 1973.

12. Wright, L. D. and Sonu, C. J.
1975 Processes of sediment transport and tidal delta development in a stratified tidal inlet. In **Proceedings of the 2nd International Estuarine Research Federation Symposium,** Myrtle Beach, South Carolina, October 16-18, 1973.

HURRICANES AS GEOLOGIC AGENTS ON THE TEXAS COAST[1]

J. H. McGowen[2] and A. J. Scott[3]

INTRODUCTION

Sand and gravel distribution and the characteristics of coarse-grained facies in coastal regions are influenced by a variety of physical processes. Tidal currents and wind waves operating under non-storm conditions play a significant or dominant role in the deposition of coarse-grained sediments in mesotidal and macrotidal areas where astronomical tidal ranges are in excess of six feet. These processes are also important in coastal areas with tidal ranges of less than six feet. However, storm-associated processes are relatively more important in the emplacement and modification of coarse-grained sediments along microtidal coasts, especially in relatively sheltered environments such as bays and lagoons.

The Texas coast is a microtidal area according to the classification of Davies (5). Astronomical tidal ranges are extremely low, being between 1.5 and 2 feet. Under non-storm conditions, large waves are inhibited from forming in shallow Texas lagoons by the presence of barrier islands.

The importance of hurricanes and tropical stors as geologic agents has been recognized by many workers. Personnel from the Department of Geological Sciences and the Bureau of Economic Geology at The University of Texas at

1. Publication authorized by the Director, Bureau of Economic Geology, The University of Texas at Austin, Austin, Texas 78712.

2. Bureau of Economic Geology, The University of Texas at Austin, Austin, Texas 78712.

3. Department of Geological Sciences, The University of Texas at Austin, Austin, Texas 78712.

Austin have been actively engaged in the study of the geologic effects of hurrianes since 1961. Hurricane Carla struck the Texas coast in September of that year. The geologic effects of this storm were impressive and launched a revival of neocatastrophism.

Hayes (9) summarized the geologic effects of Hurricanes Carla (1961) and Cindy (1963), and first recognized the relative importance of various processes during different phases of a storm and the areas affected by each. Andrews (1 and 2) described the facies of a large wash-over fan at the northern end of St. Joseph Island. He documented the fact that storm-related processes deposit large volumes of coarse-grained sediments in Texas lagoons. Scott et al. (19) reported on the effects of Hurricane Beulah that struck the south Texas coast in 1967. This storm breached the Texas barrier islands in several places and contributed sediment to many wash-over fans. The torrential rains that accompanied Hurricane Beulah flooded extensive areas of south Texas. Run-off from these rains transported large volumes of sediment and deposited them in the Texas bays. Scott et al. (19) reported that over a quarter of a million cubic yards of sand were deposited on Gum Hollow Delta. McGowen (12) detailed the depositional history and processes affecting Gum Hollow Delta and emphasized the importance of hurricane-associated run-off. In 1970 McGowen and others also described the effects of Hurricane Celia. This storm inflicted severe wind damage on structures in the Corpus Christi area in August 1970.

The Bureau of Economic Geology of The University of Texas at Austin initiated an extensive mapping program and environmental geologic study of the Texas Coastal Zone in 1969. The significance of storms was recognized, and some maps of areas flooded by Hurricanes Carla and Beulah have been published (6 and 7), while others will be included in subsequent parts of the Bureau's Environmental Coastal Atlas.

The preceding studies and subsequent experiences with Hurricanes Fern (1971) and Delia (1973) have contributed to the present paper which considers significant factors influencing the depositional effects of hurricanes and variations in storm-related deposits. Many geographic places names and physiographic features are referred to in this paper because of its regional emphasis (Fig. 1).

GEOLOGIC CHARACTERISTICS OF HURRICANES

Hurricanes vary in intensity, in track patterns, and in behavior upon crossing land, but several processes and effects are common to them all. Hurricanes are accompanied by heavy seas, high water, and wind. The relative importance of different processes and the areas affected vary with different phases of the storm as discussed by Hayes (9) and McGowen et al. (14). These differences are summarized in Figure 2.

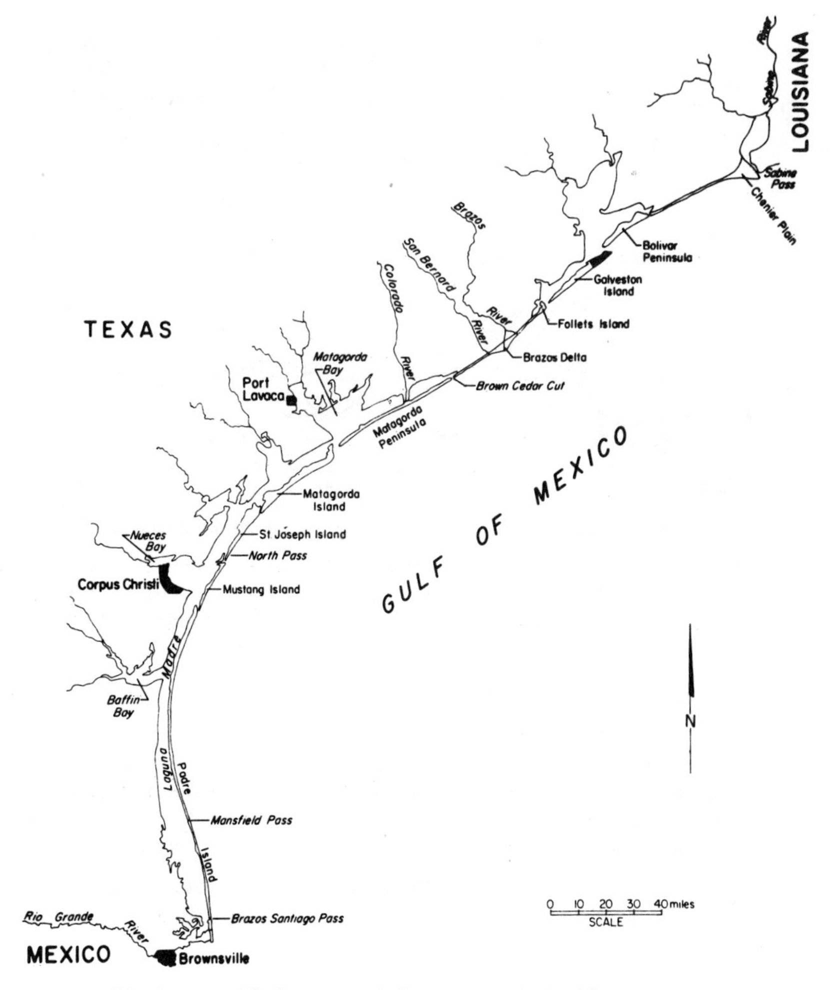

Figure 1. Physiogeographic features and place names used in this paper.

Storm approach (Fig. 2, B) is marked by rising tides and increased wind velocities. Generally, the longer a storm lingers in the Gulf, the larger the bulge of water, or surge, it pushes ashore. Storm tides are commonly higher in bays than on Gulf beaches. Hurricane Carla created 22-foot tides in Lavaca Bay at Port Lavaca (21). Large waves result from the combination of large fetch and high winds. As the storm approaches, waves first attack Gulf-facing barrier beaches. Erosion resulting from these waves and the storm-surge flood may produce breaches or channels through barrier islands and peninsulas.

As the eye of the storm passes over the shore (Fig. 2, C), the pattern of

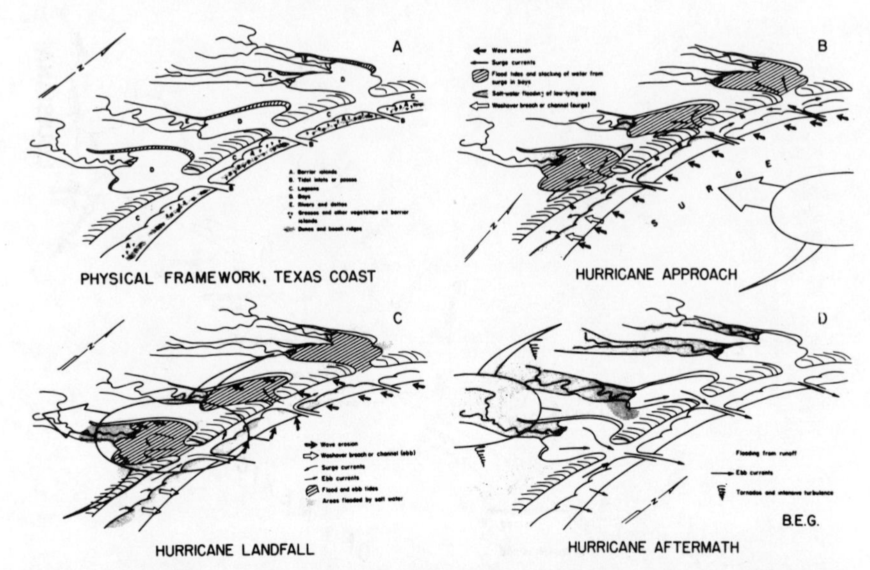

Figure 2. Schematic model of hurricane effects on the Texas coastline, after McGowen et al. (14). (A) Physical features that characterize the Texas coast. (B) Effects of approaching hurricanes. (C) Effect of hurricanes upon impact with the coast. (D) Aftermath effects of hurricanes.

current and wave attack shifts in compliance with counterclockwise winds of the hurricane. Water and sediment move out of the bays into the Gulf through passes and breaches south of the eye, while still being pushed shoreward north of the storm. Highest-intensity winds occur in the vicinity of the landfall area and may cause extensive local damage.

Moving inland (Fig. 2, D), the storm becomes more diffuse, weaker, and commonly produces numerous tornadoes. Heavy rains that normally accompany hurricanes produce large-scale run-off, inundating low-lying areas along stream courses and prograding bay-head deltas.

STORM VARIATIONS

There are significant variations among the storms that strike the Texas coast. These variations, plus the nature of the coastal segment where the storm makes landfall, determine both the geologic effects and the severity of damage to man-made structures.

The tracks of major hurricanes that have made landfall along the Texas coast during the past 12 years are shown in Figure 3. Important differences in these storm paths include: (a) approach rate and duration of the storm in the Gulf of Mexico, (b) inland path of the storm, and (c) geologic nature of the area of landfall.

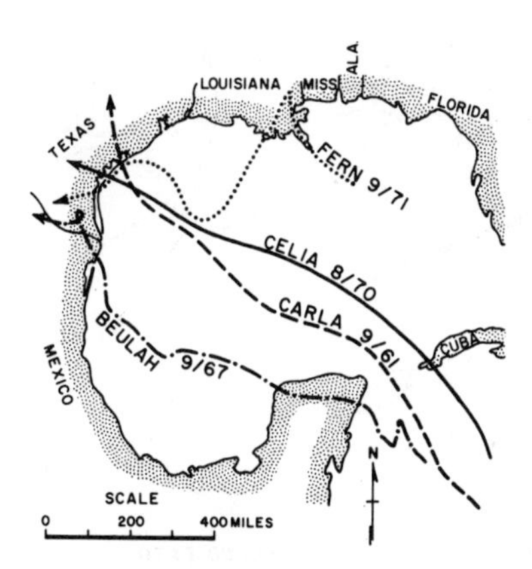

Figure 3. Tracks of major hurricanes striking Texas coast, 1961-1971.

Variations in approach rates (Fig. 4, A) greatly influence the geologic effect of hurricanes. Hurricane Carla was the largest hurricane to strike the Texas coast during the past two decades. It crossed the Gulf of Mexico in eight days, making landfall at Pass Cavallo. This slow movement toward the central Texas coast resulted in an extreme storm-surge flood.

Track 1 (Fig. 4, A) is typical of many storms that strike the Texas coast. Beulah and similar storms enter the Gulf of Mexico by crossing the Yucatan Peninsula. These storms lose forward momentum and intensity as they move over land, but regain both upon reentering the Gulf. Such storms acquire tremendous volumes of water and are accompanied by torrential rains as they drift slowly landward.

Hurricane Celia (Fig. 3) was the last major storm to strike the Texas coast. Celia became a hurricane August 1, 1970, and made landfall near Corpus Christi on August 3. As the storm neared the coast its forward motion accelerated dramatically. Upon making landfall, the eye of the storm contracted by about 40 percent, increasing wind velocity considerably. This was due to the fact that an accelerated approach rate (track 2, Fig. 4, A) maximized wind effects but greatly reduced storm-surge height. Rainfall associated with Hurricane Celia was minimal.

Hurricane Fern and Hurricane Delia are represented by track 3 (Fig. 4, A). These relatively weak storms formed in the Gulf of Mexico and drifted slowly ashore accompanied by heavy rains.

Differences in the tracks of storms after landfall (Fig. 4, B) greatly influence the extent of flooding and deltaic deposition. Hurricane Beulah stalled soon after landfall, reversed its course, and then slowly moved up the Rio Grande (track 1, Fig. 4, B). This route resulted in extremely heavy rains associated with the storm. The coastal area was flooded and continued to receive run-off waters

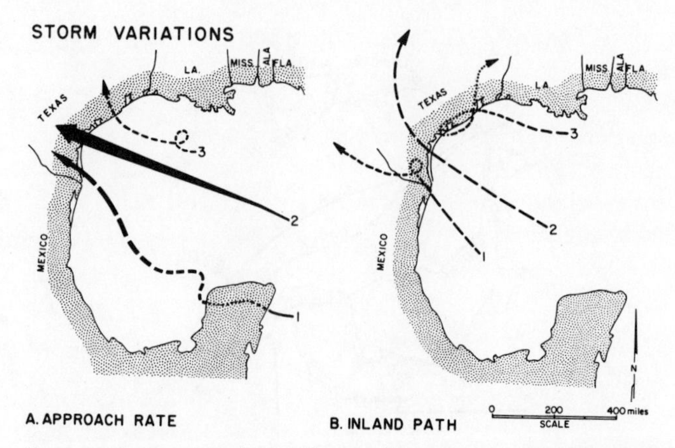

Figure 4. Storm variations. (A) Approach rate. (1) Loss of momentum and intensity while crossing Yucatan, regaining strength after entering Gulf of Mexico; (2) Acceleration of approach rate and intensity; (3) Formation in Gulf of Mexico and slow landward drift. (B) Inland path. (1) Stall and reversal of route near coast, slow movement inland along single drainage system; (2) Arcuate northwestward inland route crossing several major drainage systems; and (3) Lateral drift along coast resulting from interference of southeastward-moving cold front.

as the storm was concentrated along a single drainage system.

Path 2 (Fig. 4, B) represents the arcuate inland route of Hurricane Carla. Heavy rains accompanied this storm. However, run-off was channeled through a succession of drainage systems, as the storm moved northward across the coastal plain.

Path 3 (Fig. 4, B) is typical of the slow lateral drift of weak storms such as Fern and Delia. Heavy rains result from the stalling of such storms as they collide with southward-moving cold fronts.

FACTORS DETERMINING DISTRIBUTION OF STORM DEPOSITS

The ability of hurricanes to erode barrier islands and transport sediment through breaches to bays and lagoons depends upon the density of vegetation, the degree of fore-island dune development, and the width of barrier islands. Hurricanes most frequently breach barrier islands and other shoreline features that are poorly vegetated and lack well-developed, fore-island dunes. Sand is readily transported across narrow barrier islands, whereas most of the sand accumulates on broad barriers within a short distance of the bayward terminus of the storm channel. Geologic setting dictates whether a barrier island, peninsula, or strandplain will develop in a particular area. It also determines caliber and composition of materials making up the sedimentary body, and the rate of compactional subsidence (13). Climate and sediment availability are largely responsible for the variation in physiographic units associated with Texas shoreline features. Physiographic features of the various Gulf shoreline segments control, to a large degree, the ability of a storm to erode and deposit coarse-grained sediment.

Climate

Mean annual precipitation along the Texas coast ranges from 55 inches in east Texas to 26 inches at the Mexican border (Fig. 5, A). Mean annual temperature increases from 70°F in the east to 74°F at the Mexican border (Fig. 5, B). Cooler temperature and higher precipitation in the upper coastal zone are responsible for dense vegetation cover. Sparse vegetation is associated with the warmer temperature and less precipitation of the lower coastal zone. Pass Cavallo is the approximate boundary between well-developed, fore-island dunes to the west and south, and their absence or poor development to the northeast.

The northernmost major blowouts in dunes occur on Matagorda Island, immediately west of Pass Cavallo. Blowouts increase to the southwest in the direction of decreasing precipitation and increasing temperature. These breaks in dune vegetation weaken the fore-island dune chain and facilitate breaching by hurricanes. Where vegetation over is dense on barriers and sand-rich peninsulas storm breaching is greatly reduced.

Geologic Setting

Development of various types of shorelines (standplain, peninsula, barrier island) is controlled to a large degree by previous geologic events (13, 15). For example, standplains are situated along the seaward margin of erosional headlands which are underlain by Pleistocene or Holocene deltaic systems (Fig. 6).

Peninsulas flank deltaic headlands and are extended across drowned Pleistocene valleys and divides by spit accretion (Fig. 6). Sediment sources for peninsulas are: (a) flanking erosional headlands, (b) inner shelf and shoreface, and (c) rivers that discharge directly into the Gulf of Mexico.

Most Texas barrier islands are situated on Pleistocene drainage divides (Fig. 6). Major tidal passes that separate barriers and peninsulas overlie Pleistocene valleys. McGowen and Garner (15) have shown that the initial sediment source for barrier islands was sand-rich Pleistocene units that formed the divides. With the development of peninsulas, the longshore sediment-dispersal system was completed and sediment derived from distant sources became important to shoreline maintenance.

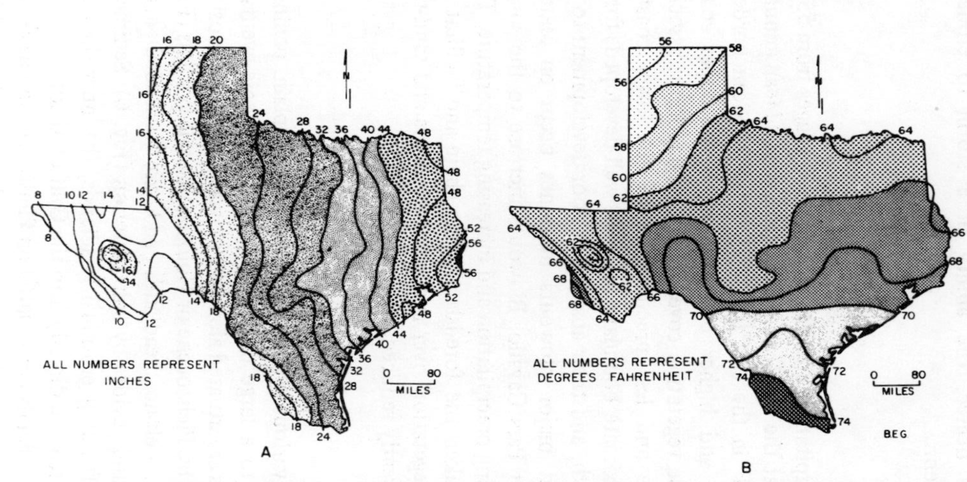

Figure 5. Regional climatic data of Texas, after Carr (4). (A) Mean annual precipitation. (B) Mean annual temperature.

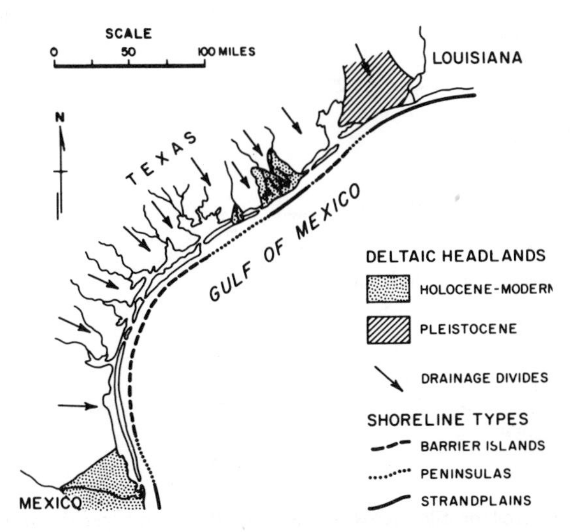

Figure 6. Distribution of modern Gulf shoreline types and major deltaic headlands.

Strandplains

Strandplains are narrow, low, sediment bodies situated upon the distal end of Pleistocene and Holocene deltas (Fig. 6). Three strandplain areas are developed on the Texas coast. These are the shoreline segments between (a) the Chenier Plain and Bolivar Peninsula, (b) Follets Island and Brown Cedar Cut, and (c) on the seaward edge of the Rio Grande delta.

The two northern strandplains are characterized by a mixture of terrigenous sand, shell, and rock fragments, derived primarily from the erosion of the associated headlands. Physiographic units of these strandplains include: (a) forebeach, (b) erosional escarpment, and (c) shell apron or ramp (Fig. 7, A). Strandplains are backed by salt and brackish marshes with attendant lakes, and tidal creeks. The narrow forebeach is a veneer of terrigenous sand, shell, and rock fragments overlying older delta-plain deposits. A one- to four-foot erosional escarpment marks the boundary between the forebeach and shell ramp. Fore-beach sediments are dominantly sand and contrast with the coarser, storm-deposited shell ramp.

Shell ramps along the northern strandplains are similar with elevations of approximately 5 to 7 feet above MSL. Breaking waves associated with tropical storms and hurricanes that have storm-surge floods in the 5-foot range construct the shell ramps. The waves spill across the shell ramp and deposit coarse shell and rock fragments up to 2 feet in diameter. The shell ramp is literally a long transverse bar with an irregular 1- to 2-foot avalanche face along the landward margin. This avalanche face forms a sharp boundary between the shell ramp and

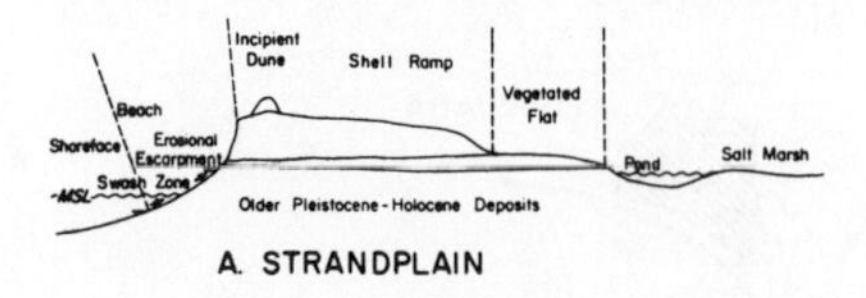

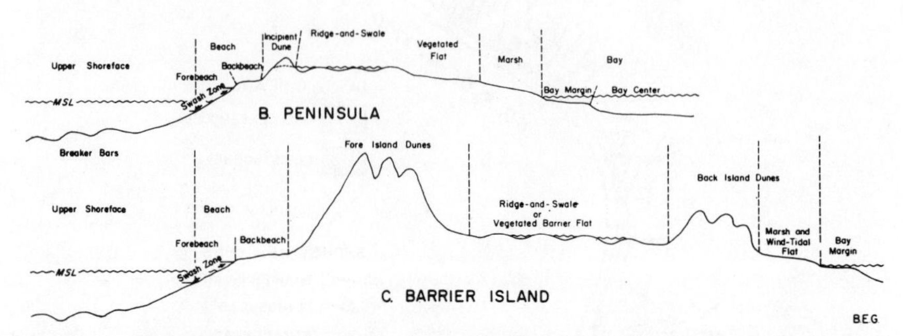

Figure 7. Generalized profiles across Texas Gulf shoreline types showing relationship of geomorphic features, from McGowen (13). (A) Strandplain. (B) Peninsula. (C) Barrier Island.

marsh. In cross section, shell ramps are trapezoid, being thickest at their seaward edge.

The Rio Grande delta is undergoing erosion and compactional subsidence, and a strandplain is developed along the seaward terminus of the deltaic plain between the river mouth and Brazos Santiago Pass (Fig. 6). Most of the delta plain is now a wind-tidal flat, a broad flat surface characterized by local ponds and clay dunes (Fig. 8). As this shoreline segment erodes, wash-over sand, dunes, and beaches are superposed upon delta plain and wind-tidal flat deposits.

Although vegetated fore-island dunes, some with heights of 30 feet, occur along this shoreline segment, several wash-over areas have formed. Breaks in the dunes range from about 260 feet to 1.3 miles in width, and wash-over sands are almost continuous from the Rio Grande to Brazos Santiago Pass (Fig. 8).

During tropical storms and hurricanes, sediment eroded from shoreface, beach, and dune areas is deposited upon wind-tidal flat, clay dune, and bay facies. Subaerial wash-over deposits are redistributed, at least in part, by aeolian processes. Wash-over sands deposited in South Bay are subjected to reworking by waves and burrowing organisms. Wash-over deposits in the south Texas strandplain area are thin (a few inches to about one foot), widespread, and consist mostly of terrigenous sand. These deposits are "sheet-like," and extend from 1400 feet to over a mile inland. In contrast, wash-overs along the upper Texas coast are sediment prisms with maximum dimension parallel to shoreline and intermediate axis perpendicular to shoreline.

Storm-surge flood is the mechanism that emplaces wash-over deposits in each

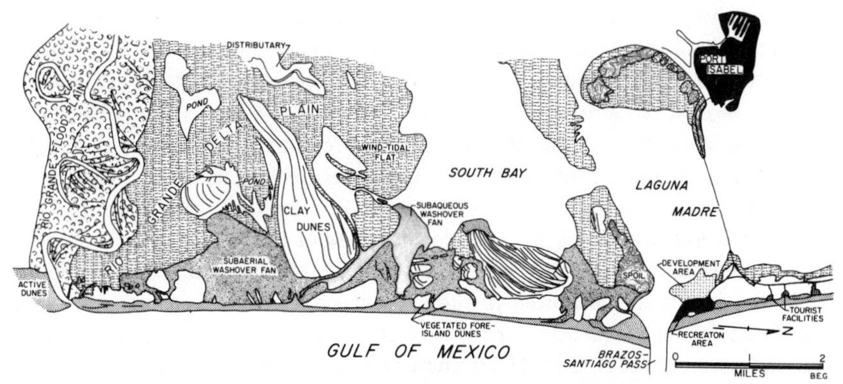

Figure 8. Strandplain developed on erosional Rio Grande deltaic headland, from McGowen (13). Strandplain deposits consist of terrigenous sand with minor amounts of shell, that have accumulated as beaches, dunes, and wash-over deposits. The Rio Grande delta plain is characterized by abandoned distributaries, ponds, and broad interdistributary areas which are now occupied by wind-tidal flats and clay dunes. The Rio Grande meanders to the Gulf of Mexico, and its flood plain is characterized by oxbow lakes and meander scrolls. Fluvial sand, near its mouth, is about 26 feet thick. Strandplain sand, near one of the storm channels, is 8 feet thick.

of the areas mentioned; the kind and intensity of process are similar in all areas. Factors that determined the geometry of the wash-over deposits are (a) fore-island dune development, (b) density of vegetation, and (c) sediment availability.

Peninsulas

Peninsulas are narrow, long sand-shell bodies that extend in a down-current direction from headlands. There are three peninsulas on the Texas coast: Bolivar Peninsula, Matagorda Peninsula, and South Padre Island (Fig. 1 and 6). Physiographic expression and facies characteristics of these areas are a function of climate, vegetation cover, and sediment availability and caliber.

Bolivar Peninsula is densely vegetated and consists chiefly of fine terrigenous sand. It is characterized by well-developed ridge-and-swale topography, and, with the exception of two prominent lobate features along its bay margin, shows no evidence of sediment being transported across the peninsula by tropical storms or hurricanes. Maximum elevations along the seaward edge of Bolivar Peninsula are about 10 feet above MSL, and several storm-surge floods have over-washed the peninsula. Dense vegetation cover has prevented scouring of channels and development of wash-over fans.

Matagorda Peninsula (Figs. 1 and 6) is about 51 miles long and is a continuous sediment body for about 48 miles. The eastern 3 miles of the peninsula are separated from the western segment by Brown Cedar Cut, a tidal pass. This pass came into existence about 1929 as a result of breaching by a hurricane (11, 17). Greens Bayou, a similar breach near the southwestern end of Matagorda Peninsula, is open only during and shortly after hurricanes. Flood deltas that extend some distance into Matagorda Bay are associated with these two passes.

Hurricanes greatly modify the appearance of these two areas, both of which are eroded and the channels are widened to about 0.65 mile and 1.1 miles for Brown Cedar Cut and Greens Bayou, respectively. Much of the sediment scoured from the passes is transported to the bay, where it accumulates as a wash-over fan. Passes are subsequently narrowed or closed by spit accretion. Greens Bayou was last severely scoured by Hurricane Carla; it is now closed. Brown Cedar Cut was scoured in 1929, and has remained open since that time, except for a period of about three years, from 1964 to 1967 (11). It has remained open since Hurricane Beulah (September 1967). Flood deltas at Brown Cedar Cut and Greens Bayou are hurricane wash-over features that are subsequently modified by astronomical and wind tides.

Matagorda Peninsula (Fig. 7, B) averages from 5 to 7 feet above MSL. Dunes are poorly developed along the peninsula. Continuous low dunes averaging 8 to 12 feet above MSL occur from the mouth of the Colorado River eastward for about 8 miles, and from Greens Bayou westward to within a mile or so of Pass Cavallo. Storm channels are common along the peninsula, and beaches that seal off their seaward ends are a maximum of 3 to 4 feet above MSL. Spring high tides and high water resulting from distant storms often over-wash the beach in the area of storm channels. Much of Matagorda Peninsula is over-washed by 5- to 7-foot, storm-surge floods. Areas of continuous dunes with heights greater than about 15 feet above MSL are not over-washed.

In addition to deposition associated with Brown Cedar Cut and Greens Bayou, major storms, such as Hurricane Carla lay down two types of wash-over deposits. These are shell ramps and wash-over fans (Fig. 9).

Shell ramps, where dunes are absent, are topographically the highest part of Matagorda Peninsula. Maximum thickness of ramps is about 4 feet. They are widest in the vicinity of Brown Cedar Cut, ranging from 180 to 2,180 feet.

Wash-over fans are associated with long narrow scour channels that cut through the peninsula. Where sediment is ultimately deposited depends upon the magnitude of the storm-surge flood. Weak storms reactivate storm channels, depositing much of the sediment within the channel, and sometimes form a narrow lunate bar or wash-over fan on the back side of the peninsula. Large storms such as Carla, with 10 to 11 feet of storm surge in the Matagorda area, cut the peninsula into numerous, small islands separated by channels scoured to widths up to 1,700 feet (Fig. 9). The shoreline was cut back as much as 800 feet. Sediment scoured from the shelf, shoreface, beach, and channel areas was

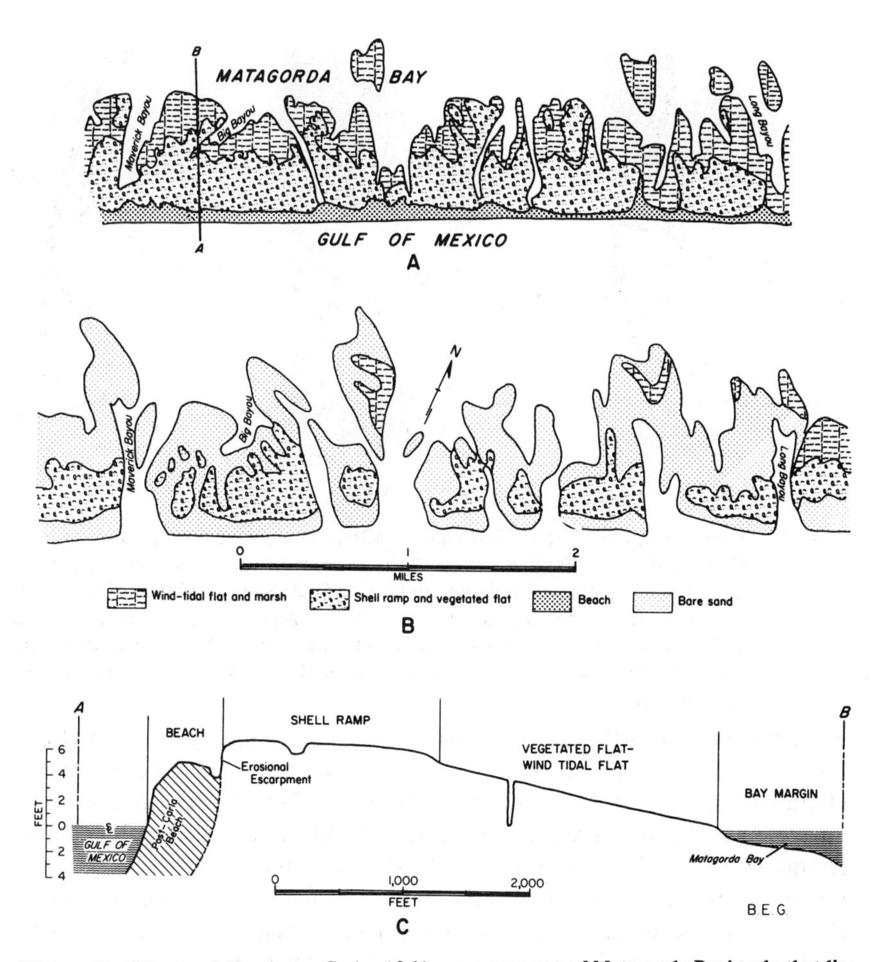

Figure 9. Effects of Hurricane Carla, 1961, on a segment of Matagorda Peninsula that lies between 1.5 to 6.0 miles west of the Colorado River, from McGowen (13). (A) Matagorda Peninsula as it appeared in 1957. (B) Matagorda Peninsula shortly after the passage of Hurricane Carla. This shoreline segment was eroded as much as 800 feet. (C) Profile across Matagorda Peninsula (May 1971). Parts of the shoreline had accreted 500 feet (300 feet landward of its position prior to Carla).

transported across the peninsula and deposited within Matagorda Bay as much as 0.5 mile beyond the bay margin. Shepard and Wanless (20) term these features wash-over deltas.

In south Texas, the strandplain developed on the distal part of the Rio Grande delta grades northward into South Padre Island (Fig. 6). South Padre Island is a peninsula which was tied to a deltaic headland (Fig. 10). South Padre Island is

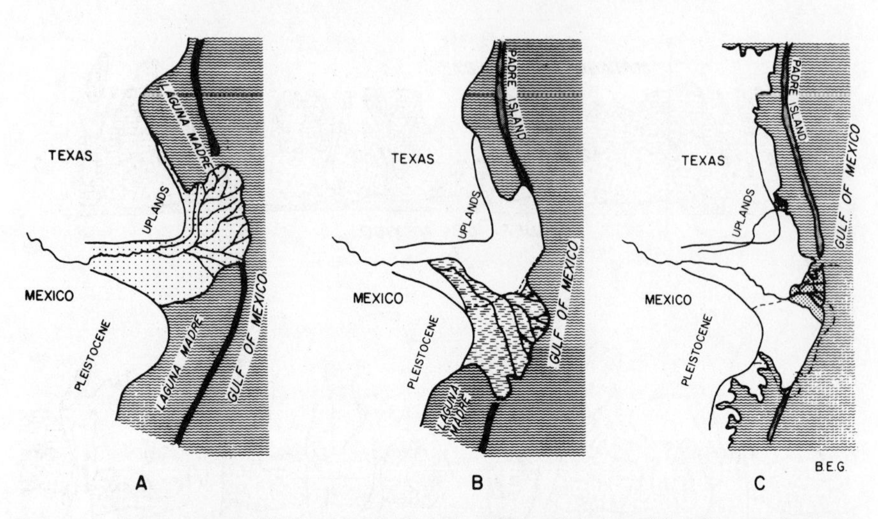

Figure 10. Development of Recent Rio Grande deltaic plain, modified from Lohse and Cook, 1958. (A) Early Recent, deltaic platform prograded seaward of present coastline. Southern peninsula tied to headland. (B) Southward shift of deltation; delta at position "A" subsides and is partially transgressed; northern peninsula is tied to headland. (C) Modern Rio Grande delta. Older deltaic lobes subsiding and being reworked; northern peninsula severed from headland by developnent of Brazos Santiago Pass.

characterized by shell beaches, sparse vegetation, and poorly developed fore-island dunes. Its morphology is a function of wind and storm activity. Hurricane-breaching by storms of the magnitude of Carla and Beulah is relatively unobstructed. South Padre Island shoreline was eroded as much as 300 feet by Hurricane Carla. Deposition occurred after the storm, but at present the shoreline has only accreted seaward about 150 feet from its storm position. Flow across the island is virtually unconfined during hurricanes, with scour channels rare and wash-over deposits poorly defined. Active dunes range in height from 5 to 25 feet above MSL, but present little resistance to flow, once a storm breach is made. Width of breaches ranges from about 0.2 to 1.0 mile.

South Padre Island is characterized by wind-tidal flats that range in width from 0.3 mile at the south to 3.5 miles near the Port Mansfield jetties. Because of the wide wind-tidal flat, sand is rarely transported by storms across the flats into Laguna Madre. Sand deposited on wind-tidal flats by storms becomes dry and is transported to the northwest by winds. This mechanism is an important contributor to the filling of Laguna Madre.

Barrier Islands

Barrier islands are elongate sand bodies that are separated from the mainland by bays or lagoons, and from each other by tidal passes. Five barrier islands are developed along the Texas coast (Fig. 6). From north to south, these islands are Galveston Island, Matagorda Island, St. Joseph Island, Mustang Island, and Padre Island (Fig. 1). A generalized profile typical of Padre Island and Mustang Island is shown in Figure 7, C.

Galveston Island is a relatively wide sand body with numerous ridges and swales (Fig. 11). Average height above sea level is about 5 feet with maximum height of poorly developed fore-island dunes being about 15 feet above MSL. Average annual rainfall is 42 to 48 inches per year. Consequently, the island is densely vegetated and hurricanes erode only the beaches and dunes, and do not affect vegetated areas of the island.

Matagorda Island is similar to Galveston Island. It is broad, has well-defined, ridge-and-swale topography, and a more or less continuous fore-island dune field from Pass Cavallo to Cedar Bayou (22). Average height above sea level is about 5 feet, with fore-island dunes averaging about 10 feet above MSL some peaks rising to 25 feet. Modern wash-over deposits are not found on Matagorda Island, but because of a decrease in average annual precipitation along this coastal segment (36 to 38 inches per year) and because of the presence of many blowouts, hurricane breaching may occur in the near future.

St. Joseph Island lies immediately southwest of Matagorda Island and also displays prominent accretionary grain. Average annual precipitation is from 34 to 36 inches. Vegetation is less dense than on Galveston Island, and blowouts are more frequent. Average elevation of St. Joseph Island is slightly more than 5 feet above MSL. Well-developed, fore-island dunes average about 15 feet above MSL with some 35-foot dunes. Active wash-overs occur at the extreme northeastern and southwestern ends of the island. The northernmost wash-over was studied by Andrews (1, 2), and the southern wash-over area by Nordquist (16). The area studied by Andrews (Fig. 12) is a large lobate feature termed a wash-over fan. This fan developed in the area of a former tidal inlet. Large, inactive wash-over fans also occur on Bolivar Peninsula and Matagorda Island. The St. Joseph fan is the only large active fan on the Texas coast.

North Pass (Fig. 1), the wash-over studied by Nordquist (16) and Scott et al. (19) was formed by a major hurricane in 1919. Nordquist estimated that approximately 9.3 million cubic yards of sediment have accumulated along the bayward terminus of North Pass as a consequence of hurricane activity, beginning with the 1919 hurricane and continuing through 1971.

Mustang Island (Fig. 1), the next island to the south, is broad, has an average elevation of 7 feet, and does not display ridge-and-swale topography. Fore-island dunes are well developed with average dune height of 15 feet and maximum dune height about 50 feet above MSL. Average annual precipitation on Mustang Island is 30 to 34 inches. Vegetation cover is less dense than on islands to the northeast, and blowouts, hurricane breaches, and wash-overs are more frequent.

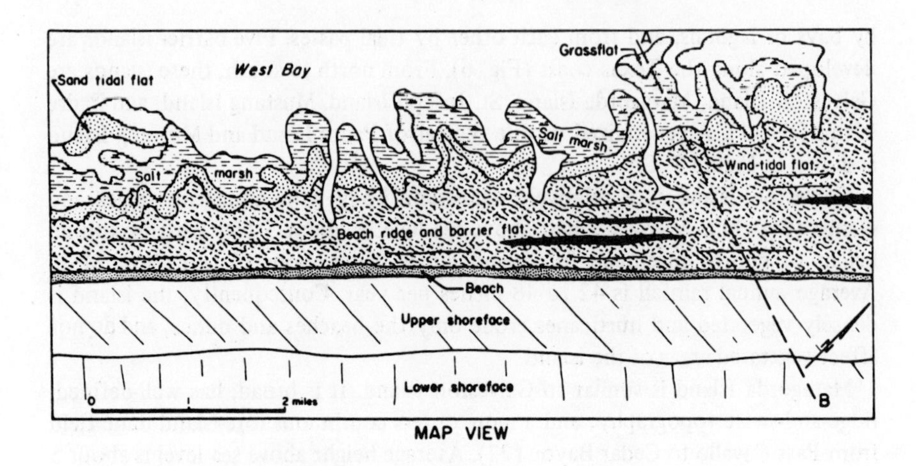

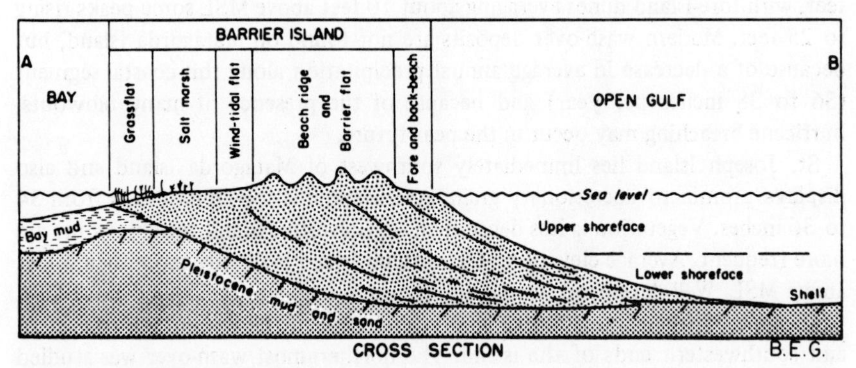

Figure 11. Modern barrier island environments and facies, Galveston Island, from Fisher et al. (6), cross section after Bernard et al. (3).

Back-island dunes occur on Mustang Island, but are restricted to the southern third of the island.

Two factors influence the occurrence of wash-over deposits on southern Mustang Island. First, rainfall decreases to the south and this is accompanied by a decrease in vegetation. Consequently, fore-island dunes are prone to blowouts. Secondly, a major tidal pass existed in this area until man's intervention during the early 1900s. Hurricanes have a tendency to breach barrier segments that are adjacent to, and on the up-current side of, tidal inlets such as North Pass on St. Joseph Island and southern Mustang Island. Evolution of the southern end of Mustang Island is summarized in Figure 13.

Padre Island is the southernmost barrier island and is district from central and upper coast barriers. Rainfall along this coastal segment ranges from about 32 inches per year at the north end to about 28 inches at the south end. As a consequence of the decrease in rainfall and increase in temperature, vegetation

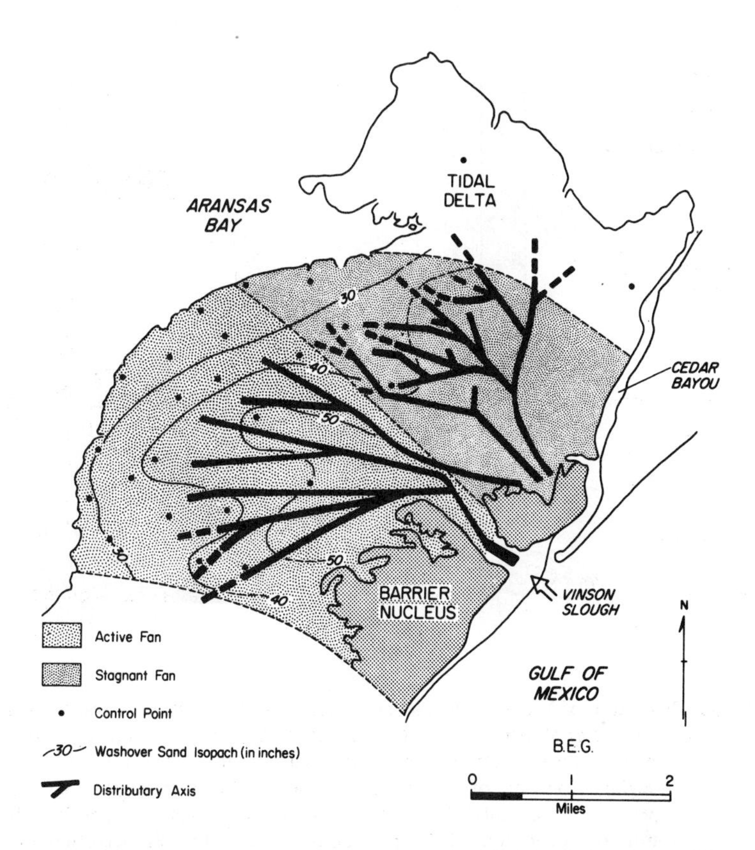

Figure 12. Hurricane wash-over fan, northern St. Joseph Island, Texas, modified from Andrews, (2).

cover is less dense here than along the central and upper Texas coast. Fore-island dunes (Fig. 7, C) are, for the most part, well developed along Padre Island southward almost to Mansfield Pass. Dune heights average about 15 feet, with maximum height being about 50 feet above MSL. In the vicinity of Mansfield Pass (Fig. 1), though, fore-island dunes are poorly developed, and hurricane storm-surge floods are virtually unimpeded.

Northern Padre Island beaches are generally broad and consist predominantly of terrigenous sand. Opposite Baffin Bay, Padre Island beaches become very shelly. As shell content and size increase, width of forebeach decreases, while berm height and elevation of back beach increase. Height of back beach increases to about 7 feet above MSL where shell is coarsest and most abundant. These exceptionally high back-beach areas provide some protection to fore-island dunes during storms. Blowouts and active back-island dunes are common along this coastal segment. However, storm channels breach the dune ridges along the

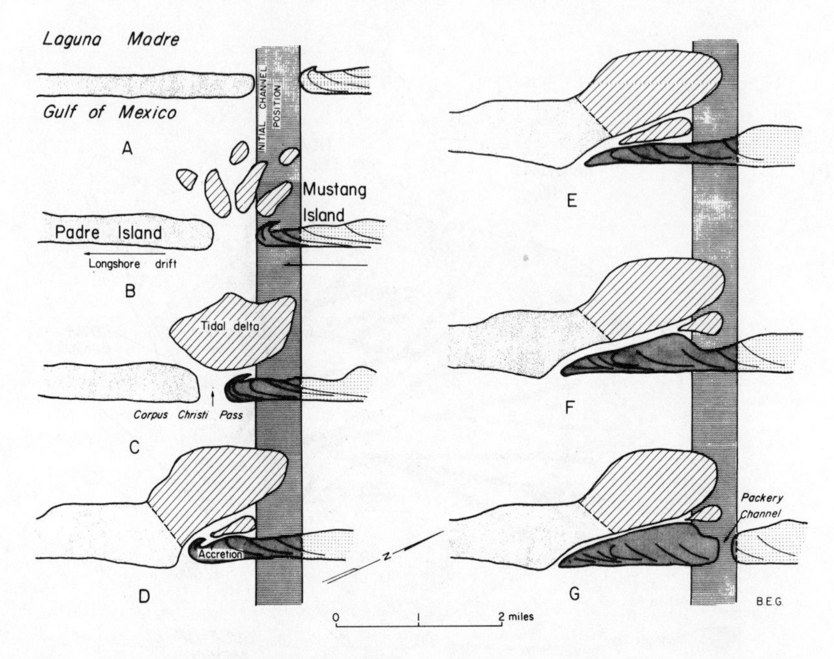

Figure 13. Evolution of Packery Channel area: northern Padre Island – southern Mustang Island, after McGowen (13). Corpus Christi Pass, a tidal inlet, closed in 1929 as a consequence of altering circulation pattern in northern Laguna Madre by dredging. Location of Corpus Christi Pass is shown in Figures A-G. Stage A represents initial inlet location shortly after stilstand. Stages B-F illustrate southward migration of Corpus Christi Pass with spit accretion on South Mustang Island and erosion of North Padre Island. Ebb tidal delta develops and becomes attached to North Padre Island at stage D. Corpus Christi Pass lengthened, became inefficient and southern tip of Mustang Island was breached by storm channel, Packery Channel, at stage G.

central Padre Island segment (Fig. 14). In this area overlapping wash-over fans form wash-over aprons. Aeolian processes redistribute the wash-over deposits forming broad wind-tidal flats that have completely filled parts of Laguna Madre.

Mainland Shorelines

Bay shorelines are also affected by hurricanes. Large waves associated with storm surge erode mainland shorelines. Cliffed shorelines have retreated as much as 100 feet during the passage of a single storm. Several streams discharge their sediment load directly into bays and estuaries. Some hurricanes and tropical storms are accompanied by excessive rainfall. Sometimes rainfall associated with

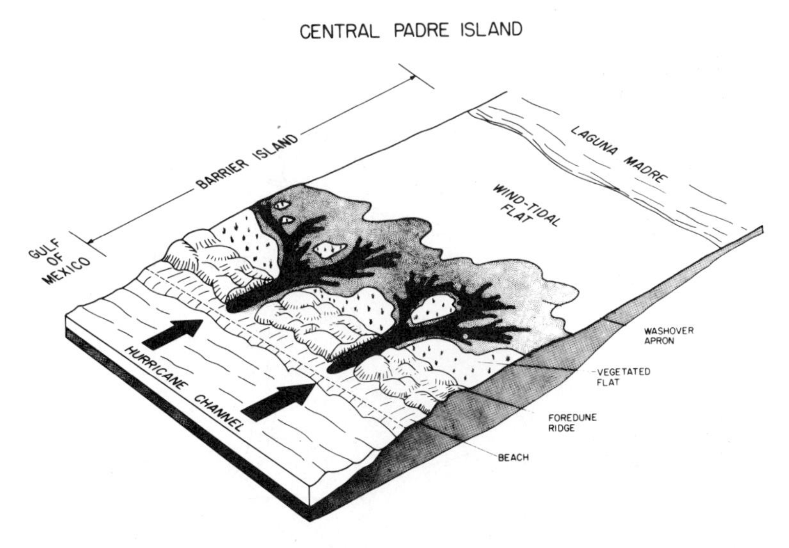

Figure 14. Sketch of storm channels, central Padre Island. Sediment eroded from beach and shoreface is transported through breaches in fore-dune ridge by hurricane surge. Sand is deposited at terminus of channels on wash-over aprons. Aeolian processes subsequently rework wash-over deposits contributing sediment to wind-tidal flats and lagoonal fill, after Scott et al. (19).

a single storm exceeds the normal annual volume of rainfall.

Deposition at the mouths of major rivers, resulting from heavy rainfall triggered by hurricanes, has not been distinguished from deposition under normal rainfall conditions. However, there are smaller drainage systems that construct fan deltas along bay margins. Because of a small drainage area and relatively high gradients, discharge through these systems is of short duration. Since there is virtually no lag between precipitation and run-off, sedimentation by these small, high-gradient streams occurs along bay margins when water level in the bays is excessively high because of storm surge and fresh-water run-off.

Gum Hollow Delta (Fig. 15) is a fan delta that is prograding into Nueces Bay along the north shoreline. The main fluvial channel that feeds the fan delta is situated along the axis of a Pleistocene distributary channel, from which the fan delta derives most of its sand. Hurricane Beulah struck the Mexican and Texas coasts while a study of Gum Hollow delta was being made (12, 19). Hurricane Beulah raised water level in Nueces Bay up to 5 feet above normal, and triggered rains in excess of 24 inches in a two-day period. The result of heavy rainfall, coupled with higher-than-normal water level, was deposition of about 250,000 cubic yards of sand upon a previously constructed fan plain. The Beulah fan (Fig. 15, C) consisted predominantly of very fine terrigenous sand. The fan surface was almost planar, broken here and there by large crescent scours, each

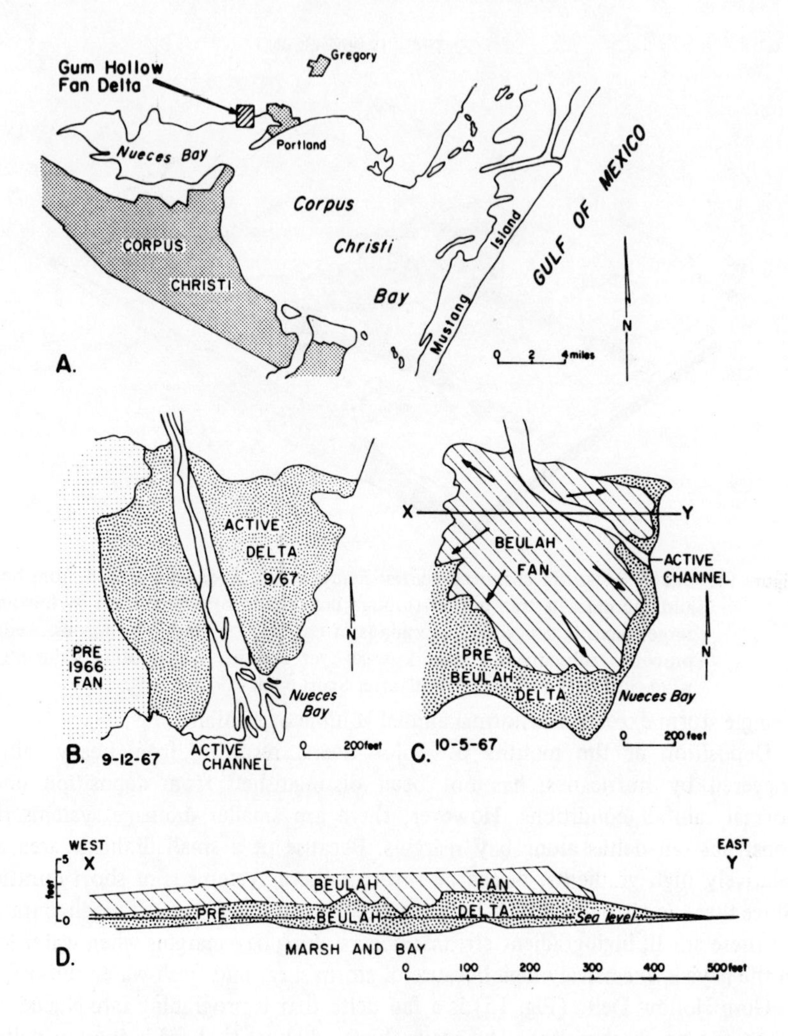

Figure 15. Storm-related deposition on Gum Hollow Fan delta, modified from Scott et al. (19) and McGowen (12). (A) Location of the delta along the north shore of Nueces Bay. (B) Subaerial fan which developed under normal water level and rainfall conditions. This fan developed within 3 weeks as a result of heavy, local thunderstorms. (C) The Beulah fan was constructed within 48 hours as a consequence of: (a) more than 24 inches of rainfall in the drainage basin; and (b) abnormally high water (3-5 feet). (D) East-west cross section across the Beulah fan and underlying older fan. Up to 6 feet of fine sand were deposited on Gum Hollow delta as a consequence of hurricane activity.

of which had a debris pile at its up-current end. The fan terminated distally with a steep avalanche face having a maximum height of about 2 feet.

Fan-delta deposits that accumulate as a consequence of hurricanes are easily recognized. They are predominantly sand, and have a limited lateral extent. They were deposited when water level was high, and, therefore, are relatively thick and rest upon an older fan delta which accumulated under normal water-level conditions. Subsequent flood events, under normal water-level conditions, rework a large volume of the storm-related deposits. Results of subsequent activity are the development of terraces underlain by the storm deposits and excessively rapid progradation because of the large volume of sand-size sediment available.

SUMMARY

Wind is an effective agent for the transportation and deposition of fine and very fine sand in, and along the margins of, Texas lagoons and bays. This is especially true in south Texas, where wind velocities are greater and vegetation is less dense because of decreased precipitation and higher temperatures. Intensity of tidal curents is relatively low; however, non-storm wave acticity is high (8, 18). Tropical storms and hurricanes play an important role in the emplacement and modification of coarse-grained lagoonal and bay deposits on the Texas coast.

Variations in approach rate and inland paths, as well as in storm size and intensity, greatly influence the geologic effects of hurricanes. In general, storms with longer durations in the Gulf of Mexico tend to be large, and slow approach rates increase flooding by storm surge. Height and duration of storm-surge conditions determine areas affected and extent of erosion by wave attack. Even relatively weak storms lacking high wind velocities or surge conditions may have significant geologic effects. If these storms stall in the coastal zone and drift inland along a single drainage system, large volumes of coarse sediment may be transported into the bays by the resulting torrential rains.

In addition to storm variations, climatic factors and the geologic nature of the affected area influence the type and volume of sediments deposited by a hurricane. Density of vegetation cover, not only controls effectiveness of aeolian processes and degree of dune development, but also determines the extent of storm scour and ability of hurricanes to breach barrier islands and peninsulas.

Development of various modern shoreline features is closely related to Pleistocene history of the specific area. Subsidence rates, sediment availability, barrier width, and foredune development are obvious factors influencing frequency and expression of storm-related features such as channels, shell ramps, and wash-over fans.

Accurate prediction of storm effects would greatly aid in minimizing property losses and damage caused by hurricanes in coastal areas. Such predictions are possible only if a broad spectrum of storm types and the detailed geologic nature of specific areas are considered.

ACKNOWLEDGMENTS

Ideas expressed in this paper have developed and evolved over a long period of time. Exchange of data, companionship in the field, and numerous lengthy discussions on hurricane effects and coastal processes with friends and colleagues have contributed much to this evolution. Notable among those who have contributed to our throughts are Miles O. Hayes, Peter B. Andrews, Richard A. Hoover, Bruce H. Wilkinson, and Ronald W. Nordquist. William L. Fisher, Director, and L. Frank Brown, Jr., Associate Director for Research, the Bureau of Economic Geology, The University of Texas at Austin, have discussed many aspects of the problem and have contributed to the editing of the manuscript. Text figures were drafted by Dan Scranton and Richard Dillon of the Bureau of Economic Geology. Final editing was done by Kelley Kennedy, and the manuscript was typed by Lillian H. Crook.

REFERENCES

1. Andrews, P. B.
1967 Facies and genesis of a hurricane washover fan, St. Joseph Island, Central Texas Coast. Doctoral dissertation. Univ. Texas, Austin, 238 p.

2. Andrews, P. B.
1970 Facies and genesis of a hurricane washover fan, St. Joseph Island, Central Texas Coast. Bur. Econ. Geology Rept. Inv. 67, Univ. Texas, Austin, 147 p.

3. Bernard, H. A., Major, C. F., Jr., Parrott, B. S., and LeBlanc, R. J., Jr.
1970 Recent sediments of southeast Texas - a field guide to the Brazos alluvial and deltaic plains and the Galveston barrier island complex. Bur. Econ. Geology Buidebook 11, Univ. Texas, Austin, 47 p. text, 97 figs.

4. Carr, J. T., Jr.
1967 The climate and physiography of Texas. Texas Water Devel. Board Rept. 53, 27 p.

5. Davies, J. L.
1964 A morphogenetic approach to world shorelines. **Zeits fur Geomorph.,** 8 (Sp. No.): 127-142.

6. Fisher, W. L., McGowen, J. H., Brown, L. F., Jr., and Groat, C. G.
1972 Evnironmental geologic atlas of the Texas Coastal Zone – Galveston-Houston area. Bur. Econ. Geology, Univ. Texas, Austin, 91 p.

7. Fisher, W. L., Brown, L. F., Jr., McGowen, J. H., and Groat, C. G.
1973 Environmental geologic atlas of the Texas Coastal Zone – Beaumont-Port Arthur area. Bur. Econ. Geology, Univ. Texas, Austin, 93 p.

8. Hayes, M. O.
1965 Sedimentation on a semiarid, wave-dominated coast (South Texas), with emphasis on hurricane effects. Doctoral dissertation, Univ. Texas, Austin, 350 p.

9. Hayes, M. O.
1967 Hurricanes as geological agents: case studies of Hurricanes Carla, 1961, and Cindy, 1963. Bur. Econ. Geology Rept. Inv. 61, Univ. Texas, Austin, 54 p.

10. Lohse, E. A.
1962 Mouth of Rio Grande, Stop VII - second day. In **Sedimentology of South Texas.** Corpus Christi Geological Society Annual Field Trip, June 8-9, 1962, p. 41-42.

11. Mason, C. and Sorensen, R. M.
1971 Properties and stability of a Texas barrier beach inlet. Texas A and M Univ., Sea Grant Program, Coastal Engr. Div. Rept. (C.O.E.) 146, 166 p.

12. McGowen, J. H.
1971 Gum Hollow fan delta, Nueces Bay, Texas. Bur. Econ. Geology Rept. Inv. 69, Univ. Texas, Austin, 91 p.

13. McGowen, J. H.
1974 The Gulf shoreline and barriers of Texas: processes, characteristics, and factors in use. Bur. Econ. Geology Rept. Inv. 78. Univ. Texas, Austin.

14. McGowen, J. H., Groat, D. G., Brown, L. F., Jr., Fisher, W. L., and Scott, A. J.
1970 Effects of Hurricane Celia – a focus on environmental geologic problems of the Texas Coastal Zone. Bur. Econ. Geology Geol. Circ. 70-3, Univ. Texas, Austin, 35 p.

15. McGowen, J. H. and Garner, L. E.
1972 Relation between Texas barrier islands and late Pleistocene depositional history (abst.). **Am. Assoc. Pet. Geol. Bull.,** 56: 638-639.

16. Nordquist, R. W.
1972 Origin, development, and facies of a young hurricane washover fan on southern St. Joseph Island, central Texas coast. M.S. thesis, Univ. Texas, Austin, 103 p.

17. Piety, W. D.
1972 Surface sediment facies and physiography of a recent tidal delta, Brown Cedar Cut, Central Texas Coast. M.S. thesis, Univ. Houston, 108 p.

18. Price, W. A.
1954 Dynamic environments: Reconnaissance mapping, geologic and geomorphic, of continental shelf of Gulf of Mexico. **Gulf Coast Assoc. Geological Societies, Trans.,** 4: 75-107.

19. Scott, A. J., Hoover, R. A., and McGowen, J. H.
1969 Effects of Hurricane Beulah, 1967, on coastal lagoons and barriers. In **Lagunas Costeras, Un Simposio, Mem. Simp. Internat. Lagunas Costeras,** p. 221-236. UNAM-UNESCO, Nov. 28-30, 1967, Mexico, D.F.

20. Shepard, F. P. and Wanless, H. R.
1971 **Our changing coastlines.** McGraw-Hill, New York, 579 p.

21. U. S. Army Corps of Engineers
1962 Report on Hurricane Carla 9-12 September 1961. U. S. Army Corps Engineers, Galveston District, 29 p.

22. Wilkinson, B. H.
1974 Matagorda Island – the evolution of a Gulf Coast barrier complex. Doctoral dissertation, Univ. Texas, Austin, 178 p.

TIDE AND FAIR-WEATHER WIND EFFECTS
IN A BAR-BUILT LOUISIANA ESTUARY

Björn Kjerfve[1]

ABSTRACT

An in-depth, fair-weather, field study in July 1972 provided information about the response of the water level of Caminada Bay, an extremely shallow, bar-built Louisiana estuary. The water surface elevation was recorded at three locations in the bay along the other parameters, an equipotential surface was established, and the time-dependent variations of a slope vector along the surface gradient were computed. It was found that the instantaneous fair-weather wind stress induced a slowly oscillating set-up around a time-averaged slope magnitude of 1.5×10^{-6} rad. This constituted less than 50% of the measured time-averaged slope. The remaining time-averaged slope is accounted for by tidal nonlinearities. The instantaneous slope vector was found to rotate or oscillate in the horizontal plane with a diurnal period. Tidal input through two entrances governed this behavior, while the wind stress and atmospheric pressure gradients served only to modify the direction of the surface slope. In general, on the diurnal scale, tidal rather than wind effects dominate the dynamics of Caminada Bay. However, the mean water level responded to the wind direction on a time-scale longer than one day. Winds parallel rather than normal to the coast controlled the water elevation, indicating an Ekman effect.

INTRODUCTION

A series of papers in the 1950's by Pritchard (17, 18, 19, 20) and by Pritchard

1. Coastal Studies Institute, Louisiana State University, Baton Rouge, Louisiana 70803. Present address: Department of Geology, University of South Carolina, Columbia, South Carolina.

and Kent (21) on coastal plain estuaries laid the dynamical framework for subsequent, intensive study efforts in various aspects of estuarine circulation and dynamics. Similarly, the study by McAlister et. al. (15) and other investigations pointed out the salient, dynamical features of fjord-type estuaries. However, the bar-built estuary, the third major estuarine type, has received virtually no attention from the point of view of physical oceanography (3). The study by Dyer and Ramamoorthy (6) is a noteworthy exception though. Apparently, there is a need for a systematic investigation of the dynamics of bar-built estuaries.

This study attempts to illustrate the relative importance and effect of astronomical tides and fair weather winds on the water surface behavior in a representative, bar-built estuary, Caminada Bay, thus shedding light on the dynamical structure of this type of coastal embayment. This is of special interest, as Collier and Hedgpeth (4) and Copeland, Thompson, and Ogletree (5) indicated that the wind effects outweighed tidal effects in controlling the surface dynamics of Laguna Madre, Texas, a bar-built estuary similar to Caminada Bay. The water surface fluctuations are, of course, intimately tied to the estuarine velocity structure.

Both coastal plain and fjord estuaries are often successfully approximated by a narrow channel, thus simplifying the analytical treatment by removing lateral variations. As bar-built estuaries are often wide, in a relative sense, with more than one opening to the ocean, the channel approximation is no longer valid.

Estuarine water surface slopes can be caused and maintained by a number of different factors, the most obvious being atmospheric pressure gradient and wind stress. Hellström (10), Haurwitz (9), Keulegan (12), Van Dorn (25), and Kivisild (13) studied wind-induced surface slopes on lakes. Van Dorn and Keulegan indicated that the wind-induced slope should be given by a linear combination of slopes produced by the surface wind shear and a wave effect. However, Saville's (22) measurement of set-up on Lake Okeechobee during storm conditions were fully accounted for by the wind stress alone, in spite of waves 3 m high. In view of this result, wind waves were ignored in this study. The effect of atmospheric pressure gradients on the water surface was measured to be comparatively small and will be treated no further.

The tide is a long wave and will of course cause a fall and rise in the water level as well as instantaneous water surface slopes. However, the tide may also maintain time-average surface slopes. Due to spatial variations in current amplitude, Cameron and Pritchard (3) showed that a mean surface slope may exist in the absence of wind shear stresses and density gradients. This effect arises from the nonlinear terms in the momentum equation. Unoki and Isozaki (23, 24) used nonlinear current effects to explain the observed mean sea level in Japanese embayments, and showed them to be similar to Longuet-Higgins' and Stewart's (14) radiation stress. Qualitatively, the mean surface is raised in regions of antinodes, usually at the head of a bay, and lowered at nodal points.

Density gradients (21) and falling rain drops (2, 25) can both cause mean surface slopes. The Caminada water density was almost homogeneous and the rain fall was limited during the study period. Therefore, these effects warrant no further discussion.

MEASUREMENTS AND DATA ANALYSIS

Study Area.

Caminada Bay is a bar-built estuary, located along the Louisiana coast, 40 km west of the active Mississippi River delta (Fig. 1). It is a portion of the Barataria basin, separating the active channel-levee system of the Mississippi River from the abandoned Bayou Lafourche channel-levee system. The bay, representative

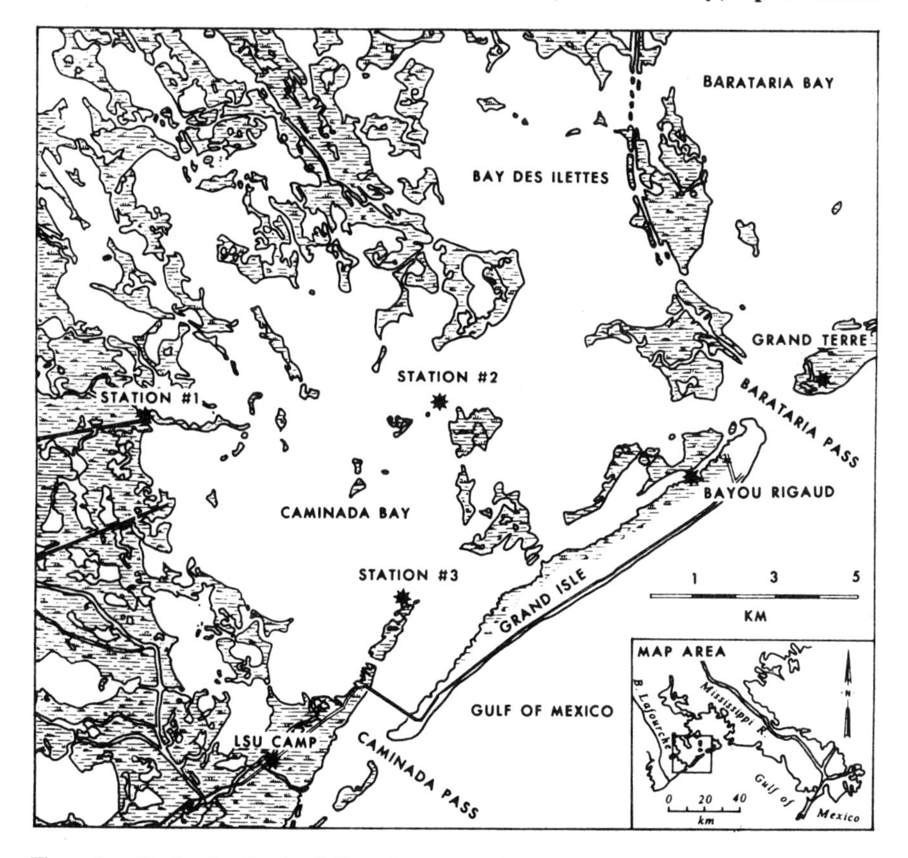

Figure 1. Study site: Caminada Bay, Louisiana, with instrument locations.

of the type of estuaries found in Louisiana, measures some 14 by 6 km, and is only on the order of one-meter deep. Surrounded by vast expanses of *Spartina alterniflora* marsh, Caminada Bay is connected to adjoining bays and the Gulf of Mexico via a series of openings.

The Gulf Coast is a microtidal environment, usually experiencing low marine energy conditions. Still, the land between the Mississippi and Bayou Lafourche is receding due to lack of an adequate sediment supply. The fresh-water run-off is largely due to rain falling over the Barataria basin, and because of dry summers, saline ocean waters encroach deep into the lower basin.

The Caminada and Gulf of Mexico tide is primarily of the diurnal kind, with a weak semi-diurnal component. Furthermore, the mean water level fluctuates on a yearly basis with a September peak and a January low. The range of this oscillation is approximately 26 cm (16). The summer weather encountered in the Caminada Bay area is dominated by the Bermuda High pressure system with a superimposed sea-land breeze circulation and locally generated thunderstorm winds. Hurricanes have struck the area occasionally. The latest direct hit on the Barataria basin was Hurricane Betsy in 1965.

Field Experiment.

The field experiment took place during a 20-day period in July, 1972, preliminary measurements having been made during six days in August, 1971. The study consisted of recordings of water surface elevation, wind speed and direction, atmospheric pressure, and temperature, along with water temperature and salinity measurements. The duration of the study was chosen to include at least one full fortnightly tidal cycle with a tropic and an equatorial tide. To facilitate presentation of the data, a time scale with origin at 0000 on July 7, 1972, was defined. The study was concluded at 492 hrs or 1000 on July 27, 1972.

Data Analysis.

All data traces were digitized with a sampling rate of 20 pts/hr and smoothed with a binomial filter. Stochastic analysis was performed on the time series to detect and describe major data features. Auto- and cross-spectra were computed for scalar and horizontal vector time series (8), using the Fast Fourier Transform (1).

Water Level Measurements.

The water level was continuously recorded at three Caminada locations, stations 1, 2, and 3 (Fig. 1), using three capacitance water level gages. A references datum was established by averaging the water level records for each station, assuming that the means define an equipotential surface. Rather than using the over-all data traces to compute these means, it was convenient to base the reference surface on the means for a three-day period during the equatorial tide. Tidal nonlinearities may then be assumed not to cause time-averaged surface slopes because of the weak currents. The computed means were further corrected for wind stress and pressure gradients. Considerations of gage response, circuit linearity, and errors in the averaging process indicated that absolute slopes greater than 1.0×10^{-6} rad could accurately be assessed.

Wind Measurements and Stress Calculations.

The horizontal wind speed and direction were recorded continuously at station 2, 6.77 m above the mean water surface. A supporting study in 1971 established a relationship between the wind speed and the surface shear stress. A highly accurate, six-level anemometer system was then used to measure the wind profile at station 2. The number of 15-min averaged profiles measured was 386. The extrapolated wind speed range at 6.77 m was from 0 to 10 m/sec. The profiles followed the logarithmic law closely; more than 90% of the profiles had a correlation coefficient in excess of 0.94.

By performing linear regression of logarithmic height above the water on wind speed, the friction velocity, U_*, (von Kàrmàn's constant divided by the regression coefficient) and the extrapolated wind speed, U, at 6.77 m were computed for each profile. Curvilinear regression of friction velocity on the 6.77 m wind speed yielded (Fig. 2)

$$U_* = 49.6 \times 10^{-4} \times U^{1.279}$$

where U_* and U are measured in cm/sec and the linear correlation coefficient is 0.96 (11). The surface wind shear stress, τ, is related to the friction velocity via

$$\tau = \varrho U_*^2$$

where ϱ is air density.

Slope Vector Calculations.

The set-up or difference in water surface elevation between any two points is a time-dependent parameter of primary importance in this study. As Caminada Bay cannot be assumed to be narrow, it is convenient to consider a two-dimensional slope vector along the water surface gradient rather than the scalar set-up.

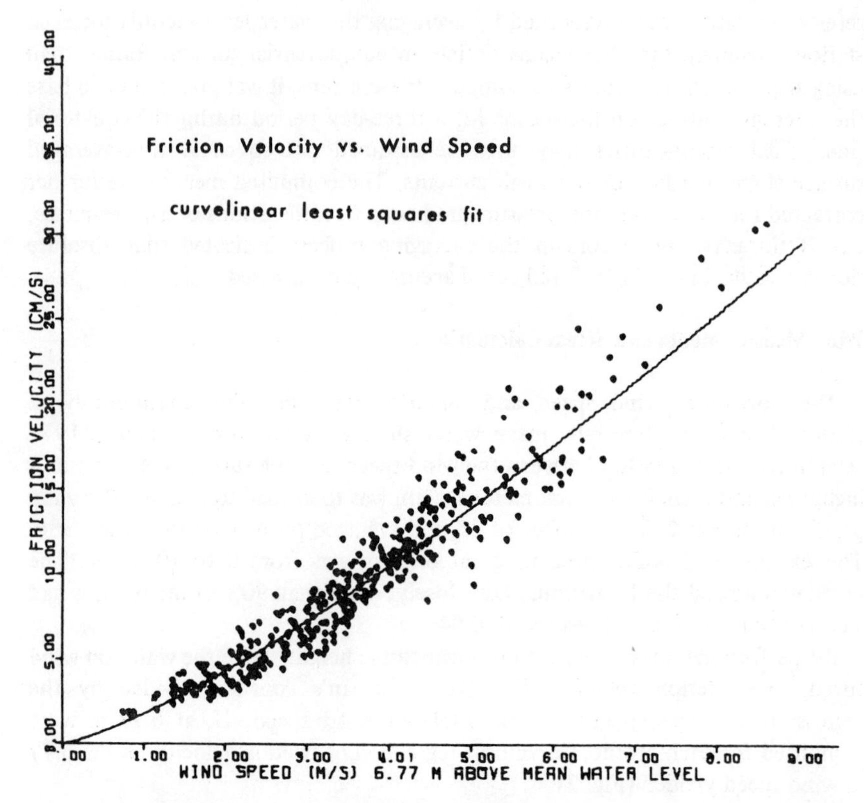

Figure 2. Curvilinear least squares fit between friction velocity, U_*, and wind speed at 6.77 m, U, based on 386 wind profiles with a correlation coefficient, r = 0.96.

Consider the Caminada water surface to be a plane and choose a circle located in this plane with its origin at the water surface at station 2. The magnitude of the slope vector is the maximum slope of the water surface at any one time, i.e., a maximized, non-dimensional set-up. The direction of the slope vector is the horizontal angle, counted clockwise from true north to the lowest elevation on the circle periphery. By analyzing the slope vector magnitude and direction time records, the three-dimensional behavior of the water surface can be induced. Of course, the approximation of the water surface by a plane is necessitated by the use of only three tide gages.

RESULTS AND DISCUSSION

Mean Sea Level.

The smoothed water surface records indicate a greatly elevated mean water level from 50 to 120 hrs and from 250 to 380 hrs at each of the three stations. The record at station 1 is presented in Figure 3. The mean water level was then 8 and 24 cm, respectively, above the filtered value for other times. As the predicted tide at Bayou Rigaud (Fig. 3) does not exhibit a similar increase in the mean elevation during these periods, meteorological rather than astronomical effects cause the surface rise.

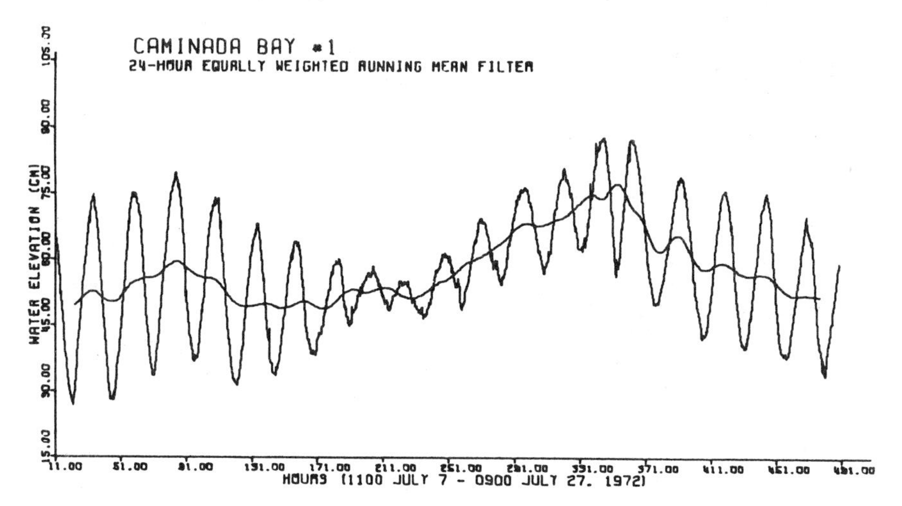

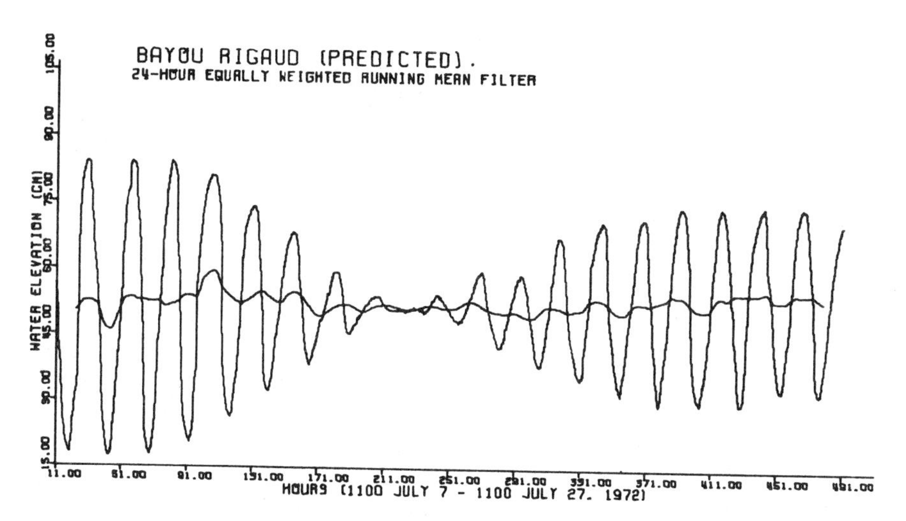

Figure 3. Comparison between measured water level at station 1 and predicted water level at Bayou Rigaud. A 24-hr equally weighted running mean filter has been used to compute the time-averaged water level. The filter frequency response is such that any oscillation longer than 48 hrs is reproduced accurately.

The alongshore component of the wind stress (Fig. 4) was most intense during the periods of high mean water. The stress was then from the east, resulting in a large alongshore (55-235°T) component from a northeasterly direction, leading at peak water level by approximately 20 hrs.

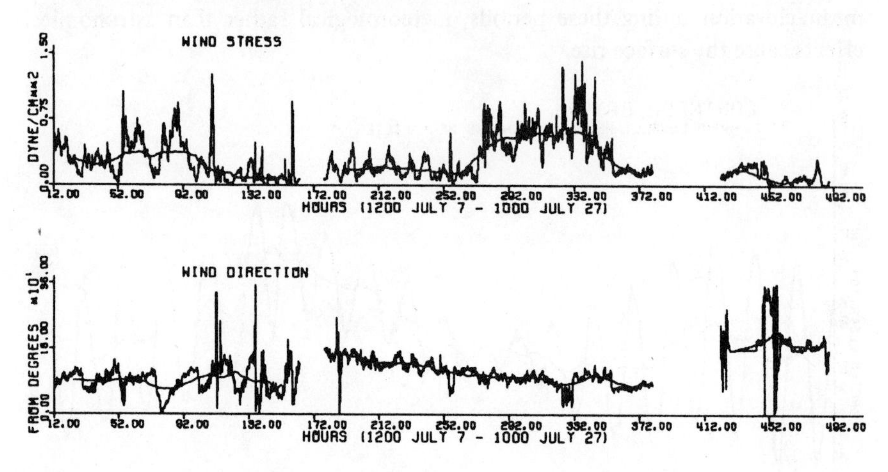

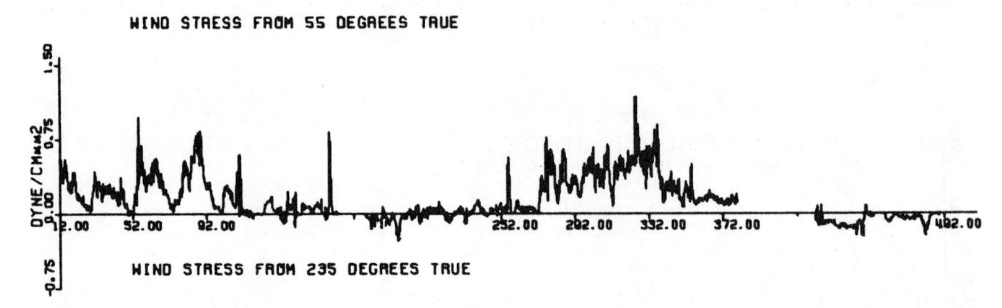

Figure 4. Measured wind stress and direction at station 2 as a function of time. The bottom graph shows the stress component along the coastline. The time-averaged magnitude and direction is a vector average, using a 24-hr equally weighted running mean filter. The wind speed varied from 0 to 14 m/sec.

According to Ekman (7), the net transport in the wind-influenced surface friction layer is in deep water 90° to the right of the down-wind direction in the northern hemisphere. This result is modified in the presence of a coastline and in shallow water with depth less than the friction layer. When the wind blows parallel to the coast with the coast to the right of the down-wind direction, a two-layered circulation may develop perpendicular to the coast, with shore-directed flow in the surface layer and an equal return transport in the bottom layer. Further, a wind regime of this kind will pile up water against the coast, causing sea surface gradients away from the land mass. In Caminada Bay, the result is an elevated mean water surface in response to the intense northeasterly wind stress and the coastal surface gradients.

Consider an average rise of water levels at stations 1, 2, and 3 to be 24 cm in 100 hrs (250-350 hrs). Take the surface area of the Bay to be 8.4×10^7 m^2. This implies that 56 m^3 of water was on the average added each second, most likely through Caminada Pass, as water flowing through Barataria Pass must primarily fill Barataria Bay. Caminada Pass has a 3×10^3 m^2 cross-section, implying a cross-sectionally averaged net inflow of 1.9 cm/sec during the rise fof the water surface. Similarly, the level fell 16 cm in the next 30 hrs in response to the decreasing wind streess. This corresponds to a 4.3 cm/sec spatially averaged net outflow through Caminada Pass. Both velocities are of reasonable magnitude and compare well with time-averaged current measurements made in Caminada Pass during a 1971 study (26).

Water Surface Slopes.

The surface slope of Caminada Bay changes both in magnigude and orientation as a function of time. The measured slope vectors are presented for a few days during one equatorial (Fig. 5) and one tropic (Fig. 6) tide.

The surface slope gradient rotated anticlockwise 360° in 24 hrs during the equatorial tide, whereas it oscillated diurnally 60° to either side of the vectorially, time-averaged value (180°) during the tropic tide. However, both direction records show similar steplike appearances, indicating long periods of constant slope direction followed by rapid direction changes. The two magnitude records are quite similar with pronounced 12-hr oscillations (recitified waves), indicating a diurnal period of the slope magnitude. This was supported by diurnal peaks in the power spectra, which were computed for vector time series (11).

At this point, it became desirable to simulate the observed slope vector behavior with an appropriate analytical model. The momentum equation was integrated vertically, the coriolis and nonlinear field acceleration terms were argued to be small in comparison to the retained terms, density was taken to be constant, and the bottom friction was made linear. The resulting set of linear partial differential equations was broken down into two sets of equations: one representing wind response and one governing tide effects. To facilitate the analytical treatment, Caminada Bay was approximated by a rectangular basin, 14x6 km, with the long axis oriented from southeast to northwest, with a uniform still-water level of 1 m.

The horizontal momentum component equations were balanced by an unsteady current term, a slope term, and the wind stress, which was assumed to be much greater than the portion of the bottom friction corresponding to the wind-induced current. As the wind stress was from the southeast (Fig. 4) during

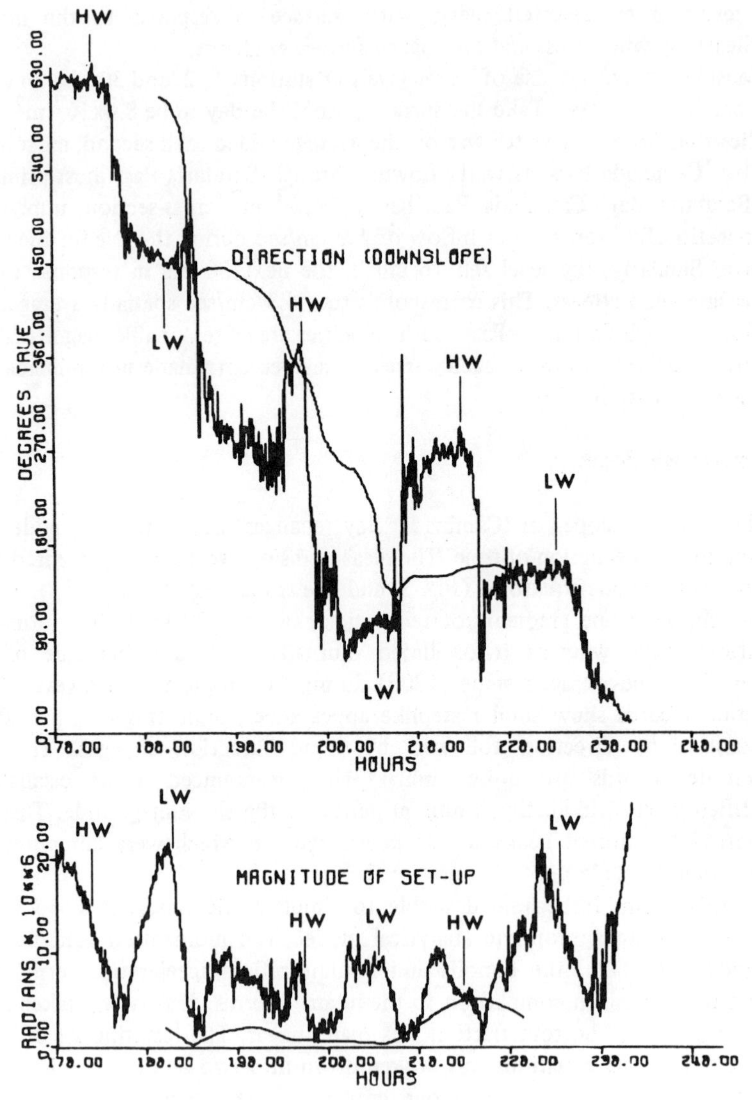

Figure 5. Measured direction and magnitude of slope vector, 1000 July 14-0113 July 17, 1972, during an equatorial tide. The smooth lines are vectorially averaged, 24-hr equally weighted running means.

both the tropic and equatorial tides shown in Figures 5 and 6, the stress direction was assumed to be constant. The spectrum of the stress magnitude indicated a dominant 24-hr peak, which appears as a regular stress oscillation in the record (Fig. 4). The wind stress was therefore simulated, assuming a linear

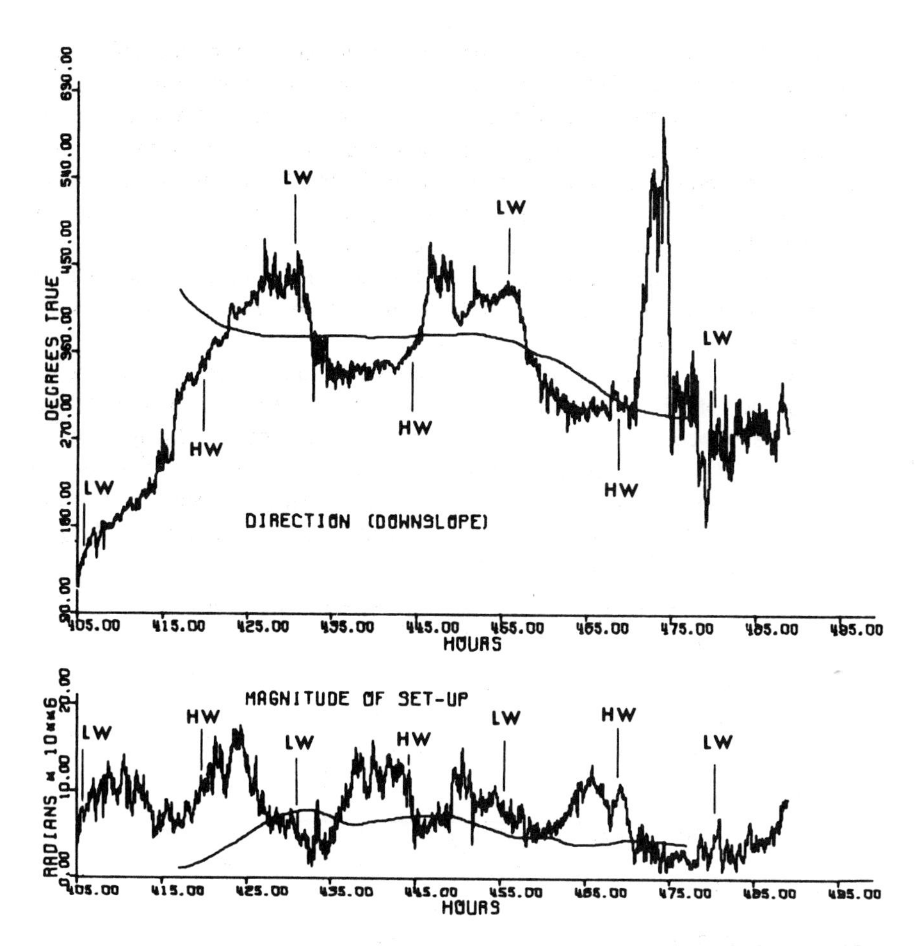

Figure 6. Measured direction and magnitude of slope vector, 2100 July 23-0900 July 27, 1972, during a tropic tide. The smooth lines are vectorially averaged, 24-hr equally weighted running means.

combination of a mean (synoptic) and a sinusoidally varying (sea-land breeze)stress. The mean was 0.18 and 0.11 dyne/cm^2 for the equatorial and tropic tide periods, respectively. The r.m.s. stress was chosen as the amplitude of the oscillatory term. Its value was 0.09 and 0.08 dyne/cm^2 for the equatorial and tropic tides, respectively. When combining the horizontal momentum component balancce with the hydrostratic equation and an appropriate mass conservation equation, it was found that the maximum wind-induced volume transport lagged 6 hrs behind the peak wind stress. Also, as can be expected, the wind-controlled water surface maintained a mean slope corresponding to the mean stress and oscillated diurnally in response to the sea-land breeze stress.

The horizontal momentum component equation for the tide effect were balanced by an unsteady, a slope, and a bottom-friction term. As a first approximation, it was assumed that two two-dimensional tidal waves enter and progress through Caminada Bay at right angles along the long and short axes of the estuary and reflect perfectly at the inside banks. Physically, the two waves may be thought of as the tidal input through Caminada and Barataria passes, respectively. The theoretical tidal surface elevations at the three stations were computed by assuming reasonable values for tidal amplitudes at the passes and a phase angle between the two waves. In general, the model was insensitive to the choice of reasonable phase angle.

The solutions of surface elevation due to wind and tide effects were added to yield a theoretical distribution of water level in space and time. A simulated surface slope vector time series was computed for an equatorial (Fig. 7) and a tropic (Fig. 8) tide in the same manner that the slopes were computed from the measured surface elevations. The qualitative agreement between the measured and simulated slope vector time series is outstanding for both equatorial (Figs. 5 and 7) and tropic (Figs. 6 and 8) conditions.

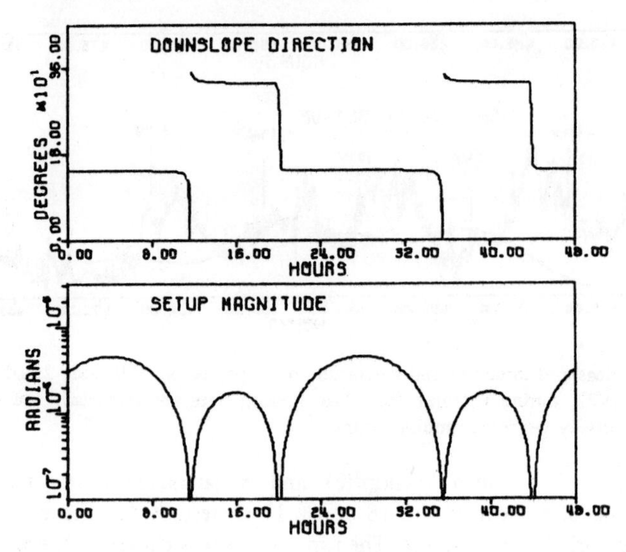

Figure 7. Simulated slope direction and magnitude for the equatorial tide, 1000 July 14-0113 July 17, 1972.

A comparison between the measured and simulated instantaneous direction series shows excellent agreement with respect to: (a) 24-hr periodicity; (b) steplike direction changes every 12 hrs; and (c) general appearance of time series, i.e., one anticlockwise rotation of the slope vector every 24 hrs during the equatorial tide and a 24 hr oscillation in direction during the tropic tide. On the

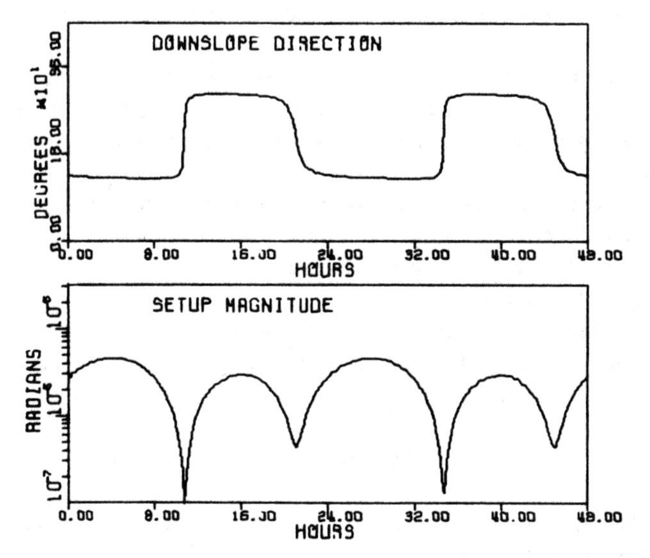

Figure 8. Simulated slope vector direction and magnitude for the tropic tide, 2100 July 23-0900 July 27, 1972.

negative side, the analytical model does not reproduce the time-averaged slope direction well. This is probably due to omission of nonlinear effects or the coriolis acceleration (11).

The agreement between the measured and simulated instantaneous slope magnitudes is also amazing, considering the simplicity of the model. A 24-hr and a 12-hr period are apparent (visually and in the spectrum) in measured as well as simulated records. The measured instantaneous magnitude is approximately a factor of 4 greater than the simulated slope. Also, the measured time-averaged slope, 4×10^{-6} rad, is underestimated by a factor of 4 in the model. This discrepancy may be explained in terms of slopes induced by the neglected nonlinear terms in the momentum equation (3, 11).

CONCLUSIONS

This study considers the response of the water elevation in a shallow bar-built Louisiana estuary to tidal and fair-weather wind effects. Also, water surface slopes were measured and are presented in a novel fashion as a vector time series. It was found that:

(a) Tidal effects dominate the dynamics (rise and fall of the water surface; currents) on the diurnal time scale.

(b) Fair-weather wind effects can cause considerable changes in mean water elevation through an Ekman mechanism on a time scale greater than the diurnal;

winds parallel rather than normal to the coast controlled the mean water level in Caminada Bay.

(c) Measured absolute water surface slopes in this microtidal bar-built estuary reached a maximum of $3x10^{-5}$ rad with a mean of $4x10^{-6}$ rad for the described fair-weather synoptic wind conditions with a superimposed sealand breeze circulation.

(d) Time series of water surface slopes can be simulated extremely well, using a linear three-dimensional analytical model.

(e) The nonlinear field acceleration terms seem to maintain surface slopes greater than wind-induced slopes.

(f) To model a "wide" estuary, e.g., Caminada Bay, which has two major tidal entrances, it is necessary to allow for lateral variations; a channel approximation does not produce useful results.

ACKNOWLEDGMENT

Financial support for this study was provided by the Coastal Studies Institute, Louisiana State University, Baton Rouge, under contract N00014-69-A-0211-0003, Project NR 388 002, with geography programs of the Office of Naval Research.

This research was part of a doctoral dissertation in the Department of Marine Sciences at Louisiana State University. The author is thankful to Drs. S. P. Murray, C. J. Sonu, S. A. Hsu, and to the technicians of the Coastal Studies Institute and the Department of Marine Sciences.

REFERENCES

1. Bendat, J. S. and Piersol, A. G.
1971 **Random Data: Analysis and Measurement Procedures.** Wiley-Interscience. 390 p.

2. Caldwell, D. R. and Elliott, W. P.
1971 Surface stresses produced by rainfall. **Jour. Phys. Ocean.,** 1(2): 145-148.

3. Cameron, W. M. and Pritchard, D. W.
1963 Estuaries. In The Sea. Vol. II, p. 306-324. (ed. Hill, M. N.) Interscience Publishers.

4. Collier, A. and Hedgpeth, J. W.
1950 An introduction to the hydrography of tidal waters of Texas. **Publ. Inst. Marine Science.,** 1(2). Univ. Texas: 125-194.

5. Copeland, B. J., Thompson, J. H., Jr., and Ogletree, W. B.
1968 The effects of wind on water levels in the Texas Laguna Madre. **Texas Jour. Science.,** 20(2): 196-199.

6. Dyer, K. R. and Ramamoorthy, K.
1969 Salinity and water circulation in the Vellar estuary. **Limnol. Oceanogr.,** 14(1): 4-15.

7. Ekman, V. W.
1905 On the influence of the earth's rotation on ocean-currents. **Arkiv för Matematik, Astronomi och Fysik.,** 2(11): 1-53.

8. Gonella, J.
1972 A rotary component method for analysing meteorological and oceanographic vector time series. **Deep-Sea Res.,** 19(12): 833-846.

9. Haurwitz, B.
1951 The slope of lake surfaces under variable wind. **Beach Erosion Board Tech. Mem.,** 25. 23 p.

10. Hellstrom, B.
1941 Wind effect on lakes and rivers. **Ingeniörsvetenskapsakademins Handlingar,** 158. 191 p.

11. Kjerfve, B.
1973 Dynamics of the water surface in a bar-built estuary Doctoral dissertation. Department of Marine Sciences, Louisiana State Univ. 91 p.

12. Keulegan, R. H.
1951 Wind tides in small closed channels. **Jour. Res. Nat. B. Stand.,** 46(5): 358-381.

13. Kivisild, H. R.
1954 Wind effect on shallow bodies of water with special reference to Lake Okeechobee. **Kungl. Tekn. Högskolans Handlingar,** 83. 146 p.

14. Longuet-Higgins, M. S. and Stewart, R. W.
1964 Radiation stresses in water waves; a physical discussion, with applications. **Deep-Sea Res.,** 11(4): 529-562.

15. McAlister, W. B., Rattray, M., Jr., and Barnes, C. A.
1959 The dynamics of a fjord estuary: Silver Bay, Alaska. Tech. Rep. 62. Oceanography. Univ. Washington. 70 p.

16. Mackin, J. G. and Hopkins, S. H.
1961 Studies on oyster mortality in relation to natural environments and to oil fields in Louisiana. **Publ. Inst. Marine Science,** 7. Univ. Texas: 1-131.

17. Pritchard, D. V.
1952 Salinity distribution and circulation in the Chesapeake Bay estuarine system. **Jour. Mar. Res.,** 11: 106-123.

18. Pritchard, D. V.
1954 A study of the salt balance in a coastal plain estuary. **Jour. Mar. Res.,** 13: 133-144.

19. Pritchard, D. V.
1955 Estuarine circulation patterns. **Proceedings of ASCE,** 81. 717/1-717/11.

20. Pritchard, D. V.
1956 The dynamic structure of a coastal plain estuary. **Jour. Mar. Res.,** 15: 33-42.

21. Pritchard, D. V. and Kent, R. E.
1956 A method for determining mean longitudinal velocities in a coastal plain estuary. **Jour. Mar. Res.,** 15: 81-91.

22. Saville, T., Jr.

1952 Wind set-up and waves in shallow water. **Beach Erosion Board Tech. Mem.**, 27. 36 p.

23. Unoki, S. and Isozaki, I.
1965 Mean sea level in bays, with special reference to the mean slope of sea surface due to the standing oscillation of tide. **Oceanogr. Mag.**, 17(½): 11-35.

24. Unoki, S. and Isozaki, I.
1966 A possibility of generation of surf beats. Proceedings of 10th Conf. Coastal Eng. (Tokyo)., 1: 207-216.

25. Van Dorn, W. G.
1953 Wind stress on an artifitial pond. **Jour. Mar. Res.**, 12(3): 249-276.

26. Walters, C.D., Jr. and Hernandez-Avila, M.
1971 Some physical observations in Caminada Pass channel. 12 p. (Unpublished data.)

PROCESSES OF SEDIMENT TRANSPORT AND TIDAL DELTA DEVELOPMENT IN A STRATIFIED TIDAL INLET

L. D. Wright[1]

and

Choule J. Sonu[2]

ABSTRACT

Flood-tidal and ebb-tidal deltas in East pass, on the northwestern coast of Florida, contrast sharply in form and absolute size. The flood-tidal delta, the more extensive, is characterized by a broad middle-ground shoal separating two diverging flood channels, whereas the ebb-tidal delta consists of a single seaward-narrowing channel flanked by subaqueous levees and having a symmetrical, crescentic, subaqueous bar at the outlet. Form differences result partially from variations in the intensities of different effluent expansion and deceleration mechanisms arising from vertical density stratification.

Over the flood-tidal delta, bayward flood flow is concentrated near the bottom beneath lighter bay water. This inflow expands as a hyperpycnal effluent under the influence of bottom friction to produce the observed configuration of the flood-tidal delta. Because flood currents attain their velocity and duration maxima in the lower layer near the bottom, bed-load transport in the channels of the flood-tidal delta is flood-dominated.

During ebb, outflow over the ebb-tidal delta is initially buoyant and is

1. Coastal Studies Institute, Louisiana State University, Baton Rouge, Louisiana 70803.

2. Tetra Tech, Inc., Pasadena, California.

restricted to the upper layer as dense sea water intrudes into the inlet near the bottom. With increasing ebb velocities, the salt wedge is forced seaward to a stationary position over the bar front, and outflow in the region between the outlet and the bar crest becomes turbulent. This sequence is similar to that which characterizes the stratified mouths of the Mississippi between low and high river stages. The morphology of the ebb-tidal delta appears to be roughly analogous to that exhibited by stratified river mouths.

INTRODUCTION

Tidal inlets may be considered analogous to river mouths in the sense that sediment transport and depositional patterns in both cases reflect varying relative contributions from outflow inertia and associated turbulence, bottom friction, buoyancy induced by vertical density stratification, and the energy regime of the receiving body of water (16). However, two basic characteristics distinguish tidal inlets from river mouths: (a) tidal inlets experience diurnal or semi-diurnal reversals of flow, and (b) tidal inlets possess two opposite-facing outlets (i.e., seaward and lagoonward openings). In tidal inlets variations in prevailing mechanisms and associated depositional forms depend on initial inlet geometry, bottom slope of the receiving basin, rate of fresh-water influx into lagoons, and tidal range and debree of tide-induced mixing.

East Pass (Fig. 1), a stratified tidal inlet located in the low tidal range environment of the northwestern Florida Gulf Coast, exhibits sediment transport patterns and depositional features which reflect the effects of two-layer stratified flow, friction-induced hyperpycnal (issuing water denser than ambient water) effluent expansion, inertiadominated turbulent diffusion, and buoyant (or hypopycnal) effluent expansion. The purpose of this paper is to examine the roles played by these processes in different regions of the inlet and with different phases of the tidal cycle.

THE STUDY SITE

East Pass (Fig. 1) is located at the extreme eastern end of Santa Rosa Island at Destin, Florida, and connects Choctawhatchee Bay with the Gulf of Mexico. The pass has a width of 300 meters at its jettied mouth on the Gulf side and a comparable width at the eastern tip of Santa Rosa Island (about 1.5 km bayward of the mouth). Maximum water depths in the East Pass channel average about 3.7 meters.

Astronomical tides in East Pass are diurnal, with a mean range of 18 cm. Maximum tidal range is 26 cm at the time of tropic tides; miniumum range at the equatorial phase is 3 cm. Ebb discharges and current velocities typically exceed those of flood tide by significant but highly variable amounts. Numerous

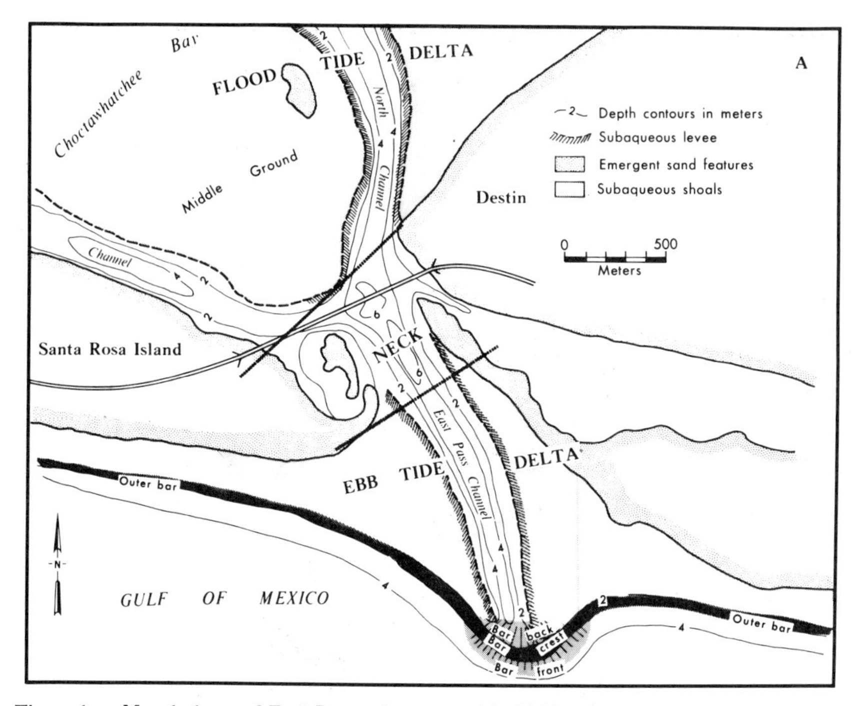

Figure 1a. Morphology of East Pass as it appeared in 1965, prior to construction of jetties.

streams entering Choctawhatchee Bay add a considerable volume of fresh water. The inflow of fresh water into the lagoon creates vertical and horizontal salinity, temperature, and density gradients within the inlet. During the observation period (July 1970), undiluted Gulf water had salinities between 33.0°/oo and 33.5°/oo and temperatures between 26° and 28°C. Salinities as low as 15°/oo and temperatures as high as 29°C were observed for bay water.

Field data were collected during July 1970. Bedform configurations were determined by ground surveys with a HYDRA hydrographic suveying system (U. S. Naval Oceanographic Office, Research and Development Branch; *see* 11) and by aerial photography using a variety of film-filter combinations. The HYDRA system consists of an Atlas-Edig AN6014 echo-sounder and a Decca Sea Fix electronic positioning system. This system provides high resolution, areal repeatability, and positioning accuracy. Water salinity and temperature were measured in the field with a Beckman portable salinometer. Values of σ_t (an index of water density; $\sigma_t = (\rho - 1.0) \cdot 10^3$, where ρ is the density) were computed directly from salinity-temperature data. Currents and circulation patterns were measured with drogues and Rhodamine dye. General outflow patterns from the inlet mouth were observed by means of thermal infrared

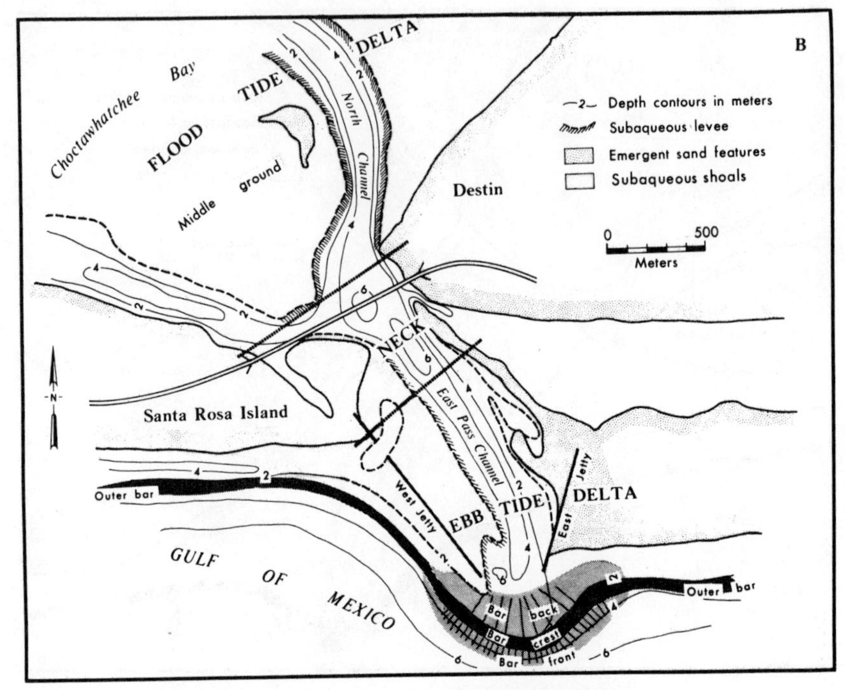

Figure 1b. Morphology of East Pass in 1970, after jetty construction.

imagery obtained through the NASA Earth Resources Aircraft Project (NASA Mission 154, on October 21, 1970, and Mission 159, on March 4, 1971).

MORPHOLOGIC CHARACTERISTICS

The present outlet of East Pass was formed in March 1929, when prolonged heavy rains and strong onshore winds caused the waters of Choctawahtchee Bay to rise 1.6 meters above normal and to breach the eastern end of Santa Rosa Island (7). The old outlet, which was situated about 2.3 km to the east of the present inlet, was subsequently filled in over a period of 6 years. Following establishment of the new inlet, the adjacent shorelines underwent progressive retreat of roughly 300 meters until about 1938, when they became stable near their present positions. Since then the position of the inlet and its shores has remained relatively constant, the exceptions being some moderate modifications resulting from construction of jetties (Fig. 1) in 1968.

East Pass presently interrupts an otherwise continuous barrier coastline composed of medium-grained quartz sand. Along the coast, wide beaches fronted by two or more offshore bars experience low to moderate wave energy and relatively persistent littoral drift to the west. These beaches and bars exhibit rhythmic patterns and have been described in detail by Sonu (9, 10). Some of the sediment in the littoral transport system is entrained into the inlet and added to the tidal delta deposits; however, much of the sediment appears to bypass the inlet by way of the inlet-mouth bar to nourish the beaches to the west.

Aerial photographs from 1935 and 1938 indicate no shoreline offsetting; however, photos from 1959, 1960, and 1965 show significant seaward offsetting of the updrift (eastern) shoreline. This is in contrast to the tendencies observed by Hayes et al. (4) for seaward offsets to occur downdrift of inlets. Following the construction of jetties in 1968, the seaward offsetting updrift of East Pass became more pronounced (Fig. 1-B).

East Pass consists of three basic morphologic units (Fig. 1): a flood-tidal delta composed of sand debouched into Choctawhatchee Bay by flood-tidal currents; a constricted "neck" region separating the barrier formations, characterized by scour and sediment trading; and an ebb-tidal delta at the seaward outlet, formed by the outflow of bay water into the Gulf during ebbing tide.

The flood-tidal delta is by far the most extensive of the units. It consists of a large "middle-ground" bar (Fig. 1) flanked by two bifurcating flood channels. The bar is similar to the middle-ground bars which develop in the Mississippi Delta at the outlets of crevasses and distributaries which enter shallow bays (e.g., 13, 16). This type of bar appears to be a response to rapid effluent expansion induced by bottom friction (1, 16). The associated bifurcating channels average about 3 meters in depth with symmetrical cross sections and are bound on either flank by linear, sharp-crested ridges (Fig. 1) similar in form and function to the subaqueous natural levees of prograding delta distributaries. The channels widen bayward at a linear rate.

The inlet "neck" (Fig. 1) is a single channel about 1 km long and 600 meters wide and is the deepest portion of the inlet, averaging 4.5 meters. In the vicinity of the bridge, scour holes locally exceed 12 meters in depth. This region serves largely as a conduit for the bidirectional transport of sand and undergoes no net accretion by tidal action.

The ebb-tidal delta consists of a single channel which narrows seaward between pronounced subaqueous levees and a single crescentic inlet-mouth bar (Fig. 1). It is smaller in area than the flood-tidal delta. This composite form is roughly similar to the idealized morphodynamic model of a stratified (buoyant) river mouth proposed by Wright and Coleman (16). Prior to jetty construction the subaqueous jetty deposits of the ebb-tidal delta had forms similar to the "ramp-margin shoals" described by Oertel (6) and were surmounted by wave induced megaripples.

The crescentic inlet-mouth bar is made up of three units (Fig. 1, A and B): the seaward-ascending bar back, the bar crest, and the seaward-descending bar front. Each of these units is related to a different set of morphodynamic mechanisms, to be discussed shortly. The bar crest and bar front are essentially continuous with the outer bars of the adjacent coast and serve as the avenue of littoral bypassing. The eastern half of the bar is shoaler and contains a larger sand accumulation than the western half, probably owing to the fact that it is directly exposed to incident waves and lies in a region of convergence between the littoral drift and inlet outflow. A fathometer profile of the bar is shown in Figure 2. Over the bar back, water depths decrease progressively from about 8.5 meters in the scour hole that prevails at the outlet to 2.5 meters at the bar crest. This depth decrease takes place over a distance of 400 meters to produce a mean slope of 0.86°. Over the bar front the depth increases rapidly and the bottom descends at an angle of 2°. The steepness of the inlet bar front distinguishes it from most river-mouth bar fronts, which normally slope more gently than the associated bar backs. The steep front of the East Pass bar probably reflects the effects of wave shoaling and littoral drift.

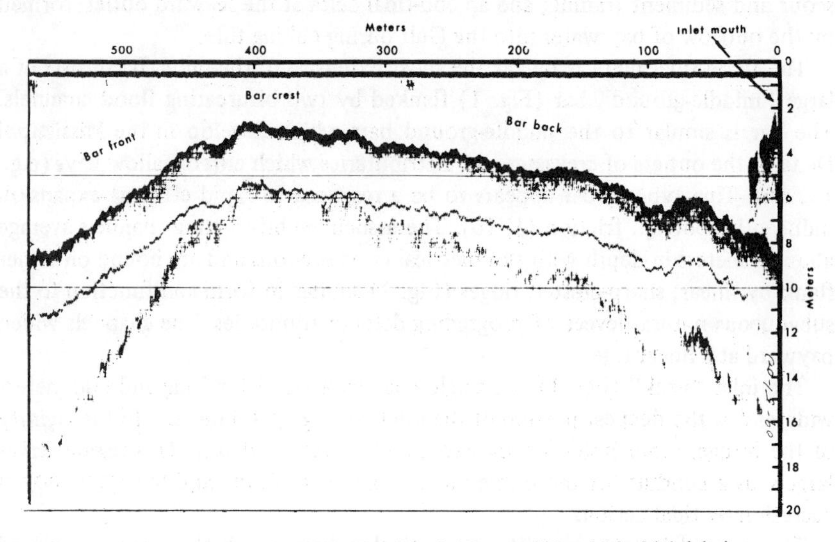

Figure 2. Fathometer profile of the East Pass inlet-mouth bar (ebb-tidal delta).

FLOOD-TIDE PROCESSES

With the onset of flood tide, dense Gulf water intrudes into the inlet and flows bayward beneath lighter bay water remaining from the previous ebb. Flood currents initially enter the inlet along the eastern jetty, while weak ebb flow continues on the western side of the outlet. As the flood phase becomes fully established, inflow eventually prevails across the entire width of the inlet. Drogue studies conducted on July 14, 1970, during flood tide revealed that over the bar crest and bar front flow was westward, bypassing the inlet.

Within about 2 hours after the beginning of flood tide, sea water completely displaced bay water over the ebb-tidal delta and in the inlet neck region, occupying the entire depth and width of the East Pass channel and exhibiting negligible stratification. Bayward flood currents in the neck and near the inlet mouth averaged 0.46 m sec^{-1}, and there was little vertical variation except in immediate proximity to the bottom.

Over the flood-tidal delta (Fig. 1), the flooding marine water converged with low-density bay water at an abrupt density front. Bay water extended farthest seaward over the shallow middle-ground shoal, whereas within the adjacent channels Gulf water intruded about 1 km farther bayward. This caused the front to have a distinctive "V" shape which was observed during tidal food throughout the study period (17). The same configuration is evident on aerial photographs taken during a flooding tide in 1965.

In cross section the front exhibited extremely steep density gradients (Fig. 3), across which density changed as much as 8-10 σ_t units within a distance of less than 1 meter. Beneath the surface boundary, or "foam line", the front extended abruptly downward to depths of 1 meter. However, within a short distance bayward the boundary was nearly horizontal, separating an upper layer of brackish water from underriding salt water. Water depths in the vicinity of the front were on the order of 3 meters. The two-layer flow convergence pattern shown in Figure 3 was evidenced from drogue studies and dye experiments conducted on four separate occasions (17).

As the flood-tide currents approached the frontal boundary, surface velocities diminished; in the lower layers, flow accelerated slightly to about 0.6 m sec^{-1}. Bayward of the front, weak seaward drift prevailed at the surface; but beneath the density interface Gulf water continued to flow bayward with progressively decreasing velocities. In plain view, flood-tidal currents in the intruding Gulf water were swiftest in the channel, and sea water at any given depth penetrated farther bayward by way of the channel.

Because flood currents over the flood-tidal delta are swiftest near the channel bottoms, flood-dominated bed-load transport prevails in the channels. Figure 4 is a longitudinal bathymetric profile of North Channel showing well-developed megaripples, which consistently characterized the channel beds. The megaripples averaged about 15 meters in chord and 1 meter in height and were highly asymmetrical; their steeper (lee) slopes faced bayward (Fig. 4). These orientations persisted during both flooding and ebbing tide and during both tropic and equatorial lunar phases throughout the study period, and are evident on aerial photographs dating back to 1935. They testify to the predominance of sediment transport by flood-tidal currents. Aerial photographs disclosed that less pronounced bedforms over the shoals on either side of the channel and over most of the subaqueous middle ground were in conformity with seaward-flowing ebb currents. The levee-like subaqueous ridges served as boundaries between the

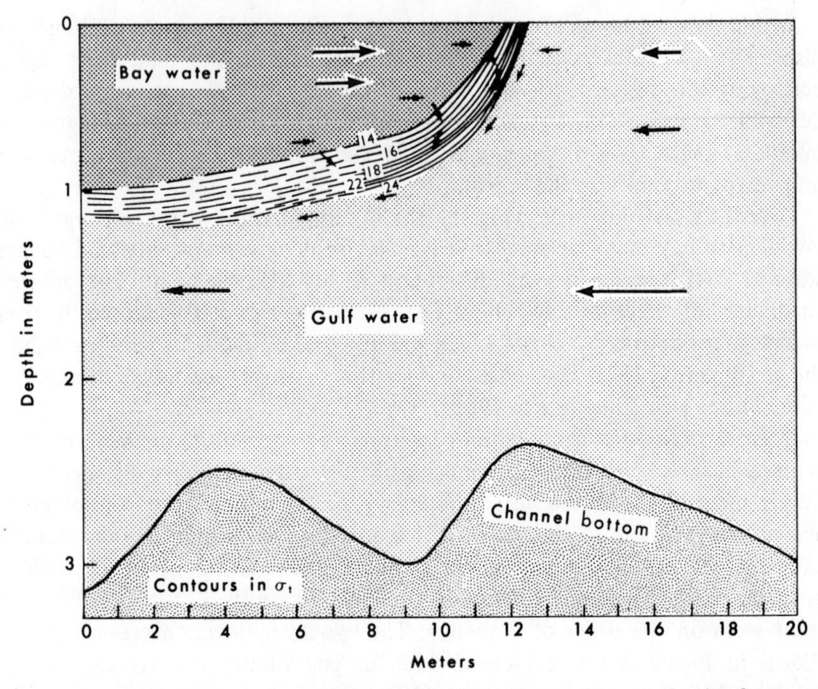

Figure 3. Density (σ_t) cross section and generalized flow pattern at density front over flood-tidal delta during flooding tide (July 7, 1970). (Arrows indicate approximate relative flow velocity.)

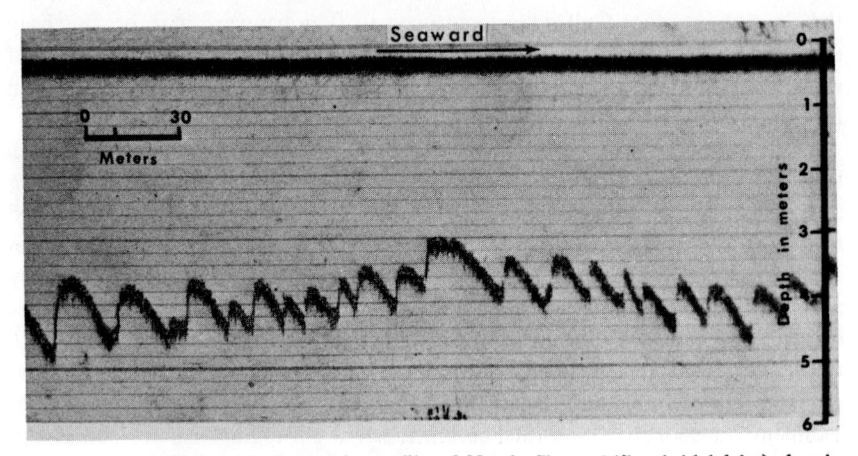

Figure 4. Longitudinal bathymetric profile of North Channel (flood-tidal delta) showing flood-oriented bedforms (July 20, 1970).

two directions of bed-load transport.

Bayward of the inlet neck, inflowing Gulf water expands as a hyperpycnal effluent beneath the fresher bay water. Maximum velocities and velocity gradients are concentrated near the bottom, where the effects of frictional deceleration and expansion are enhanced. It has been demonstrated that increases in flow velocity near the bottom and decreases in water depth increase the effects of friction, causing increases in the rates of effluent deceleration and lateral expansion (1, 16). The flood-tidal delta, with its characteristic middle-ground shoal and divergent channels, is a response to friction-dominated hyperpycnal effluent expansion combined with net bayward transport of bed load.

EBB-TIDE PROCESSES

With the onset of tidal ebb, seaward flow in the upper brackish layer over the flood-tidal delta underwent an abrupt velocity increase. Simultaneously, the two arms of the "V"-shaped front began to converge and to advance seaward. Surface velocities immediately behind the advancing front attained values as high as 0.6 m sec^{-1} within one-half hour after the beginning of ebb tide. Initially, saline Gulf water beneath the interface continued to flow into the bay but with progressively diminishing velocities. There was generally a lag of about 2 hours between the beginning of tidal ebb and complete current reversal across the water column.

Figure 5 shows density and current cross sections parallel to the inlet axis during the initial, intermediate, and advanced stages of the ebb cycle. Within 1 to 2 hours after the beginning of ebb the low-density surface layer reached the inlet mouth above bayward-flowing Gulf water, while above the flood-tidal delta the upper layer underwent a substantial increase in thickness which was accompanied by a decrease in the degree of stratification (Fig. 5-A). During this stage the inlet was characterized by a two-layer "salt-wedge" stratification and circulation pattern similar to that which prevails in the passes of the Mississippi River during low and normal river stages (14, 15, 16). Seaward ebb currents were concentrated in the upper layer.

As the ebb currents accelerated they displaced the toe of the salt wedge seaward and resulted in increased vertical mixing and decreased stratification; seaward flow prevailed across the entire column over the flood-tidal delta and in the inlet neck, but the highest velocities remained in the upper layer (Fig. 5-B). Between the inlet mouth and the crest of the ebb-tide bar, seaward ebb flow remained confined to the upper layer above a salt wedge that intruded a short distance into the seaward end of the pass. This wedge continued to migrate seaward with progressing ebb until it became stationary just seaward of the bar crest, where it remained for the rest of the ebb cycle (Fig. 5-C). This situation is

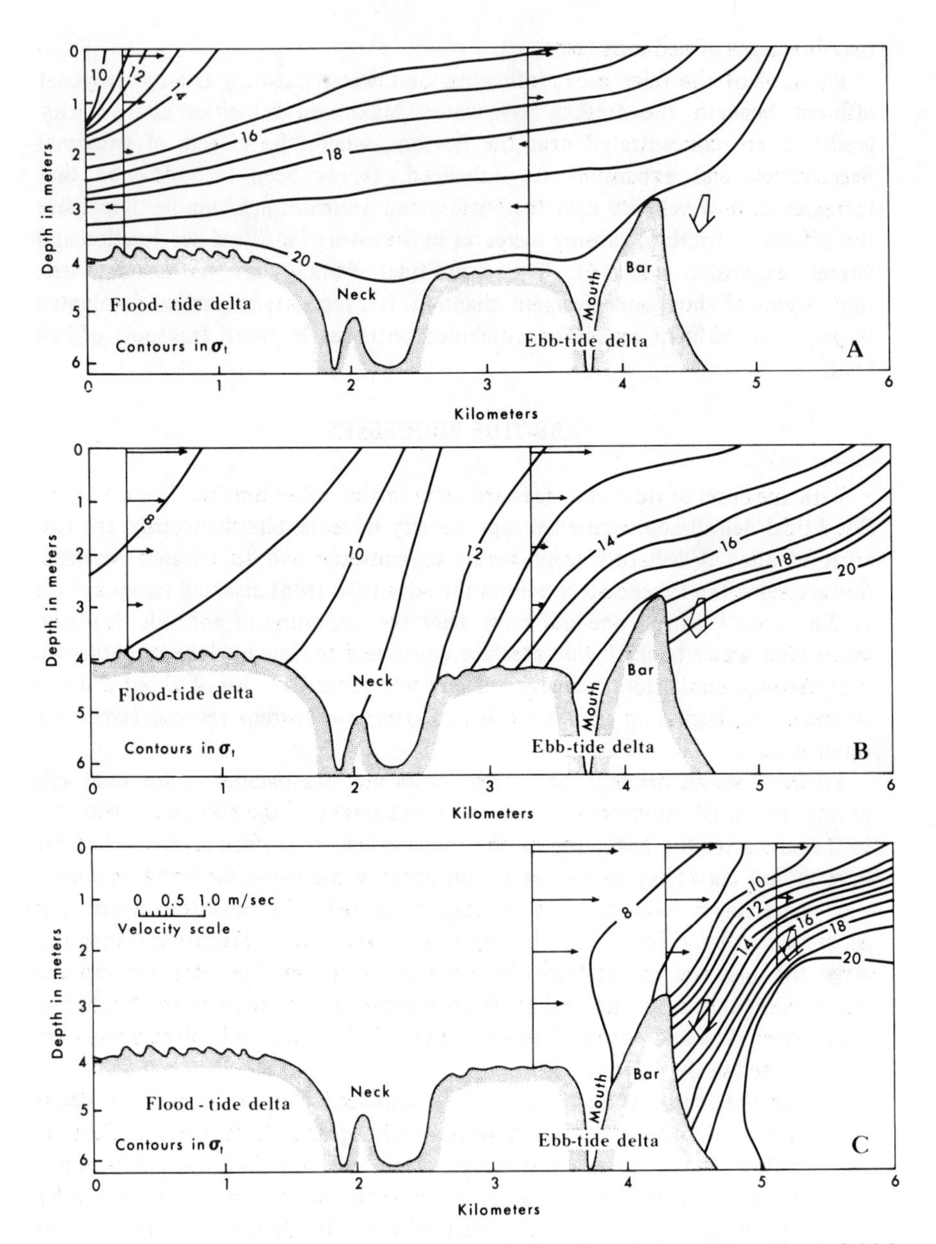

Figure 5. Stratification and current sequence in East Pass during ebbing tide. A. Initial stage, 1-2 hours after onset of ebb. B. Intermediate stage (+2-3 hours). C. Established ebb outflow.

similar to that observed at the mouths of the Mississippi under conditions of river flood (16).

During this established ebb phase, surface velocities at the inlet mouth averaged 1.2 m sec^{-1}; at a depth of 3 meters ebb velocities of 0.45 m sec^{-1} were comparable to those observed at that depth during flooding tide. Velocities at all depths decreased with increasing distance seaward of the outlet. This deceleration can be considered responsible for the accumulation of the bar. Strong seaward flow in the surface layer persisted well beyond the crest of the bar, and surface velocities as high as 0.4 m sec^{-1} were observed more than 1 km from the outlet. However, seaward-directed ebb flow at depths of 2 meters and greater ceased near the bar crest upon encountering the nearly stationary salt-water interface. Below the interface, over the bar front flow turned westward to parallel the coast in association with the prevailing littoral drift. Hence, density stratification appears to protect littoral drift near the bottom from ebb outflow, thus effecting sand bypassing.

The sequence of ebb-tidal effluent expansion patterns and mechanisms is illustrated in Figure 6, which is based on ground observations and analyses of thermal infrared remote-sensing imagery. During the early stages of ebb outflow, when a salt wedge intruded into the inlet, bottom friction and inertial effects were negligible and effluent expansion was dominated by buoyancy (Fig. 6-A). The outflow expanded radially at a rate of 15 m min^{-1}. Initially the densimetric Froude number F', which is the ratio of inertia to the forces of buoyancy, had values less than unity. ($F' = u/[(1 - \rho_1/\rho_2)\, gh']^{1/2}$, where u is the average velocity of the upper layer, ρ_1 and ρ_2 are the densities of the upper and lower layers, respectively, g is the acceleration of gravity, and h' is the depth of the density interface.) Studies of Mississippi River outflows have shown that buoyant expansion is characterized by dual helical circulation, exhibiting flow divergence at the surface and flow convergence beneath the interface (12, 16). Flow convergence in the lower layers inhibits the divergence of subaqueous levees and may play a role in accounting for the configuration of the ebb-tidal delta.

As ebb velocities increased, inertial effects became more important and F' exceeded unity, and the region of radial buoyant expansion was displaced seaward. With the increased inertia, flow near the outlet became turbulent and large-scale lateral eddies developed (Fig. 6-B).

Finally, the buoyant expansion region was dispalced to a position seaward of the bar crest above the stationary density interface. Between the outlet and the bar crest, buoyancy effects became negligible in contrast to inertia and the flow over the bar back behaved as a turbulent jet with linear expansion (Fig. 6-C). Intensification of lateral eddies during this stage probably enhances lateral flow convergence toward the tips of subaqueous levees. Flanking the eddies were zones of buoyancy-induced lateral expansion which propagated along the adjacent shores. It is during this stage of tidal ebb that bottom shear is

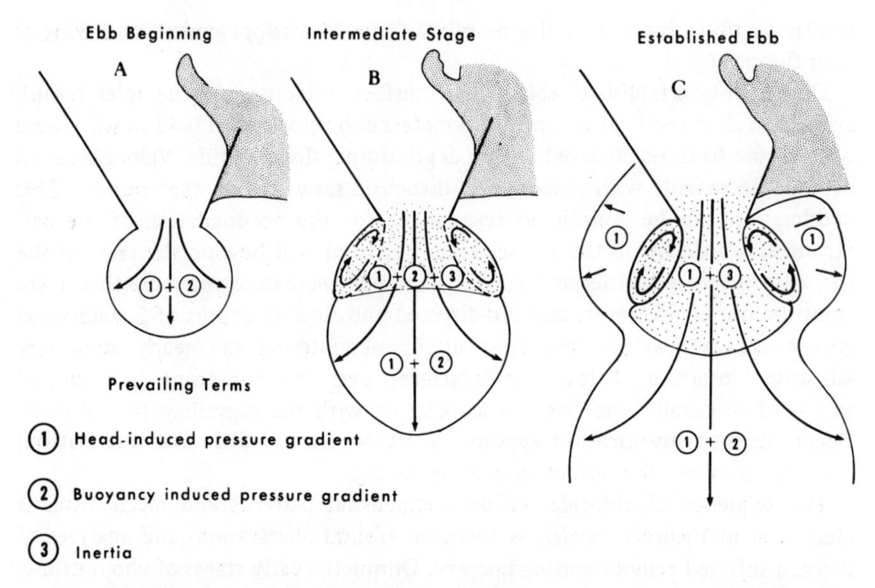

Figure 6. Sequence of effluent spreading and diffusion at moutth of East Pass during ebbing tide. A. Subcritical buoyant spreading during initial stage of outflow. B. Turbulent expansion and incipient generation of lateral eddies near outlet; byoyant spreading region displaced seaward. C. Turbulent expansion between the outlet and the bar crest with intensification of lateral eddies; buoyant expansion seaward of bar crest.

significant and seaward transport of bed load takes place over the bar back. This situation is analogous to high-river-stage outflow from the passes of the Mississippi (16).

CONCLUSIONS

Sediment transport in East Pass is almost entirely in the form of bed load; therefore, it is the flow conditions near the bottom that determine transport and depositional patterns. The tendency for flood- and ebb-dominated flows to favor separate courses is a common trait of tidal inlets (2, 3, 5, 8); in East Pass, density stratification causes vertical rather than horizontal segregations of maximum flood and ebb velocities. Hence, over the flood-tidal delta, flood flows are concentrated near the bottom, where they can move the most sediment and where they experience frictional expansion, whereas ebb flows favor the upper layers in terms of velocity and duration. Sediment therefore experiences a net thrust bayward, where it accumulates as a broad middle-ground shoal between diverging channels.

During ebb, seaward outflow over the ebb-tidal delta expands under the combined influences of buoyancy and turbulence. Maximum velocities and expansion rates occur in the upper layer. Accordingly, the form of the ebb-tidal delta is similar to that developed at the mouths of stratified rivers such as the Mississippi.

ACKNOWLEDGMENTS

This study was supported by the Geography Programs, Office of Naval Research, under contract N00014-69-A-0211-0003, Project NR 388 002, with Coastal Studies Institute, Louisiana State University.

REFERENCES

1. Borichansky, L. S. and Mikhailov, V. N.
1966 Interaction of river and sea water in the absence of tides. In **Scientific problems of the humid tropical zone deltas and their implications,** p. 175-180. UNESCO.
2. Bruun, P. and Gerritsen, F.
1960 **Stability of coastal inlets.** North Holland, Amsterdam, 123 p.
3. DaBoll, J. M.
1970 Holocene sediments of the Parker River estuary, Massachusetts. Coastal Res. Group, Dept. Geol., Univ. Massachusetts, Amherst, Massachusetts. Contrib. 3-CRG, 138 p.
4. Hayes, M. O., Goldsmith, V., and Hobbs, C. H.
1970 Offset coastal inlets. **Proceedings of the 12th Coastal Engineering Conference,** Washington, D.C., p. 1187-1200.
5. Ludwick, J. C.
1970 Sand waves and tidal channels in the entrance to Chesapeake Bay. Inst. of Oceanography, Old Dominion Univ., Norfolk, Virginia, Tech. Rept. 1, 79 p.
6. Oertel, G. F.
1972 Sediment transport of estuary entrance shoals and the formation of swash platforms. **Jour. Sed. Pet.,** 42(4): 858-863.
7. Park, R.
1939 Study of East Pass channel, Choctawhatchee Bay, Florida. U. S. Engineers Office, Mobile, Alabama, 37 p. (Unpublished report.)
8. Price, W. A.
1963 Patterns of flow and channeling in tidal inlets. **Jour. Sed. Pet.,** 33(2): 279-290.
9. Sonu, C. J.
1972 Field observation of nearshore circulation and meandering currents. **Jour. Geophys. Res.,** 77: 3232-3247.
10. Sonu, C. J.
1973 Three-dimensional beach changes. **Jour. Geol.,** 81(1): 42-64.
11. Spinning, J. N., Paradis, M. G., Dixon, D. G., Fletcher, J. P., and Von Nieda, D. G.
1970 Test and evaluation of the HYDRA survey and data processing system. U. S. Naval Oceanographic Office, Informal Rept., IR 70-3, 38 p.

12. Waldrop, W. R. and Farmer, R. C.
Three-dimensional flow and sediment transport at river mouths. Louisiana State Univ., Coastal Studies Inst., Tech. Rept. 150, 137 p. (In press.)

13. Welder, F. A.
1959 Processes of deltaic sedimentation in the lower Mississippi River. Louisiana State Univ., Coastal Studies Inst., Tech. Rept. 12, 90 p.

14. Wright, L. D.
1971 Hydrography of South Pass, Mississippi River delta. Am. Soc. Civil Engrs. Proc., **Jour. Waterways, Harbors and Coastal Eng. Div.**, 97: 491-504.

15. Wright, L. D. and Coleman, J. M.
1971 Effluent expansion and interfacial mixing in the presence of a salt wedge, Mississippi River delta. **Jour. Geophys. Res.**, 76: 8649-8661.

16. Wright, L. D. and Coleman, J. M.
Mississippi River mouth processes: effluent dynamics and morphologic development. (In preparation.)

17. Wright, L. D., Sonu, C. J., and Kielhorn, W. V.
1972 Water-mass stratification and bedform characteristics in East Pass, Destin, Florida. **Mar. Geol.**, 12: 43-58.

ORIGIN AND PROCESSES OF CUSPATE SPIT SHORELINES

Peter S. Rosen[1]

ABSTRACT

Cuspate spits and cuspate forelands result from a shoreline being reoriented into dominant wave approaches. Cuspate spits form in elongate lagoons where the basin shape acts as a selective filter on the wave spectrum, so dominant wave approaches are at a high angle to the shoreline.

Coatue Beach, the northern shoreline of Nantucket Harbor, Massachusetts, has been modified into a series of six *abrasional cuspate spits* (20). The long axis of the harbor is parallel to two opposing wind directions, the dominant northeast and prevailing southwest, causing nearly equal longshore transport in opposite directions. Sediment is eroded from each of the concavities between the cuspate spits and transported to the spit ends, where it is deposited as subaqueous bars. These bars are eroded by the tidal currents, which act as the destructive agent of the cuspate spit-shoreline reorientation process. The tidal currents and opposing longshore drift have resulted in an abrasional shoreline approaching dynamic equilibrium.

Comparisons with other areas show that the abrasional or accumulative form of the spits is a function of tidal range. The tendency for cuspate spits to segment a lagoon occurs only with accumulative spits in nontidal areas. Cuspate spits form only in microtidal areas.

A section of beach oriented obliquely to dominant wave-approach directions reacts by forming several reorientation features, including beach cusps, beach protuberances, looped spits, cuspate spits, and cuspate forelands. The minor forms, cusps and protuberances, may act as the initial phase of shoreline reorientation.

1. Virginia Institute of Marine Science, Gloucester Point, Virginia 23062.

INTRODUCTION

There is a family of beach-accumulation forms generated by longshore processes as the result of waves approaching a shoreline at a high angle. These forms include cusp-like structures, cuspate spits, looped spits, and cuspate forelands. Lewis (10) stated that a shoreline will tend to orient into the dominant wave-approach directions as a result of net longshore transport.

Cuspate spits are defined by Shepard and Wanless (17) as, "a cuspate projection into an enclosed or semi-enclosed lagoon". The origin of cuspate spits has been discussed by Gilbert (4), Shaler (16), Gulliver(6), Johnson (8), and Jones (9). These investigations concluded that such spits are the result of scour and deposition by eddies of tidal currents forming circulation cells in a lagoon. Price and Wilson (12) suggested seiche, or harbor oscillations, as the spit-building agent.

Based on studies of the Chukotsky Peninsula, U.S.S.R., Zenkovitch (20) considered cuspate spits to be a result of reorientation of the shoreline to dominant wave approach (Fig. 1). In an elongate lagoon where dominant wave approach is perpendicular to the shore, there will be a point where longshore sediment transport becomes constant. If the sediment load is obstructed in any way, the capacity will decrease due to the decreased wave-approach angle,

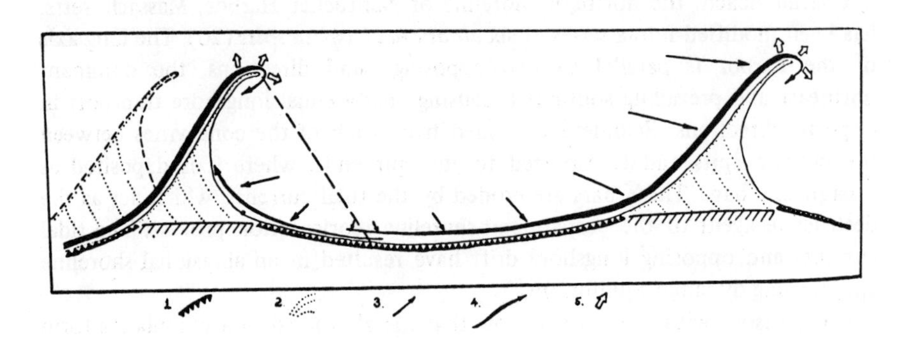

Figure 1. Processes of cuspate-spit development, as outlined by Zenkovitch. This model concurs with the processes in Nantucket Harbor, except the cuspate shape is formed by the abraded shore. (1) The abraded shore. (2) The dune ridges. (3) The relative magnitude and orientation of local wave resultants. (4) The longshore debris streams, with the width of the arrows showing the stream's relative capacity. (5) The part of sediment going away to the deep bottom.

causing deposition. A spit will begin to accrete out from the shore at an angle of about 45°. Waves from the opposite direction will act similarly on the opposite side of the spit, resulting in the cuspate form. The growing spit will form a wave shadow and the process will be repeated along the lagoon shore. As the process continues, spits in favored positions with relation to the wave regime will grow at the expense of others. Studies of the processes of the cuspate spit shoreline in Nantucket Harbor, Massachusetts, support Zenkovitch's model.

NANTUCKET CUSPATE SPIT PROCESSES

Physical Setting

Nantucket Island is located 30 km south of Cape Code, Massachusetts. Nantucket Harbor, on the north side of the island, is an elongate lagoon trending approximately northeast-southwest (Fig. 2). The harbor is bordered on the south by Pleistocene material and on the north and east by the Holocene sand spits, Coatue and Haulover beaches.

Coatue Beach is approximately 10 km long with a maximum relief of 3 m. The topography consists of dune ridges, most of which are parallel to each other and oriented to the northwest. Six regularly-spaced cuspate spits project into Nantucket Harbor from Coatue Beach. Recurved dune ridges are preserved on the two southwesterly spits (First and Second Points), but truncated dune ridges on the four northeasterly spits demonstrate the erosional form of most of the shoreline (Fig. 3). The tides in Nantucket Harbor are semi-diurnal and slightly mixed. The mean range is 1 m.

The wind diagram for Nantucket (Fig. 4) shows that the southwest wind blows 25 percent of the year and its average velocity is over 7 m/s. The northeast wind blows over 10 percent of the year and has an average velocity of nearly 8 m/s. Both major wind components parallel the long axis of Nantucket Harbor, obtaining the maximum fetch.

Longshore currents and shoreline changes

Figure 5 shows the longshore currents and wave parameters measured simultaneously at the stations on 30 June, 1971, a day with a steady southwest wind of 13 m/s. Wave heights showed a marked increase from the concavity centers to the northwest (downwind). The low wave heights in Stations 5 to 3 were due to the "wave shadow" created by the upwind spits. Longshore currents showed a concomitant increase from the centers of the concavities to the northeast (Stations 3 to 1). Sediment was thus transported from the center of the concavities to the ends of the spits, where it was deposited as bars projecting from the end of each spit. Similar measurements under northeast winds showed

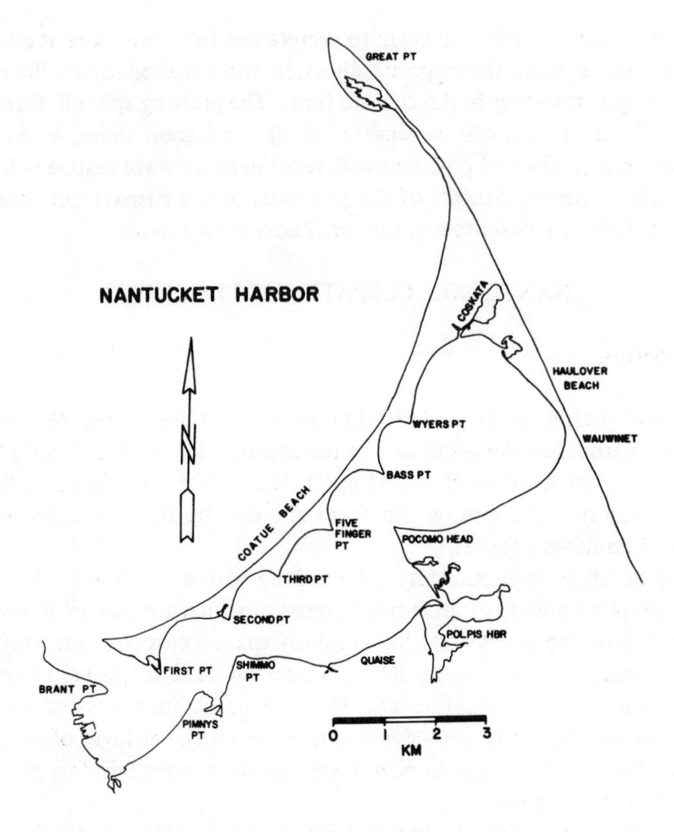

Figure 2. General orientation map of Nantucket Harbor, Massachusetts.

the reverse of this process, i.e., longshore transport to the southwest. The data show that under both conditions, longshore transport was from the centers of the concavities to the spit ends.

This process of longshore transport from the centers of the concavities to the ends of the cuspate spits was reflected in the change in characteristics of the foredune ridges along the length of Coatue Beach (Fig. 6). Steeply cut erosional scarps occurred in the centers of the four concavities between First and Bass Points. The Wyers-Bass concavity showed the opposite trend, i.e., dune-ridge stability in the concavity center.

A comparison of the shoreline of 1781 (map courtesy of Library of Congress, 1972) and that of 1971 (Fig. 7) reveals remarkably little change in location of each cuspate spit. The relatively greater long-term net stability of the Wyers-Bass concavity shows this section of shoreline has come closest to equilibrium

Figure 3. Aerial photograph of Coatue Beach. Note truncated dune ridges on cuspate-spit shoreline, and preserved recurved dune ridges on First and Second Points.

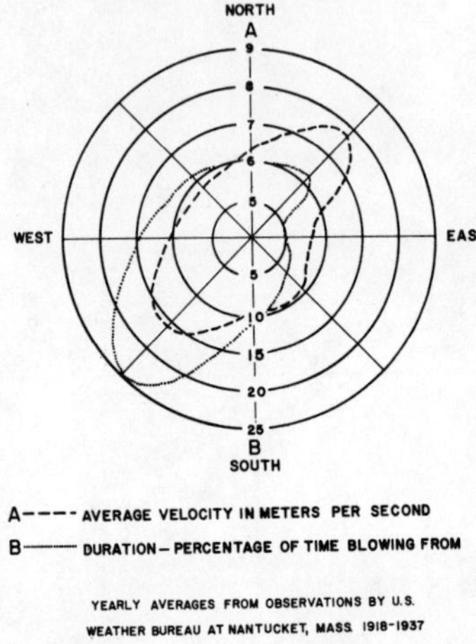

Figure 4. Wind diagram for Nantucket Island, Massachusetts. Note major southwest and northeast wind components that align with the long axis of Nantucket Harbor (curves based on eight compass points).

conditions. In general, there has been erosion in the centers of the four westerly concavities, which concurs with the process studies and shoreline characteristics.

Hydrography

A hydrographic study of Nantucket Harbor (14) showed the lack of circulation cells, or any other tidal effect that might assist in forming a cuspate spit in Nantucket Harbor. Tidal currents were strongest near the spit ends where water in the concavity joins the main volume of flow in the harbor. Measurements of the velocity of tidal currents along the cuspate spit shoreline were made on 24 July 1971 during a dead calm, so there was no influence of wind- or wave-generated currents. The maximum tidal currents occurred on the tips of the spits and tended to erode the spits and straighten the shoreline rather than to develop cuspate spit forms.

Ridge-and-runnel systems

Well-developed ridge-and-runnel systems were present continuously through the summer of 1971 on the harbor side of Coatue Beach from Coskata to 1 km

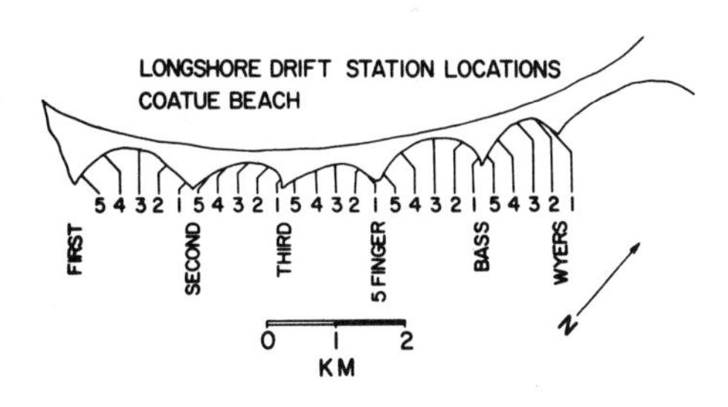

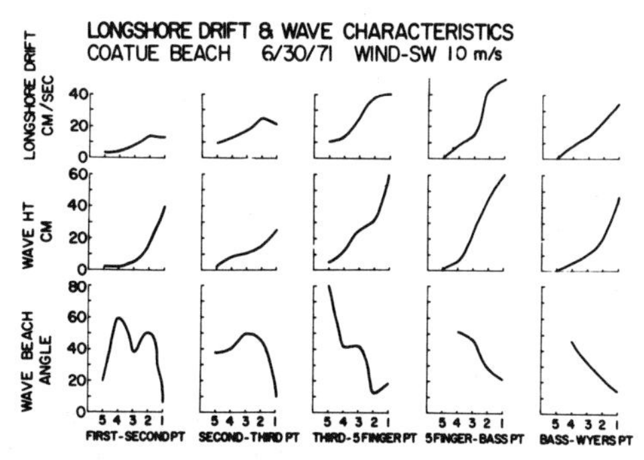

Figure 5. Longshore drift and wave characteristics on Coatue Beach. Each graph suggests that sediment is transported from the center of the concavity to the spit ends.

west of Five Finger Point. They occurred intermittently to Third Point, and not at all between Third and First Points. Ridge-and-runnel systems occur in intertidal areas where there is no eelgrass (*Zostera* sp.). Since 1940, the eelgrass colonized from the lower harbor and migrated toward the east (J. C. Andrews, personal communication). Dense eelgrass adjacent to the low-water line terminates east of Five Fingers Point, where continuous ridge-and-runnel systems begin.

The height of the ridge varied with differing shoreline orientations. The southwest- and northeast-oriented ridges had consistently lower heights than those at the center of the concavity. The ridges in the centers of the concavities had a distinct form with a well-developed slip face, and were higher. The ridges oriented into dominant winds were lower and had irregular crests with numerous runnel outlets.

The ridge on the straight beach located ½ km east of Wyers Point in July, 1971, had the well-developed slip face and higher amplitude characteristic of ridges at the centers of concavities. It was, however, exposed to the higher waves and increased longshore drift, similar to the ridges at the spit ends. This ridge had the same shoreline orientation and high angle of wave-approach as the ridges

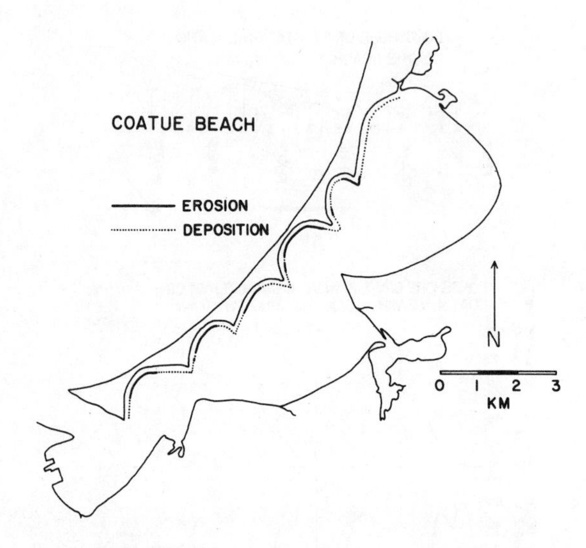

Figure 6. The erosional-depositional characteristics of the foredune ridges on Coatue Beach. The four westerly concavities are erosional in the centers, while the Wyers-Bass concavity is depositional in the center.

in the centers of the concavities. Wave-approach angle is a significant factor in determining the form of beach ridges on Coatue Beach.

The larger volume of sediment provided to the centers of the concavities by onshore migration of higher ridges is a factor in the dynamic equilibrium process of the shoreline, especially between Wyers and Bass points. The ridges provide much of the sediment that is later transported by longshore currents to the spit ends.

Initial form of cuspate spits

The relatively straight beach between Wyers Point and Coskata (Fig. 2) served as a model for the initiation of the cuspate spit form. Throughout the summer of 1971, periods of maximum southwest wind fetch of several days' duration caused several shoreline reorientation features to develop. Non-periodic, cusp-like accumulations of sediment formed in the intertidal zone. The continuous ridge-and-runnel system reoriented into the dominant wave-approach, forming a series of small transverse bars projecting into the harbor at a 10°-20° angle with the strike of the beach. The reorientation of the ridge-and-runnel systems usually formed an accumulation of sediment in the intertidal zone where connected to the beachface. This entire feature migrated downdrift up to 4 m/day.

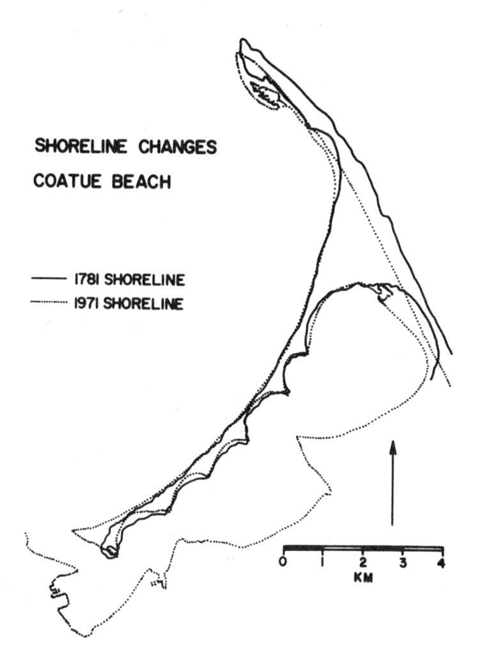

Figure 7. Comparison of shorelines of Coatue Beach, 1781-1971. Stability of the spit ends and erosion in the four westerly concavities is evident (map of 1781 courtesy of Library of Congress, 1972).

None of these minor reorientation features on Coatue Beach established long-term equilibrium, but they did demonstrate how a saturated longshore sediment stream can effect a shoreline reorientation without a pre-existing irregularity in the beach.

Harbor segmentation

Zenkovitch (20) proposed a model for the evolution of cuspate spit shorelines. The end product of the cuspate spit process is segmentation of the lagoon. Raisz (13) also described this process. Nantucket Harbor began to evolve in this manner between 1896 and 1908, when there was an inlet in Haulover Beach. The same tide level occurs at both the ocean side of Haulover Beach and the main harbor entrance, so the opening at Haulover caused a decrease in the velocity of tidal flow. Barnard (1) observed, "The entrance of the tide through the (Haulover) inlet has caused a decided increase in the five narrow bars which extend like finger points from the shore of Coatue Beach." Without the tidal scour governing the growth of the bars, the spits began to evolve toward segmentation. The growth of the bars was terminated with the closing of Haulover Inlet in 1908.

DISCUSSION

Spit Processes

The cuspate spits in Nantucket Harbor are an equilibrium shoreline, as evidenced by a lack of long-term shoreline changes. The wind and waves act through longshore currents as the spit-building agent, which approximately equals the tidal currents as the primary erosional agent. This concurs with Gierloff-Emden's (3) observations that cuspate spits form only in areas of low tidal range.

The process in Nantucket Harbor fits closely to Zenkovitch's process model (20). With two dominant winds in opposing directions the longshore processes act in both directions, each removing sediment from the center of the concavity. This accounts for the symmetry of the spits. The Nantucket spits represent a reorientation of the shoreline into dominant winds. Many cuspate spits in other areas (Table 1) are not oriented into dominant winds. An elongate basin selectively restricts fetch so that dominant wave-approaches are generally perpendicular to the shoreline, regardless of the wind rose. The cuspate spits at First and Second Points on Nantucket probably were initiated by the projection of earlier recurved dune ridges, while the beach protuberance growth on the Wyers-Coskata beach served as a general model for the initiation of this beach form.

The rhythmic form of the Nantucket Harbor cuspate spits is not fully understood, but the spacing appears to represent the distance necessary for waves to gain enough longshore competency for the cuspate spit processes to work, as outlined by Zenkovitch (20). Tanner (18) believes spit-spacing is a function of available fetch. It is more meaningful to view the spacing as a function of average wave size. Larger waves would necessitate longer distances to reform and reach longshore sediment saturation. None of the cuspate spit shorelines surveyed (Table 1) showed rhythm comparable to that of Nantucket Harbor.

Zenkovitch listed two types of cuspate spits, accumulative and abrasional. On accumulative spits the cuspate form is composed primarily of redeposited material; on abrasional spits, it has been eroded from previously deposited material. No cuspate spit can be purely abrasional or accumulative, since the formative process involves both erosion in the concavities and deposition at the spit ends. Zenkovitch's process model is for an accumulative spit. The Nantucket cuspate spits are abrasional, as shown by the truncated beach ridges found on

TABLE 1.

Survey of Cuspate Spit Shorelines

Area	Mean Tidal Range (Meters)	Shoreline Trend	Major Wind Components	Harbor Segmentation	Form	Rhytimicity	Reference
Nantucket Harbor, Massachusetts	1	NE-SW	NE-SW	No	Abrasional	Yes	Rosen (1972)
Waquoit Bay, Massachusetts	1	N-S	NE-SW	No	Abrasional	No	—
Pond, South and Monomoy Island, Massachusetts	0	E-SW	NE-SW	Yes	Accumulative	No	Goldsmith (1972)
Chappaquidick Island Marthas Vinyard, Massachusetts	0	N-S	NE-SW	No	Accumulative	No	—
Sengenkontacket Pond, Marthas Vinyard, Massachusetts	1	NE-SE	NE-SW	No	Abrasional	No	—

TABLE 1 (continued).

Area	Mean Tidal Range (Meters)	Shoreline Trend	Major Wind Components	Harbor Segmentation	Form	Rhytimicity	Reference
Chuktotsky Peninsula, U.S.S.R.	0	—	—	Yes	Accumulative	Yes	Zenkovitch (1959)
St. Lawrence Island, Alaska	0	NW-SE & E-W	N-E-SW	Yes	Accumulative	Yes	Fisher (1955)
Coskata Pond, Nantucket Massachusetts	1	NE-SW	NE-SW	No	Abrasional	No	—
Zululand Coast, South Africa (Kosi Bay, Lake Sibayi, Lake St. Lucia, Richards Bay)	0	NE-SW	NE-SW	Yes	Accumulative	No	Orme (1972)
Chika Lake, India	0.8	NE-SW	NE-SW	Yes	Accumulative	No	Venkatarathnam (1970)
Mobile Bay, Alabama	0.6	E-W	N-SW	Yes	Mixed	Yes	Shepard & Wanless (1971)

each spit. All the spits, however, have made slight lateral readjustments in position by the accretion of dune ridges facing the southwest or northeast. The cause for either an accumulative or abrasional form is due to tidal-current velocities in the lagoon which are generally functions of tidal range. Sediment is transported from the center of each concavity by longshore currents until slowed by deeper water and deposited on the bars projecting from each spit. In Nantucket Harbor, the tidal currents scour the sediments, preventing the bar from forming a sub-aerial spit. An abrasional cuspate spit has also formed in Waquoit Bay in southern Cape Cod, Massachusetts, an area also with a 1-m tidal range. In non-tidal settings, such as St. Lawrence Island, Alaska (2), the ponds and lakes of southeastern Massachusetts (13), and other areas (Table 1), the cuspate spits are accumulative, since material transported from the concavities is deposited as a spit between the concavities.

Accumulative cuspate spits occasionally occur in pairs projecting from opposite sides of the lagoon. Segmentation involves the joining of the spit ends. The origin of this pairing is probably accretion of the bar at the end of a primary spit so that it extends across the lagoon, forming an irregularity on the opposite shore sufficient to initiate the cuspate spit longshore processes.

Cuspate forelands

Cuspate forelands appear to be genetically identical to cupate spits, except that basin shape does not act as a selective filter on the wave spectrum. The similarity in shoreline form between the Nantucket cuspate spits and the Carolina capes on the southeastern coast of the United States is not surprising, since the shore of the southeastern United States is parallel to opposing dominant winds (15). Shepard and Wanless (17) considered the effects of the Gulf Stream to be of vital importance in developing the cuspate forelands.

The barrier-island coast of northwest Alaska has been modified into a series of three cuspate forelands: Icy Cape, Point Franklin, and Point Barrow. The spacing of these capes (approx. 129 km) is similar to that of the Carolina capes. As there are no strong oceanic currents in this area, the role of these currents in the process of this shoreform is dubious. The seas are frozen in this area nine months of the year, and a unique situation exists during thaws. The ice near the coast melts, breaks into floes, and drifts away from the shore to varying degrees (7). This creastes a situation similar to lagoonal cuspate spits: an elongate basin is created with the icefields acting as the northern boundary. The basin shape restricts fetch to favor waves perpendicular to the shoreline, as in Nantucket Harbor. The wind rose for Barrow, Alaska, shows dominant and prevalent directions perpendicular to the shoreline.

Tidal range is not a significant factor on cuspate foreland shorelines because there are no constrictions to form strong tidal currents parallel to shore. All cuspate forelands examined in a world survey were accumulative forms.

The following definitions serve to differentiate between cuspate spits and cuspate forelands:

Cuspate spit: A cuspate projection of a beach into a body of water with fetch limited by basin shape, representing a reorientation of the shoreline to dominant wave-approach.

Cuspate foreland: A cuspate projection of a beach on a shoreline with no fetch limitations, representing reorientation of the shoreline to dominant wind/wave approach.

CONCLUSIONS

(a) Net longshore transport on the Nantucket cuspate spit shoreline is from the center of each concavity to the ends of each cuspate spit. This is the constructive process in cuspate spit development.

(b) The end product of the cuspate-spit process is harbor segmentation. The Nantucket cuspate spits are not segmenting the harbor because tidal currents erode the ends of the spit and projecting bars.

(c) The Nantucket Harbor cuspate spits are a shoreline in dynamic equilibrium where longshore currents, the spit-building agent, approximately equal the destructive tidal currents.

(d) Features similar to beach cusps and transverse bars are short-term shoreline reorientation features in Nantucket Harbor. These features may serve as the initial form of the cuspate spit process.

(e) The accumulative, or abrasional, form of cuspate spits is a function of tidal range. Accumulative spits form in non-tidal areas; abrasional spits form in areas of high tidal range.

(f) Ridge-and-runnel systems, when present, provide an increased volume of sediment to the centers of the concavities. This form of ridges is a function of shoreline orientation.

(g) Cuspate spits represent a reorientation of shorelines into dominant wave-approach directions.

ACKNOWLEDGMENTS

Most of the funds for this project were provided by grants to the University of Massachusetts (Miles O. Hayes, Principal Investigator) by the Coastal Engineering Research Center, U. S. Army Corps of Engineers, (contract DACW 72-67-C-0004) and the Office of Naval Research, Geography Branch (contract N00014-67-A-230-0001). Thanks are extended to Victor Goldsmith for helful discussions and suggestions during the preparation of the manuscript. Virginia Institute of Marine Science contribution number 545.

REFERENCES

1. Barnard, C.
   1900 Some recent changes in the shoreline of Nantucket. **New York Acad. Sci. Annals,** 12: 683-684.
2. Fisher, R. F.
   1955 Cuspate spits of St. Lawrence Island, Alaska. **Jour. Geol.,** 63(2): 133-142.
3. Gierloff-Emden, H. G.
   1961 Nehrungen und lagunen. **Petermanns Geographischen Mitteilungen,** Quartalsheft 3: 161-176.
4. Gilbert, G. K.
   1885 The topographic features of Lake Shores. **U. S. Geol. Survey, Fifth Annual Report,** 69-123.
5. Goldsmith, V.
   1972 Coastal processes of a barrier island complex and adjacent ocean floor: Monomoy Island-Nauset Spit, Cape Cod, Massachusetts. Doctoral dissertation, Univ. Massachusetts, 417 p. (Unpublished.)
6. Gulliver, F. P.
   1896 Cuspate forelands. **Geol. Soc. America Bull.,** 7: 399-422.
7. Hume, J. D. and Schalk, M.
   1964 The effects of ice-push on Arctic beaches: **Am. Jour. Sci.,** 262: 267-273.
8. Johnson, D. W.
   1925 **The New England-Acadian shoreline.** Hafner Publishing Co., New York. 608 p.
9. Jones, W. F.
   1938 Report on Nantucket Harbor and its improvement. 31 p. (Unpublished manuscript.)
10. Lewis, W. V.
    1938 The evolution of shoreline curves. **Proceedings of Geol. Assoc.,** 49: 107-127.
11. Orme, A. R.
    1973 Barrier and lagoon systems along the Zululand Coast, South Africa. In **Coastal Geomorphology,** p. 181-218. (ed. Coates, D. R.) Publications in Geomorphology, S.U.N.Y., Binghamton, New York. 404 p.
12. Price, W. A. and Wilson, B. W.
    1956 Cuspate spits of St. Lawrence Island, Alaska, a discussion. **Jour. Geol.,** 64(1): 94-95.
13. Raisz, E. R.
    1937 Rounded lakes and lagoons of the coastal plains of Massachusetts. **Jour. Geol.,** 42: 839-848.
14. Rosen, P. S.
    1972 Evolution and processes of Coatue Beach, Nantucket Island, Massachusetts: A cuspate spit shoreline. M.S. thesis, Univ. Massachusetts, 203 p. (Unpublished.)
15. Rudee, G. T.
    1923 Shore changes at Cape Hatteras. **Assoc. Amer. Geogr. Annals,** 12: 87-95.

16. Shaler, N.
1889 The geology of Nantucket. **U.S. Geol. Survey Bull.** 53, 55 p.

17. Shepard, F. P. and Wanless, H. R.
1971 **Our changing coastlines.** McGraw Hill, New York, 579 p.

18. Tanner, W. F.
1962 Reorientation of convex shores. **Am. Jour. Sci.,** 260: 37-47.

19. Venkatarathnam, K.
1970 Formation of the barrier spit and other sand ridges near Chilka Lake on the east coast of India. **Mar. Geol.,** 9: 1165-1190.

20. Zenkovitch, V. P.
1959 On the genesis of cuspate spits along lagoon shores. **Jour. Geol.,** 67(3): 269-277.

MOVABLE-BED MODEL STUDY OF GALVESTON BAY ENTRANCE

by

F. A. Herrmann, Jr.[1]

In 1960, a model of the Galveston Bay entrance area (Fig. 1) was constructed at the U. S. Army Engineer Waterways Experiment Station (1). At the time of the model study, there were three major problems within the jetty channel at the entrance to Galveston Bay. The first of these was concerned with the extremely sharp turn located at approximately the inner end of the jetty channel. Large ships, especially tankers, had difficulty negotiating this turn at or near the strength of the tidal currents. The second problem was concerned with the extremely deep water immediately alongside the north jetty, in the area where the navigation channel was very close to the structure. There was considerable concern that, during a severe storm, a section of the jetty might slough into the channel and block navigation. The third problem was concerned with shoaling in the inner bar and outer bar parts of the navigation channel. For the depth of -11.6 m of the inner bar and -12.2 m of the outer bar (including 0.6 m allowable over-depth dredging), the average annual shoaling rates were 346,000 cu m and 560,000 cu m, respectively. As a part of the improvement desired, it was anticipated that the depth of the entrance channel would be increased throughout its length by 1.2 m and an increase of shoaling in both the inner and outer bars could logically be expected.

THE MODEL

The movable-bed model of the Galveston Bay entrance, with scale ratios of 1:500 horizontally and 1:100 vertically, reproduced about 452 sq km of

1. Assistant Chief, Hydraulics Laboratory, U. S. Army Engineer Waterways Experiment Station, P. O. Box 631, Vicksburg, Mississippi 39180.

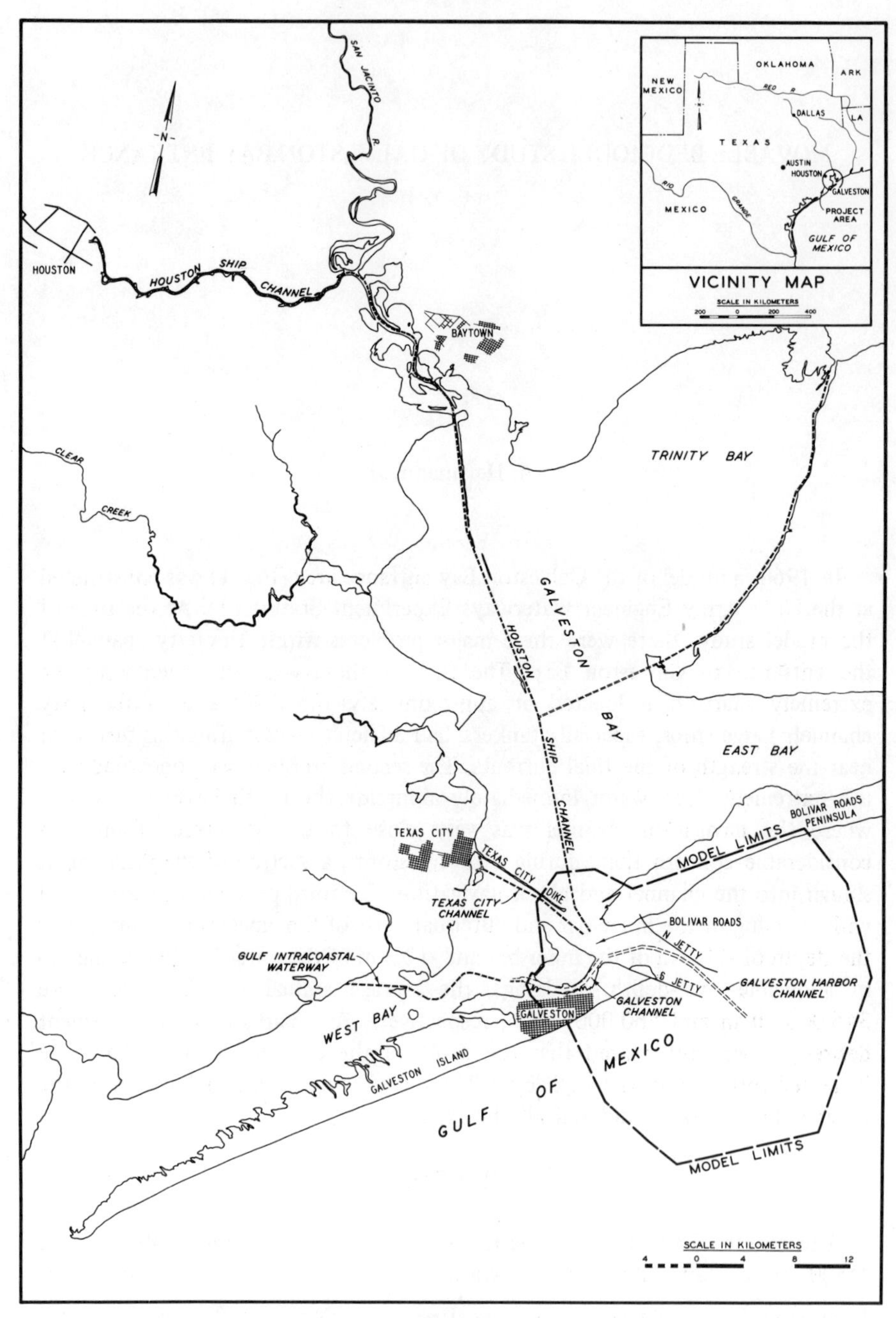
SAN JACINTO R.
N
HOUSTON
HOUSTON SHIP CHANNEL
BAYTOWN
CLEAR CREEK
TRINITY BAY
GALVESTON BAY
HOUSTON SHIP CHANNEL
EAST BAY
BOLIVAR ROADS PENINSULA
MODEL LIMITS
TEXAS CITY
TEXAS CITY DIKE
TEXAS CITY CHANNEL
BOLIVAR ROADS
N JETTY
S JETTY
GULF INTRACOASTAL WATERWAY
GALVESTON HARBOR CHANNEL
GALVESTON
GALVESTON CHANNEL
WEST BAY
GALVESTON ISLAND
GULF OF MEXICO
MODEL LIMITS
SCALE IN KILOMETERS
4 0 4 8 12
NEW MEXICO
OKLAHOMA
ARK
RED R
DALLAS
LA
TEXAS
AUSTIN
HOUSTON
GALVESTON
RIO GRANDE
MEXICO
PROJECT AREA
GULF OF MEXICO
VICINITY MAP
SCALE IN KILOMETERS
200 0 200 400

Figure 1.

prototype area, including a small part of Galveston Bay and a part of the Gulf of Mexico extending 13 km north of the north jetty, 10.5 km south of the south jetty, and offshore to about the 15-m depth contour. Tests were conducted to: (a) develop plans for relocation and stabilization of the jetty channel on an alignment and at a depth suitable for the safe passage of large tankers; (b) determine means for protecting the north jetty from the undermining action of tidal currents; (c) determine the shoaling characteristics of the relocated and deepened inner bar part of the jetty channel, and develop plans for minimizing shoaling in the relocated channel; (d) determine the shoaling characteristics of the deepened outer bar part of the jetty channel; and (e) determine the best locations for additional anchorage areas within the jetty channel or in Bolivar Roads.

From the linear scale relations, it was possible to calculate the following scales using the Froude scale law: velocity 1:10, time 1:50, discharge 1:500,000, and volume 1:25,000,000. The computed time scale applied only to the hydrodynamics of the system and had no relation to the time required to reproduce hydrographic changes. Based on the knowledge that prototype velocities in the problem area were relatively low and on experience gained from previous model studies, it was determined that the movable bed should be molded of crushed coal. This material was of proper weight and size to permit known hydrographic changes of the prototype to be accurately simulated in the model. The coal used had a specific gravity of 1.4, a grain-size range of 0.1 to 5.0 mm, and a median grain diameter of 1.4 mm.

MODEL VERIFICATION

The value of any model study depends entirely upon the proven ability of the model to produce with a reasonable degree of accuracy the results that can be expected to occur in the prototype under given conditions. It is essential, therefore, before any model tests of proposed plans of improvement are undertaken with a view to predetermining their effects in the prototype, that the required similitude first be established between the model and prototype and that all scale relations between the two be determined.

Since it is presently beyond the state of the art analytically to determine the appropriate bed material to use in the model or the time scale for bed movement, it is necessary to resort to a trial-and-error procedure to determine appropriate material for the model bed and the required model scales. This procedure is variously referred to as verification, adjustment, calibration, tuning, etc.

Verification of a movable-bed model is an intricate cut-and-try process of progressively adjusting the various hydraulic forces (tide, discharge, and wave characteristics) and model operating techniques (duration, wave sequence, and bed material) until the model will accurately reproduce hydrographic changes that are known to have occurred in the prototype between certain dates. In this manner, the accuracy of the functioning of the model is established and certain of the scale relations with the prototype are determined. Verification of a movable-bed model usually consists, in general, of the following steps: (a) two prototype surveys of past dates are selected – the time between these dates being known as the verification period – and the movable bed of the model is molded to conform to the earlier survey; (b) the hydraulic phenomena which occurred in the prototype during the verification period are simulated in the model to the proper time scale, all regulative measures undertaken in nature during that period being reproduced in the model at their proper time; and (c) the movable bed of the model is surveyed at the end of this period, and the model is considered to be satisfactorily verified only when this survey is an accurate reproduction of the later prototype survey. It should be pointed out that, until the verification has been completed, the "proper time scale" is not known and must be guessed for each trial run.

The verification period initially selected for the Galveston Bay entrance model was the 10 years between prototype surveys of 1950 and 1960, since these two surveys were the only ones available that provided complete coverage of the movable-bed part of the model. Figure 2-P illustrates the changes in the

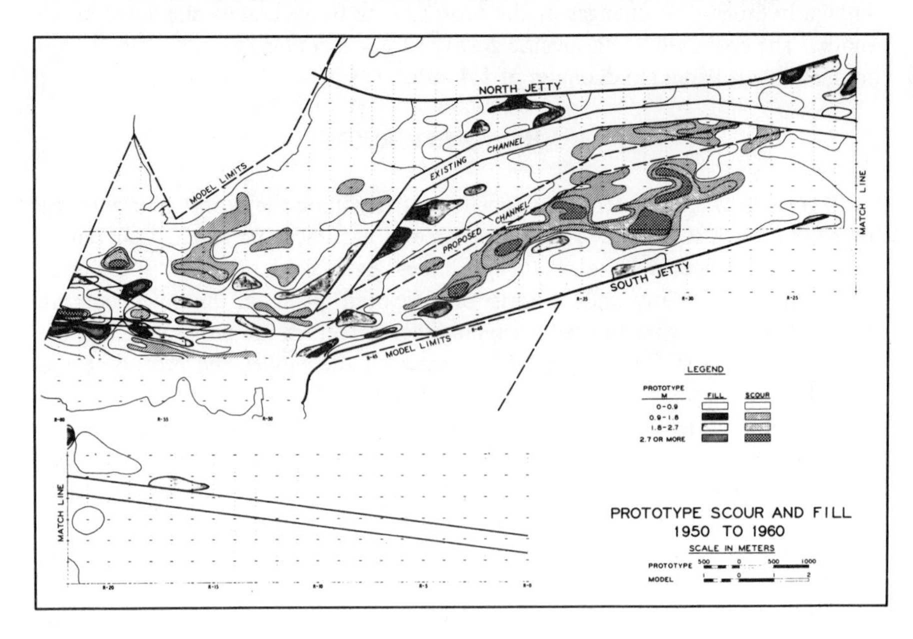

Figure 2-P.

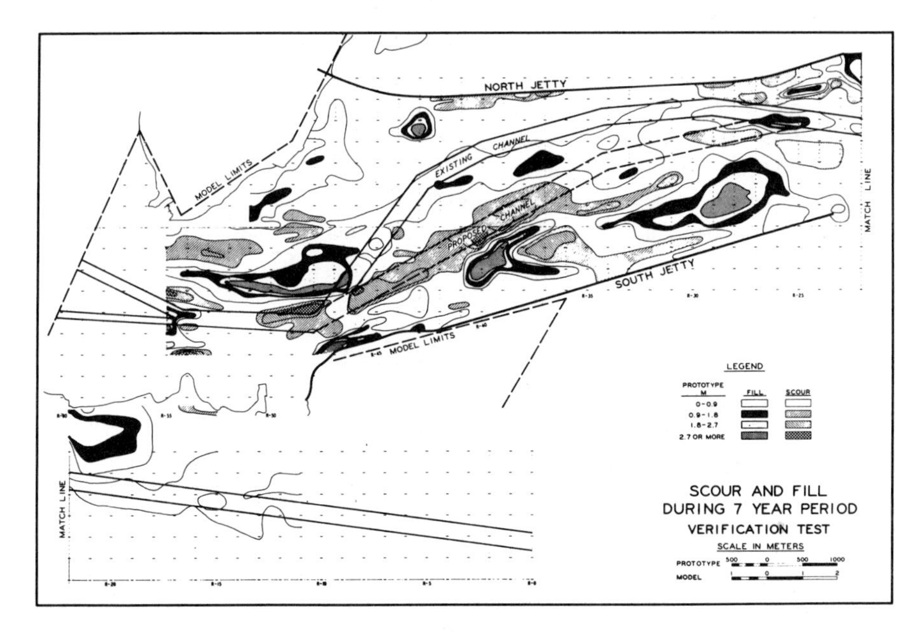

Figure 2-M.

prototype bed configurations that occurred between these two surveys. A number of major storms occurred during this same period and sufficient data to define their effects on entrance conditions were not available. Neither were there complete records of channel-dredging performed during this period.

It was subsequently determined that the scour and fill characteristics for the 10-year period from 1950 to 1960 were a satisfactory guide to the long-term trends of deposition and scour outside the navigation channel, and that detailed dredging records for 1957-1961 could be used to define channel-shoaling characteristics.

The scour and fill patterns achieved in the model after a 7-year verification test are shown in Figure 2-M. Comparison with Figure 2-P indicates that the model reproduced with good accuracy the long-term trends of changes in prototype bed configurations. The following tabulation presents the results of the channel-shoaling verification.

	Prototype		Model	
	Inner Bar	Outer Bar	Inner Bar	Outer Bar
Annual Shoaling Rate (cu m)	346,000	560,000	219,000	470,000
Percentage Distribution	38	62	32	68

Although the yearly average of total model shoaling was about 24 percent less than the prototype quantity, the distribution of material between the model inner and outer bar channel areas was within 6 percent of the prototype distribution. The volume scale determined from the Froudian relations (1:25,000,000) was used for determining model shoaling volumes. The movable-bed model verification was considered to be satisfactory. The empirical time scale for bed movement developed during the verification procedure was 7.5 hours in the model to 1-year prototype (1:1167), as opposed to the Froudian time scale for hydraulic phenomena of 1:50.

MODEL TEST RESULTS

A number of improvement plans were then tested in the model, and the plan recommended for construction in the prototype is shown in Figure 3-M. This plan involved relocating essentially the full length of the jetty channel for the principal objectives of easing the sharp bend at the inner ends of the jetties and shifting the channel an appreciable distance away from the north jetty in the area where undermining by tidal currents was threatened. It was further recommended that, in constructing this plan in the prototype, a large percentage of the spoil from the new work be deposited alongside the north jetty to reduce the depth and the possibility of undermining and to prevent any increase in the cross-sectional area of the entrance as a whole. The realigned channel was tested with the desired 1.2-m increase in depth, and the model test results indicated that the deeper channel on the new alignment could be maintained with essentially the same annual maintenance dredging required for the lesser channel depth on the old alignment. The results also indicated that most of the inner bar shoal would be eliminated and that shoaling in the outer bar and entrance channel would be increased. It was also determined that the inner bar channel would be unstable during the construction program, some shoaling of the inner bar could be expected during and possibly for the first year following construction, and the total shoaling rate would probably accelerate during the first year after construction.

POSTCONSTRUCTION CONFIRMATION STUDY

The recommended plan was subsequently constructed in the field, and a confirmation study is being conducted to evaluate the accuracy of the model predictions (2). Such studies are valuable to field engineers because they provide a measure of the degree of reliance that can be placed on model predictions and to laboratory engineers because they provide guidance for improving modeling techniques.

Actual construction required two and one-half years, as opposed to two years used in the model test, and 4,470,000 cu m of spoil was placed in the abandoned channel, as compared to 3,550,000 cu m in the model test. Otherwise, the construction sequence was very similar to that used in the model. Figures 3-M and 3-P show the channel conditions in model and prototype at the end of construction. It can be seen that the prototype channel was dredged to a considerably deeper depth than was the model channel, in which the dredged depth could be controlled more carefully. On an overall basis, the prototype channel was 0.6 m deeper than the model channel. In addition, a considerable amount of dredging in the prototype exceeded the allowable 0.6 m over-depth dredging. Adjusting the prototype dredging quantity to compensate for the excessive over-depth dredging (1,900,000 cu m), the model construction dredging (9,940,000 cu m) was only about 6 percent less than was required in the prototype (10,610,000 cu m).

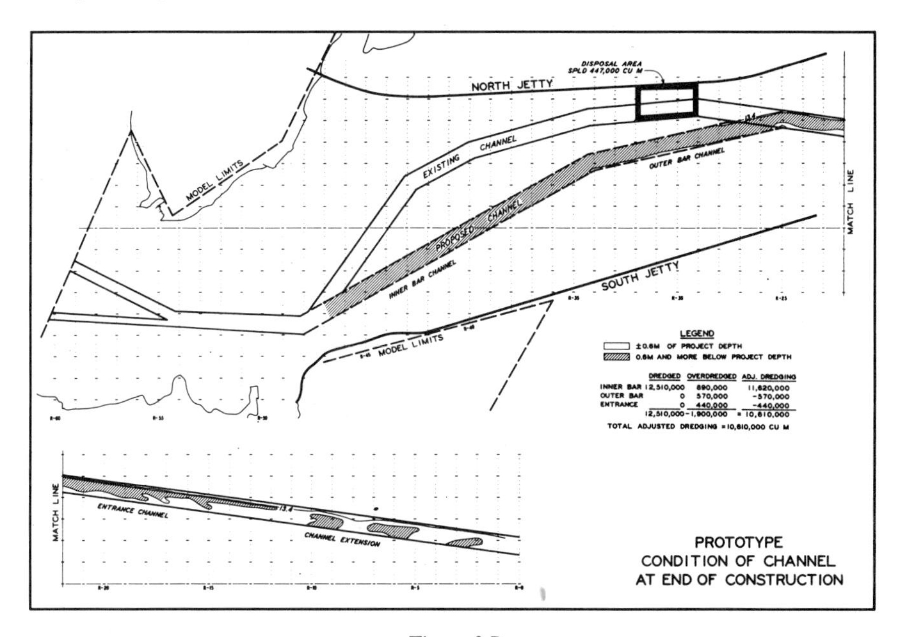

Figure 3-P.

Figures 4-M and 4-P show the condition of the model and prototype channels, respectively, one year after construction of the project and before any maintenance dredging was performed. Both model and prototype condition surveys indicate the possible need for some maintenance in the inner bar, outer bar, and the entrance channel. No dredging was done in the model inner bar, however, because the depths were essentially at project elevation.

Figures 5-M and 5-P show the condition of the model and prototype

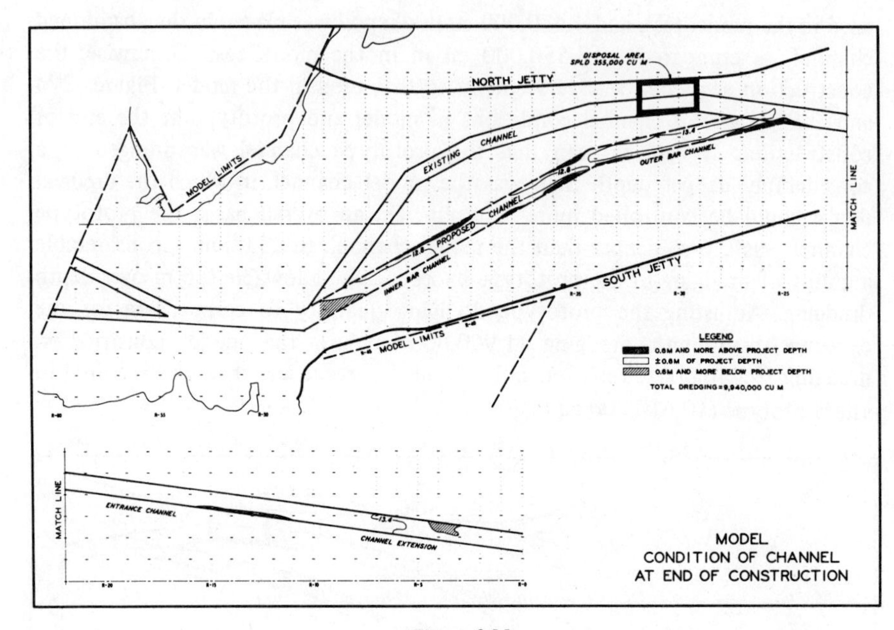

Figure 3-M.

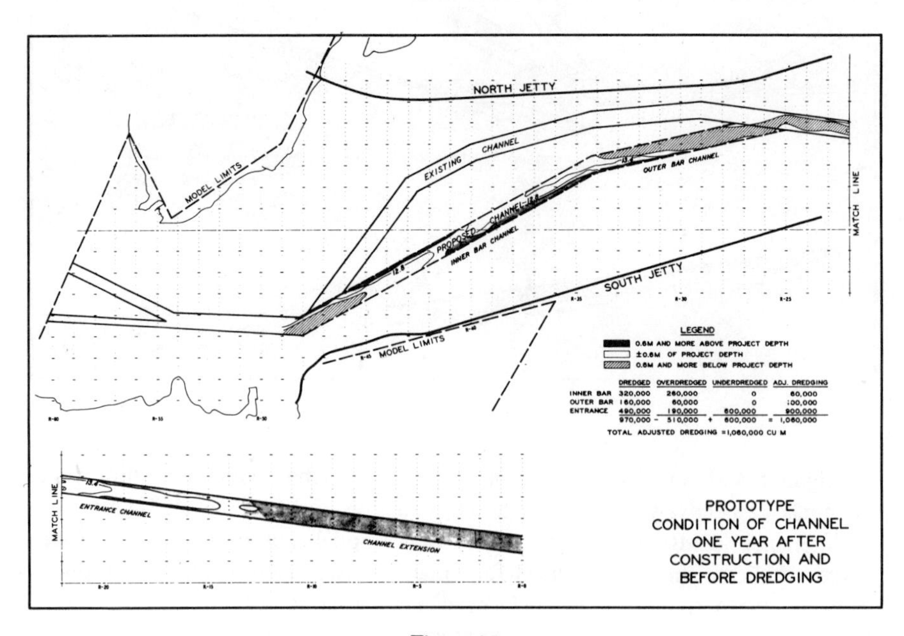

Figure 4-P.

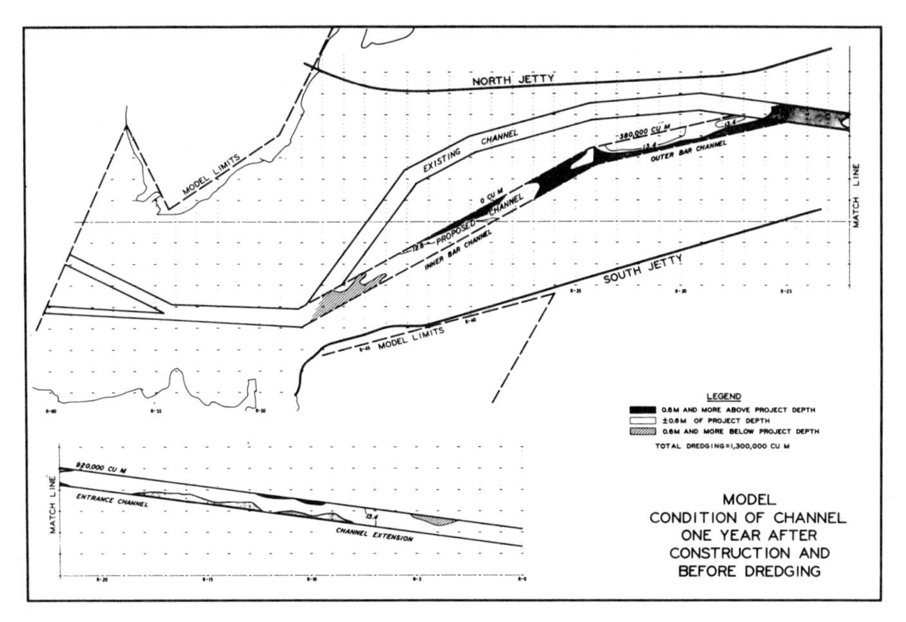

Figure 4-M.

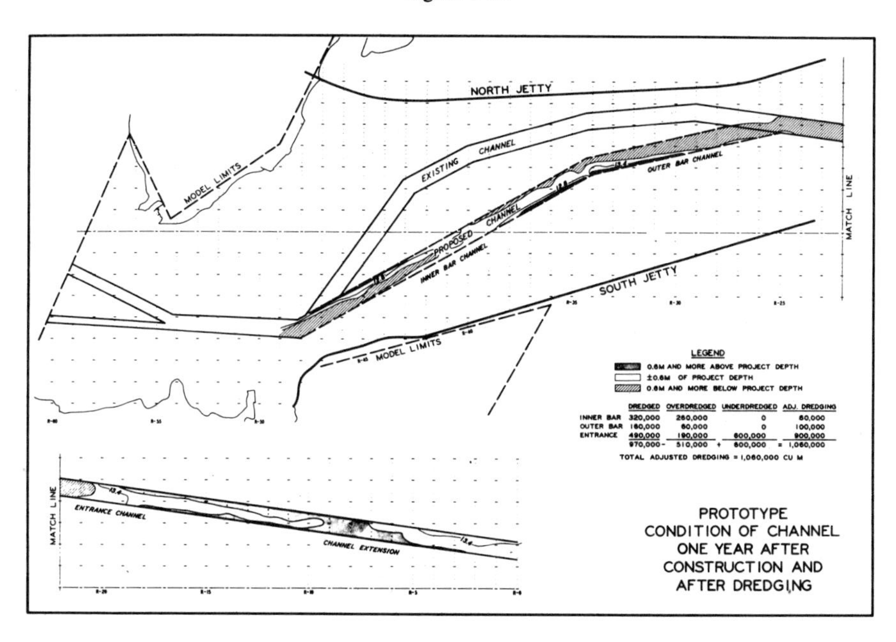

Figure 5-P.

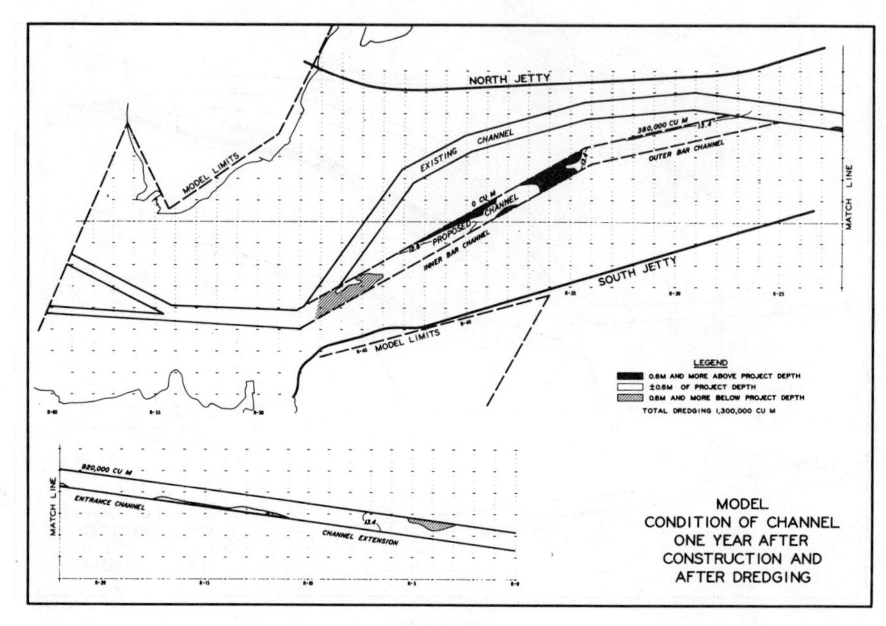

Figure 5-M.

channels, respectively, following the first maintenance dredging, which was done approximately one year after construction was completed. The amounts of total maintenance dredging performed in model and prototype were 1,300,000 and 970,000 cu m, respectively (Table 1). However, some of the prototype dredging was performed in areas already having depths in excess of the allowable 0.6-m over depth, while other areas having depths less than the project depth were not dredged. To make a valid comparison between model and prototype, it was, therefore, necessary to adjust the prototype dredging data to compensate for these factors. The details of this adjustment are shown in Table 1, and the adjusted prototype dredging quantity was 1,060,000 cu m. Thus, at the end of the first year of operation, the model predictions were fairly well borne out by prototype experience, in the sense that total maintenance dredging for the enlarged channel on the new alignment was somewhat greater than that for the channel of lesser depth with the old alignment (906,000 cu m), and the inner bar was no longer a critical shoaling area.

Figures 6-M and 6-P show the condition of the model and prototype channels, respectively, at the end of the second year following construction and before maintenance dredging was begun. On an overall basis, the agreement between the model prediction and the recorded condition of the prototype channel for this point in time is quite close.

Figures 7-M and 7-P show the condition of the model and prototype

TABLE 1.

Maintenance Dredging After Construction

		Prototype					Model	
Years After Construction	Channel	Yearly Maintenance (cu m)	Over-Dredged (cu m)	Under-Dredged (cu m)	Adjusted Dredging (cu m)	Accumulated Total Maintenance (cu m)	Yearly Maintenance (cu m)	Accumulated Total Maintenance (cu m)
1	Inner	320,000	260,000	0	60,000		0	
	Outer	160,000	60,000	0	100,000		380,000	
	Entrance	490,000	190,000	600,000	900,000		920,000	
	Totals	970,000	510,000	600,000	1,060,000	1,060,000	1,300,000	1,300,000
2	Inner	570,000	570,000	0	0		0	
	Outer	300,000	200,000	0	100,000		140,000	
	Entrance	870,000	250,000	0	620,000		340,000	
	Totals	1,740,000	1,020,000	0	720,000	1,780,000	480,000	1,780,000

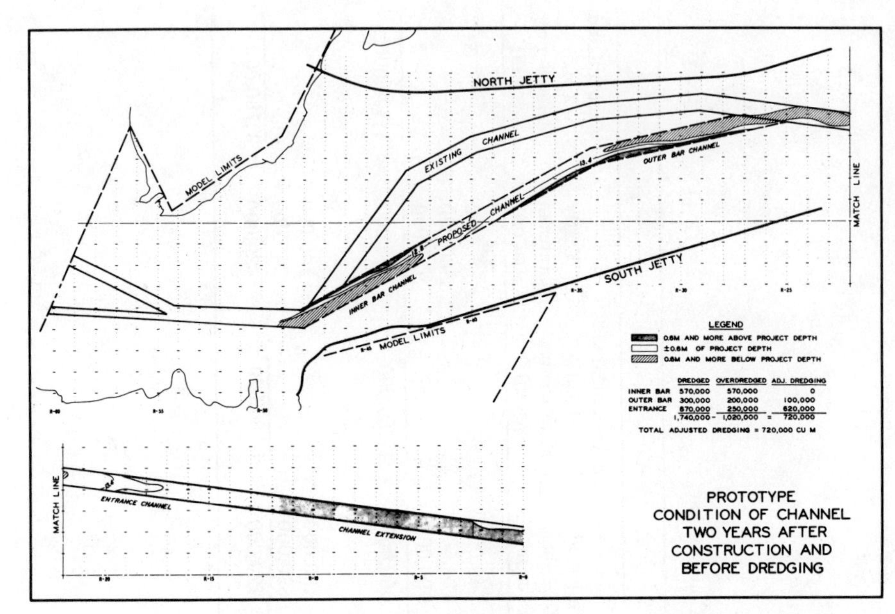

Figure 6-P.

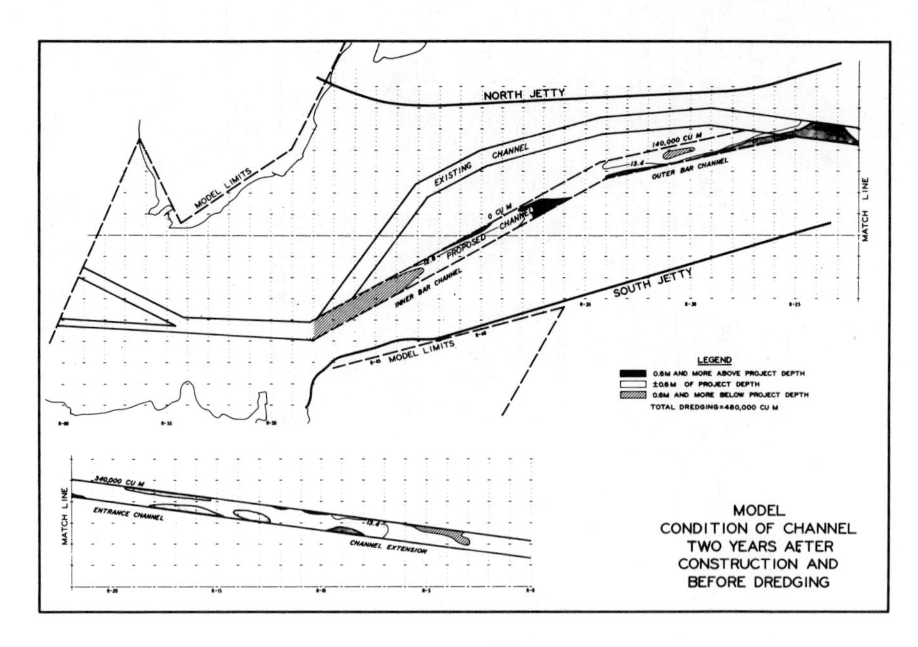

Figure 6-M.

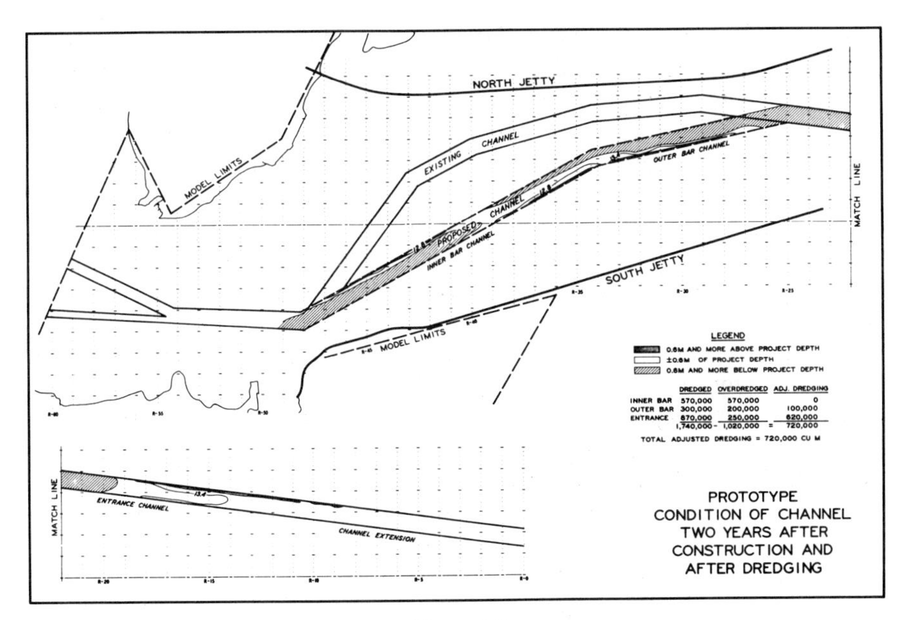

Figure 7-P.

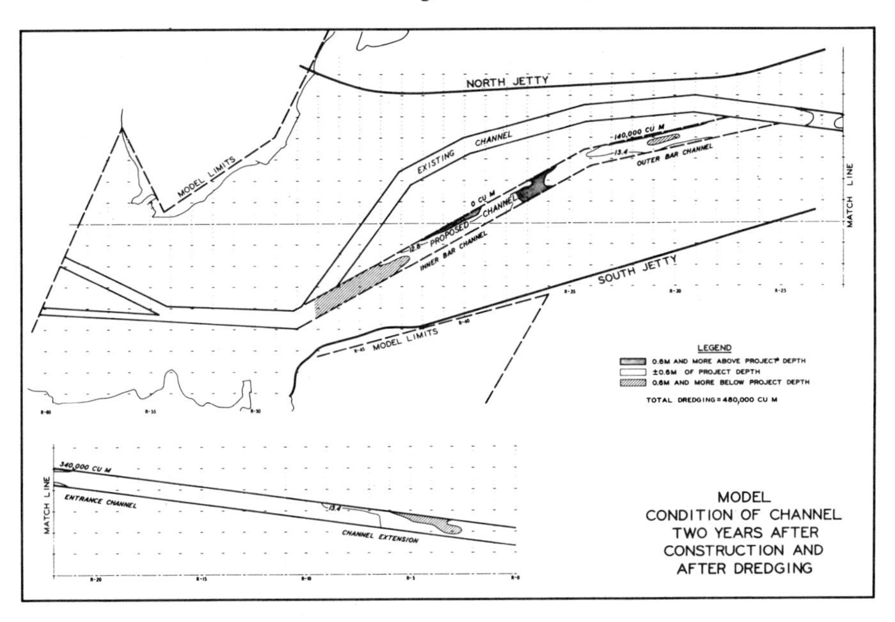

Figure 7-M.

channels, respectively, at the end of the second maintenance-dredging operation. It will be noted that the condition of the model channel is very close to project depth plus allowable over-depth, while that of the prototype channel is well below the allowable over-depth throughout most of the channel. Thus, a much greater amount of dredging was actually accomplished in the prototype than that required to restore the channel to project dimensions. This occurred because the dredge operates on the basis of a time schedule developed from a prior prediction of the anticipated shoaling quantity. Although shoaling was substantially less than anticipated, the dredge operated for the full time scheduled.

During the second maintenance-dredging operation, the quantities dredged in model and prototype were 480,000 and 1,740,000 cu m, respectively (Table 1). After making the adjustment for dredging accomplished below the project depth plus allowable over-depth (Table 1), the adjusted prototype maintenance was 720,000 cu m. The total dredging requirements for the two years following completion of construction for both the model and prototype (adjusted) were 1,780,000 cu m. Therefore, based on two years of experience following completion of construction, it is quite obvious that the model predictions with respect to both shoaling requirements and location of shoaling in the realigned channel were borne out almost exactly by experience in the field. The inner bar is no longer a serious shoaling problem, and total maintenance-dredging requirements for the realigned channel are actually appreciably less than for the original channel alignment at lesser depth.

Considering the jetty entrance as a whole, and excluding the navigation channel, which has already been discussed in detail, Figures 8-M and 8-P show conditions throughout the jetty channel in model and prototype, respectively, two years after completion of construction. It will be noted that, in the part of the abandoned channel used for disposal of dredge spoil, depths have been reduced from about 13 m to approximately 9 m in both model and prototype. Prototype depths immediately adjacent to the jetty in the area are significantly greater than indicated in the model because the hopper dredge could not be safely operated close to the jetty. On the other hand, dredge spoil in the model was placed throughout the disposal area shown in these figures. It is interesting to note that a shoal having depths less than 9 m in both model and prototype developed across the abandoned channel, whereas major shoaling formerly occurred in the inner bar channel. On an overall basis, it appears that shoaling of the abandoned channel has been slightly more extensive in the prototype than was indicated from the model tests. One possible explanation for this difference is the fact that considerable dredging was done in the prototype to create an anchorage area just north of the inner bar part of the relocated channel, and this dredging may have resulted in the more extensive filling of the abandoned channel. This anchorage area was tested separately in the model, and while the

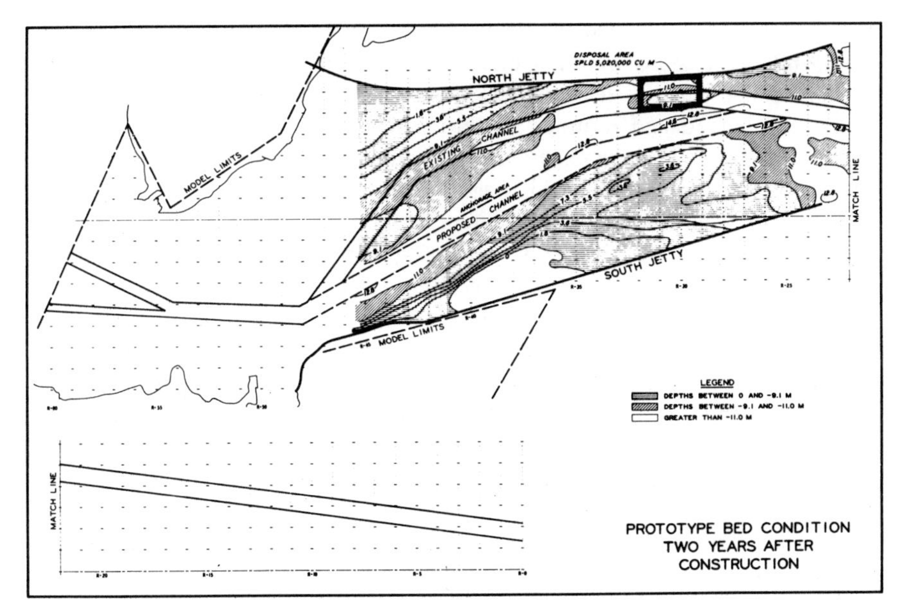

Figure 8-P.

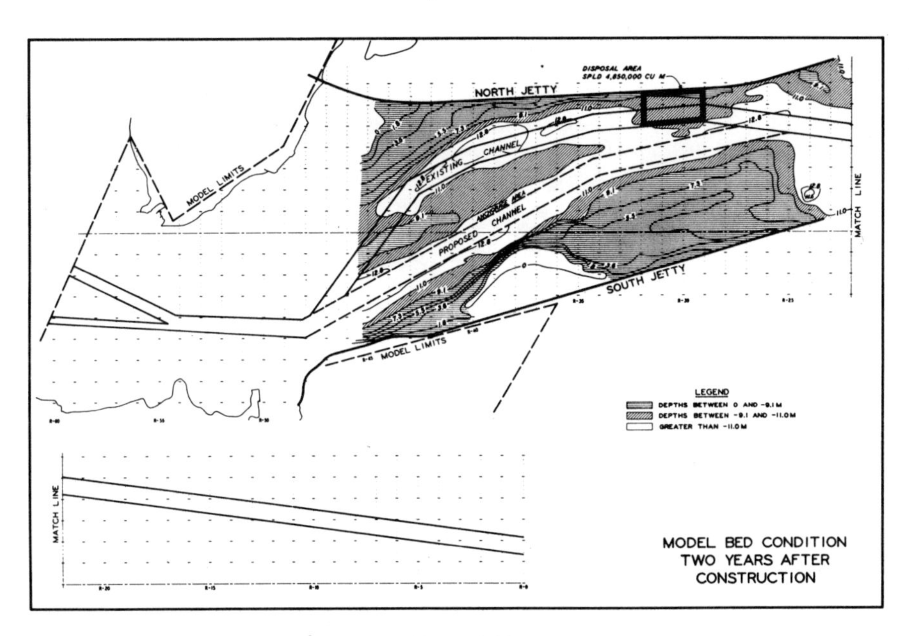

Figure 8-M.

results indicated no adverse effects on the realigned channel, possible effects on the abandoned channel were not evaluated. Concerning dredge-spoil disposal in the abandoned channel, fears were expressed by some that this material would be rapidly eroded and returned to the realigned channel. However, both the model tests and prototype experience indicated that no serious problem would be created by spoil disposal in this area.

Figures 9-M and 9-P show the extent of scour and fill that has occurred in model and prototype, respectively, during the period from beginning of construction to the end of the second year following construction. The navigation channel has been eliminated from this comparison, because the realigned channel was constructed and two maintenance-dredging operations therein were performed during this period of time. It will be noted that major changes in depth are concentrated in the abandoned channel, and especially in that area used for dredge-spoil disposal. In this comparison, one can readily see the increased tendency for shoaling of the abandoned channel in the prototype as compared to the model predictions. In the area between the abandoned channel and the relocated channel, the model indicates fill and the prototype indicates scour; however, all of this indicated scour is attributable to the anchorage area dredged in the prototype. South of the abandoned channel, the

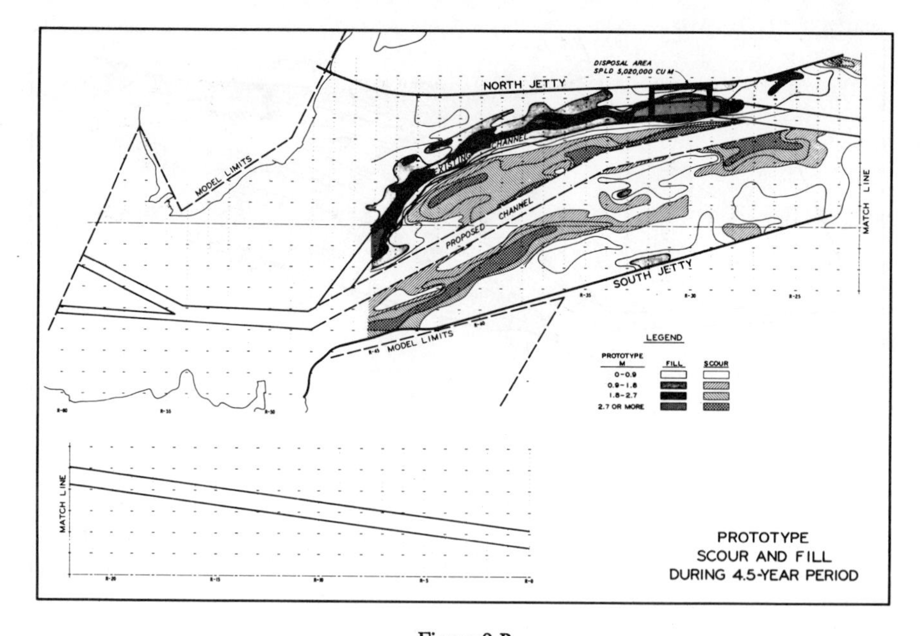

Figure 9-P.

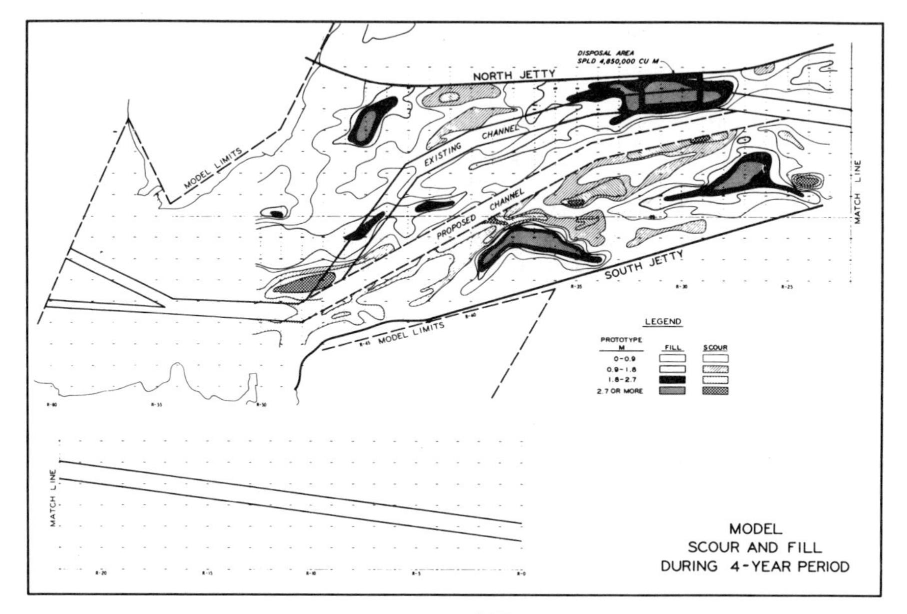

Figure 9-M.

model indicates two areas of rather substantial fill that are not borne out by prototype experience. The inner of these areas is in extremely shallow water and is typical of the unnatural model bed formations developed in shallow water when using crushed coal as bed material and with exposure to significant wave action. No immediate explanation is apparent for the outer fill area; however, it is probably due to unnatural effects on wave defraction and refraction in the jetty entrance caused by the distorted scales of the model. Some filling in this general area is indicated in the prototype surveys, but such filling was much more extensive in the model tests.

On an overall basis, it appears that tests in the Galveston Bay entrance model predicted with good accuracy all significant effects of the plan developed and constructed in the prototype, and especially in the navigation channel. It is planned, however, to continue to watch carefully future developments in this area, since a two-year period following completion of construction may not be adequate completely and accurately to document the performance of the improvement plan.

ACKNOWLEDGMENT

The study described and the resulting data presented herein were obtained from research conducted under the Civil Works Program of the U. S. Army Corps of Engineers by the Waterways Experiment Station and the Galveston

District. Permission was granted by the Chief of Engineers to publish this information.

REFERENCES

1. Simmons, H. B. and Boland, R. A. Jr.
1969 **Model study of Galveston Harbor entrance, Texas; hydraulic model investigation.** Technical Report H-69-2, U. S. Army Engineer Waterways Experiment Station, CE, Vicksburg, Mississippi.

2. Simmons, H. B.
1971 Model-prototype confirmation. Paper presented at Coastal Engineering Research Center Coastal Inlet Short Course. U. S. Army Engineer Waterways Experiment Station, CE, Vicksburg, Mississippi. (Unpublished data.)

SIMULATION OF SEDIMENT MOVEMENT FOR MASONBORO INLET, NORTH CAROLINA

by

William C. Seabergh[1]

INTRODUCTION

The prototype

Masonboro Inlet is an entrance through a sandy barrier beach into a channelized bay that has very little fresh-water inflow. The inlet, located on the North Carolina coast about 40.2 km north of Cape Fear, is known to have existed since 1738 and has held a relatively stable position since 1928 (5). However, improvements were needed, since channel depths over the bar were shallow and the channel alignment was continually changing.

In 1965 and 1966, a single jetty (1200 m long), a deposition basin (5.2 m deep), and a channel (4.3 m deep by 122 m wide) were constructed as shown in Figure 1. The inner portion of the jetty, designated as having a weir, was built of concrete sheet piling with a crest elevation of mean sea level. The weir part of the jetty would permit long shore transport from the north to pass over the jetty and settle in the basin rather than enter and shoal in the channel. When the basin filled, a hydraulic dredge could pump the sand upbeach or downbeach, depending on where the need for beach restoration was greatest. The location of the jetty-basin system on the north side of the inlet was based on considerations of the wave climate. Observations indicated that strong northeasters brought the largest waves and thus the greatest littoral drift from the northerly direction.

The project, as designed and built, functioned well for about a year, the

1. U. S. Army Engineer Waterways Experiment Station, P. O. Box 631, Vicksburg, Mississippi 39180.

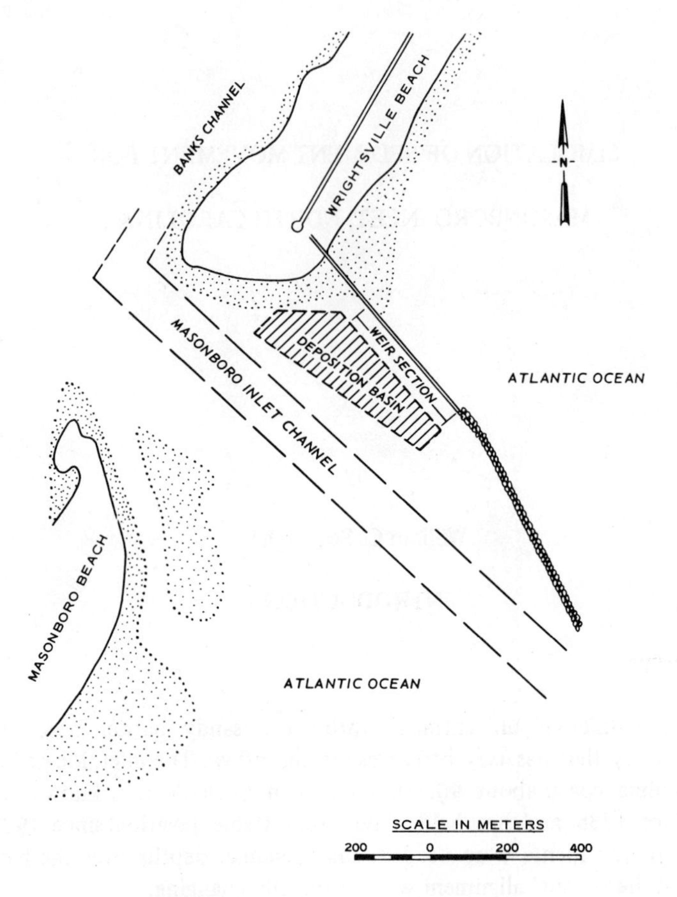

Figure 1. Plan of project design.

deposition basin filling with sand. Early in 1968, however, the channel relocated to the north, toward the jetty and deposition basin. The channel has since deepened and moved adjacent to and along the entire length of the jetty, presently reaching depths over 9 m near the structure (7).

The model

Under a Corps of Engineers research program, the General Investigation of Tidal Inlets, a fixed-bed model study of the inlet is underway. This model (Fig. 2) had a dual purpose: first, it would be a design tool to find solutions to problems at the inlet; second, it would be an investigative tool, used to determine whether a model could find the causes of the problems at the inlet and whether a model study could have aided in the prevention of the shifting

Figure 2. The model.

channel. As a part of the program to meet these purposes, sediment movement on the bar and near the inlet was simulated.

The model was constructed to a horizontal scale of 1:300 and a vertical scale of 1:60. Prototype velocity and tidal height at the inlet were measured. In a process called "verification," the model was adjusted by trial and error to reproduce those data. The hydraulic verification indicated that the vertical and horizontal water measurements were in agreement with the prototype and that the velocity distributions of the prototype were reproduced in the model.

VERIFICATION OF SEDIMENT MOVEMENT

Preliminary tests

With the use of wave machines installed in the model, trial tests were conducted to determine the ability of the model to simulate sediment movement. In order to select the most suitable material for the tests, various materials were placed on the concrete bed of the model and allowed to move under the influence of waves and tidal currents. The materials tested included sand (S.G. = 2.65; d_{50} = 0.23 mm), expanded shale (S.G. = 2.1; d_{50}=1.6 mm), naturalite (a ground pumice with S.G. = 1.7; d_{50} = 0.95 mm), and 2 plastic materials with S.G. = 1.18 and 1.05, both cubical in shape, about 3 mm on a side. The heavier materials moved well on the ocean shoal regions, but did not respond to tidal currents in deeper regions of the model where maximum velocities of about 16 cm/sec were observed. The lightweight plastic (S.G. = 1.05) was too light when placed on shallow regions and was washed away. The plastic of S.G. = 1.18, known commerically as Tenite Butyrite, proved to be the most suitable material to use under the influence of both waves and tidal curents.

Method of study

The model region to be studied for sediment movement was approximately 4.5 m wide by 6 m long and included the inlet and the ocean shoal. A uniform layer of plastic averaging 5.7 mm thick was placed on the region in a blocked grid. The area of each block was measured so that a known amount of material could be proportioned for placement in each block. The placement procedure included bagging material for each block in separate packets, then carefully spreading the material in its appropriate block during a period of low-water slack, attempting to complete placement as quickly as possible, usually within a 5-minute period. A photograph of the grid after a test is shown in Figure 3.

Figure 3. After a shoaling test.

With the material and the techniques for its use established, a historical period with which to verify the sediment movement was chosen. Ideally, at least a one-year history of bed movement is needed in order to realize the effect of a full-wave season. The model bed was molded in concrete to the September 1969 hydrography, so the period of verification began in 1969. A two-year verification period, 1969-1971, was chosen to take advantage of the greater accumulation of bed movement that took place in the two-year period. A fill and scour map of the 1969-1971 period (Fig. 4) shows the sedimentation pattern that the model was required to duplicate to demonstrate a proper adjustment.

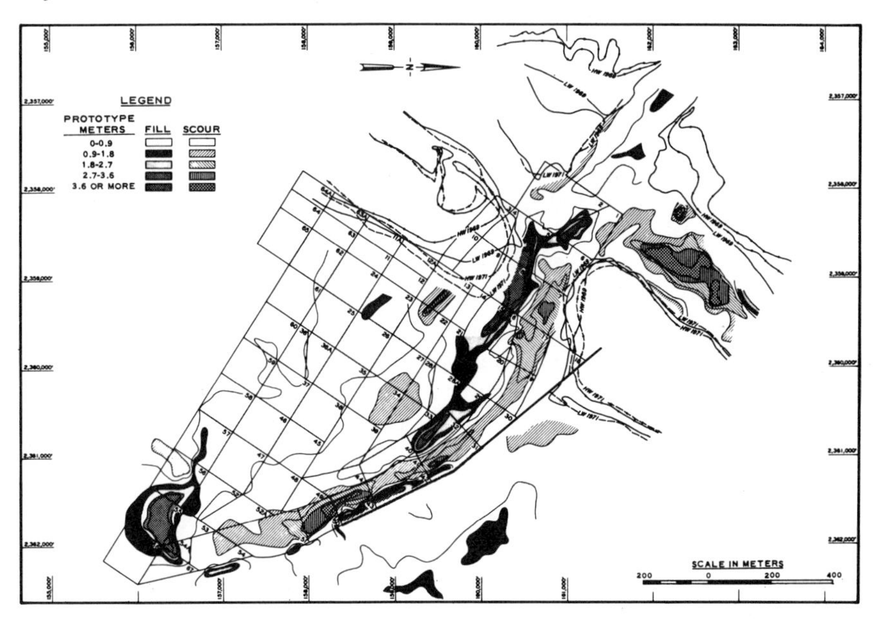

Figure 4. Masonboro inlet scour and fill.

Adjustment of the model

Many adjustable model parameters act together to produce the right balance of forces to simulate the prototype movement of sediment. These include: wave height, wave period, wave direction, duration of wave activity, duration of the test, tidal range, and mean tidal elevation. These parameters must be adjusted by trial and error until they, in concert with the response of the modeling material, produce model sediment patterns similar to those produced by the prototype during the selected verification period.

Since two wave generators were available for use and because of the difficulty in moving them during any test run, two wave directions were chosen and fixed for the duration of the tests. The choice was based on limited wave-observation data for the area – mostly ship-wave observations. Essentially the directions selected, S16°E and N84°E, represent the approaches of longer period, summer swell from the South Atlantic, and shorter period, steep winter waves from North Atlantic storms.

With the wave directions determined, the next step concerned selection of wave characteristics. Since the model was of distorted scale, it was not possible to reproduce refraction and diffraction simultaneously. Refraction in distorted-scale models is governed by the depth scale and diffraction by the horizontal scale. The refractive effect was considered the most important to model, since in the region of the inlet the offshore bar causes considerable refraction (2). Model wave characteristics were therefore based on the vertical scale. It is important to note that the refraction pattern will be accurately reproduced; however, the wave kinematics will not be similar (4) since the time scale for the wind-wave period is not the same as that for the tidal wave. This point, however, is not critical since the time scaling for the sediment movement is found by trial and error, and does not necessarily correspond to the wave kinematics in a distorted model.

A series of 30 verification tests were run to determine the proper combination of height, period, and duration of waves. Wave heights were relatively easy to choose since the plastic material could not endure too high a wave without being completely eroded from the shallower shoals. Heights chosen for study varied from 0.38 m to 1.2 m prototype, i.e., 0.6 cm to 2.0 cm in the model. The tests showed that wave heights which generally followed prototype trends of higher waves from the north and lower waves from the south actually gave shoaling trends similar to those of the prototype. The wave period for waves from the north was chosen as 7.4 sec prototype or 1.0 sec in the model, and the wave period for waves approaching from the south was 11 sec prototype or 1.42 sec in the model.

Once wave magnitudes and periods had been chosen, the test duration could be so adjusted that the energy input from all hydraulic parameters gave reasonably good results. A duration of six tidal cycles (two hours in the model) was found to be sufficient to give enough material movement to simulate a two-year period.

The hydraulic variables that were considered possible controls were the tide range and its mean level. The mean range of tide (1.16 m) was used throughout, but it was noted that, if the mean level was raised 0.15 m (prototype) above mean sea level, better shoaling patterns were achieved. This change was by no means a drastic change to the verified hydraulics.

Since in nature there is usually a continuous input of material to the inlet system, the technique of feeding material into the system to simulate littoral drift material was used. Trial and error determined a method of feeding material from the south along Masonboro Beach and determined what amount it was necessary to use.

Another parameter whose wide range of variations could be adjusted for the verification study was the duration of the attack of a particular wave. At first, a 2:1 proportion of north waves to south waves was tried. This meant that the north waves were run for two model tidal cycles, the south waves were run for one model tidal cycle, then the sequence was repeated to complete the test. Shoaling patterns in the model did not show good agreement with the prototype with this technique. Another variation was to generate 10-sec wave bursts each minute of the 19-minute tidal cycle. The idea behind this was that the wave turbulence would suspend sediments and they could then be more easily transported by tidal currents. This technique gave poor results also, possibly because the current patterns that were set up by the mass transport caused by the continuous wave train were lost. Finally by alternating north and south waves through six model tidal cycles, a fill-and-scour trend similar to that in the prototype was achieved.

Interpretation of verification results

The model scour-and-fill pattern is shown in Figure 5. The numbers appearing in each block of Figure 5 are percentage increases or decreases (fill or scour) of the original amount of material placed in the block at the beginning of the test. Regions of extremely heavy fill have percentages in the hundreds, but the maximum scour indicated can be only 100 percent, since once material has entirely vacated a block the concrete bed is reached and no more scour can occur. It was obviously not possible to determine corresponding percentages for the prototype.

A comparison between Figure 5, the model verification, and Figures 4 and 6, the prototype patterns, shows how well sedimentation was simulated. Blocks 29 and 32 show a heavy degree of fill, as does the prototype. Generally, material of the more southern shoal blocks moved over to locations alongside the present channel, filling in the southern edge of this channel. Also, blocks 3 and 10 are typical of the region along the south side of the inlet gorge which shows increased filling, thus shifting the deep gorge farther northward. In blocks 53, 54A, and 67, regions well beyond the jetty, it can be seen that material in the model is filling blocks just as in the prototype. The main channel shows scour (blocks 30, 31, 41, 42, 43, 50, and 51) and continues to deepen in the prototype; this is indicated in the model by the heavy scour in these blocks. The shoaling verification was thus considered to be satisfactory, although the model sedimentation tests must be considered to be qualitative rather than quantitative.

1964 CONDITION TESTING

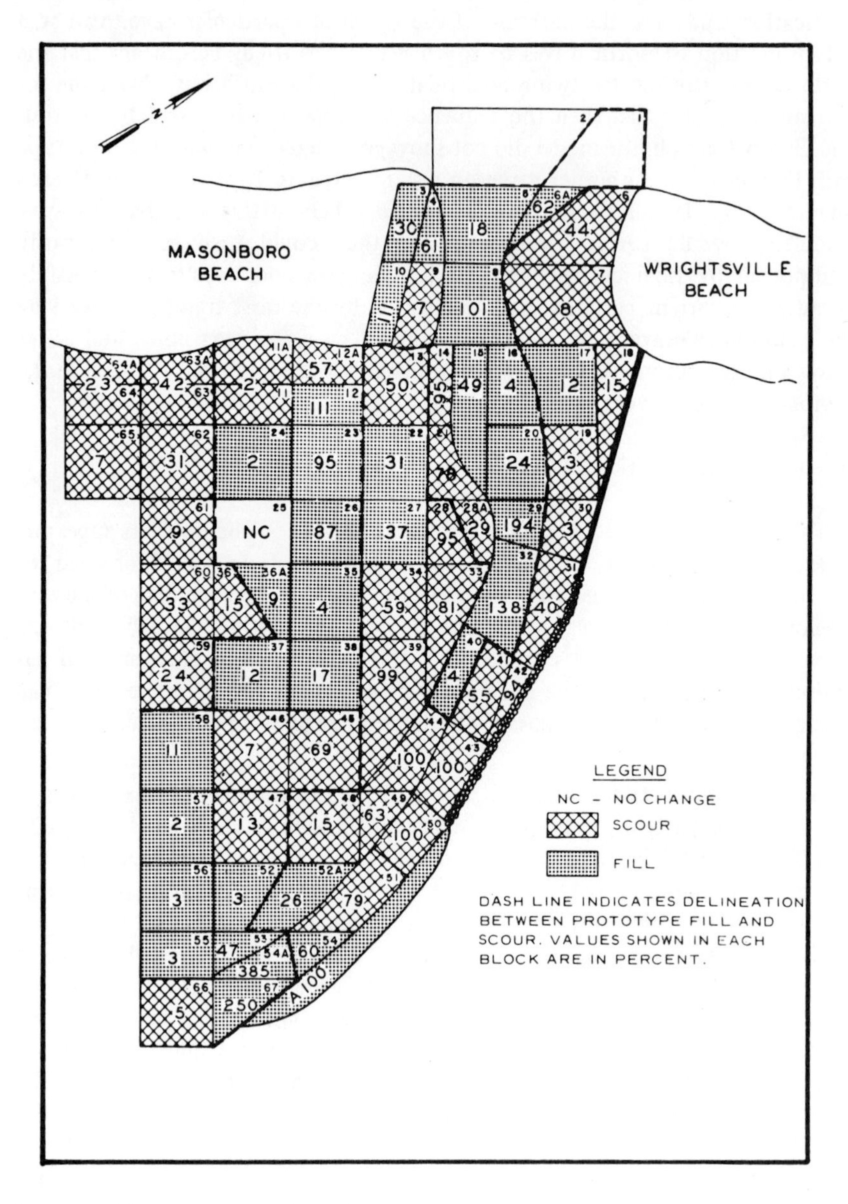

Figure 5. 1969 verification of shoaling.

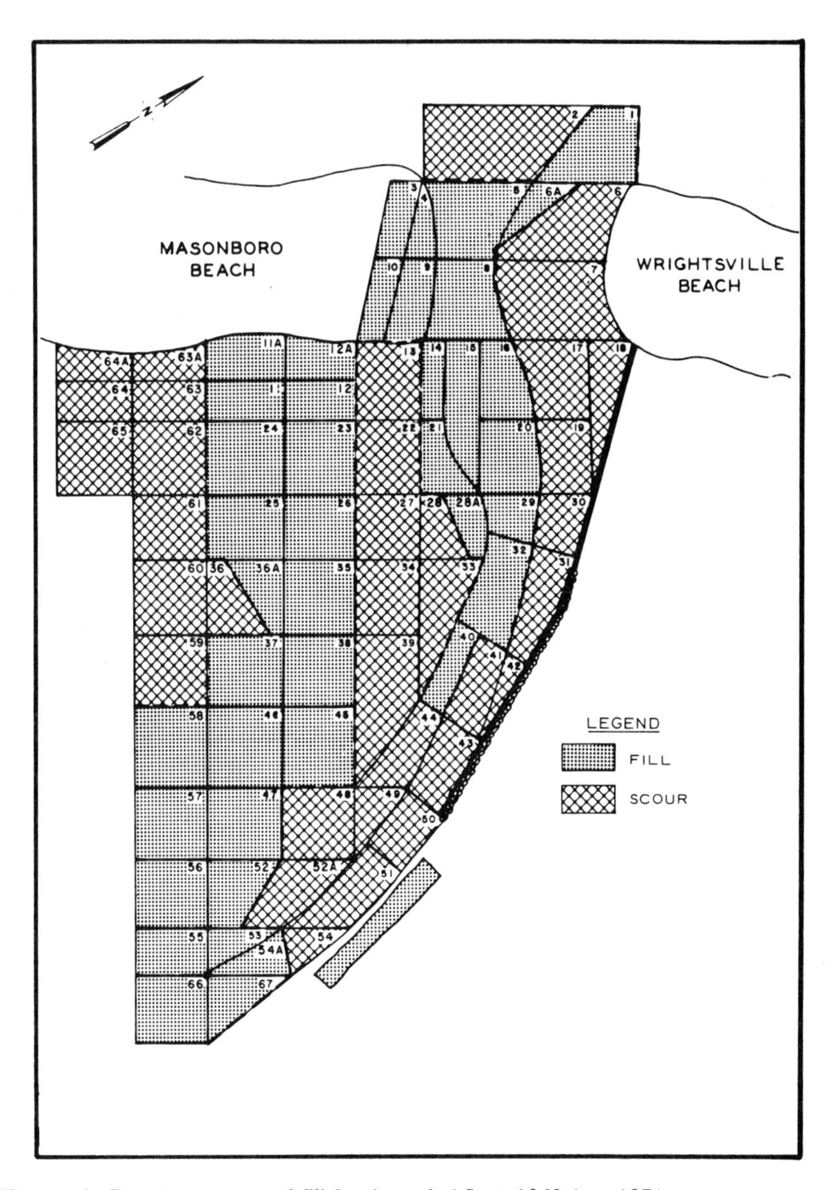

Figure 6. Prototype scour and fill for the period Sept. 1969-Aug. 1971.

At this point in the study the hands of time were turned back, and the model inlet was remolded to its 1964 pre-jetty hydrography. The objective of this step was to take the point of view of an investigator who would test a project plan in a model before it was constructed. It can be assumed that the 1964 conditions were verified by a method similar to that used for the 1969-1971 data. Next, a base shoaling test was run for the inlet as it existed in 1964 using a testing technique identical to that developed during the sedimentation verification. Then a plan was installed, i.e., a channel and deposition basin were dredged and a jetty constructed in the model. The test results shown in Figure 7 indicate that sediment would encroach from the south side of the inlet, that the channel would fill, lose its identity, and possibly shift northward to the scoured locations near the jetty, and that the deposition basin would fill. Since all these results did occur in nature, it appears that a model study would have been helpful in the design of the project for this particular inlet.

1966 CONDITION TESTING

The inlet region was next remolded to its 1966 conditions and a shoaling test was again run. The 1966 condition corresponded to that time just after the design project had been completed. Results were similar to those for the 1964 conditions, except that more severe scour appeared nearer the jetty, as shown by fill-and-scour patterns in Figure 8. These differences may be attributable to a sharply defined ridge that separated the deposition basin from the navigation channel for the 1964 condition but not for the 1966 condition. This ridge may have diverted ebb currents from the jetty and deposition basin for the earlier tests.

The 1966 shoaling test can also be compared to what occurred in the prototype between 1966 and 1968, since the test is simulating a two-year period. The model predictions of sediment motion in the form of fill and scour for 1966-1968 are shown in Figure 8. The fill-and-scour patterns for the prototype for the period 1966-1968, using the block layout scheme, are shown in Figure 9. Patterns are again very similar with a few variations of one from the other.

PATTERNS OF SEDIMENT MOVEMENT

From the above, it can be seen that fixed-bed model shoaling tests can be an aid to the design of various coastal projects. Also, through these tests certain detailed mechanisms of sediment movements were noted and will now be discussed.

Two diagrams, Figures 10 and 11, illustrate the general patterns of sediment movement for ebb and flood flows, respectively, as observed in the model. These

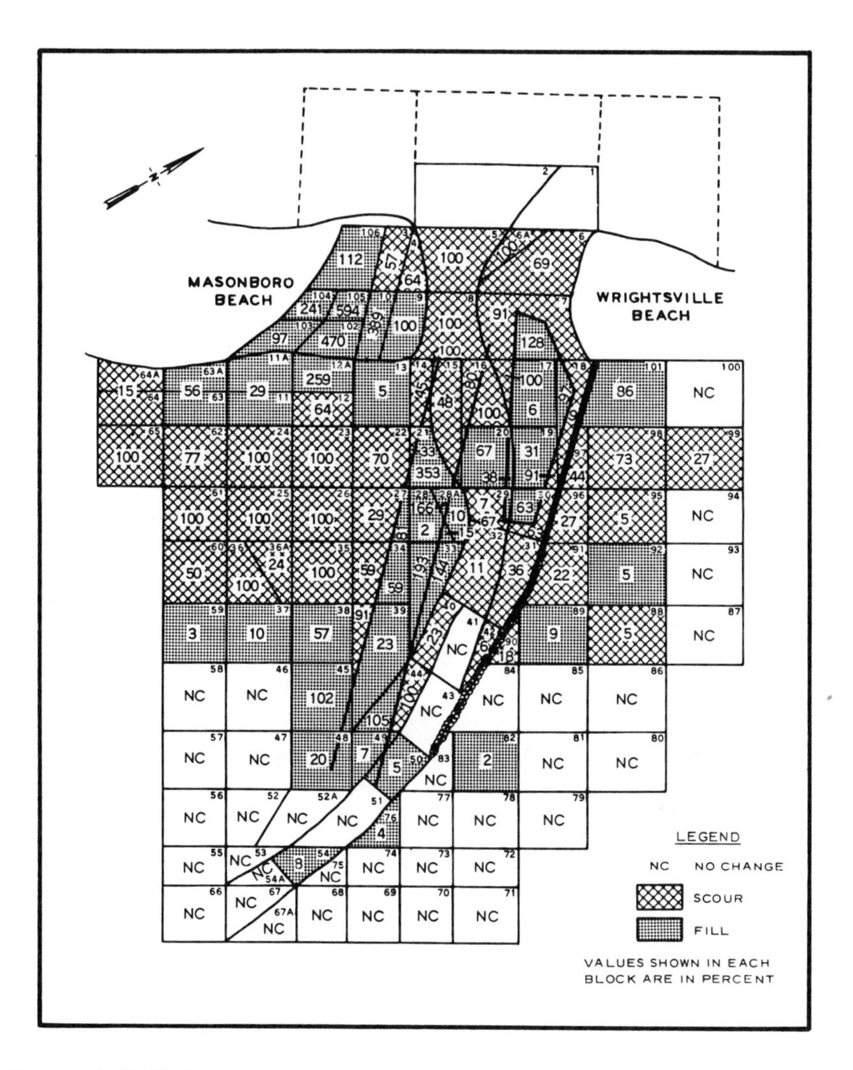

Figure 7. 1964 plan test.

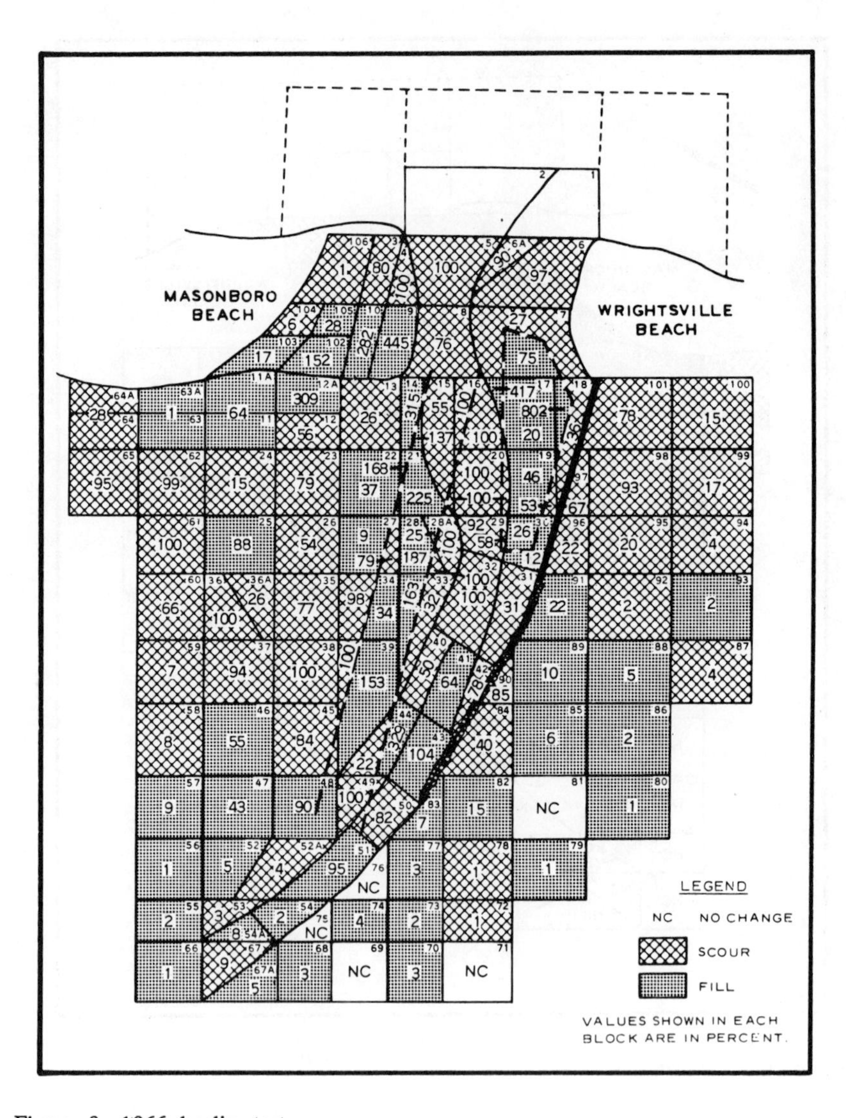

Figure 8. 1966 shoaling test.

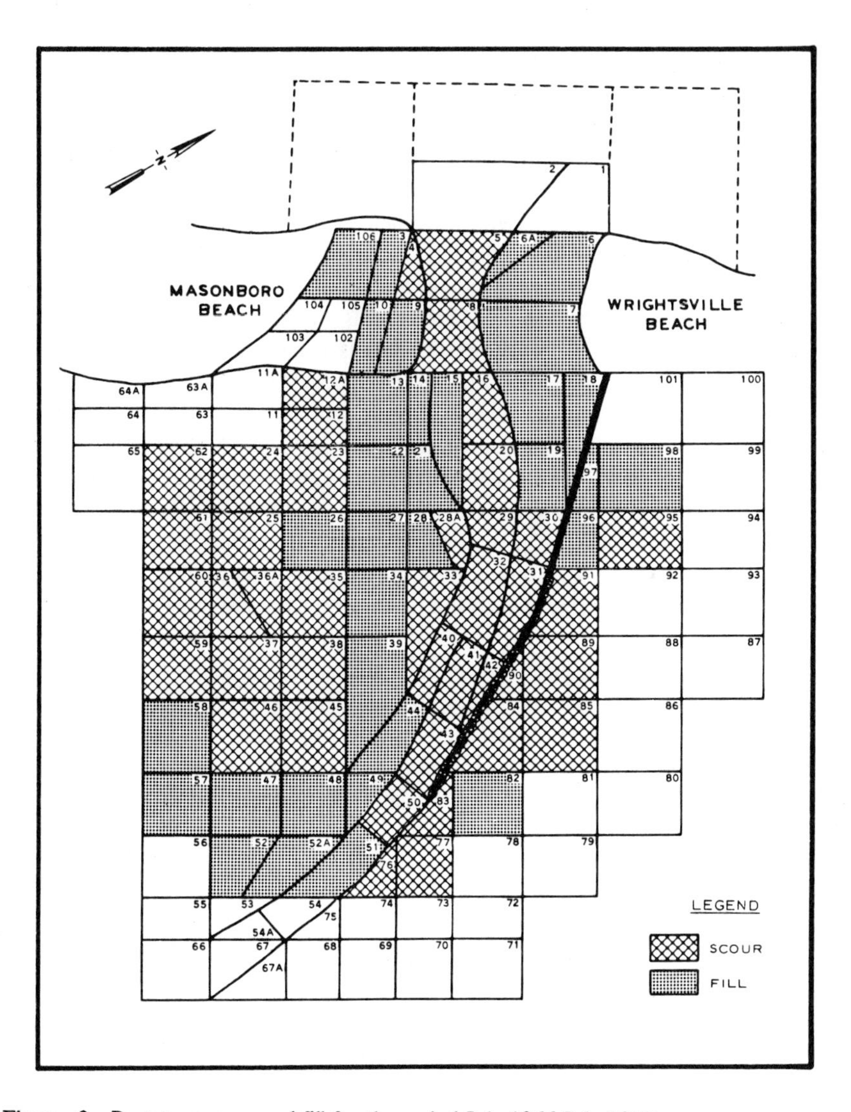

Figure 9. Prototype scour and fill for the period July 1966-July 1968.

patterns are general and, for a given tide, exist for almost any direction of offshore wave approach, since the ocean bar refracts waves so that their approach at the shorelines near the inlet is approximately the same, no matter what their original direction.

The ebb patterns (Fig. 10) are generally away from the inlet and out over the bar. Sediment also travels down the channel along the jetty where some deposits oceanward of the tip of the jetty keeps building the bar farther outward. In 1964, the bar had not reached the end of the jetty, but by 1969 it had built out 450 m past the jetty.

A very interesting ebb-sediment-movement pattern occurs where sediment moving out of the channel along the jetty is directed somewhat southward when it meets waves. On the oceanward side of the shoal opposite the jetty, which is a region in the lee of ebb currents, material is pushed up on the shoal in a direction opposite to the ebb currents. This is a mechanism that continually builds the shoal and occurs for both the south and the north wave approaches.

Material carried seaward during ebb flow is probably the same as that moved toward the inlet during flood, as shown in Figure 11. The flood currents over the shoal, aided by refracted waves, carry material towards the inlet entrance from both sides of the inlet. In the model, material passed over the landward edge of the weir along the beach. In the prototype, one would suspect that the turbulence north of the weir would be great enough to suspend sand along most of the weir region and carry it into the basin, or into the north side of the inlet. Sediment carried into the interior channels during flood flow finds areas of settlement just inside the entrance on each side of the inlet, as shown in Figure 11. It can be noted that the ebb channels hug the banks opposite these deposition regions, so that material tends continually to fill these regions.

Another factor that contributes to the overall sediment patterns is related to a complex phenomenon of wave-breaking and the mass transport of the water from the breaking wave. Waves break on the shallow south shoal, then push water towards the inlet. There is some interaction with currents at the inlet gorge, but generally the water from the breakers tends to flow towards the deposition basin of the jetty. The current then proceeds to flow out along the southerly side of the jetty. This circulation exists, no matter what the wave approach, and its effect is superimposed on the tidal currents. Its pattern relates to the sediment movement by increasing ebb velocities; thus, the wave-enhanced scouring power of the ebb current is also a possible contributor to the movement of the channel towards the jetty by helping to align the ebb flow close to the jetty. Model ebb-velocity measurements near the crook or elbow in the jetty with waves being generated indicated significantly increased velocities compared to velocity measurements without waves in the model.

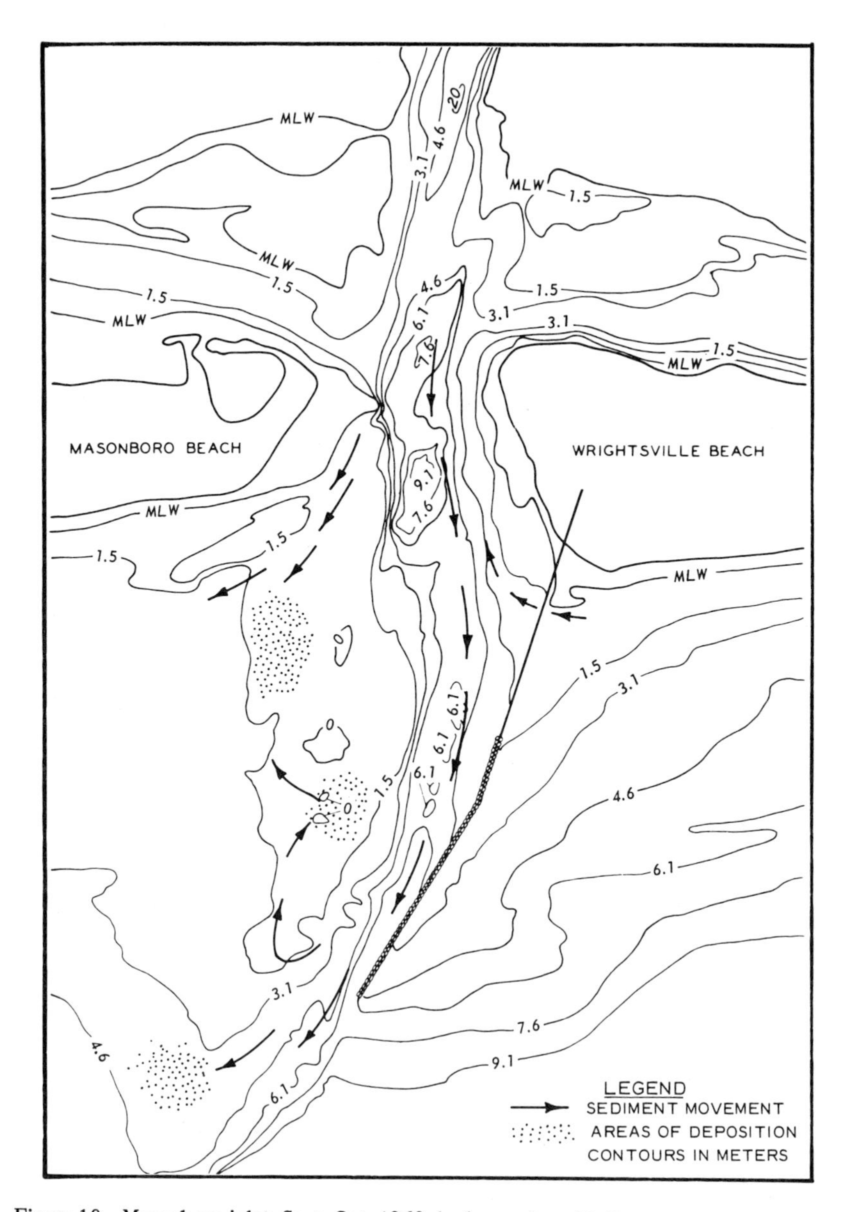

Figure 10. Masonboro inlet, Sept.-Oct. 1969, hydrography, ebb flow.

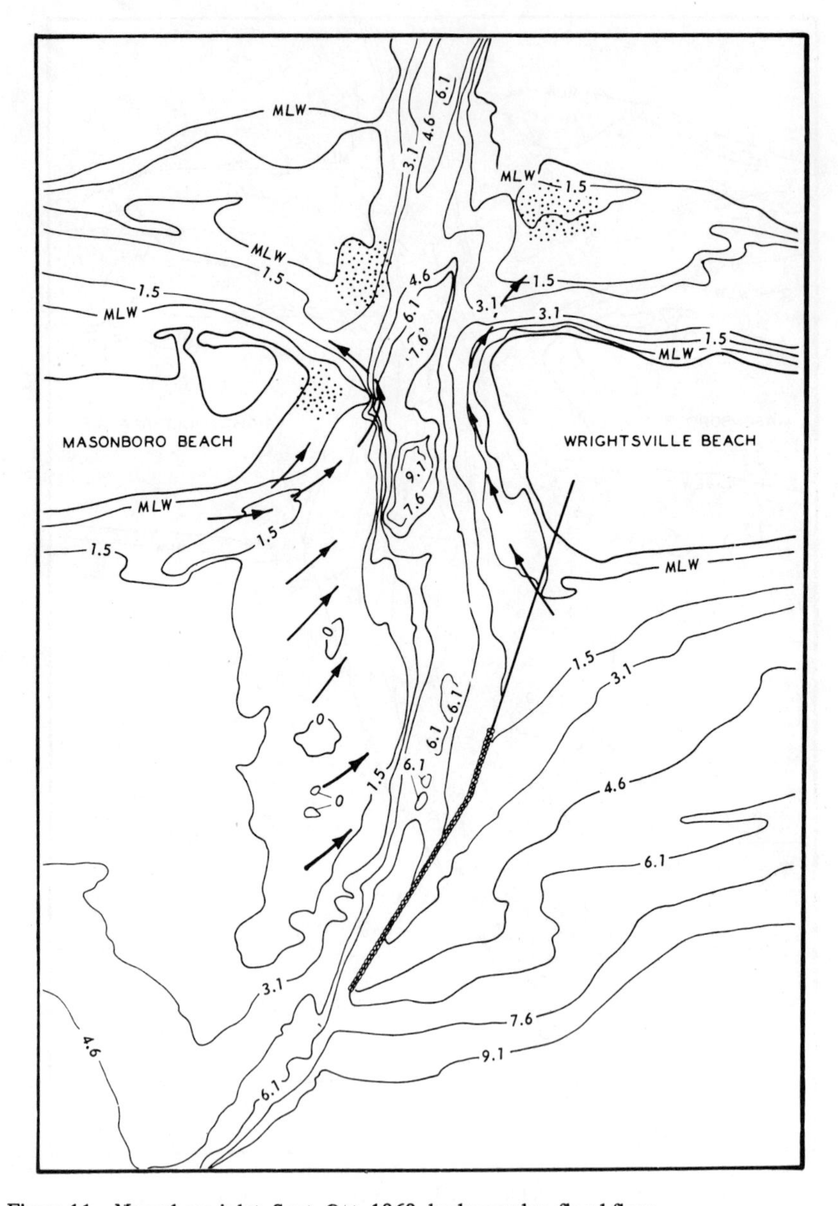

Figure 11. Masonboro inlet, Sept.-Oct. 1969, hydrography, flood flow.

CONCLUSIONS

A fixed-bed model study can give important aid in evaluating coastal projects. It can be helpful to an understanding of complex regions of sediment movement, and can simulate this movement in a qualitative sense. This was seen in all phases of the study. The verification showed that the model produced fill-and-scour regions similar to those of the 1969-1971 period it was simulating. Also the 1964 predictive tests pointed out sediment-movement trends that did occur in the prototype after the project was completed in 1966.

ACKNOWLEDGMENT

Data in this paper were obtained from research conducted by the U. S. Army Engineer Waterways Experiment Station under the Civil Works Research and Development Program of the Army, Corps of Engineers. Permission of the Chief of Engineers to publish this information is appreciated. The findings of the paper are not to be construed as representing the official position of the Department of the Army unless so designated by other authorized documents.

REFERENCES

1. Fisackerly, G. M.
1970 Estuary entrance, Umpqua River, Oregon. Technical Report H-70-6, U. S. Army Engineer Waterways Experiment Station, CE, Vicksburg, Mississippi.
2. Hayes, M. O., Goldsmith, V., and Hobbs, Carl H.
1970 Offset Coastal Inlets. In **Coastal Engineering–1970 Proceedings.** Vol. II, p. 1187-1200.
3. Herrmann, F. A., Jr.
1972 Model studies of navigation improvements, Columbia River Estuary Report 2, Entrance Studies, Section 3, Report No. 2-735, U. S. Army Engineer Waterways Experiment Station, CE, Vicksburg, Mississippi.
4. Keulegan, G. H.
1968 St. Lucie Inlet Model. (Unpublished data.)
5. Magnuson, N. C.
1967 Planning and design of a low-weir section jetty. In **Jour. Waterways and Harbors Division, American Society of Civil Engineers.** Vol. 93, No. WW2, May 1967, p. 27-40.
6. U. S. Army Engineer District, Wilmington, CE
1965 Masonboro Inlet, North Carolina – North Jetty; design memorandum. Wilmington, North Carolina.
7. U. S. Army Engineer District, Wilmington, CE
1970 Masonboro Inlet, South Jetty Report, Wilmington, North Carolina.

SEDIMENT TRANSPORT PROCESSES IN THE VICINITY OF INLETS WITH SPECIAL REFERENCE TO SAND TRAPPING[1]

by Robert G. Dean and Todd L. Walton, Jr.[2]

ABSTRACT

The hydraulic and sedimentary processes in the vicinity of an inlet govern the evolution of adjacent barrier islands and accumulations in the outer shoals. The wave-energy level can be an important agent in limiting the ultimate growth of these outer shoals or, in cases of low wave energy, in contributing to the ultimate closure of the inlet for nonmigrating inlets. The stabilization of an inlet, either by maintenance dredging or by jetties, may contribute to the continual accumulation of shoal material and intensified beach erosion. The latter is especially true in cases of bar dredging and disposal at sea.

A method is described and applied for calculating the materials accumulated in the outer shoals adjacent to inlets. The calculated outer shoal accumulations are presented for twenty-three Florida inlets; these volumes range from approximately 1 million cubic yards for Jupiter Inlet to more than 200 million cubic yards for Boca Grande Inlet. Calculated volumetric accumulations in inner bars and shoals are presented for four Florida inlets. This procedure contains inherent errors associated with changes of horizontal and vertical datums between earlier and later surveys.

1. This work is a result of research sponsored by NOAA Office of Sea Grant, Department of Commerce, under grant no. 04-3-158-43. The U. S. government is authorized to produce and distribute reprints for governmental purposes notwithstanding any copyright notation that may appear hereon.

2. Coastal and Oceanographic Engineering Laboratory, University of Florida, Gainesville, Florida 32611.

One conclusion of this study is that stabilization of inlets, particularly in areas of low wave energy, results in an offshore shoal of continuously increasing volume. The relatively high wave energy and strong currents on these shoals would winnow out the fine materials, leaving only coarser material suitable for beach-nourishment purposes. The volumes of these shoals are very significant when evaluated in terms of the erosion of adjacent barrier islands. As an example, for the estimated present total outer shoal volumes in Florida and the present rate of erosion along Florida's shoreline, these outer shoals contain enough material to forestall erosion for 76 years.

INTRODUCTION

Sedimentary transport processes in the vicinity of inlets are a complex phenomenon and are the combined results of tidal-current flows, wave action, effects of vegetation trapping and stabilizing material on the inner shoals, and possible effects of salinity and thermal stratification resulting in a net inward flow at the bottom of the channel. These factors are further complicated by the extreme variation in wave climate encountered, ranging from very mild wave conditions to hurricane conditions that may transport large volumes of sand to the inlet. A knowledge of these processes and the transported and accumulated quantities involved is important to a general understanding of coastal changes and to the functional design of inlets including maintenance dredging and such modifications as the addition of jetties.

Sediment transport in the vicinity of inlets is probably of greatest concern to coastal engineers because it is at these locations that the possibility exists for rather substantial losses of high quality beach sand from the littoral system. These losses are manifested by beach erosion which in extreme cases in Florida have amounted to average horizontal retrogressions of 40 feet per year over a period of 50 years.

This paper presents a qualitative discussion of the hydraulic and sedimentary processes in the vicinity of inlets, including classical concepts, and recent developments. In Section II attention is focused on the quantities of material accumulated in the inner and outer shoals of inlets and a method is presented, based on a single survey, for readily calculating the outer-shoal volumes for inlets that satisfy certain geometrical requirements. This method is applied to twenty-three Florida inlets and the volumes are presented and interpreted in terms of their significance with regard to the prevailing beach-erosion problem in the state. The calculation of the inner-shoal volumes is much more tedious and must be based on successive historical surveys. Calculations of inner-shoal volume are presented for four Florida inlets.

HYDRAULICS AND SEDIMENTARY PROCESSES

Inlet Configuration

As most inlets are shaped by nature to the idealized form of a nozzle, the ebb-flow patterns at an inlet are similar to those of a turbulent jet of water issuing from a nozzle, whereas the flood-flow patterns are similar to a more streamlined flow into the bell-shaped entrance of the nozzle.

On ebb flow, the momentum of the exiting water is confined to the deeper portions of the outer shoals, due to a lower water level, and forms a jet directed seaward. The high velocities in the central core of the jet can carry sand a considerable distance seaward. Lateral transfer of the jet's momentum causes an entrainment of adjacent waters and a consequent eddy formation (Fig. 1a). These eddies move water toward the inlet on ebb tide. On flood tide (Figure 1b) the water converges toward the inlet from all sides, especially in the swash channels (often called "flood" channels) which have a greater depth at this time. This idealized picture thus leads to alongshore currents on both sides of the inlet carrying sand toward the inlet at all times.

This effect is somewhat of a self-stabilizing process since sand that is deposited in the inlet causes increased velocities in the main channel section,

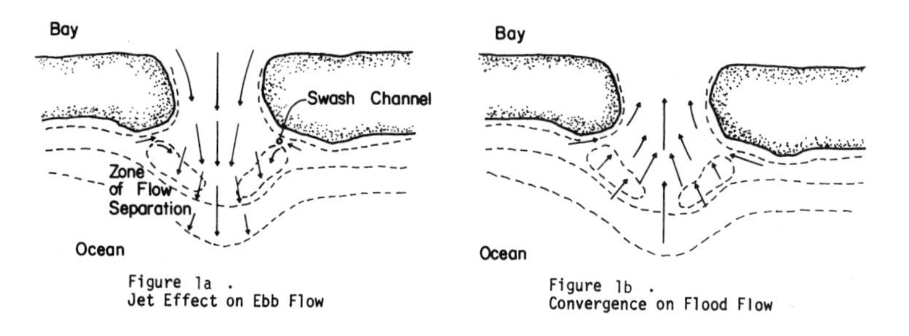

Figure 1. Idealized inlet flow patterns for ebb and flood.

which then widens the throat area of the inlet by erosion and decreases the currents again.

This configuration effect occurs mainly on the ocean side of inlets because of the breaking wave action on the inlet which "molds" the encroaching sand into a nozzle shape. Swash channels, though, are often seen on both sides of an inlet.

A good example of the effects mentioned is seen in Figure 2, showing Redfish Pass on both flood and ebb tides. The velocity patterns shown were determined by photographing stationary dye packets in the water during the maximum flows of the inlet.

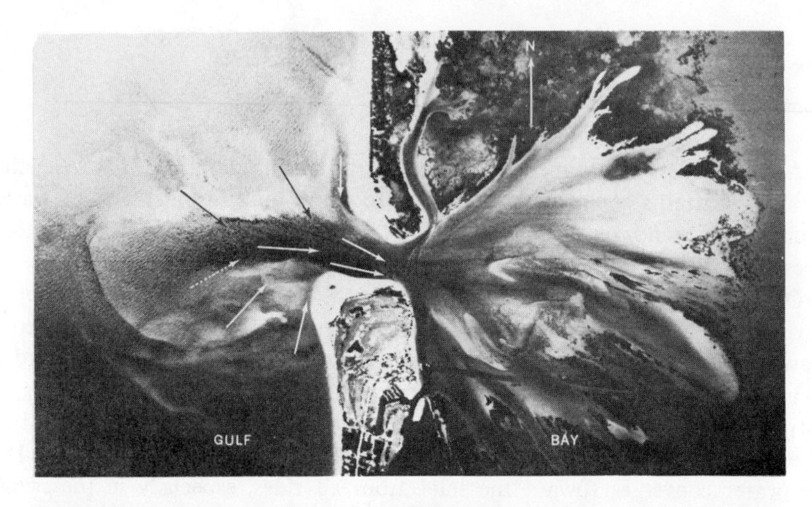

Figure 2a. Redfish Pass flow patterns on flood current.

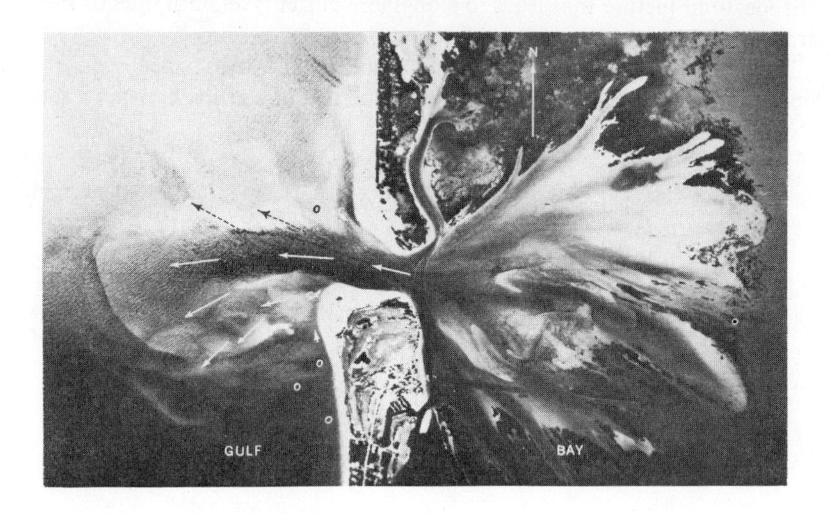

Figure 2b. Redfish Pass flow patterns on ebb current.

Refraction and Sheltering Effects Produced by Outer Bars

It has been noted in the literature (1 and 5) that refraction plays a major role in sand-trapping processes around inlets. This effect is very apparent at inlets where the dominant waves approach almost perpendicular to shore. Refraction of the waves around the outer bar causes a longshore current directed toward the inlet on both sides of the inlet, thus working in concert with the ebb-jet effect.

The outer bars also provide sheltering from the wave activity, thus producing a zone of low energy in which alongshore currents can deposit their material. This effect, combined with the refraction effect noted above, can produce an "offset" inlet configuration (2), where the dominant waves approach from a preferred direction. Such is the case of Redfish Pass (Fig. 2) which has considerable offset on the downdrift (south) side of the inlet.

The net effect of the above processes is to drive sand into the inlet, or to the edges of the channel, whence the tidal currents can then move the sediment to the bay shoals or outer shoals.

The Effect of Wave Energy on Limiting Shoal Volumes

Consider the inlet and outer-shoal features presented in Figure 1. The "forces" acting upon sediment which is brought within the influence of the inlet include both the effects of ebb- and flood-tidal flows and, particularly on the outer shoal, include the effect of waves that tend to drive the material back toward shore. To some degree then it is valid to regard the material in the outer shoal as affected by two opposing forces: (a) the tidal forces, which act predominantly offshore and, (b) the wave forces, which act to return the material toward the inlet. In a situation where these two forces are exactly balanced, the shoal has achieved an equilibrium condition in which over some reasonable time period there is neither growth nor reduction of the shoal volumes.

It is of interest to consider two areas for which the levels of wave energy are markedly different. In an area of high wave energy the forces tending to return the sand ashore will be significant. The result of these high shoreward forces is an equilibrium shoal volume that is probably relatively small. This is particularly true in a situation where there is a small net littoral drift and consequently there is not a requirement for the shoal to bypass large quantities of material. On the other hand, in an area where the wave energy is much lower, for example the west coast of Florida, the forces that tend to return the sand ashore are relatively small and therefore a large offshore shoal region would develop. Furthermore, if there is a small net littoral drift the tendency will be for the shoal feature to be composed of a permanent central channel between two shoal flanking regions. Recalling that the ebb flow occurs predominantly during periods of low ocean tide, for a small net littoral dirft the shoal material is transported farther and farther seaward through a relatively deep channel and the "equilibrium shoal volumes" would be much larger than would be the case in a higher wave energy region.

Figure 3 presents the bathymetry for Boca Grande Inlet, Florida. This entrance is in an area of relatively low wave energy and the offshore bathymetry

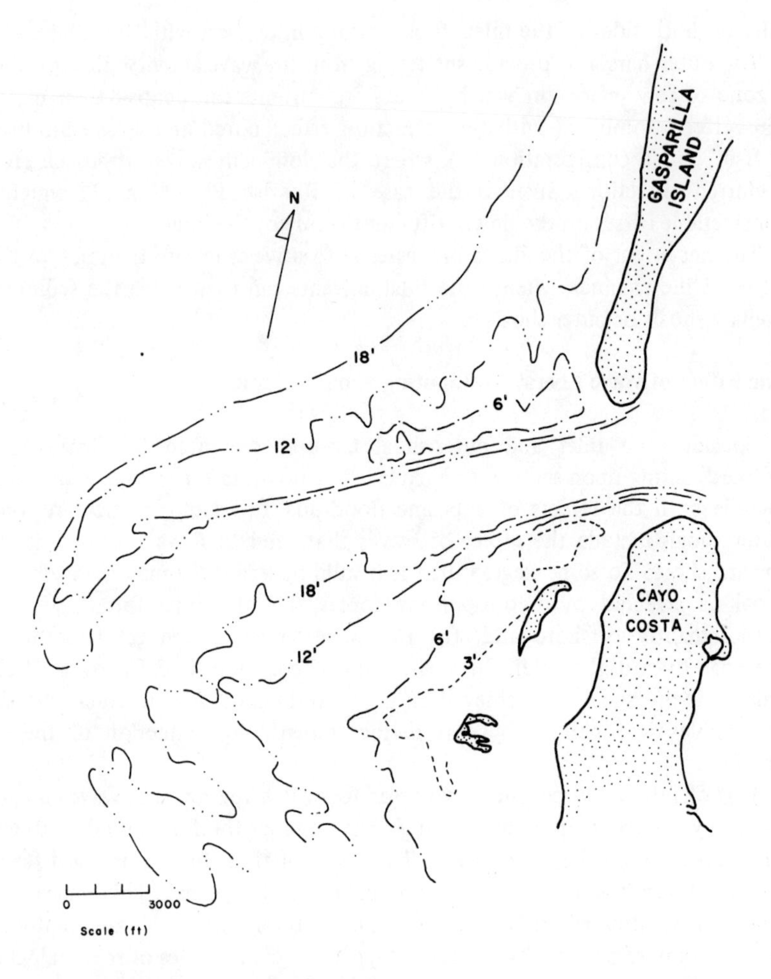

Figure 3. Showing volumes accumulated in outer shoals adjacent to Boca Grande Inlet, Florida (low energy shoreline).

includes a relatively deep channel flanked by two massive shallow depositional features.

Effect of Inlet Migration on Equilibrium of Inner and Outer Shoal Volumes

It is worthwhile to consider the effects of inlet migration on the equilibrium volumes of inner and outer shoals. The effects to be discussed will be more significant in the vicinity of an inlet in a region where wave energy is low. As described previously, if the inlet is stationary then the accumulation of material in the flank shoals will tend to direct the ebb currents toward deeper water, hence the material transported will be deposited in ever deeper water. The results of this are that if an equilibrium volume of the outer shoal is indeed reached, then the associated volumes are very high and would only be achieved through deposition of sand in deep water. The ultimate limit of such a process may occur when the channel is so long and its hydraulics are so inefficient that the inlet is not able to compete with other openings occurring as a result of breakthrough during storms. These competing inlets will survive whereas the old inlet will close and, over a period of decades or centuries, the shoal material will tend to be returned to shore. Much of this material that is deposited in depths greater than the wave base, or just marginally within the wave base, will not be returned to shore or, if returned, the process will be very slow.

Next consider the case of a migrating inlet. Various inlets such as Aransas Pass have been noted to migrate at rates approaching 200 feet per year (6). For the case of a migrating inlet, the channel would tend to be directed obliquely offshore in a direction opposite to that of inlet migration, and the outer bank of this channel would tend to be breached during periods of high wave action and spring tide. The equilibrium volumes associated with this offshore shoal would be much smaller than those for a stationary inlet because the material would not be transported to as deep water as for the former case. The equilibrium can be described as a process in which the "trailing" shoals are being continually driven ashore, whereas the material being driven offshore is occurring within the immediate vicinity of the inlet and is being derived, at lease in part, from the erosion of the shore in the direction of the inlet migration. In cases where habitation and development of a barrier island or shoreline exist, the allowance of inlet migration is not an acceptable approach to preventing large volumes of high quality beach sand from being tied up in the outer shoals.

Turning attention to the accumulation of material within the inner shoals, it is seen that there is no limit to equilibrium volume for a migrating inlet. Generally speaking, the inner shoal may play a primary role in causing an inlet to migrate. This is particularly true in the case of a strong net littoral drift. As the material accumulates on the updrift side, the channel tends to become smaller than the equilibrium size required by the interior bay plan area and the ocean tidal range. Therefore some erosion of the inlet throat occurs. Because there is a supply of sand on the updrift side, the effect of this erosion will be to cause a small migration toward the downdrift side. In addition, the inner shoals will result in an accumulation that tends to be located on the updrift side, i.e., near the source of supply. As these shoals grow, the resulting hydraulics become less efficient, and hence the competition of channels located further downdrift becomes more effective (3). It is clear that the shoals deposited on the interior of the migrating inlet will not be returned to the outer shoreline. However, in the case of a barrier island, the shoals may ultimately become part of the outer shore through continual migration of the barrier island and erosion of the outer shore due to loss of material (3). On the other hand, if the inlet does not migrate but is stationary, the inner shoals might develop to a stage where they would be in equilibrium. In essence, an upward bottom slope develops from the inlet toward the inner bay and, in the equilibrium stage, the downward seaward gradient produces a force which opposes those acting to cause additional material to be transported into and deposited within the bay.

Effects of Inlet Improvements with Examples

Inlet improvements effect the sedimentary processes around inlets and modify the sediment budget in the vicinity of an inlet. Generally speaking, there are two types of inlet improvements: (a) jetty construction, and (b) dredging to maintain a navigable depth. Improvements to any particular inlet may include one or both of these.

Jetty construction

The origin of the word "jetty" reflects the intended of the structure, i.e., to confine the current and to cause sand deposited near its tip to be "jetted" out into deeper water by the ebb currents. Additional important purposes of a sandtight jetty are the blocking of sand actively transported in shallow water from being deposited in the channel and the stabilization of an inlet against migration. In areas of moderate wave action, where the net littoral drift is substantial and the navigation channel is greater than 20 feet, natural processes are not then likely to be effective in reestablishing the natural bypassing after the jetty construction, and the only possible consequence (if provision is not made for artificial bypassing) is accretion on the updrift side and erosion on the downdrift side. If bypassing were to be reestablished naturally, then it follows that, because the sand-transporting capacity of waves is small in depths greater than 20 feet, an extensive plateau of sand at this depth would be required to provide an area sufficient for the total transport. The amount of sand that would comprise such a plateau would be very large and would be derived from and represent a loss from the nearshore region. Again the only possible effect is a net erosion of the shoreline. Therefore if one or more jetties are constructed in an area of significant net littoral drift, and if substantial beach erosion cannot be tolerated, then provision must be made for effective artrficial bypassing.

Weir jetties are a relatively recent innovation and include a low section close to shore in the updrift jetty. Ideally, the net drift passes over the weir and settles in a deposition basin which is located adjacent to but separate from the navigational channel. This design is best suited to areas in which there is a low tidal range and a strong predominant drift. If the tidal range is too great, strong currents and waves can pass over the weir section at high tide, thereby reducing the effectiveness of the jetty in protecting the channel area from wave action. Currents in conjuction with the waves may act to transport sand beyond the deposition basin and perhaps cause deposition in the navigation channel. If the net drift is not a large portion of the downdrift, then there is a danger that all of the downdrift will be trapped, thereby resulting in a deficit of material during updrift conditions and a consequent erosion updrift of the inlet.

The first and perhaps most effective weir jetty system was installed at Hillsboro Inlet, Florida (Fig. 4). This area is uniquely suited for such an installation, because there is a low reef which angles seaward from the shore at the mouth of the inlet. The reef, which is at Mean Low Water near the shore, serves as the weir and also provides a solid base for the seaward jetty. The navigational channel has been cut through the reef seaward of the jetty tip. Figure 4 shows that the Hillsboro weir is oriented more or less parallel to shore and the deposition occurs as an extension of the beach berm and "spills over" the weir into the deposition basin. A weir oriented parallel to shore appears to have some desirable features compared to a shore-perpendicular orientation.

A second weir jetty system has been installed at Masonboro Inlet, North Carolina (Fig. 5). The north (updrift) jetty was completed in 1966 (4). No south jetty has been constructed, although one is planned. One interesting feature of the behavior of this system is the extensive shoal formed immediately to the south of the inlet and the associated encroachment of the navigational channel into the deposition basin. Model studies being conducted at the Waterways Experiment Station indicate that this shoal may be the combined result of (a) the jetty's sheltering of this area from the predominant northeast waves, thereby creating a sand trap, and (b) a large clockwise eddy carrying sand toward the inlet on both the flood and ebb currents due to the mechanism described earlier. Although it has not been definitely established, it appears that significant bypassing may be occurring by natural means, with the ebb currents transporting sand from the deposition basin to the outer shoals where waves and currents cause it to be transported shoreward and deposited in the shoal. If this is the case, then it would appear that there is a possibility of simply dredging the shoal and discharging during the season and at a distance where the southerly dirft is effective. Furthermore, the shoal would provide protection for the dredge, if the dredging pattern were such that a partial shoal barrier was left seaward of the dredged area. Also, it would be possible to control the encroachment of this shoal on the navigational channel by leaving a narrow ridge of the shoal adjacent to the channel with the further removal to be effected by waves and currents. It should be noted that the existing natural bypassing of sand past the channel seems to be due to a combination of a concentrated ebb flow through the deposition basin area and a relatively high wave action which is capable of transporting the material shoreward.

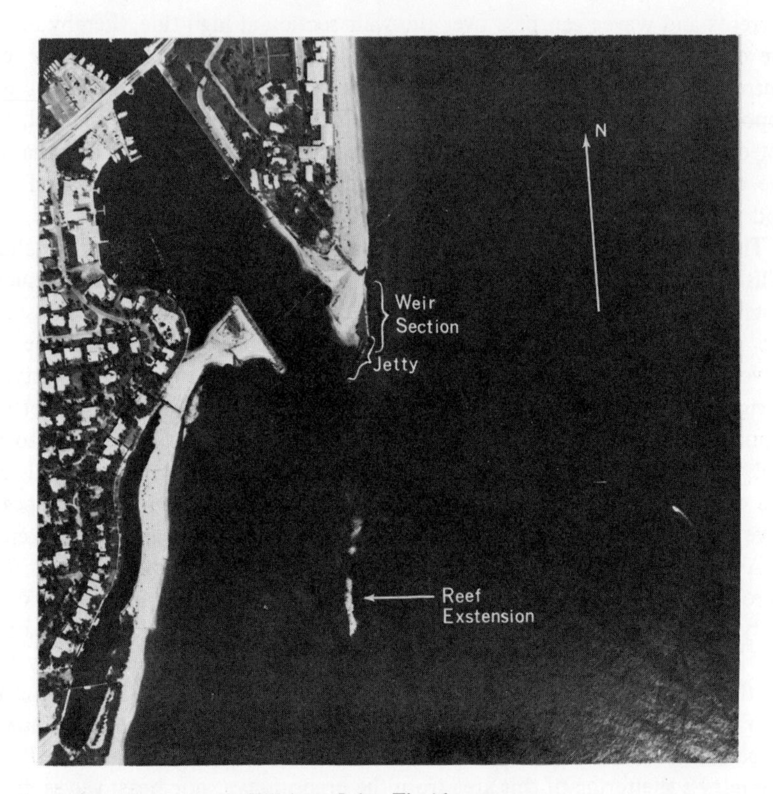

Figure 4. Weir jetty system at Hillsboro Inlet, Florida.

Dredging

Dredging of the outer bar is a common practice in maintaining navigable depths through the bar area. The role of this bar in effecting bypassing is well known and has been noted previously. In cases where there is a strong predominant littoral drift, not only does increasing the depth in the vicinity of the bar interrupt the net drift, but it will eventually cause some lowering of the general elevations of the entire bar formation. Although this is difficult to demonstrate quantitatively due to the complexity of the process, the case can be argued qualitatively from two different considerations. First, by lowering the elevation of any point in the bar formation, the gradient of sand, and hence the sand transport toward that point, will be increased. An analogy is provided by heat flow in a solid in which if the temperature is decreased (i.e., heat is extracted) then the temperature in the vicinity of that point is lowered and there is a flow of heat toward that point. Second, although the sand-transport processes are complex, it is evident from observations that waves tend to return

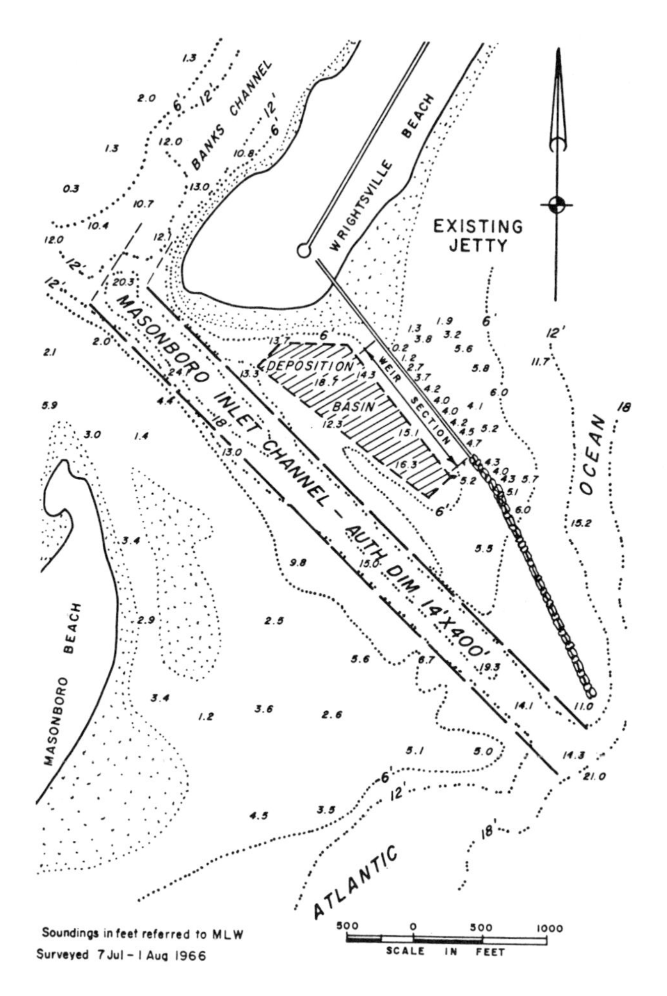

Figure 5. Masonboro, North Carolina, weir jetty design. (From reference 4).

sand from the outer bar to the shore along the shoals adjacent to the main channel. If the elevation and availability of sand are reduced in the outer-bar area, then the effectiveness of these onshore transport mechanisms will be reduced.

OUTER- AND INNER-SHOAL ACCUMULATION VOLUMES

Outer-Shoal Volumes

A total of twenty-three Florida inlets were investigated to estimate quantities of sand residing in outer shoals of inlets on sandy coasts. The inlets spanned a wide range of tidal prisms and geomorphological types (i.e., tidal river entrances, lagoonal entrances, bay entrances).

The procedure followed in calculating the volume of sand in the outer bar is summarized in Figure 6, an idealized case, and Figure 7, an actual calculation for John's Pass (Gulf Coast). First, available National Ocean Survey (formerly U.S.C.G.S.) Hydrographic Sheets (scales 1: 10,000 and 1: 20,000) or N.O.S. navigation charts (scales 1: 40,000) were pieced together to obtain hydrographic charts of the inlet and its surrounding shores. Figure 7 is a chart of John's Pass pieced together from four separate N.O.S. hydrographic sheets surveyed at a common date; the small sounding numbers are due to a reduction of scale necessary for one of the charts.

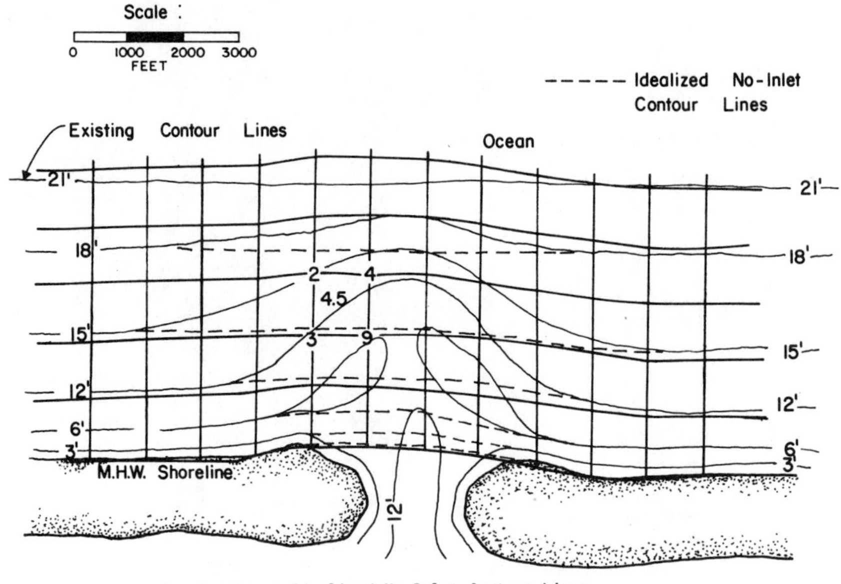

1. Construct Idealized No-Inlet Contour Lines
2. Impose 1000 foot square grid system on chart and calculate differences between actual depth and idealized no-inlet depth at grid line intersections (see example block)
3. Average depth differences at intersections and record in center of block (see example block)
4. Compute volume of sand in outer shoal by summing averaged block depth differences and multiply by 10^6 feet^2

Figure 6. Steps in calculation of accumulated volume of sand in the outer bar. Procedure illustrated for idealized inlet.

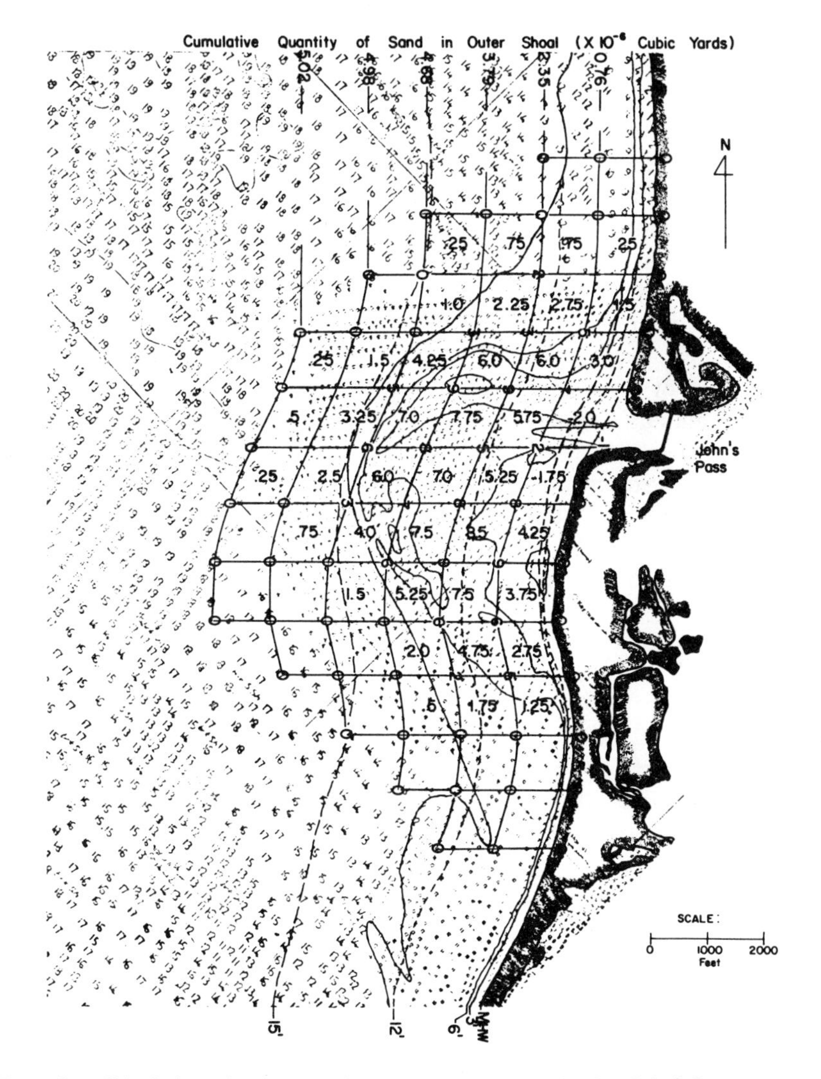

Figure 7. Calculation of accumulated sand volume in outer shoal at John's Pass.

The parallel contour lines upcoast and downcoast of the inlet were assumed the natural topography of the coast without the inlet. Contour lines of a common depth upcoast and downcoast were then joined together generally paralleling the shoreline. These lines are dashed on Figures 6 and 7. A 1000-foot-square grid system was then superimposed on this combination chart of idealized no-inlet hydrography and actual hydrography. This grid system is shown in Figures 6 and 7 by the heavy, solid-black lines.

Depth differences between the actual existing bathymetry and the idealized no-inlet bathymetry were then calculated to the nearest foot and plotted at the intersections of the grid, interpolating where necessary. Depth differences were then averaged for the grid square (shown centered in each square on Figure 7) and the total difference in sand volume between the idealized no-inlet shoreline and the existing inlet shoreline was calculated. In some cases the calculations were made from the 3-foot or 6-foot contour line outward, due to irregularity in the actual MLW shoreline. This made a relatively small difference in the total volume of sand in the outer shoals.

To evaluate the magnitudes of possible different interpretations of the natural no-inlet shoreline and bathymetry and the consequent differences in sand volumes, eight of the twenty-three inlet outer-shoal volumes were calculated by each of the authors. Differences in the sand volumes in almost all instances were found to be under 30 percent. On the inlets with the most sand accumulations, the differences were usually less than 10 percent. A listing of the inlets and the calculated volumes of sand in the outer shoals are summarized in Table 1. As a final note it should be mentioned that both authors thought themselves conservative in their final estimates of accumulated material.

Inner-Shoal Volumes

The volumes of sand accumulated in inner-shoals were calculated for four inlets. In three of the inlets only two surveys gave reasonably comprehensive coverage, therefore one volume accumulation of sand was calculated. The fourth inlet, St. Lucie Inlet, had six (6) surveys of inner-shoal areas over its history which provided a wealth of information on its shoaling areas and trends.

The calculated volumes of sand in the interior shoals of the inlets are listed in Table II. These volumes represent the amount of sand accumulated in the period between the first and last (most recent) survey of the inlet.

Two of the inlet interior sand volumes (Ft. Pierce and St. Lucie) were calculated by comparative cross sections spaced at 600-1000 feet. The remaining two inlets, Redfish Pass and Blind Pass, were calculated by drawing contour lines and planimetering the contoured areas to determine the volume of water beneath the plane of reference (MLW) in the interior bay area. The difference in this volume between the two comparative surveys represents the shoaled volume of sand. This latter method requires a higher degree of accuracy in that both total planimetered plan areas must be exactly equal; but is believed to be an improved method of calculation for bay tidal deltas in which complex shoaling patterns are represented on small-scale survey sheets.

Both methods have an inherent error due to the fact that the absolute elevation of the plane of reference (MLW) may be changing. This datum-plane

TABLE 1.

Volumes of Material Deposited in Outer Inlet Shoals

Inlet	Calculated Volume of Shoaled Material in Outer Bar (cubic yards)		Relative Error[1]	Critical Cross Section Area of Inlet ($ft.^2$)	Survey Year
	by Walton	by Dean			
Nassau Sound	49.4×10^6			58,200	1871
Nassau Sound	55.89×10^6			69,400	1924
Nassau Sound	52.52×10^6	54.83×10^6	$\pm$ 2%	49,000	1954
Jupiter Inlet	1.00×10^6			6,180	1967
Redfish Pass	4.25×10^6	4.33×10^6	$\pm$ 1%	15,000	1956
Captiva Pass	11.28×10^6			28,500	1879
Captiva Pass	12.35×10^6			33,250	1956
Boca Grande Pass	206.60×10^6			119,200	1956
Midnight Pass	1.39×10^6	0.69×10^6	$\pm$ 33%	2,800	1955
Venice Inlet	0.93×10^6	0.84×10^6	$\pm$ 5%	1,600	1954
Longboat Pass	7.95×10^6			13,340	1954
New Pass	7.74×10^6			8,040	1954
John's Pass	7.16×10^6	5.24×10^6	$\pm$ 15%	6,280	1926
John's Pass	5.02×10^6			8,720	1952
Clearwater Pass (formerly Little Pass)	3.11×10^6			31,850	1926
Clearwater Pass	4.85×10^6			25,200	1950

TABLE 1. (continued)

Inlet	Calculated Volume of Shoaled Material in Outer Bar[1] (cubic yards)		Relative Error[2]	Critical Cross Section Area of Inlet (ft.^{2})	Survey Year
	by Walton	by Dean			
Dunedin Pass (formerly Big Pass)	13.65×10^6	12.70×10^6	± 7%		1879
Dunedin Pass	6.40×10^6	12.70×10^6	± 33%	9,400	1926
East Pass (Dog Isl.)	17.78×10^6			76,000	1935
West Pass	32.77×10^6			79,400	1874
West Pass	54.44×10^6	48.60×10^6	± 6%	79,200	1943
Destin Pass (also East Pass)	7.04×10^6			17,750	1941-47
Pensacola Harbor Entrance	12.38×10^6			86,700	1940

[1]Relative error = $\frac{\text{calculated volume - average of calculated volumes}}{\text{average of calculated volumes}}$

TABLE 2.

Volumes of Material Deposited in Inlet Bay Shoals

Inlet	Calculated Volume of Shoaled Material in Bay Shoals (cubic yards)	Estimate of Error[1] due to changing MLW Datum Plane	Survey Years
Redfish Pass	3.76×10^6	-6.9%	1879 - 1956
Blind Pass	1.11×10^6	-9.4%	1879 - 1956
Fort Pierce	7.49×10^6	-1.2%	1882 - 1930
St. Lucie	7.96×10^6	-3.1%	1882 - 1930

1 Calculated error = (plan area of bay used in shoaling calculation) x (estimated difference of elevation in tidal datum planes[2])

2 Estimated difference in datum planes in 0.5 feet for Redfish and Blind Passes, and 0.32 feet for Fort Pierce and St. Lucie Inlet

change was not taken into account in the present calculations. For example, if the MLW reference plane had risen 0.5 feet in the 50 years between surveys, then the soundings on the recent survey should be reduced by 0.5 feet or the soundings on the old survey increased by 0.5 feet. The error due to a changing reference plane (in times of rising sea level) is thus conservative in the sense that actual accumulated volumes of sand would be greater than the calculated volumes. An estimation of the magnitude of this error is presented in Table II for the four inlets considered. On the inlets calculated, this error was less than 10 percent. In all cases though, since recent hydrographic surveys (N.O.S.) give soundings to the nearest foot, this error may not be of great importance.

Another source of error not considered is the volume of sediment entering the region from other outside sources (i.e., drainage creeks, channel spoiling, tidal-river deposits). It is believed that these sources of shoaling material are relatively small in comparison with the sand entering through the inlet for the four inlets considered. Of the four inlets, only Blind Pass on the lower Gulf Coast of Florida was open at the time of the first survey. Survey areas covered in the calculations of the shoaled sand generally covered the total area of shoaling in the inlets. In the case of St. Lucie Inlet, (Figure 8), some of the surveys did not cover the total shoaled area, thus Region 1 on Figure 8 covers a smaller area in the vicinity of the inlet while Region 2 covers a much larger area, the entire area of shoaling. Figure 9 presents a history of the St. Lucie Inlet interior shoaling volumes for Regions 1 and 2. It appears from this figure that inlets generally shoal rapidly in their earlier years and gradually approach a much smaller "equilibrium" shoaling rate as represented by the slope of the right-hand sides of the Figure 9 curves. Much more data than now exist are needed to investigate this important aspect of sand-trapping in inlets.

III. INTERPRETATION OF SHOALED VOLUMES IN RELATION TO BEACH-EROSION RATE

One of the more effective means of countering beach erosion is through artificial nourishment with good quality (coarse) sand. Although there are substantial sources of sand offshore, some recent experience has shown that ample quantities of good quality sand located nearshore may not be economically available. Because of the high current and wave energy conditions under which the offshore shoals exist, the quality of material in these shoals should be generally good. It is therefore of interest to relate the volumes of these outer shoals to the current erosion rate in Florida.

The total outer-shoal volume presented in Section II for twenty-three inlets is approximately 400 million cubic yards. Based on the number of inlets in Florida, it is estimated that this represents approximately one-half of the total

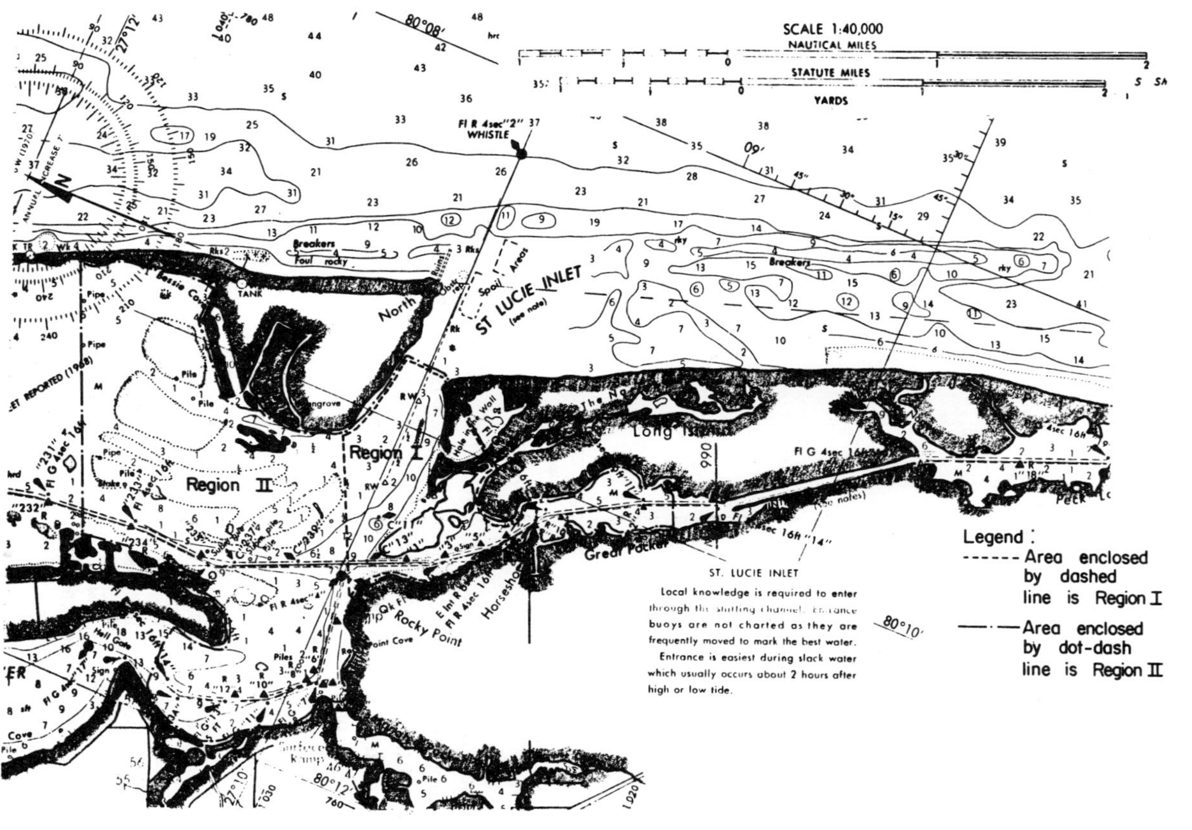

Figure 8. Area covered in sand accumulation calculations, inner shoals of St. Lucie Inlet.

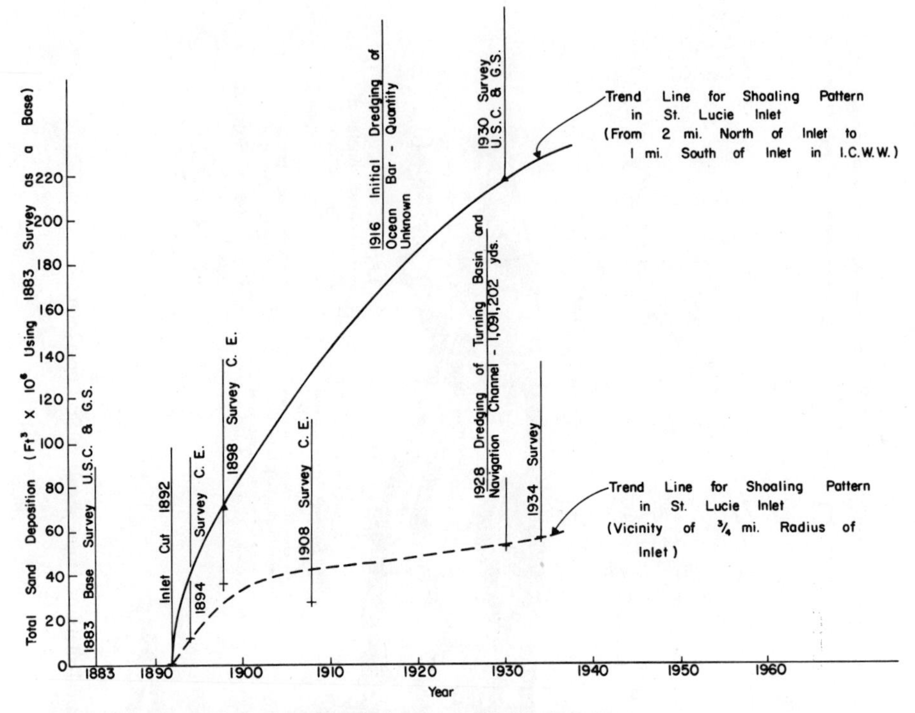

Figure 9. Deposition of sand in the interior of St. Lucie Inlet.

outer shoal quantities (i.e., total outer shoal volumes $\simeq$ 800 x 10^6 yd^3). If the total outer sandy shoreline length of the state is taken as 1000 miles and if this is eroding at the generally accepted rate of 2 ft per year, or 2 yd^3/ft of beach front per year, then the outer-shoal volumes contain sufficient quantities of material to forestall erosion for

$$\frac{800 \times 10^6}{5280 \times 1000 \times 2} \simeq 76 \text{ years.}$$

This calculation indicates that the amount of material residing in the outer shoals adjacent to inlets is very significant in terms of the present rate of beach erosion. Furthermore, because this material was generally derived from beach erosion and resides in a high-energy environment, it is in close proximity to areas of beach erosion and is expected to be of good quality.

IV. SUMMARY AND CONCLUSIONS

The hydraulic and sedimentary processes in the vicinity of inlets are such that they tend to trap material in the interior and exterior shoal features. The volumes and quality of this trapped material in the outer shoals are of special significance relative to beach-erosion problems.

ACKNOWLEDGMENTS

The investigation leading to this paper was carried out under NOAA Office of Sea Grant support to the State University System of Florida. The funding and encouragement provided to the program are appreciated.

REFERENCES

1. Bruun, P.
1966 **Tidal inlets and littoral drift.** H. Skipnes Offsettrykker, Trondheim, Norway.

2. Hays, M. O., Goldsmith, V. and Hobbs, C. H.
1970 Offset coastal inlets. **Proceedings, Twelfth Conference on Coastal Engineering,** p. 1187-1200. Washington, D.C.

3. Johnson, D. W.
1972 **Shore Processes and Shoreline Development.** (Reprinted from original 1919 Edition) Hafner Publishing Company, New York.

4. Magnuson, N. C.
1967 Planning and design of a low-weir section jetty. **ASCE Jour. Waterways and Harbors,** Vol. 93, No. WW2, 27-40.

5. O'Brien, M. P.
1969 Equilibrium flow areas of inlets on sandy coasts. **ASCE Jour. Waterways and Harbors,** Vol. 95, No. WW1, 43-52.

6. Price, W. A.
1963 Patterns of flow and channeling in tidal inlets. **Jour. Sed. Pet.,** 33: 279-290.

A RECENT HISTORY OF MASONBORO INLET,

NORTH CAROLINA

by Limberios Vallianos[1]

INTRODUCTION

Masonboro Inlet is located along the southeastern ocean shoreline of North Carolina and, with respect to the general Atlantic coast of the United States, is located at a point approximately equidistant from Boston, Massachusetts, and the southern tip of the Florida peninsula. The morphology of the coastal margin of North Carolina (Fig. 1) is characterized by three cuspate forelands known as Cape Hatteras, Cape Lookout, and Cape Fear. The ocean shoreline is comprised of a narrow band of low, sandy barrier islands whose widths are generally on the order of 250 meters. These barrier islands are separated by a total of 22 inlets which couple the ocean with the interconnected estuarine waters of the State. Though a barrier island-inlet chain is typical of the entire coastline, there is a marked difference between the northern and southern halves of the State's coastal zone in terms of estuarine areas. In the northern half, the estuaries are expansive bodies of water known as Pamlico Sound and Albemarle Sound. These are the largest embayments on the Atlantic coast contained between a barrier island system and the continental land mass. The southern half of the State's coastal zone is characterized by narrow elongated lagoons comprised of salt marshes and circuitous networks of tidal-flow channels. The width of most of these elongated lagoons does not exceed 1.5 kilometers. The coastal setting at and in the vicinity of Masonboro Inlet embodies the general features of the southern half of the State's coastal zone, as evidenced by Figure 2.

1. Chief, Coastal Engineering Studies Section, U. S. Army Engineer District Wilmington, P. O. Box 1890, Wilmington, North Carolina 28401.

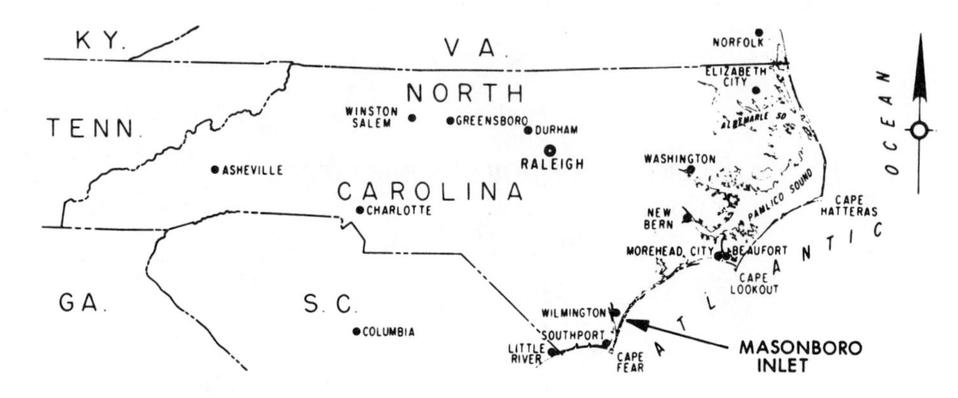

Figure 1. Morphology of coastal North Carolina.

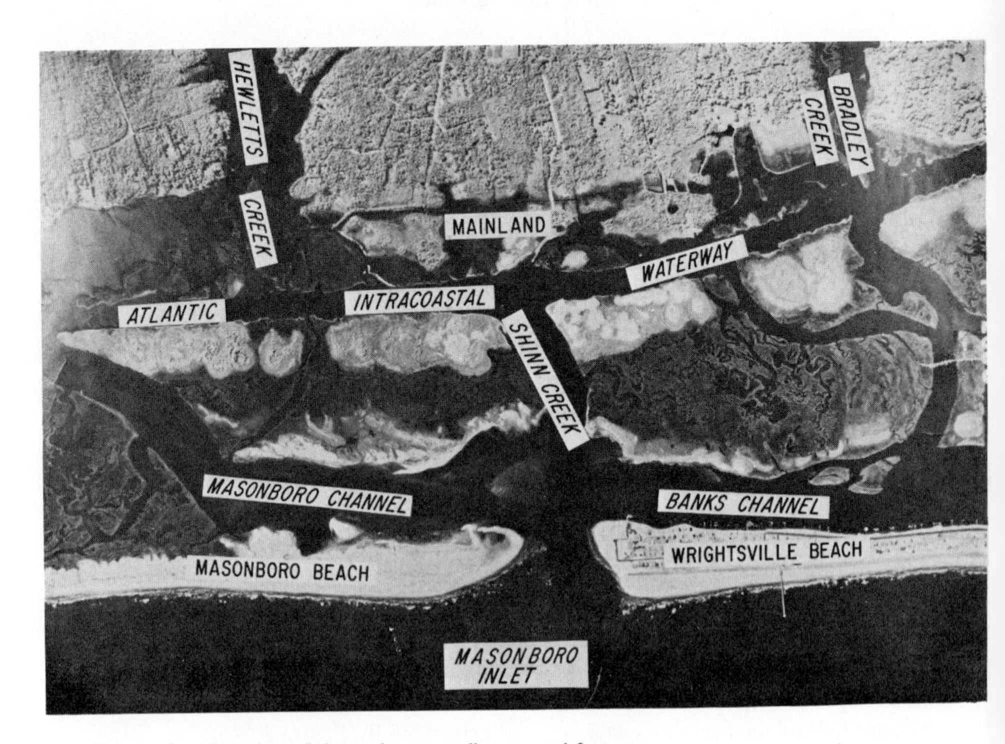

Figure 2. Masonboro Inlet and surrounding coastal features.

ENVIRONMENTAL FACTORS AND INLET CHARACTERISTICS

As shown in Figure 2, Masonboro Inlet is bounded by two barrier beaches – on the north by the Town of Wrightsville Beach, North Carolina, and on the south by the undeveloped barrier island known as Masonboro Beach. The shores adjacent to the inlet are gently sloping beaches composed primarily of a mixture of sand and shell fragments ranging in mean particle sizes between 0.2 and 0.5 millimeters. The slopes of the active beach profiles are approximately 1V to 10H on the foreshore to a depth of 1.5 meters below mean low water (mlw); 1V to 35H between 1.5 and 7.0 meters below mlw; and 1V to 400 H between 7.0 and 9.0 meters below mlw. The area is exposed to large oceanic fetches from directions south to northeast, and meteorological-oceanographic conditions that range from calm to hurricane intensity; however, the preponderance of waves affecting the area have a height range of 0.3 to 1.5 meters with periods varying from 5 seconds to 10 seconds. Because of the wide wave direction exposure, the shores adjacent to Masonboro Inlet experience significant reversals in the direction of alongshore littoral transport. Therefore, over any particular period, the predominant direction of alongshore material transport may be southerly or northerly. However, the long-term predominant direction of littoral transport is determined to be from north to south in accordance with currently estimated average annual alongshore littoral transport rates of 92,000 and 168,000 cubic meters in a northerly and southerly direction, respectively.

The astronomical tide ranges at the inlet are 1.2 meters and 1.4 meters for mean and spring conditions. Inlet geometry and flow conditions have varied significantly with time in accordance with natural and man-induced changes. The variations at the inlet are reflected in the data presented in Table 1.

TABLE 1

Cross-Sectional Area and Flows
Masonboro Inlet, North Carolina

Date	Inlet area at mlw (sq. m.)	Average Maximum instantaneous flow velocity (m./s.)		Average maximum instantaneous discharge (cu. m./s.)		Total flow 10^6 (cu. m.)	
		Flood	Ebb	Flood	Ebb	Flood	Ebb
June 1937	930	0.79	1.25	1,050	1,420	14.8	23.3
June 1938	410	0.61	1.10	490	610	5.8	10.1
Sept. 1969	1,030	0.98	1.49	1,190	1,250	18.1	18.3

HISTORY OF INLET CHANGES AND RESULTING BEHAVIOR

The existence of Masonboro Inlet has been documented by surveys dating back to 1733 and, since that time, there is no record indicating its possible closure. Accordingly, it can be considered a permanent coastal feature; however, this inlet has from time to time changed positions (migrated) in a manner common to many inlets on sandy coasts which experience significant alongshore material transport. These movements notwithstanding, the net long-term change in Masonboro Inlet's location has been small due to changes in the inlet's direction of migration, probably resulting from significant reversal in alongshore littoral transport. This conclusion was reached from a comparison of plane table surveys conducted in and around the inlet in 1857, 1909, and 1929 and controlled aerial photographs dating from 1933. A summary of the distances and directions of the inlet's centerline movements with respect to the 1857 centerline position is given in Table 2.

Distances and Directions of Masonboro Inlet's
Movements with Respect to 1857 Position

Date of survey	Distance from 1857 position (meters)	Location with respect to 1857 position
1909	1,190	south
1929	0	(coincident)
1933	290	south
1959	200	south
1969 to present	160	south

The first man-made influence imparted to Masonboro Inlet is related to the excavation of the Atlantic Intracoastal Waterway (Fig. 2), through the lagoonal marshland in that vicinity during the period from 1930 to 1932. Unfortunately, we have no record of changes (or lack thereof) occurring at the inlet immediately following this work. The next episode involving a man-induced change at the inlet occurred in 1947. At that time an elongated spit developed on Masonboro Beach which extended northward into the mouth of Banks Channel creating a constriction and flow orientation that resulted in a rapid erosion of the Wrightsville Beach shoulder of the inlet (Fig. 3). At the request of local interests, the Wilmington District, U.S. Army Corps of Engineers, developed a plan designed to eliminate the erosion, primarily through tidal-flow division. This was to be accomplished by cutting a channel, having a bottom width of 46 meters and depth of 2.5 meters, through the south end of the spit. This would allow the tidal exchange through Masonboro Channel to be effected

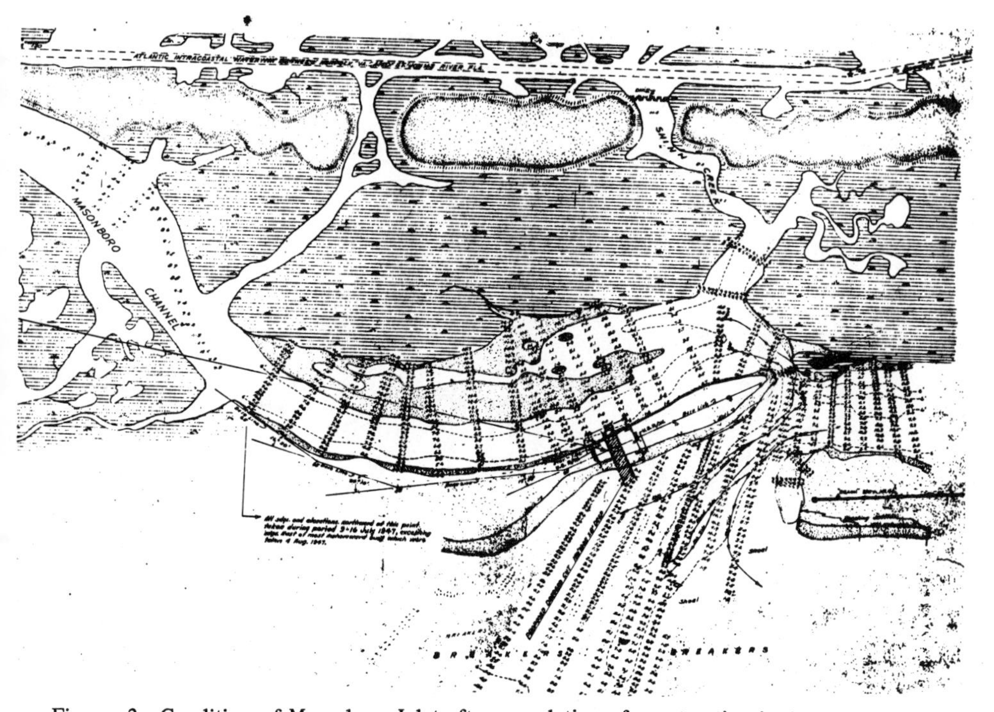

Figure 3. Condition of Masonboro Inlet after completion of construction in April 1947.

without making it necessary for that part of the total flow to follow a route through the existing inlet and around the north end of the spit. Further to assist in this partial flow diversion, a small earthen dike was to be placed across Shinn Creek, near its confluence with the Atlantic Intracoastal Waterway. The dike was also to serve to eliminate recurrent shoals in that vicinity of the waterway. Additionally, the plan involved the construction of two groins at the south end of Wrightsville Beach and three groins on the south end of the separated spit, in order to stabilize that side of the cut and to force any natural expansion of the cut to occur in a southerly direction. This plan had been implemented by April 1947, and was highly successful in eliminating the erosion at Wrightsville Beach. Figure 3 shows conditions surveyed at the completion of construction in April 1947.

Conditions surveyed one year later, in April 1948, are shown in Figure 4. In Figure 3, the road cul-de-sac on Wrightsville Beach is approximately 25 meters from the high-water line, whereas in Figure 4, the high-water line is shown at a position 85 meters from the cul-de-sac, where it has essentially remained since. In fact, the progradation of the high-water line at the end of Wrightsville Beach occurred within about six months after the construction was completed. Comparison of Figures 3 and 4 also shows a large southerly expansion of the cut

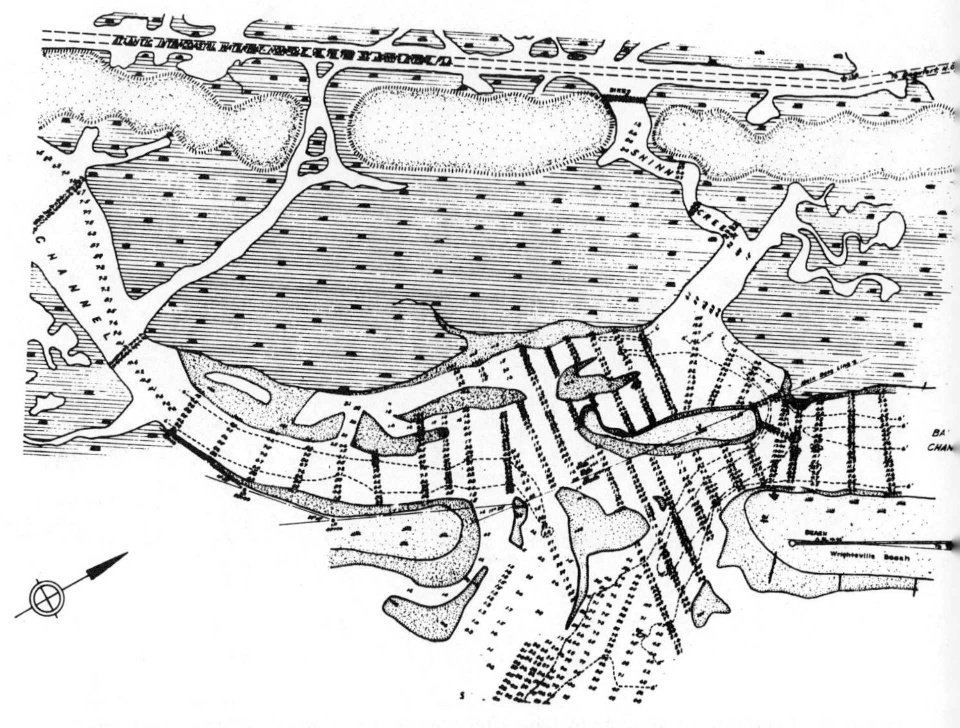

Figure 4. Condition of Masonboro Inlet one year after completion of construction, April 1948.

through the elongated spit. As previously mentioned, the original cut was 46 meters wide and 2.5 meters deep. By April 1948, the cut had expanded naturally to a width of approximately 460 meters and its depths varied from 0.3 meters to 1.5 meters. It will be noted in Figure 4 that the three small groins constructed on the north side of the original cut are awash. These short structures were flanked and severed from their land anchorage by a winter storm that caused tides and waves to overtop the low, separated spit segment.

In 1949, the congress of the United States authorized improvements to the navigability of Masonboro Inlet in the interest of commercial fishing and recreational boating. These improvements included: (a) interior access channels to the Atlantic Intracoastal Waterway via Banks Channel; (b) an ocean entrance channel through the inlet's outer shoals; and (c) two jetties that would flank the ocean entrance channel; however, these structures were to be built only if it were proved that the entrance channel could not be economically maintained by dredging operations.

The interior access channels were dredged in 1957. Excavation of the ocean entrance channel, whose natural controlling depths were from 1.8 meters to 2.4

meters (6 feet to 8 feet), was accomplished in the period between 15 April and 1 June 1959. (English measurements are shown parenthetically for ease of reference to surveys presented in Figures 7-A through 7-H.) During this operation the entrance channel design dimensions of a 122-meter (400-foot) bottom width at a low-water depth of 4.3 meters (14 feet) were attained through the removal of 265,000 cubic meters of ocean shoal materials. This work constituted the third attempt to impose on the inlet a man-made change. In planning the excavation of the entrance channel, it was considered necessary to confine the entire tidal flow in a single inlet channel, in order to optimize flushing action. Therefore, the materials removed from the entrance channel were placed on the remnants of the southern inlet channel created by the 1947 partial flow diversion. Under this plan, any spit development on Masonboro Beach that might again place the end of Wrightsville Beach in jeopardy of erosion would be removed during routine maintenance of the entrance channel. Figures 5 and 6 show conditions at Masonboro Inlet prior to and after placement of material on the south side of the inlet. Figure 5, an aerial photograph taken May 1958, shows the flanks of the inlet in the state that had generally existed since October 1954, following the impact of Hurricane Hazel on the area. This

Figure 5. Conditions prevailing prior to placement of material on the south side of Masonboro Inlet.

Figure 6. Conditions prevailing after placement of material on the south side of Masonboro Inlet.

particular storm devastated the southeastern coast of North Carolina. At Wrightsville Beach, the storm surge attendant upon Hazel reached an elevation of 3.7 meters above low-water datum. The earthen dike placed across Shinn Creek in 1947 was overtopped by the tides and waves of Hurricane Hazel and was destroyed.

The general history of inlet behavior following initial excavation of the entrance channel on 1 June 1959 is represented by the series of displays designated Figures 7-A through 7-H. Each of these displays shows the boundaries of the excavated channel, the shaded areas representing shoals whose water depths are less than the channel design depth of 4.3 meters (14 feet) at low water. Though the design depth was set at 4.3 meters (14 feet), the initial dredging of the entrance as well as a subsequent restoration of the channel in October 1959 attained actual average channel depths of approximately 4.9 meters (16 feet). As can be readily seen by comparing Figures 7-A and 7-B, shoaling in the entrance channel was quite rapid. By the beginning of September 1959, only three months after initial dredging, the amount of shoaling had reached major proportions and, in October 1959, a dredge was detailed to the site, where it removed 84,000 cubic meters of material in restoring project

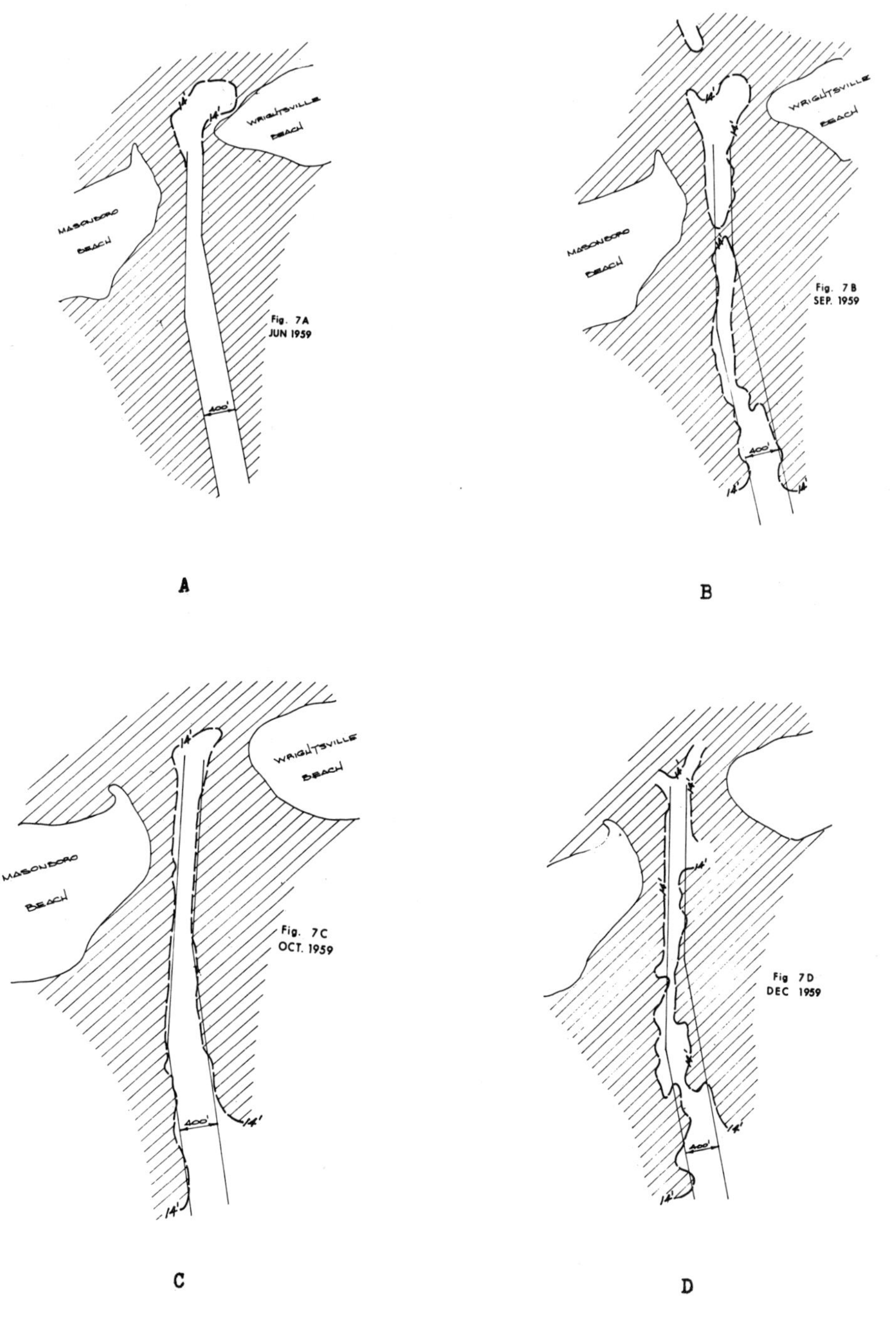

Figure 7. General history of Masonboro Inlet behavior following initial excavation of the entrance channel on 1 June, 1959.

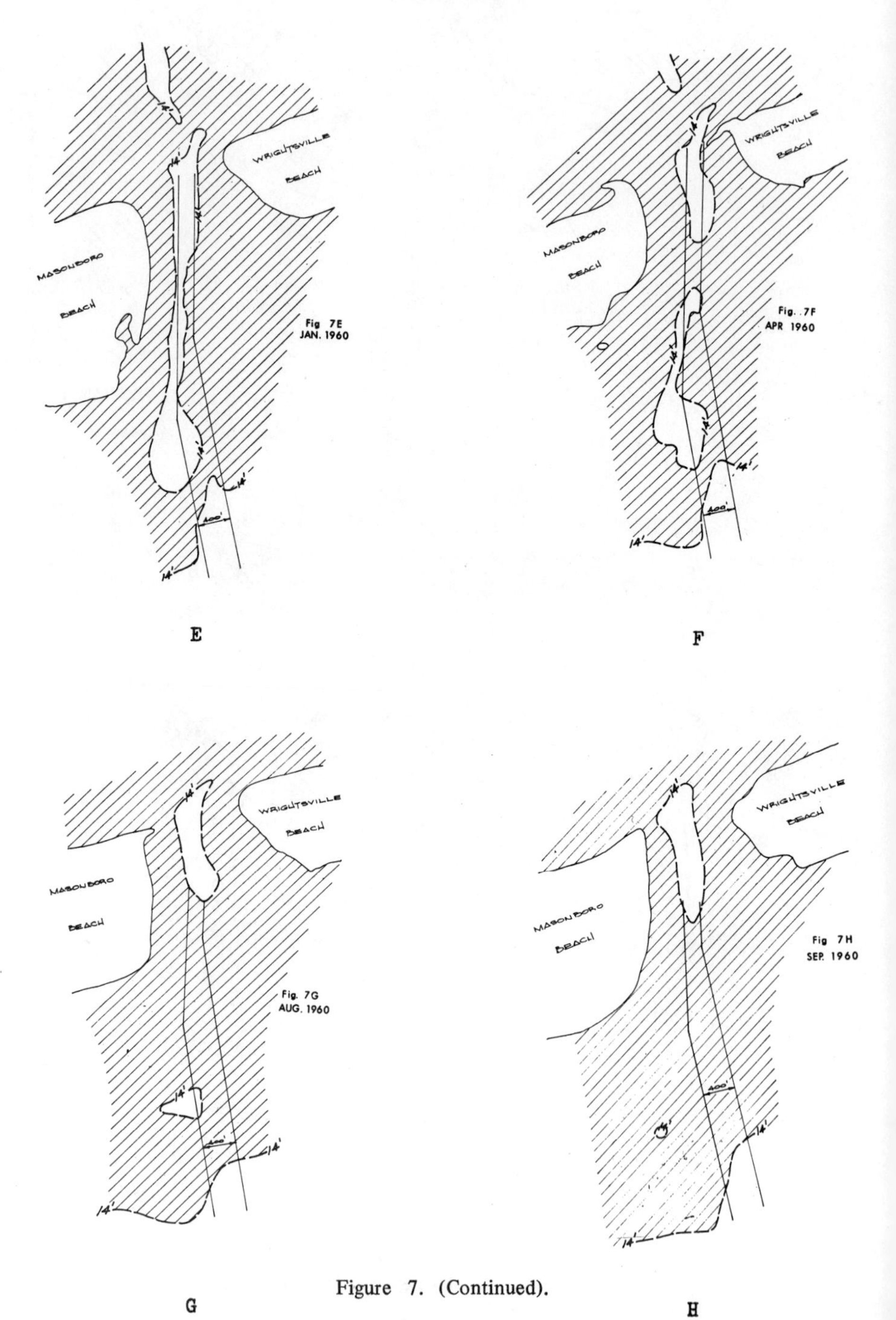

Figure 7. (Continued).

dimensions (Fig. 7-C). However, immediately following this second dredging effort, the same pattern of shoaling as previously experienced again manifested itself. By December 1959 (Fig. 7-D) only three months after project restoration, the channel was again in a rather severely shoaled condition. It is interesting to compare Figures 7-B and 7-D. The reader will note that these two displays are almost identical both in terms of the character of shoaling and the period of time (three months) after channel excavation. Figures 7-E through 7-H demonstrate that the shoaling was a continuous phenomenon characterized by the apparent influx of sediments to the channel from the north side of the inlet.

In view of the obvious futility of attempting to maintain the entrance channel by dredging, the Wilmington District, Corps of Engineers, recommended construction of the jetties that had been conditionally authorized for Masonboro Inlet in 1949. These recommendations were accepted and the work began in July 1965; however, due to budgetary constraints, only the north jetty on the apparent long-term updrift side of the inlet was built. This was the fourth and, to date, last major man-made change to the inlet. The north jetty, shown in plan view on Figure 8, was designed as the prototype of the so-called weir-type jetty

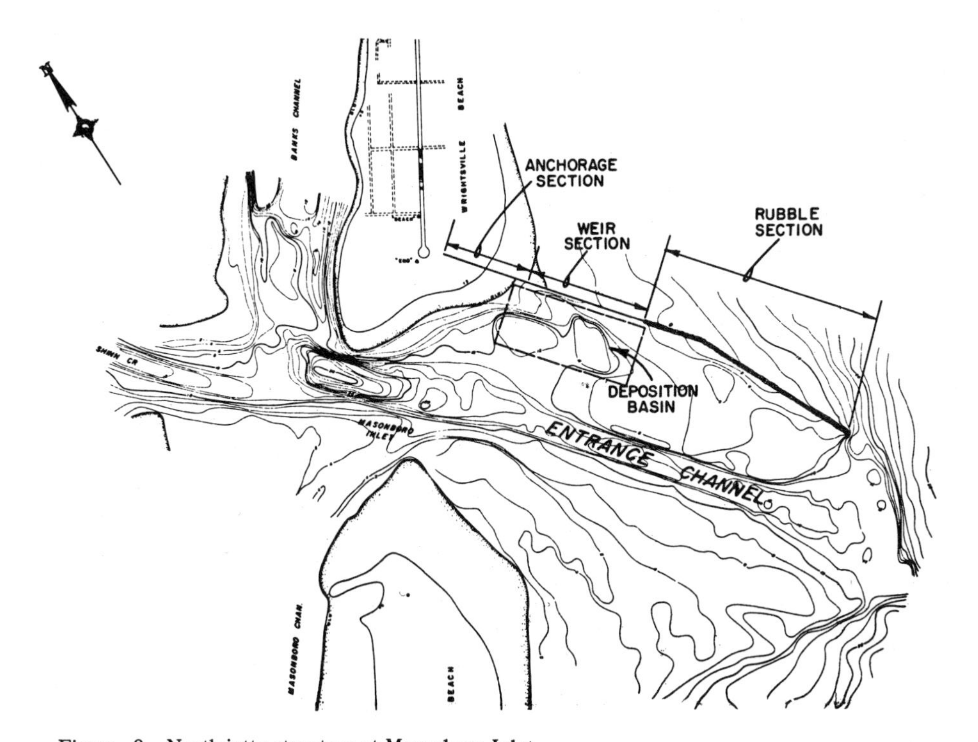

Figure 8. North jetty structure at Masonboro Inlet.

structures which provide a means for mechanically by-passing sediments across a controlled inlet. This is accomplished by building a low inner section (weir) along the updrift jetty profile over which littoral materials can be transported by wave and tide action and, thence, deposited in a preexcavated deposition basin or sediment trap on the immediate leeward side of the structure. The deposited material can then be retrieved from the deposition basin by conventional dredging equipment and transported hydraulically to the downdrift beach – in this case, Masonboro Beach.

The north jetty at Masonboro Inlet was constructed over the period July 1965 to June 1966. Its total length is 1,110 meters comprised of a concrete sheet-pile anchorage section 195 meters long, a 335-meter weir section of concrete sheet-pile with a top elevation at mean sea level, and a 580-meter outer or seaward section constructed of stone (rubblemound) with a crest elevation of about 2.1 meters above low water. The deposition basin was excavated to a volume of 288,000 cubic meters. Additionally, the entrance channel was reestablished through the ocean shoal. The inlet as it now appears is shown on Figure 9.

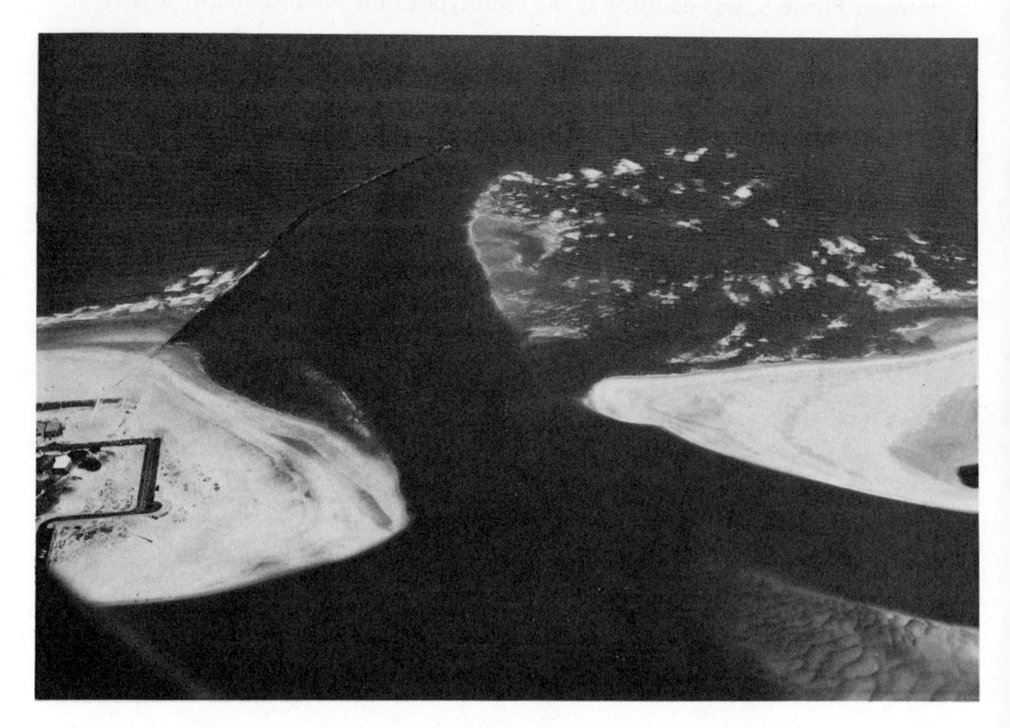

Figure 9. Present condition of Masonboro Inlet.

The history of the inlet behavior since the jetty was built is shown in the series of displays designated Figures 10-A through 10-G. In these displays, the plain flow areas represent depths equal to or greater than the channel depths of 3.0 meters (10 feet), whereas the shaded areas represent depths less than 3.0 meters (10 feet). A comparison of Figures 10-A through 10-C reveals that during the first year (July 1966 to June 1967) following jetty construction, the basic system of jetty, deposition basin, and entrance channel functioned as generally planned. During this period, approximately 115,000 cubic meters of sediment were transported by natural forces across the weir section of the jetty and deposited in the lee-side deposition basin; also, average depths in the entrance channel exceeded the design depth of 4.3 meters (14 feet) and the channel orientation remained favorable to navigation. However, by January 1968 (Fig. 10-D), the channel was shoaled considerably with controlling depths being only about 2.7 meters (9 feet). Furthermore, there had developed along the lee side of the seaward end of the structure substantial scour trough, which caused concern with respect to possible structural failure through undermining. Therefore, in March and April 1968 the outer shoal was dredged to widen and deepen the outer entrance channel, in order to reduce flow velocities. A total of 47,000 cubic meters of material were removed in this operation. It will also be noted in Figure 10-D that the inlet channel had encroached into the deposition basin, thereby rendering it inoperable.

The history of the inlet since January 1968 is characterized by a progressive northward intrusion of its south shoal, concomitant with a northward shift of the entrance channel to juxtaposition with the jetty structure. It will be noted that, notwithstanding the dredging effort undertaken in March and April 1968 survey (Fig. 10-E) revealed an increasingly worsening condition, with scour depths next to the structure reaching depths of 6.1 meters (20 feet). Accordingly, further dredging was done early in November 1968 and 81,000 cubic meters were removed from the south side of the entrance channel. However, as the January 1969 and September-October 1969 surveys show (Fig. 10-F and 10-G), the threat of the structure being undermined had not been eliminated. In fact, by March 1969, there was evidence of some instability in the structure which required that, as a safety measure, side-slope protection be placed along the entire lee side of the stone part of the jetty. Since the end of 1969, the outer 460 meters of entrance channel have maintained a width of approximately 60 meters (200 feet). The remaining landward section of the entrance channel has continued to shift in a northward direction under the influence of the inner south shoal, which progressively intrudes into the channel. This has made it necessary for bottom side slope protection to be placed against the lee side of the jetty's weir section.

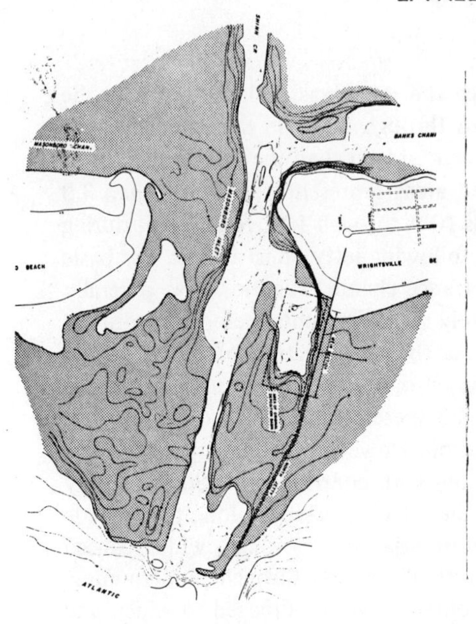

A - July, 1966

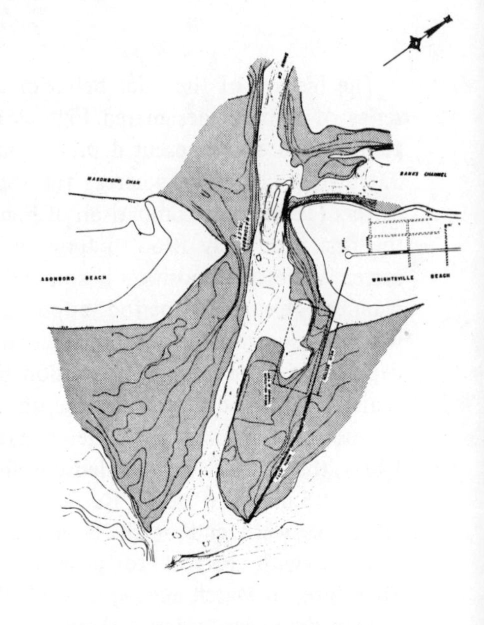

B - December, 1966

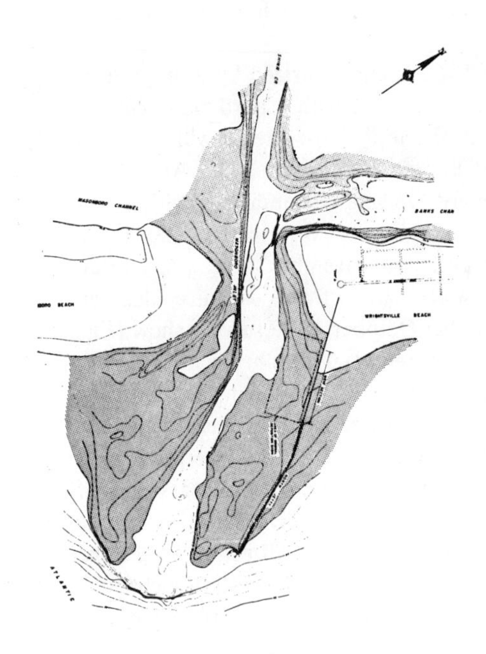

C - June, 1967

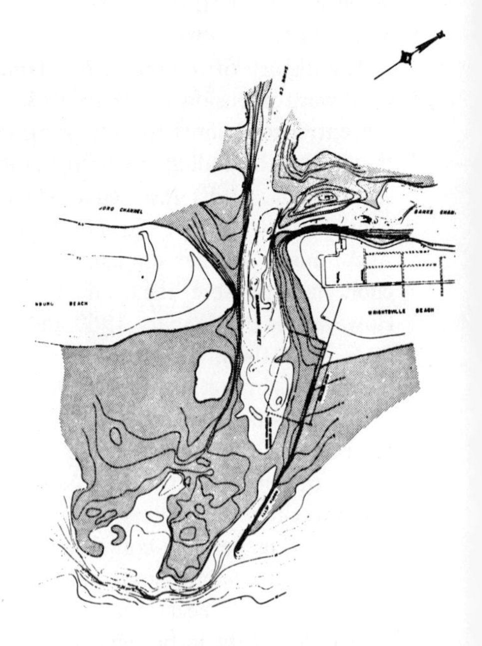

D - January, 1968

Figure 10. General history of Masonboro Inlet behavior following completion of the north jetty structure.

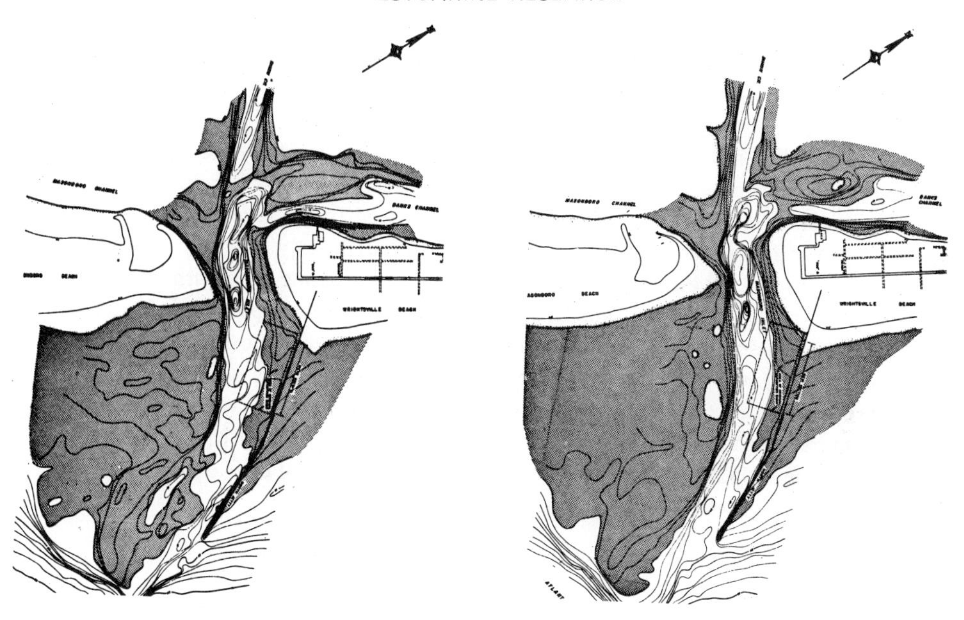

E - July, 1968

F - January, 1969

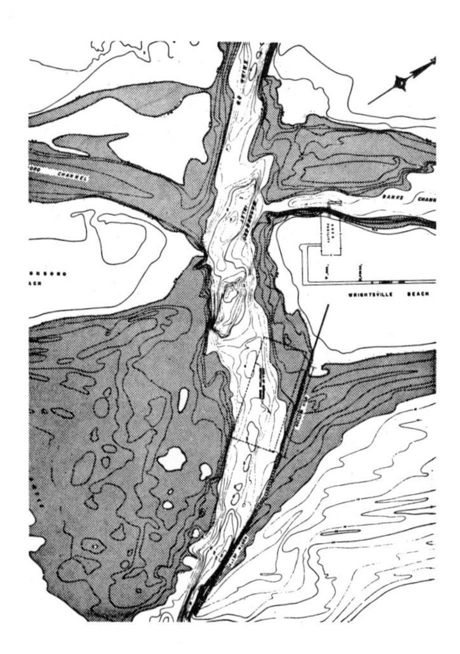

G - October, 1969

Figure 10. (Continued).

PLANNED FUTURE CHANGES

Uncontrolled northward drift, ebb-flow orientation caused by interior channel alignment, particularly at Shinn Creek, and wave refraction-diffraction at the inlet under all wave-approach conditions are apparently the three major factors that have influenced the inlet behavior since the north jetty was built. Detailed hydraulic model investigations for the purpose of designing the south jetty are currently in progress at the U. S. Army Waterways Experiment Station, Vicksburg, Mississippi. Construction of the south jetty will follow model studies as soon as funds are made available. This will provide a wider and safer navigation entrance; additional security for the north jetty, which could be undermined during hurricane conditions, notwithstanding the bottom protection that has been placed to date; and a more controlled littoral drift deposition for by-passing purposes.

THE RECENT HISTORY OF WACHAPREAGUE INLET, VIRGINIA[1]

J. T. DeAlteris and R. J. Byrne[2]

ABSTRACT

The configuration of Wachapreague Inlet has been traced since 1852 to the present from bathymetric surveys and aerial photographs. During the last 120 years this offset tidal inlet has migrated to the south at a rate of 1 meter per year. A cyclic growth and decay of the lateral ramp margin shoals has been documented over the last 24 years. These variations are not likely due to variations in littoral drift along the adjacent islands. A study of the net long-term sand-volume changes on the ebb-tidal delta shows no significant long-term change in the storage of sand.

The mobile sediment distribution of the inlet was investigated with respect to spatial variations over the entire inlet complex and temporal variations in the deep inlet throat. The sediment distribution correlated well with the various depositional environments ranging from gravels in the deep inlet throat to silty sands on the flood-tidal delta. Changes in the inlet throat sediment distribution were monitored over a 3-month period. Short-term fluctuations in the inlet cross-sectional areas were correlated with overall changes in the bottom sediment characteristics in the inlet throat.

Investigation into the geomorphology of the inlet orifice shows the north flank to be a sandy spit extending south from the barrier island, while the south flank is a firm cohesive lagoonal mud. Thus, as Wachapreague Inlet migrates south in response to a predominantly southerly littoral drift, it leaves in its path a wedge of sand (the only sand sink in the system) and erodes into firm marine lagoonal deposits.

1. Virginia Institute of Marine Science contribution no. 573.

2. Virginia Institute of Marine Science, Gloucester Point, Virginia 23062.

INTRODUCTION

Wachapreague Inlet, a natural tidal inlet located on the eastern shore of Virginia, is typical of several inlets with a similar bathymetric structure along a 30-km expanse of the southeastern Delmarva Peninsula coast (Fig. 1). Wachapreague Inlet is an "offset coastal inlet" similar to those described by Hayes et al. (4); it is offset to the downdrift side. The Wachapreague Inlet Complex is composed of the inlet throat channel, Parramore Island to the south, Cedar Island to the north, a crescent-shaped, ebb-tidal delta to the east, and finally a system of lagoons and tidal channels to the west. The geometry of the inlet throat cross-section is unusual, in that the southern flank has an average slope of 15° and reaches a maximum slope of 45°. In contrast, the northern flank has a gradually sloping wall with an average inclination of 5°. In order to explain these unusual characteristics, a study was undertaken to trace the recent morphometric history of the inlet and to determine what geological and sedimental controls have influenced its stability and evolution.

HISTORICAL CHANGES IN THE INLET COMPLEX CONFIGURATION

U.S.C. & G.S. Hydrographic Survey Sheets for 1852 (Fig. 2), 1871, 1911, and 1934 were compiled and contoured at 0.913-m (3-ft.) intervals. An accurate bathymetric survey of the entire complex was made by the authors in 1972 (Fig. 3). Comparison of the charts showed that the axis of the inlet channel has migrated to the south at a rate of 1 m per year during the last 120 years (Fig. 4), and has rotated counterclockwise from a southeastern axial orientation to a more easterly orientation. In its migration the channel flow has eroded the northern flank of Parramore Island, leaving a wedge of sand as the south tip of Cedar Island. In addition, the northeastern face of Parramore Island has accreted seaward, while the southern end of Cedar Island has migrated landward, thus accentuating the apparent offset. The long-term cross-sectional area of the inlet throat has been relatively stable since 1871 at about 4,400 m^2, (less than 10% variation from mean); however, between 1852 and 1871, the cross-sectional area increased from 1,845 m^2 to 4,473 m^2 (Table 1). Historical evidence indicates that the configuration of the interior marsh-lagoon system has changed very little since 1852, thus the potential tidal prism appears to have remained unchanged. Flow gaging at the inlet throat indicates the present channel cross-sectional area and the spring tidal prism follow the linear relationship noted by O'Brien (5). At present, the ocean and lagoonal tidal ranges are the same. There is insufficient tide information for the 1850 period to determine if the reduced cross-section admitted a smaller tidal prism.

The length of the inlet throat channel (based on the 12-m contour) has

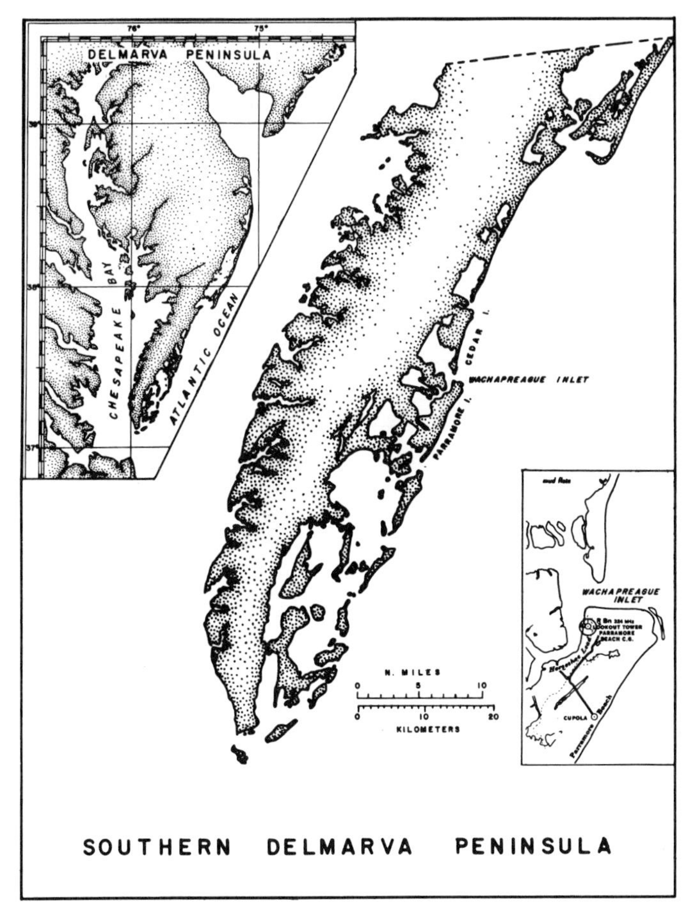

Figure 1. Wachapreague Inlet, located on the Virginia eastern shore of the Delmarva Peninsula.

increased from 1662 m in 1852 to 3,046 m in 1972 (Table 2), significantly increasing the frictional characteristics of the inlet. Various hydraulic radii of the inlet throat cross-sections were calculated based on (a) an unmodified cross-section; (b) a modified cross-section (long shallow tails removed); and (c) a modified and normalized cross-sectional area (expanded to a uniform area). The results are tabulated in Table 3 and the trend is similar for all three techniques, an increasing hydraulic radius until the turn of the century, then a decreasing one. Thus, with a steadily increasing channel length, and a decreasing hydraulic radius, Wachapreague Inlet appears to be evolving toward a less efficient channel.

To investigate the possibility that the entire inlet complex is serving as either a source or a sink of sand to the littoral drift moving down the barrier island coast, the volume of sand, to a base of 21 m below MLW, was calculated for each of

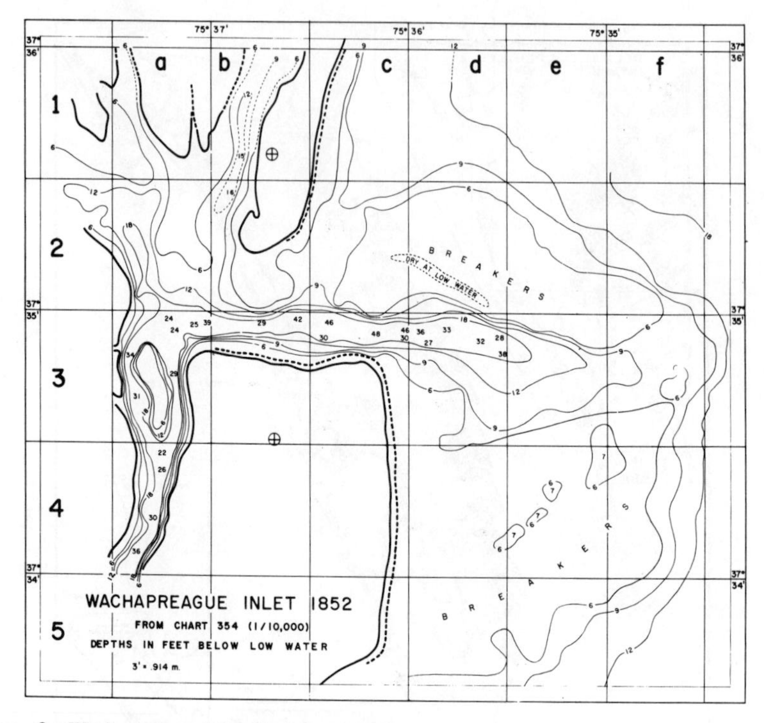

Figure 2. Wachapreague Inlet Complex, 1852.

the survey charts from 1852 through 1972 (no data for 1871). The net change in total material gained or lost during 120 years was a loss of 7 x $10^6 m^3$. In addition, there was a nebligible change in the sand volume stored on the ebb-tidal delta. As noted earlier, the inlet channel has migrated to the south during the last 120 years; consequently the northern flank of Parramore Island has lost material, while the south tip of Cedar Island has gained sand in the form of a sand spit extending from the barrier island. The increase in volume of this sand wedge extension from 1852 to 1972 is 4.5 x $10^6 m^3$. Thus the migration of the inlet channel has served to develop a localized sand sink, while the system as a whole has experienced a small, overall net loss of material.

Short-term changes in the geometry of the barrier islands flanking the inlet and of the lateral ramp margin shoals were studied from 1949 to 1973 with the aid of aerial photography. A photograph taken in 1957 (Fig. 5) shows the inlet complex at a critical time in its life history; note the breakthrough inlet on Cedar Island. Tidal prism being lost through the breakthrough inlet represents Wachapreague Inlet's reduced capability to flush sand from its main channel, and thus a threat to its existence.

More significantly, note the wedges of sand on the northeastern face of

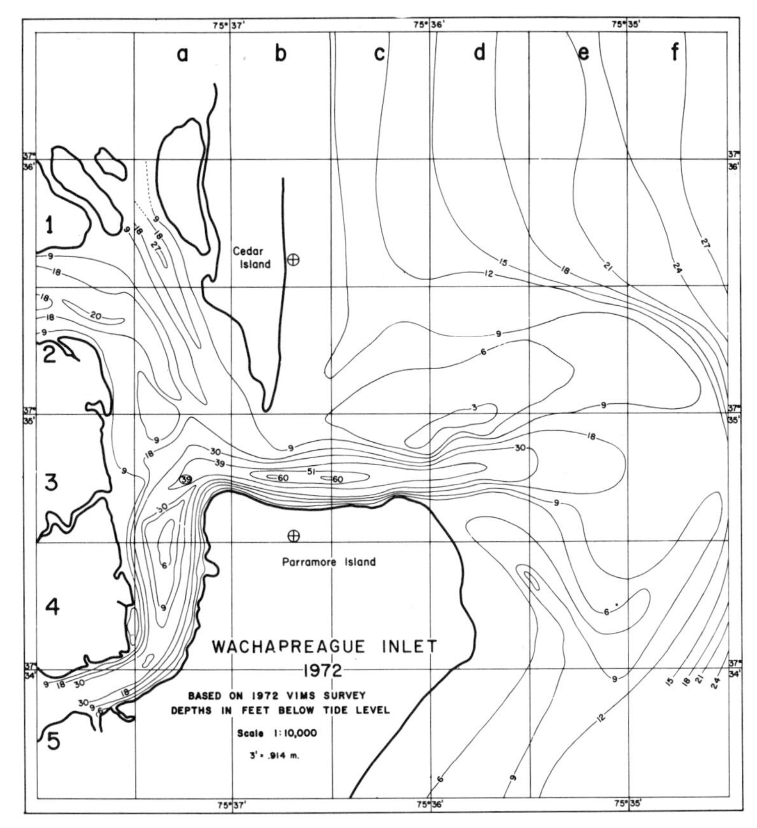

Figure 3. Wachapreague Inlet Complex, 1972.

Parramore Island and on the southern tip of Cedar Island. These accretional features represent 3.3 x 10^6m^3 and 2.4 x 10^6m^3 of sand respectively, while the northern shoal represents 1.5 x 10^6m^3 of sand (based on the U. S. Army Corps of Engineers' rule of thumb that "One square foot of beach is equivalent to one cubic yard of sand," (6)). In addition to the main channel, note also the well-developed northern channel. In April of 1962 (Fig. 6), after the March "Ash Wednesday" storm, the shoal disappeared, a loss of 1.5 x 10^6m^3 of sand. the southeastern tip of Cedar Island, although elongated, lost 0.6 x 10^6m^3 of sand, and the northern face of Parramore Island gained about 0.5 x 10^6m^3 to a total volume of 3.8 x 10^6m^3. Note also, that the northern shoreline of Parramore is now straight. In February of 1970, a single northern shoal existed, with a volume of 2.9 x 10^6m^3 of sand, Cedar Island, narrowed and lengthened, shows no net change in sand. By June 1971 (Fig. 7) the north shoal was disected with a volume-reduction of 1.4 x 10^6m^3 of sand. Note the presence of a concavity of the northern shoreline of Parramore Island. This is due to diffraction of waves approaching

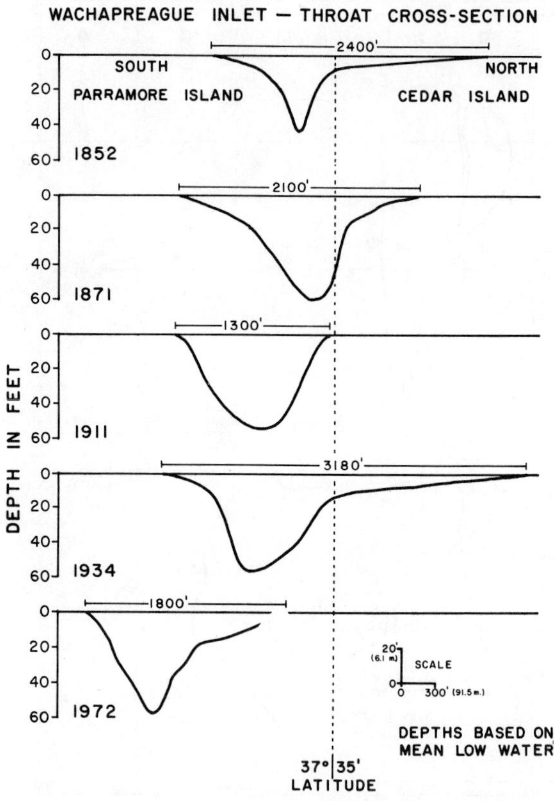

Figure 4. Migration and changes in geometry of inlet throat cross-section over the last 120 years.

TABLE 1.

Historical Cross-sectional Areas of Wachapreague Inlet Throat

Year	Area (m^2)
1852	1845
1871	4473
1911	4737
1934	4572
1972	4047

TABLE 2.

Channel Length (based on 12-m contour)

Year	Length (m)
1852	1662
1871	no data
1911	1701
1934	1909
1972	3046

TABLE 3.

Hydraulic Radius (m)

Year	Unmodified	Modified	Modified and Normalized
1852	2.5	4.3	6.7
1871	6.9	10.3	9.6
1911	9.6	9.6	10.0
1934	4.7	9.6	8.9
1972	6.1	7.4	7.4

from the northeastern sector and then passing through the channel between Cedar Island and the northern shoals. In September of 1972 the northern shoals again disappeared, but there was no corresponding accretion on either Cedar Island or Parramore Island. By November of 1972, another northern lateral ramp margin shoal had begun to accrete, and by July of 1973 it had developed further, to a volume of 0.9 x $10^6 m^3$ of sand.

Beaches on the eastern shore of Virginia are shallow, narrow and apparently sand-starved. Byrne et al. (2) estimate that the net drift from the north of the inlet does not exceed 450,000 m^3/yr. During the period from February 1970 to September 1972 a shoal of 2.9 x $10^6 m^3$ of sand disappeared. Yet by July 1973 another shoal reappeared with a volume of 0.9 x $10^6 m^3$ and an accretion of 1.9 x $10^6 m^3$ occurred on the northeastern face of Parramore Island. Changes of this magnitude cannot be reasonably related to fluctuations in littoral drift, but more likely correspond to cyclic short-term changes on the ebb-tidal delta. For example, a 1-m change in the depth over the area of the ebb-tidal delta (4 x $10^6 m^3$) will yield a volume change of 4 x $10^6 m^3$.

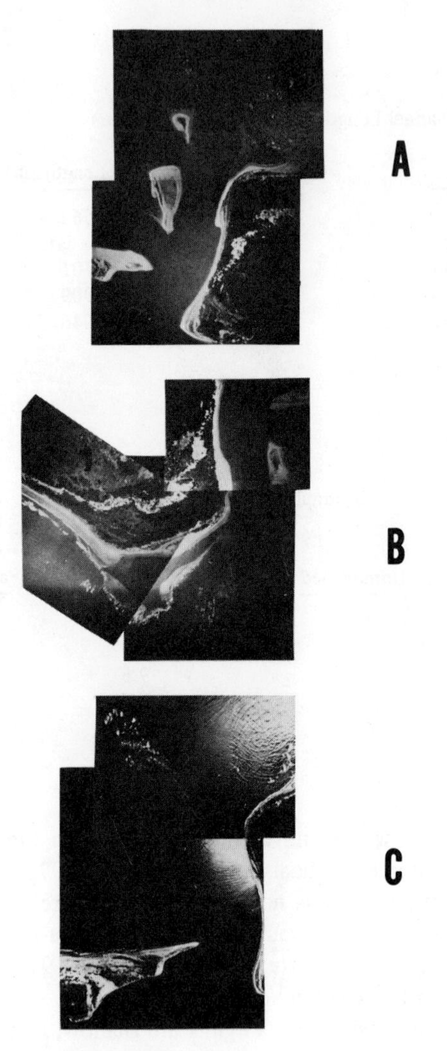

Figure 5. Wachapreague Inlet, 1957.

PRESENT MOBILE SEDIMENT DISTRIBUTION OF WACHAPREAGUE INLET COMPLEX

The mobile-sediment distribution was investigated with respect to both spatial variations over the entire inlet complex and temporal variations in inlet throat channel. The results of the spatial-sediment distribution survey are summarized in Figure 8. The sediments varied from a veneer of very coarse sediments,

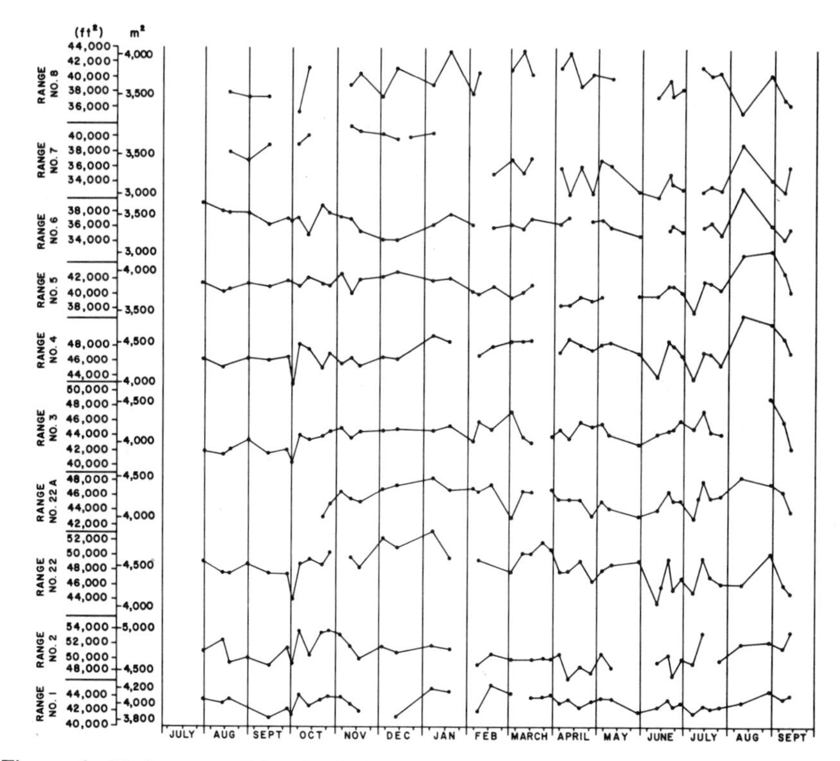

Figure 6. Wachapreague Inlet, April 1962.

composed of shell debris, cobbles, and gravels overlying a stiff cohesive sandy, clay substrate in the deep inlet throat channel, to well-sorted medium-to-fine sand surrounding the inlet throat to very fine silty sands both inside and outside the immediate area of the inlet channel. The sediment distribution appears to correlate well with the various depositional environments. Several very interesting points came to light as a result of the survey. The apparent flood-tidal deltas or bathymetric highs are in fact relative topographic highs of lagoonal sediments, overlain by a thin veneer of fine sand. Secondly, the northern flank of Parramore Island, on the steep wall adjacent to the inlet, is an exposure of very firm lagoonal deposits. Finally, there appears to be a swath of fine sand ($2.0\,\phi < \text{mean size} < 2.5\,\phi$) that intersects the coarser sediments in the inlet axial channel. Perhaps this is a pathway for sand to bypass the inlet. The slope of the channel sides in this area is 2°.

The bottom-sediment distribution in the throat of the inlet was sampled fortnightly for a period of three months at various high and low slack waters. Sample stations were located at the deepest part of each of eleven transects that cross the throat on the inlet. The loose sediments recovered from the bottom

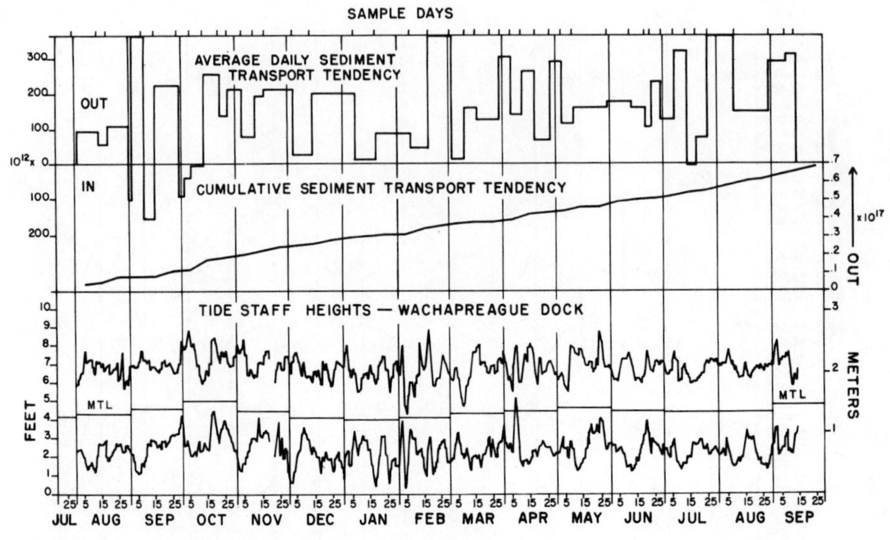

Figure 7. Wachapreague Inlet, June 1971.

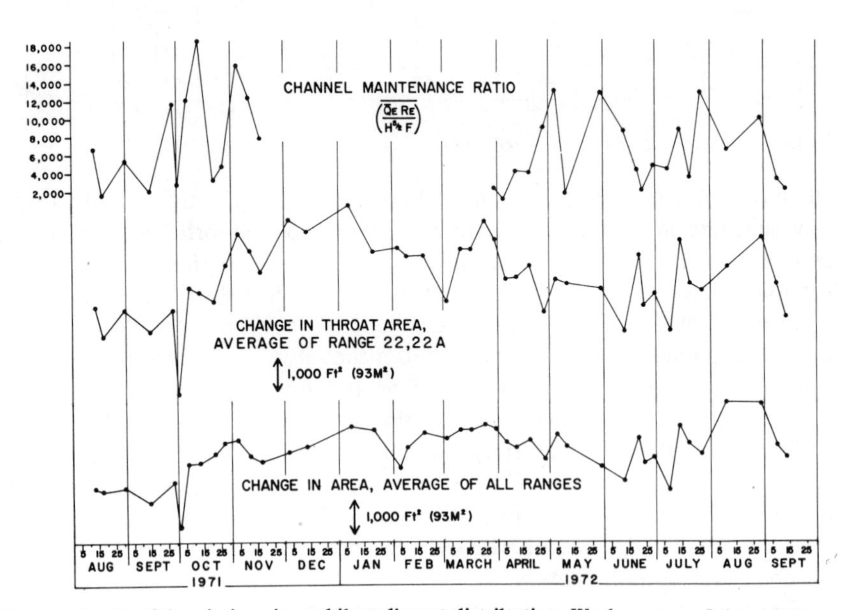

Figure 8. Spatial variations in mobile-sediment distribution, Wachapreague Inlet, 1972.

included medium- and coarse-grain sands, gravels, boulders (up to 15 cms in diameter), shell debris of various sizes and shapes, and rounded chunks of hard mud. These mud chunks proved to be identical to the substrate material along the southern flank and bottom of the inlet throat.

No obvious re-sorting pattern between high and low slack was observed during the sampling period. In the deepest parts of the inlet throat, below 15 m, the loose bottom sediment usually consisted of gravels and large shell debris (*Mercenaria* sp. and *Crassostera* sp.). Toward the eastern and western extremities of the throat channel in depths ranging between 12 m and 15 m, the mobile bottom sediments usually varied between coarse sand and smaller shell fragments.

The bottom-sediment distribution did reflect measured fluctuations in the cross-sectional area of the inlet's throat during the sample period. That is, during the last week in May 1972 and the first two weeks in June 1972, appreciable amounts of sand were recovered from most of the transects across the gorge, perhaps indicating a choking or filling-in of the throat. Later, this was correlated with an overall decrease in the cross-sectional areas of the transects across the inlet throat (over a 15% decrease at one transect). In mid July 1972, mud clumps (rounded chunks of lagoonal mud) were the principal sediment type recovered at almost all sample stations in the inlet. This indicated erosion in the inlet of the southern flank and the bottom, which later was verified by a significant overall increase in the cross-sectional areas.

Migration of shell debris was visually verified by divers who went down shortly after a slack water period. Thus it was determined that the inception of shell motion occurred at a relatively low current velocity in the inlet. These shifting coarse sediments appear to be abading into the hard bottom substrate, as evidenced by scour pits observed in the bottom.

THE GEOMORPHOLOGY OF THE INLET COMPLEX

The geomorphology of the inlet complex was studied from samples, cores, and observations taken by scuba divers, a 22-m well recently drilled on Parramore Island, and sub-bottom profiles made across the inlet throat and in Horseshoe Lead, landward of Parramore Island. The first realization that Wachapreague Inlet was different from the typical sandy trough described for inlets on sandy coasts (1), came as a result of the mobile-sediment distribution survey. Further observations and samples made by divers along the inlet bottom and southern flank confirmed that underlying the mobile coarse sediments on the deep inlet bottom there was a stiff sandy-silty clay substrate, with interspersed layers of gravels, and coarse sands. Samples taken of the south wall of the inlet (6-9 m below MSL) showed it to be composed of lagoonal deposits with a mean grain size of 4.8 ϕ.

Data taken by sub-bottom profiling across the inlet throat are shown in Figure 9 and their interpretation is shown in Figure 10. Note the horizontal reflectors below 20 m; these underlie the sedimentary deposits both to the north and to the south. The sloped reflectors on the north side between 20 and 15 m represent the recent sand deposits of the southern tip of Cedar Island as it extends southward. On the south side of the inlet, the reflectors are parallel and horizontal from below 20 m to a depth of 15 m; but note the two strong reflectors between 15 and 16m. Between 15 m and 11 m on the southern flank, the reflectors are again inclined toward the bottom, indicating either recent sand deposits or the deposits along the flank of an older channel. From 11 m to 6 m the reflectors are again parallel and horizontal.

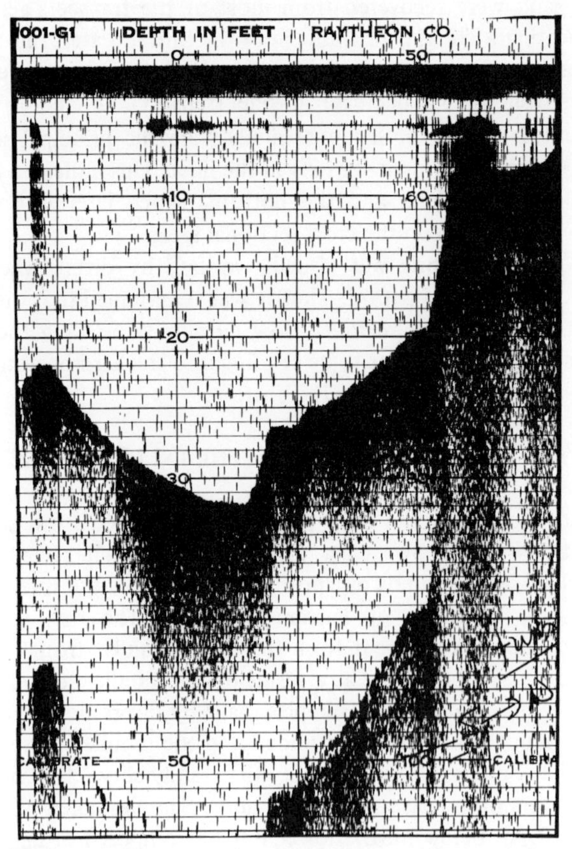

Figure 9. Sub-bottom profile across the throat of Wachapreague Inlet, from north to south.

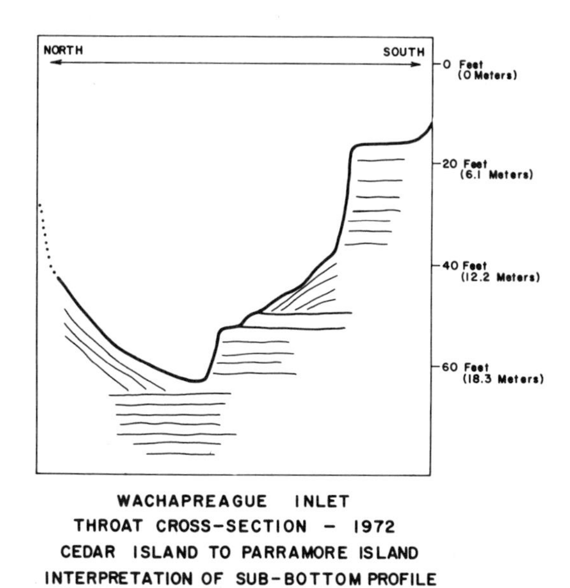

Figure 10. Interpretation of sub-bottom profile data, Wachapreague Inlet.

In order to be able to correlate the various reflectors shown in Figures 9 and 10 with specific geologic strata, a well was drilled on Parramore Island and continuously split-spoon-sampled from 0 to 22.4 m below MSL. A summary of this well log is in Table 4. There are some inconsistencies between the samples taken along the mud exposures on the northern flank of Parramore Island in the inlet and the well log. The very fine silty sand (mean size 3.5 ϕ) found from 9.0 m to 14.5 m were not found in the samples taken from the south wall of the inlet. However, immediately below that horizon, the coarse sands, shells, and gravels do correlate well with the two strong reflectors found in the inlet between 15 m and 16 m and found also in Horseshoe Lead between 15 m and 16 m. Below this, there are alternating layers of medium sands, fine sands, gravels, and finally at 22.4 m, a layer of very stiff, silty clay.

SUMMARY AND CONCLUSIONS

The inlet channel has migrated to the south at a rate of 1 meter per year. Since 1871, the cross-sectional area of the inlet throat has remained relatively constant at about 4400 m^2. The system has not served as either a significant long-term source or sink of sand. The main channel has progressively evolved to a less efficient configuration. In contrast, the short-term changes in the inlet are dramatic. The volume of sand involved in the growth and decay of the lateral ramp margin shoal, and the sand wedge on the northwestern face of Parramore

TABLE 4.

Summary of Parramore Island Well Log

Depth Below MTL (meters)	Mean Grain Size (phi units)	Comments
0 - 8.9	5.0	firm lagoonal mud, shells and rhizomes
9.0 - 14.5	3.5	very fine silty sand, shells (Littorina)
14.6 - 15.3	1.71	medium sand, shells and shell fragments (Crassostrea)
15.4 - 15.7	transition zone	no samples
15.8 - 16.4	5.8	firm silty clay mud, small shell (Littorina)
16.5 - 17.4	1.74	medium sands, shells and gravels
17.5 - 19.0	2.40	clean fine sands, small shells
19.1 - 20.4	1.93	medium sand, shells fragments
20.5 - 20.9	2.87	fine sands
21.0 - 21.3	transition zone	no samples
21.4 - 21.6	-4.0	gravels, shell fragments
21.7 - 21.9	1.89	medium sands, shell fragments
22	-4.0	gravels
22.1 - 22.3	5.25	very stiff silty mud

Island is quite large when compared to estimates of littoral drift. Sand must be moving on and off the ebb-tidal delta and the offshore area for these features to appear and disappear. It is interesting to note also, that, while the inlet channel has migrated south, the ebb-tidal delta has moved very little, giving rise to an apparent counterclockwise rotation of the channel axis.

The geology of the area has had an influence on Wachapreague Inlet. DeAlteris, et al. (3), described "A Geological Control of a Natural Inlet", Wachapreague Inlet. The firm cohesive lagoonal muds on the southern flank of the inlet have had a stabilizing effect on the inlet both by slowing the rate of migration and by allowing the inlet cross-section to assume a more efficient geometry at least on the south side.

ACKNOWLEDGMENTS

This work was supported by the Office of Naval Research, Geography

Programs, contract N00014-71-C-0334, Task No. NR388-103. The aerial photography of the study area was provided by NASA, Wallops Island, under contract NAS 6-1902. We would like to acknowledge Mr. Michael Castagna, Scientist in Charge of the Virginia Institute of Marine Science, Wachapreague Lab, for his assistance in the inlet program and Mrs. Cindy Otey for typing the manuscript. Particular thanks are due to D. G. Tyler and Ray O'Quinn for assistance in the field.

REFERENCES

1. Brown, E. I.
   1928 Inlets on sandy coasts. **Proceedings of A.S.C.E.,** 54(2): 505-553.
2. Byrne, R. J., Bullock, P., and Tyler, D. G.
   1973 Response characteristics of a tidal inlet, a case study. **Proceedings of 2nd Inter. Estuarine Research Conf.,** Myrtle Beach, South Carolina.
3. DeAlteris, J. T. and Byrne, R. J.
   1973 A geological control of a natural inlet. **Geol. Soc. Am., N. E. Section Meeting, 21-24 March 1973 (Abs.).**
4. Hayes, M. O., Goldsmith, V. and Hobbs, C. H.
   1970 Offset coastal inlets. **Proceedings of A.S.C.E., Twelfth Coastal Eng. Conf.,** p. 1187-1200.
5. O'Brien, M. P.
   1969 Equilibrium flow areas in inlets on sandy coasts. **Proceedings of A.S.C.E., Waterways and Harbors Div.,** 95(WW 1): 43-51.
6. U. S. Army Corps of Engineers
   1966 Shore Protection, planning and design. **Tech. Rpt. no. 4, C.E.R.C.,** Washington, D.C.

[illegible] No. NR 388-102. The aerial photography of the study area was provided by NASA, [illegible] contract NAS [illegible]. We would like to acknowledge Mr. Michael [illegible], Scientist in Charge of the Virginia Institute of Marine Science's [illegible] for his assistance in the pilot program and Mrs. [illegible] for typing the manuscript. Additional thanks are due to [illegible] and [illegible] O'Connor [illegible] and [illegible] for their help.

REFERENCES

[illegible]

[illegible] Proceedings [illegible] Conf. [illegible]

[illegible] March 1973 [illegible]

[illegible] Proceedings of [illegible]

[illegible] Proceedings [illegible]

[illegible] Waterways and Harbors Div. [illegible]

[illegible]

THE INFLUENCE OF WAVES ON THE ORIGIN AND DEVELOPMENT OF THE OFFSET COASTAL INLETS OF THE SOUTHERN DELMARVA PENINSULA, VIRGINIA[1]

Victor Goldsmith, Robert J. Byrne,

Asbury H. Sallenger, and David M. Drucker[2]

ABSTRACT

Comparisons of the bathymetric surveys of 1852 and 1934 indicate that during that 82-year interval these barrier islands became substantially offset (up to 1 km) seaward on the downdrift side of the inlets. The inlets migrated southward while the ebb-tidal deltas remained stationary. The offshore bathymetry underwent concomitant changes within the same 82-year interval, most notably in the ridge-and-swale bathymetry, which deepened in the troughs and built upward on the crests. This and other detailed analyses of the bathymetry has encouraged high confidence in the older bathymetric survey.

Using standard computational wave-refraction techniques and the older bathymetry, it was determined that in 1852 the shorter-wavelength northeast waves (T = 4-6 secs) tended to concentrate wave energy at the south ends of these islands, whereas longer northeast waves (T = 12 secs) tended to concentrate wave energy at the north ends of the islands. Moreover, the longer waves approached the shore with their wave orthogonals closer to the perpendicular of the shoreline than did the shorter waves. Thus, the more accretional waves built up the shoreline on the downdrift sides of the inlets,

1. Virginia Institute of Marine Science, contribution no. 579.

2. Virginia Institute of Marine Science, Gloucester Point, Virginia 23062.

while the shorter erosional waves eroded the shoreline on the updrift sides. This effect was amplified by a feedback mechanism: the more the inlet offset the greater the refraction of the longer waves, which resulted in more buildup and a decrease in littoral drift, especially to the north. This computed wave behavior is consistent with both the long-term volumetric stability and the extreme volumetric fluctuations observed annually by DeAlteris and Byrne (4) and Byrne, et al. (2) for these offset inlets. However, since 1852 there has been a tendency for the shoreline wave-energy distribution to become more uniform along any one of these barrier islands, which suggests that when the wave-energy distribution reaches equilibrium the growth of the inlet offsets will cease, and the inlets will become more stable.

INTRODUCTION

The 90-km coastline of the southern Delmarva Peninsula, extending from Chincoteague to Fisherman's Island, is composed of numerous inlets and extensive marsh systems with a thin veneer of sand forming the ocean barrier islands (Fig. 1). The northern two-thirds of this 50-mile section of coastline contains numerous inlets with large downdrift offsets; i.e., the net downdrift side of the inlet protrudes appreciably more seaward than does the updrift side. Though longshore sediment might be expected to accumulate on the updrift side of the inlets and result in updrift offsets, downdrift-offset inlets are so common on such barrier islands coastlines as New England, New Jersey, and the southeast coast of Alaska (8) that they appear to be the rule rather than the exception.

Historical studies of the southern Delmarva inlets by Byrne (1) and DeAlteris and Byrne (4) show that since 1852 Hog and Cobb islands have become offset on their north ends and eroded on their south ends, resulting in clockwise rotation and development of the downdrift inlet configuration. Wachapreague Inlet, between Cedar and Parramore islands, is the most offset of these inlets. Since 1852 the south end of Cedar Island has recessed landward approximately 0.5 km, while the north end of Parramore Island has accreted seaward 0.5 km, resulting in a net growth in this downdrift offset of approximately 1 km in the last 120 years. Concomitant changes include a slight southern migration of the inlet and a counterclockwise rotation of the main inlet channel because of the positional stability of the ebb-tidal delta.

Despite these major long-term morphological changes at Wachapreague Inlet, DeAlteris and Byrne (4) have determined that the total volume of material in this inlet system has remained nearly constant, with a slight loss of seven million cubic meters between 1852 and 1972. However, Byrne, et al. (2) have shown that dramatic, storm-associated, short-term fluctuations in total sand volume of up to two million cubic meters can occur in one year within the inlet channel. They suggest that such volumetric changes are too large to be simply attributed

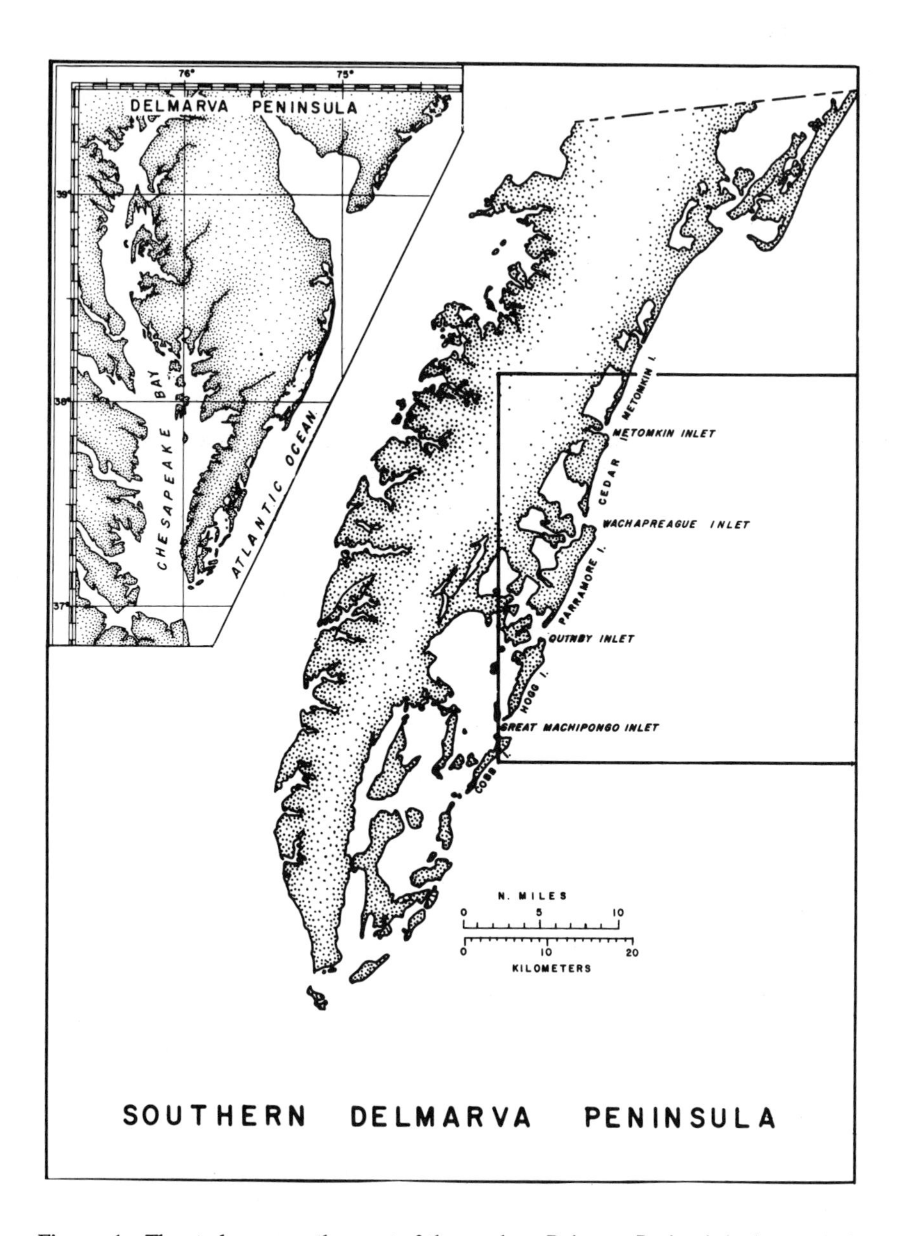

Figure 1. The study area on the coast of the southern Delmarva Peninsula is shown within the depth grid, 46 km on a side, and includes Metomkin, Wachapreague, Quinby, and Great Machipongo inlets.

to littoral drift processes alone.

The purpose of this paper is to report on our study of the offset inlets and to examine the influence of waves on the origin and development of these inlets from two points of view, the long-term historical changes and the short-term volumetric fluctuations. This will be accomplished primarily through detailed wave-refraction studies of the waves in the vicinity of these offset inlets; and in comparisons between the wave patterns of 1934 and those of 1852.

The question to be answered is "Did the wave behavior active since 1852 drive the system towards its present configuration?".

BATHYMETRY

The older wave patterns were determined by applying our present wave-climate models and using depths from the 1852 hydrographic survey in the vicinity of Metomkin, Wachapreague, and Quinby inlets. These surveys extend approximately 18 km offshore to depths of 25 to 30 m. Depths were accumulated from original U.S.C. & G.S. hydrographic surveys (Table 1) on a square grid 46.25 km by 46.25 km in size, at 0.46 km (0.25 nm) intervals for both the 1852 bathymetry and the 1934 (most recent) bathymetry. This has resulted in a total of 10,000 depths for each grid, which are contoured in Figures 2 and 3.

TABLE 1.

Hydrographic Charts Used in This Survey

Survey No.	Year	Survey Method	Navigation
348	1952	Lead-line	Sextant sitings on shore.
5673	1934	Lead-line	With distance from shore:
5674	1934	Lead-line	1. Sextant sitings on shore.
5703	1934	Lead-line	2. Sextant sitings on shore-located buoys.
5704	1934	Lead-line	3. Radio Acoustic Ranging (RAR)
5770	1934	Lead-line	
5771	1934	Lead-line	

In order to test the accuracy of the older hydrographic lead-line survey, and to delineate the areas where the comparisons would be too weak to use (e.g., at some distance from shore), the two surveys were compared by computing the differences between each of the depths at 0.46-km intervals, and then contouring these differences at 2-m intervals, starting at ± 1 m. This wide contour interval was chosen in order to allow for the inevitable "slack" in precision in even the modern bathymetric surveys (Fig. 4).

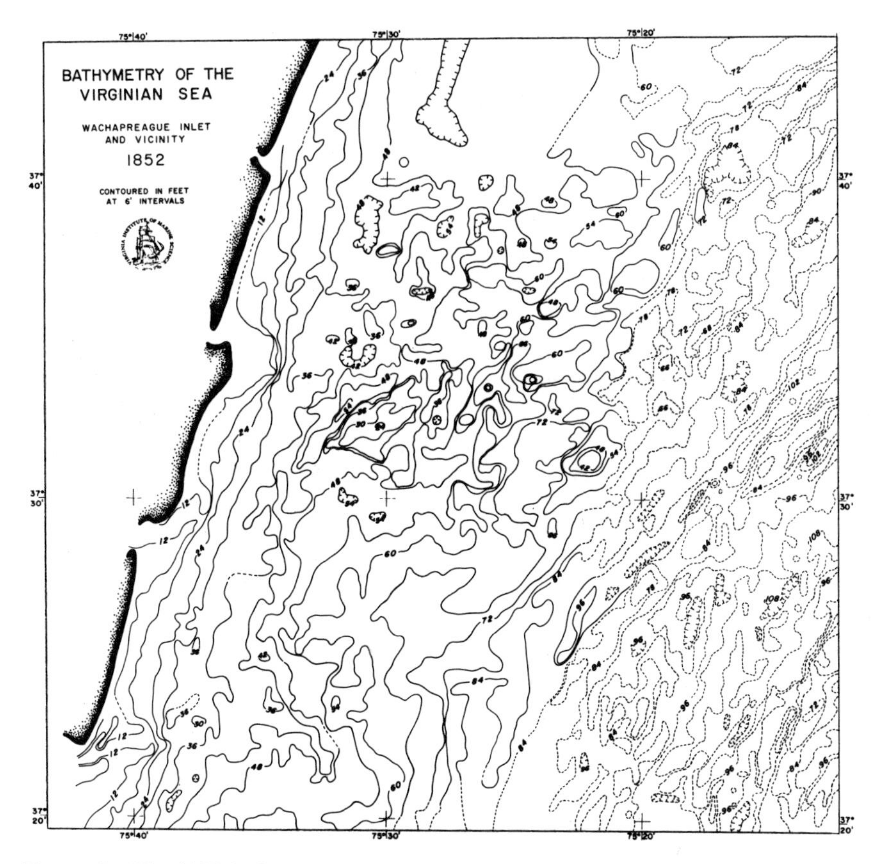

Figure 2. The 1852 bathymetry was read off U.S.C. & G.S. Hydrographic Survey 348 at 0.46 km intervals and contoured at 2-m (6-ft.) intervals. The dashed lines at the seaward edge (right) are based on the 1934 surveys. Note the complex ridge-and-swale bathymetry in the center of the grid and the low-relief areas to the south and north.

Several very interesting results are apparent in this bathymetric comparison figure, all of which give confidence in this approach:

(a) There is no apparent relationship between the amount of bathymetric change (i.e., comparison of 1852 and 1934 bathymetry) and distance from shore. Instead, the amount of change appears to be directly related to the complexity of the bathymetry (Figs. 2, 3, and 4).

(b) Definite trends in bathymetric changes, such as the erosional area immediately east of Cedar and Metomkin islands, give added credence to the accuracy of both surveys.

Also, in order to determine if the greater bathymetric changes in the areas of complex bathymetry were due to misalignment of track lines resulting from

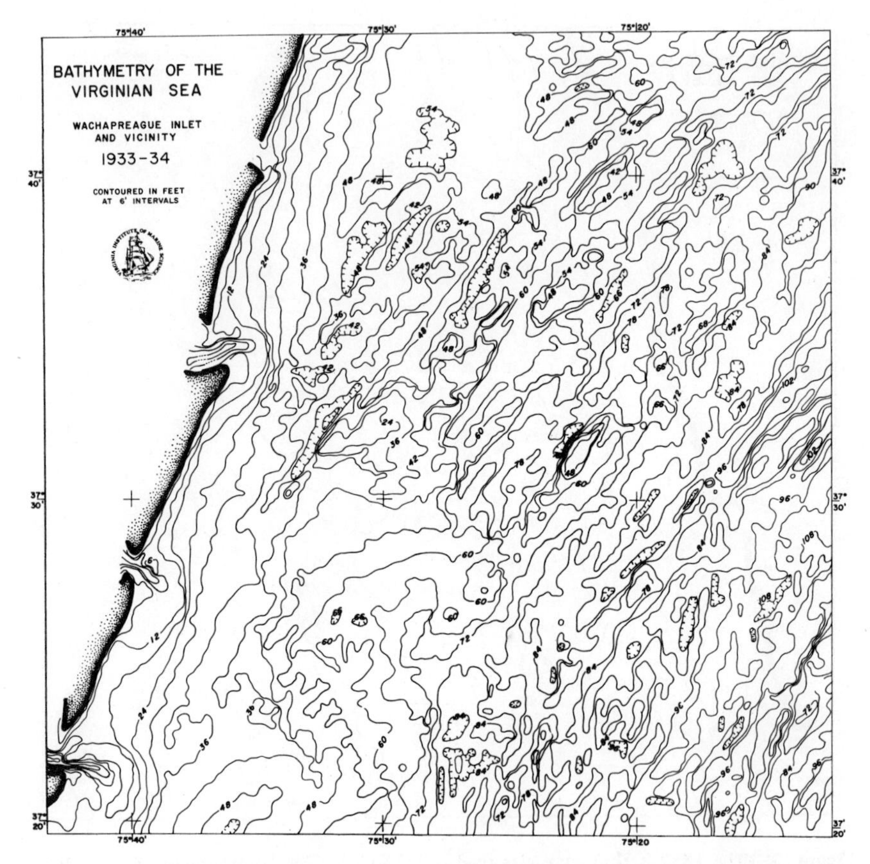

Figure 3. The 1934 bathymetry was read off several U.S.C. & G.S. Hydrographic Surveys (Table 1) at 0.46-km intervals and contoured at 2-m (6-ft.) intervals. Note the complex ridge-and-swale bathymetry at the center and the southeast corner of the grid, and the low-relief areas to the south and north. Also note the growth of the downdrift inlet offsets.

navigational inaccuracies, 100 profiles for each of the two grids were computer-plotted in an east-west direction. The profile comparisons showed that in the areas of complex bathymetry, many of the ridges had shallowed (i.e., grown upwards) and the swales had deepened between 1852 and 1934 (Fig. 5). Ship-track misalignments resulting from navigational inaccuracies would have the opposite effect; that is, ridges would decrease in size and swales would become shallower, resulting in a decrease in relief. Moreover, the scale of most of these bathymetric changes in the areas of complex bathymetry ($\pm$ 3 to 5 m) appear too large to be attributed only to navigational and sounding inaccuracies. This subject is treated much more thoroughly in Sallenger et al. (9).

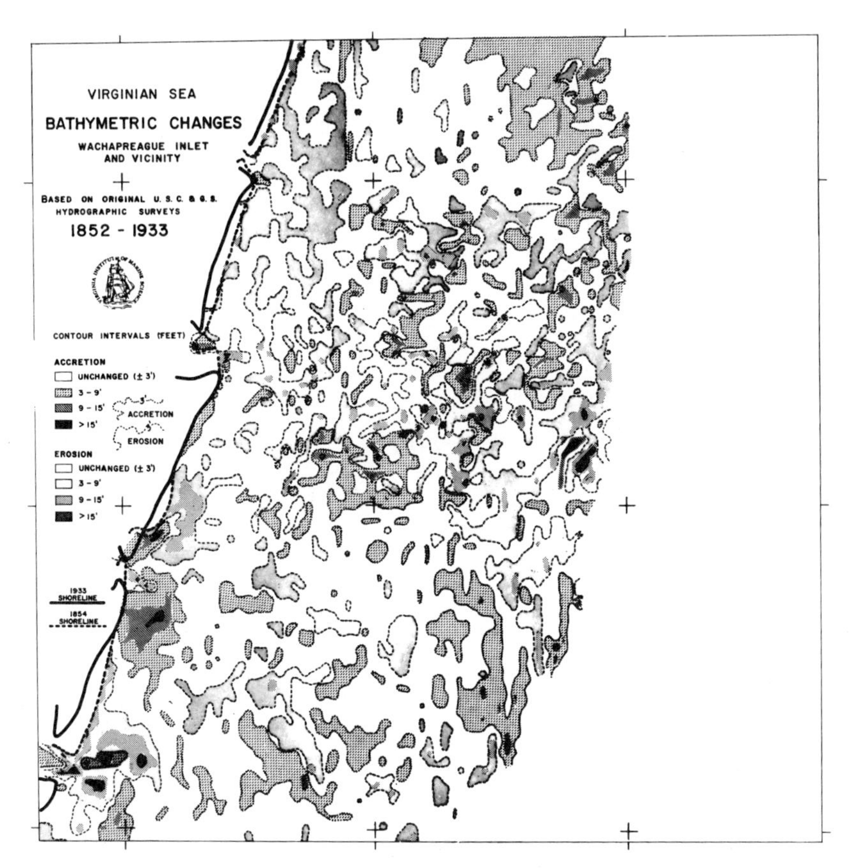

Figure 4. The contoured comparison of the 1852 and 1934 bathymetric surveys was determined by computing the depth differences at 0.46-km intervals. Note that the largest changes occurred in the highest relief areas (Figs. 2 and 3), and that these large changes show no relationship with distance from shore. Also note the large bathymetric changes located near inlets, and the erosional area adjacent to the shoreline in the northern half of the grid.

The significance of the results of the 1852-1934 bathymetric comparisons is that definite changes in the inner shelf bathymetry have occurred within this time interval, and that these changes are in sufficiently shallow depths to affect most incoming waves. A wave with a period of four seconds will begin to be appreciably affected at depths of 17 m, and a nine-second wave at 34 m depths.

WAVE REFRACTION

Wave input for this second-order grid adjacent to the southern Delmarva Peninsula came from the first-order Virginian Sea wave climate model (Fig. 6), in

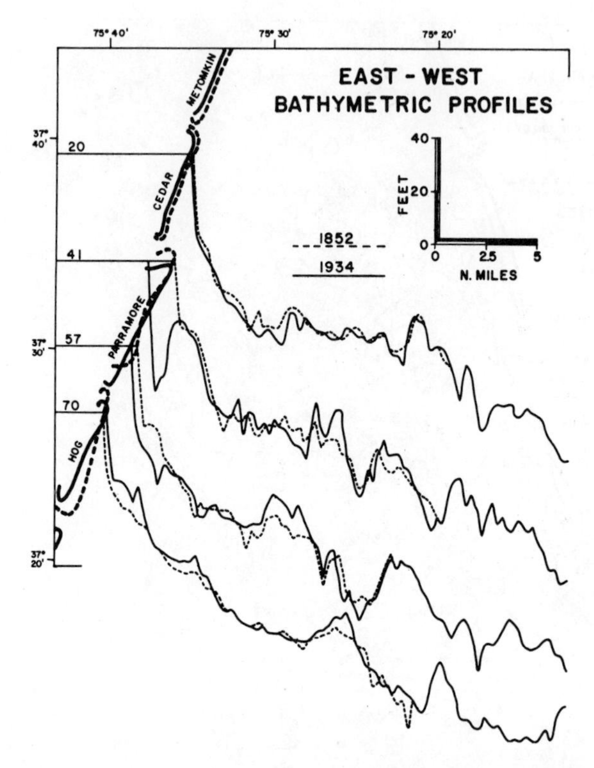

Figure 5. Comparison between 1952 and 1934 of four of the 100 east-west bathymetric profiles (Numbers 20, 41, 57, and 70) that were plotted from each of these surveys. Note the erosion in the swales and accretion on the crests. Such "realistic" changes give confidence in the navigation and depth accuracy of the surveys (see text).

which approximately 126 different wave conditions were propagated landward from deep water using standard linear wave theory at a .92-km depth grid (5, 7). The strengths, weaknesses, and applications of such an approach, which employs a dense grid of depths based on original hydrographic surveys as the basic data input, is discussed by Colonell et al. (3). In general, such an approach has been found by experience to be quite satisfactory. Wave input data were taken from the first-order grid and then input into the second-order grid at its seaward margin. The 39 wave conditions that were computed are listed in Table 2.

Output consists of wave-refraction diagrams and a printout of computations of 18 different wave parameters for each ray at approximately 1-km intervals as the wave propagates landward (6).

Wave-refraction diagrams for both 1852 and 1934 are presented for northeast waves with wave periods of 4 seconds (Figs. 7-A and 7-B) and 12 seconds (Figs.

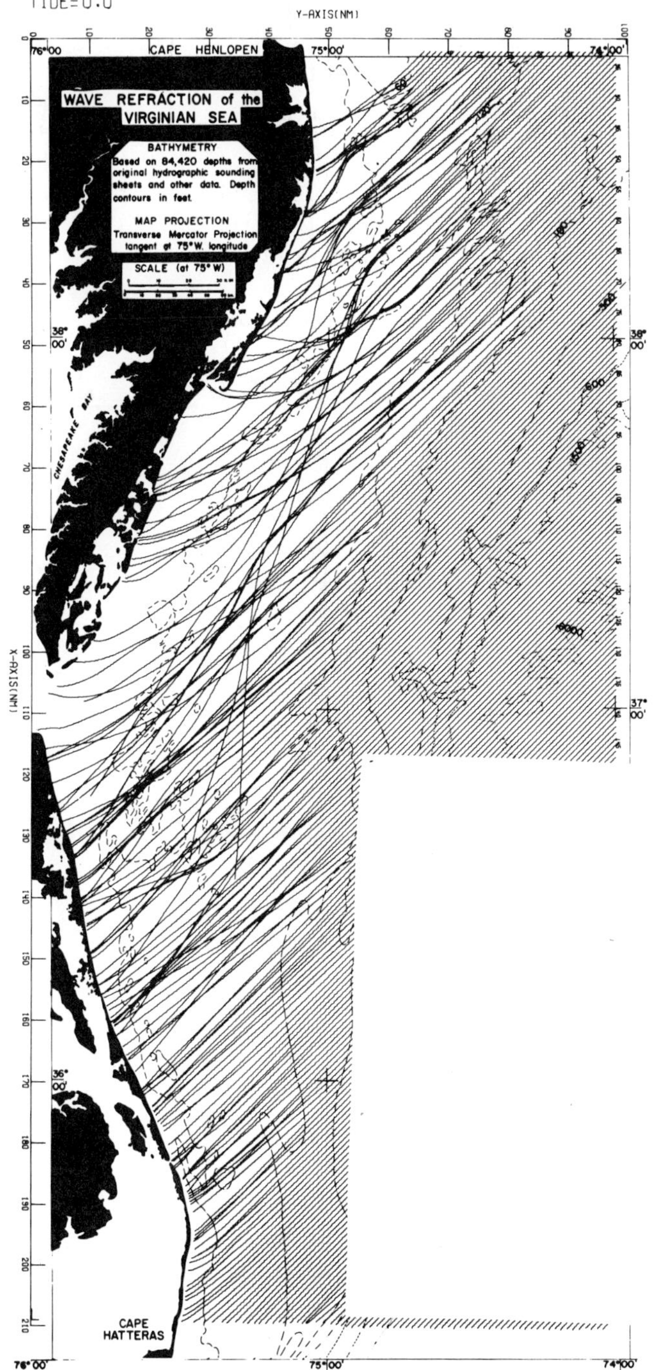

Figure 6. Wave-refraction diagram of the Virginian Sea, Cape Henlopen to Cape Hatteras, for waves from the northeast with period equal to eight seconds, height equal to 0.67m, at low tide. Wave rays were initially spaced at 2-km intervals. Wave output from diagrams such as these first-order depth grids is used as input for the second-order depth grids (Figs. 7-9).

TABLE 2.

Tide (m)	Direction	Wave Period (sec)	Wave Ht (m)
0	NE	4, 6, 8, 10, 12	0.7, 2
	E	4, 6, 8, 10, 12, 14	0.7, 2
	SE	4, 6, 8, 10, 12, 14	0.7, 2
+1.3	NE	4, 6, 8, 10, 12	0.7

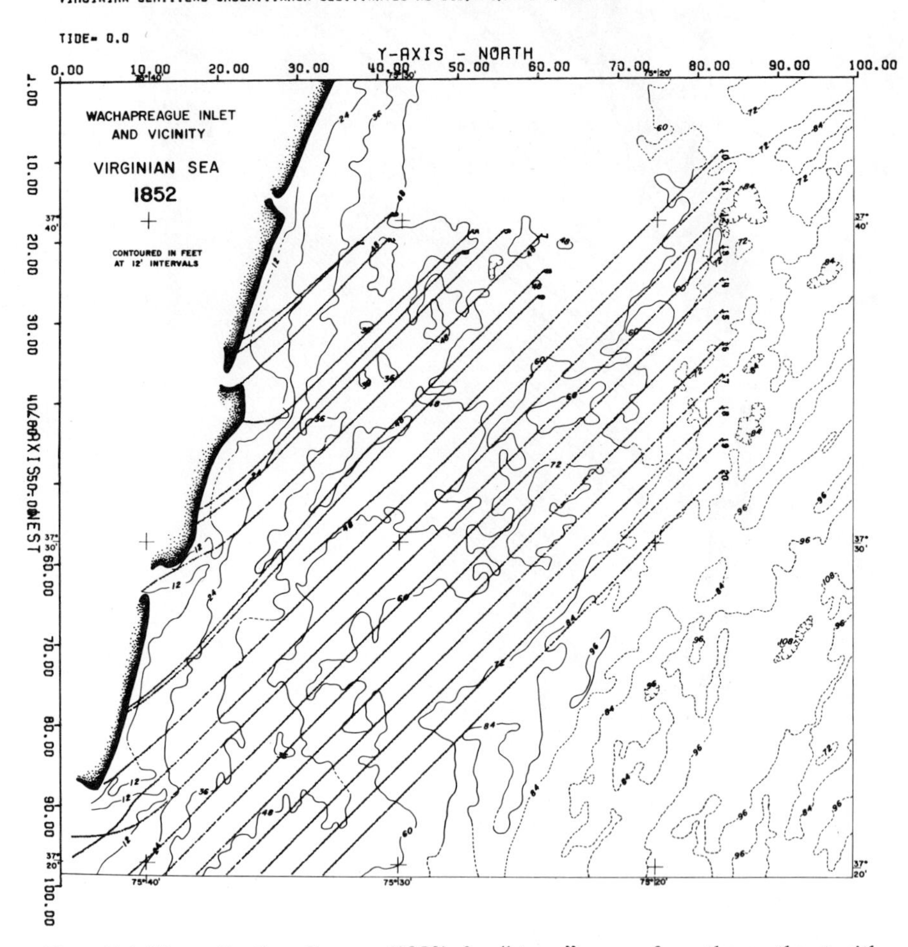

Figure 7-A. Wave-refraction diagram (1852) for "storm" waves from the northeast with period equal to 4 seconds, height equal to 0.67 m, at low tide. Note the convergence of wave rays, indicative of wave-energy concentrations, at the south ends of the barrier islands. This has resulted in relatively higher rates of shoreline erosion at these locations.

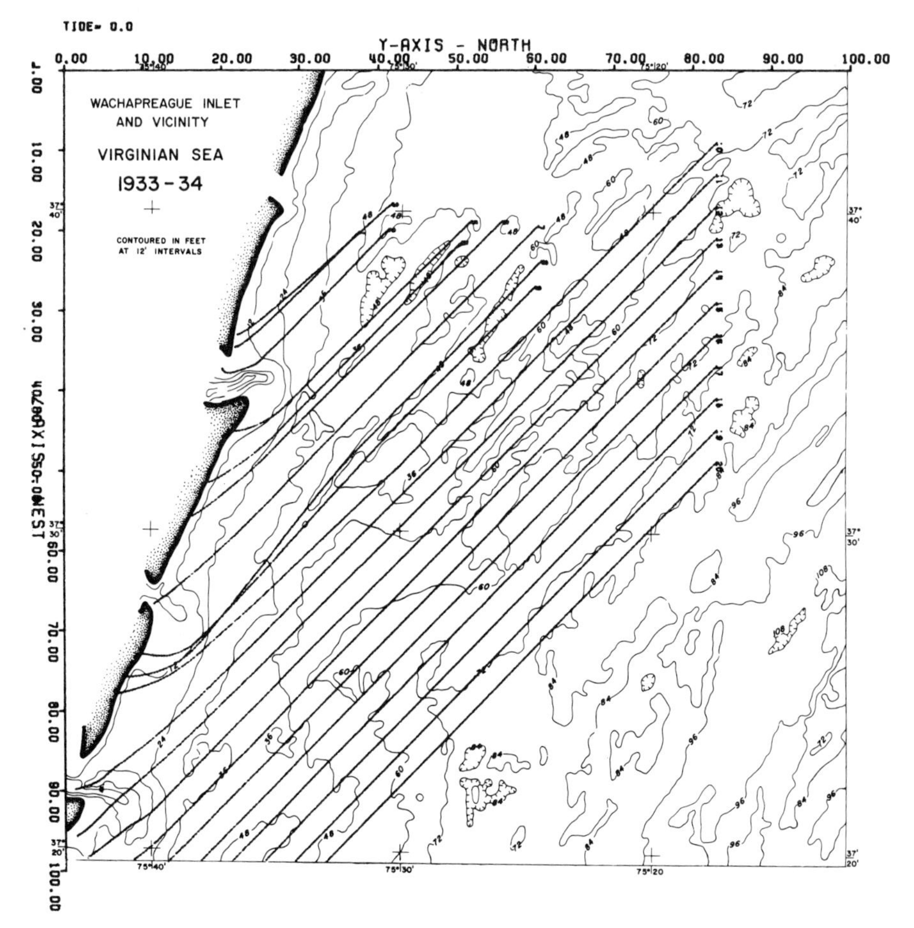

Figure 7-B. Wave-refraction diagram (1934) for "storm" waves from the northeast with period equal to 4 seconds, height equal to 0.67 m, at low tide. Note that the convergence of wave rays, indicative of wave-energy concentration, are not quite as concentrated at the south ends of the barrier islands as they were in 1852.

8-A and 8-B); all at low-tide conditions, and for 10-second southeast waves (Figs. 9-A and 9-B).

Computed values of wave energy for northeast waves of 4- and 12-second periods are shown in Figure 10 and compared with the strandline migration of these barrier islands.

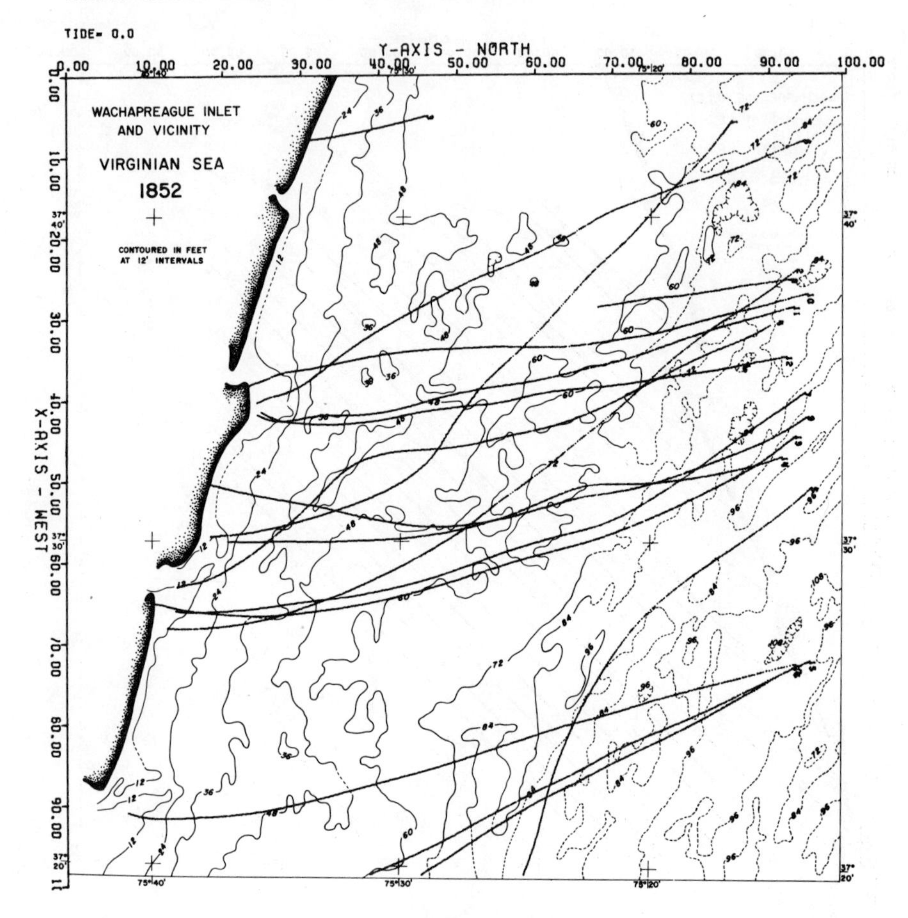

Figure 8-A. Wave-refraction diagram (1852) for "fair weather" waves from the northeast with period equal to 12 seconds, height equal to 0.67 m, at low tide. Note the convergence of orthogonals, indicating wave-energy concentration, at the north ends of the barrier islands. Note also that these wave rays form a higher angle with the shoreline than the shorter waves (Figs. 7a and 7b). This has resulted in greater rates of accretion in these areas, with concomitant decrease in littoral drift.

CONCLUSIONS

(a) In 1852 the shorter-wavelength northeast waves (T = 4-6 secs) tended to concentrate wave energy at the south ends of these islands, whereas longer

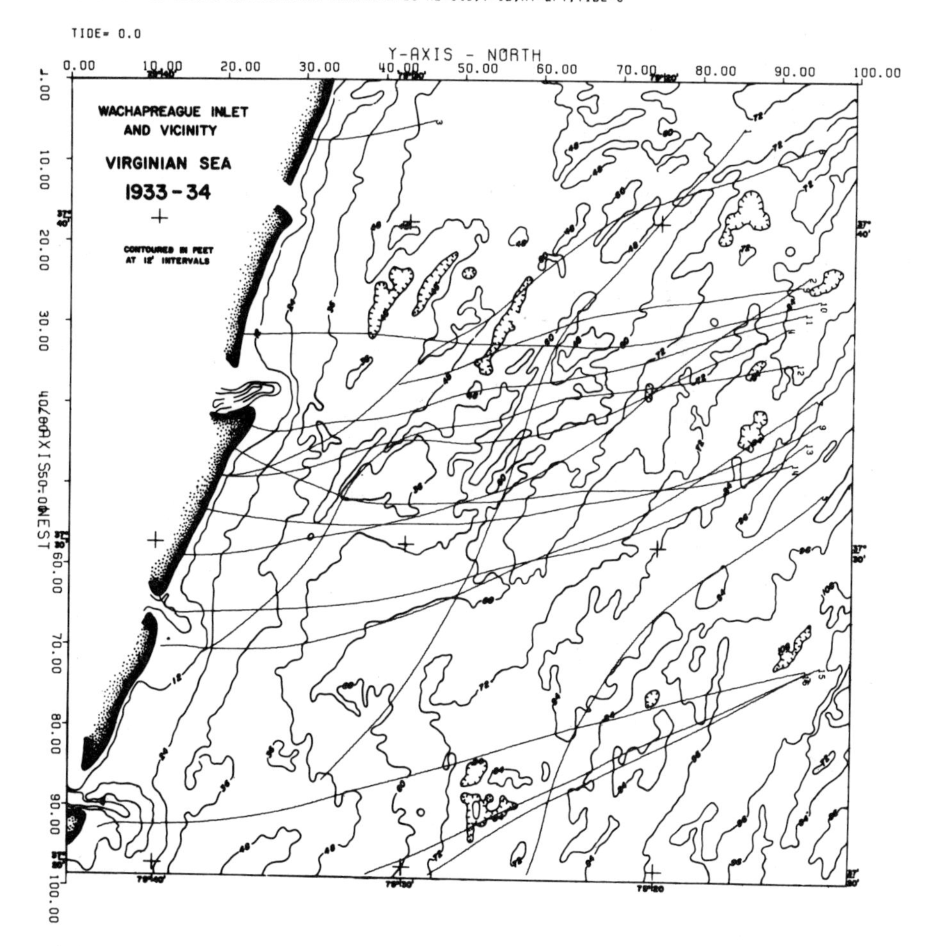

Figure 8-B. 1934 Wave refraction diagram for "fair-weather" waves from the northeast with period equal to twelve seconds, height equal to 0.67 m, at low tide. Note that the wave ray convergences at the north ends of the barrier islands are less pronounced with the growth of the downdrift offsets since 1852.

northeast waves (T = 12 secs) tended to concentrate wave energy at the northends of the islands. Moreover, the longer waves approached the shore with their wave rays closer to the perpendicular of the shoreline than did the shorter waves. Thus, the more accretional waves built up the shoreline on the downdrift sides of the inlets, while the shorter erosional waves eroded the shoreline on the updrift sides. This effect was amplified by a feedback mechanism: the more the inlet offset the greater the refraction of the longer waves, which resulted in more

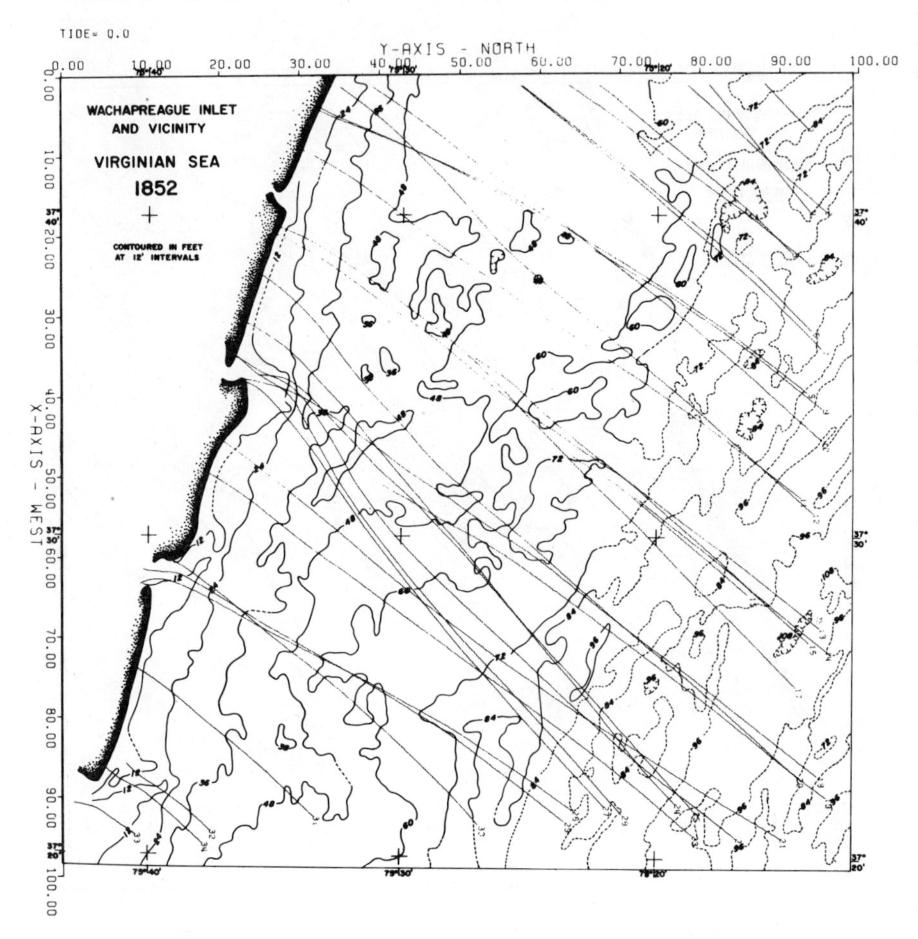

Figure 9-A. 1852 wave refraction diagram for "fair-weather" waves from the southeast with period equal to ten seconds, height equal to 0.67 m, at low tide. Note the very striking wave ray convergences at the inlets and the high angles of the wave rays with the shoreline. Southeast waves, therefore, do not contribute significantly to the northward littoral drift along this shoreline.

buildup. The greatest wave-energy concentrations in 1852 were along those parts of the islands that had the largest rates of shoreline erosion, resulting in the observed clockwise rotation of the islands between 1852 and the present (Fig. 10).

(b) The southeast wave-refraction diagram clearly shows that wave refraction increased between 1852 and 1934, especially at Wachapreague Inlet. The

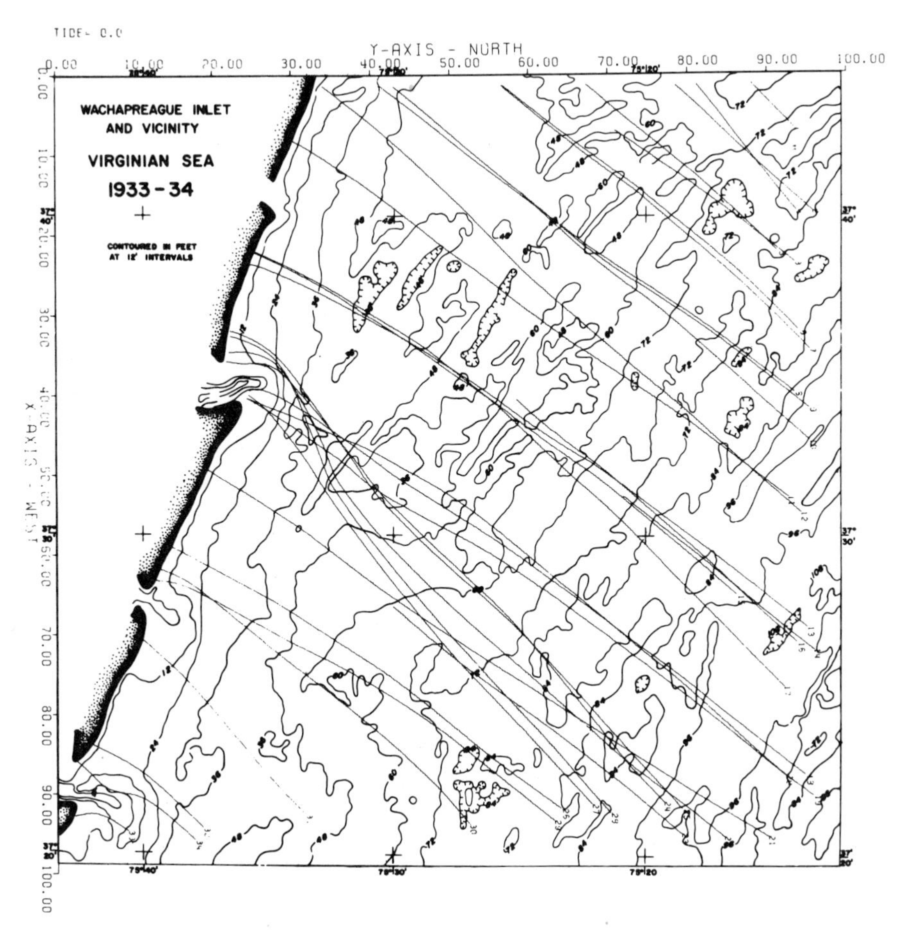

Figure 9-B. 1934 wave refraction diagram for "fair-weather" waves from the southeast with period equal to ten seconds, height equal to 0.67 m, at low tide. Note that wave ray convergences along the shoreline, especially at Wachapreague, have become more pronounced since 1852. This increased wave refraction has resulted in a further decrease in northern littoral drift since 1852.

increased tendency for the wave rays to approach at a higher angle with the shoreline has resulted in a "downgrading" of the importance of littoral drift along this coastline, especially in the amount of drift to the north.

(c) The low littoral drift around the inlet mouths and the fact that both the northeast and southeast longer-period waves tend to lock and enhance the downdrift buildup explains the observed increased offset. In addition the associated decrease in longshore drift is compatible with the general permanency

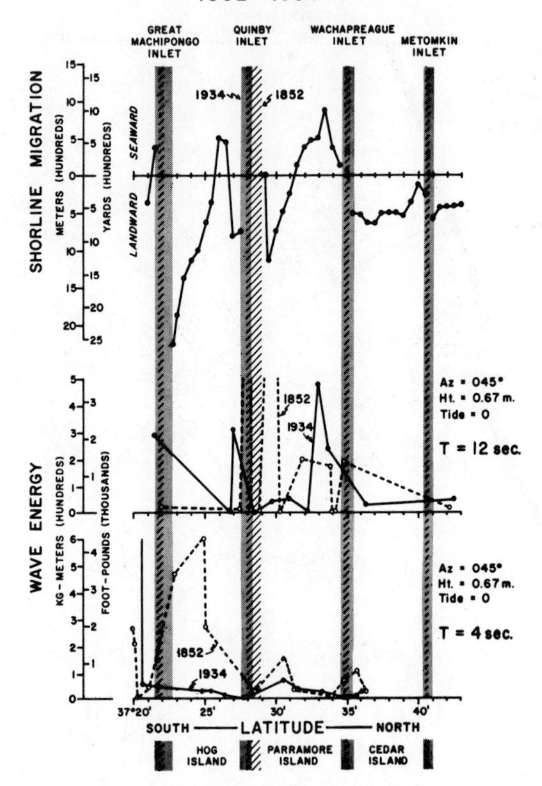

Figure 10. Shoreline changes (top) compared with wave-energy changes, between 1852 and 1934, for northeast waves of 4-second (bottom) and 12-second (middle) periods. Note the very striking correlation between erosion (i.e., landward shoreline migration) at the south ends of the islands and the location of the wave-energy peaks for 4-second waves, and a similar comparison between accretion at the north ends of the islands and the location of the wave-energy peaks for the 12-second waves. Note also, the strong tendency for the wave-energy distribution along the shoreline to become more evenly distributed between 1852 and 1934 coincident with the growth of the downdrift offset inlets and the other morphological changes of this barrier island chain.

of the total amount of material in the inlet system (4) and the interpretation that the short-term changes in sand volume in inlet channel and north soals is, at least in part, due to adjustments within the inlet complex (2).

FUTURE CHANGES

In answer to the question posed in the introduction to this paper, the data clearly indicate that the wave climate active since 1852 did drive the system to its present configuration. What about future changes?

An examination of Figure 10 shows that along any given barrier island, the wave energy of the 4-second wave tended to approach closer to a horizontal line, i.e., to reach equilibrium along the barrier shoreline, between 1852 and 1934. This is most notable at Parramore Island, between Wachapreague and Quinby inlets. Therefore, it is safe to predict that when the wave-energy distribution becomes evenly distributed along the barrier island shoreline, the strongly offset inlet configuration will have approached equilibrium. The *short-term* stability of these inlets should then increase appreciably.

However, it appears that these coastal inlet configurations are continuously changing, largely in response to the changes in incoming wave-energy distribution that results from bathymetric changes on the hydraulically active shelf. As a result of these offshore changes, *long-term* stability of these naturally occurring inlets in unlikely.

ACKNOWLEDGMENTS

The support by Sea Grant project 5-72 under P.L. 891-688 is gratefully acknowledged. We would like cordially to thank Lieutenant John J. Almy, U.S.N., and his crew of the U. S. Coast Survey, for their excellent bathymetric survey in 1852, without which this study would not have been possible. Appreciation is also extended to Mrs. J. S. Davis and Mr. K. Thornberry for their meticulous and talented efforts in preparing the graphic aspects of this study.

REFERENCES

1. Byrne, R. J.
   1973 Recent shoreline history of Virginia's barrier islands (Abs.). Va. Academy of Sciences Annual Mtg., Williamsburg, April 1973.
2. Byrne, R. J., Bullock, P., and Tyler, D. G.
   1973 Response characteristics of a tidal inlet: a case study. 2nd International Estuarine Research Conf.
3. Colonell, J. M., Farrell, S., and Goldsmith, V.
   1973 Wave refraction analysis: aid to interpretation of coastal hydraulics. A.S.C.E. **Hydraulics Div. Conf., Hydraulic Engineering and the Environment,** Montana State Univ., Bozeman, Aug. 15-16, 1973.
4. DeAlteris, J. and Byrne, R. J.
   1973 Recent history of Wachapreague Inlet, Virginia. 2nd International Estuarine Research Conf.

5. Goldsmith, V., Morris, W. D., Byrne, R. J., and Sutton, C. H.
1973 The Virginian sea wave climate model: a basis for the understanding of shelf and coastal sedimentology (Abs.) Northeastern G.S.A. Annual Mtg., March 1973, 5(2): 167.

6. Goldsmith, V., Morris, W. D., and Byrne, R. J.
1973 Wave climate model of the mid-Atlantic continental shelf and shoreline. VIMS SRAMSOE No. 38. (In press.)

7. Goldsmith, V., Sutton, C. H., and Davis, J. S.
1973 Bathymetry of the Virginian sea-continental shelf and upper slope, Cape Henlopen to Cape Hatteras. Bathymetric Map, VIMS SRAMSOE No. 39.

8. Hayes, M. O., Goldsmith, V., and Hobbs, C. H.
1970 Offset coastal inlets: a discussion. In 12th Coastal Engr. Conf., p. 1187-1200. Washington, D.C., 2286 p.

9. Sallenger, A. H., Goldsmith, V. and Sutton, C. H.
1973 Historical comparisons of bathymetric data on the inner continental shelf: evaluation and applications. (In press.)

RESPONSE CHARACTERISTICS OF A TIDAL INLET:

A CASE STUDY[1]

R. J. Byrne, P. Bullock, and D. G. Tyler[2]

ABSTRACT

Wachapreague Inlet, Virginia, was monitored over a 13-month period to document the cross-sectional area response of the channel to short-term variations in wave activity and hydraulic inputs. The offset inlet channel, whose length is 1.5 km, width 500 m, and maximum depth 20 m, was surveyed 46 times in the 13 months. The significant conclusions are:

(a) The complex basin-storage characteristics result in an ebb-dominated net sand transport over the long term.

(b) The channel responds very quickly to storm events and large tidal prisms with changes of 10-15 percent in cross-sectional areas occurring within a few days.

(c) Although the channel-area response is reasonably correlated with the ratio of ebb-tidal power to wave power, the short-term net sediment transport characteristics play a large role.

(d) The volume of sand injected into the channel cannot reasonably be attributed to longshore drift alone: a significant but undetermined portion of it is due to transfer of material on and off the ebb delta.

(e) The steric changes in sea level play a stong role in inlet maintenance where complex storage systems are involved.

1. Virginia Institute of Marine Science contribution no. 580.

2. Virginia Institute of Marine Science, Gloucester Point, Virginia 23062.

INTRODUCTION

Although the behavior of tidal inlets has been widely studied, there are many knowledge gaps that preclude predicting the details of behavior for any given natural inlet (2, 3, 7, 13). In general, the cross-sectional stability of an inlet is understood to represent a balance between the advection of sand by littoral drift and the scouring capability of the hydraulic currents generated at the inlet. Most of the previous work has cast the problem in terms of regional littoral-drift rates and gross hydraulic characteristics. This report presents the results of a 13-month field study wherein the objective was to relate the response in the inlet channel cross-sectional area to the variations in wave activity and tidal prism that arise from storm activity and fortnightly variations in tide range (11, 12).

Wachapreague Inlet, in the barrier island chain of the lower Delmarva Peninsula (Fig. 1), was selected because it has had a relatively stable and well-defined channel for the past century and it appears to be a good example of the offset inlets common to this barrier chain.

METHODS

In order to ascertain the changes in cross-section areas at different positions in the inlet channel, range lines were established on the north shore of Parramore Island at intervals of 200 meters (Figs. 1, 2). During operations, the sounding boat progressed across the inlet on a range line, while distances from the shore were recorded as horizontal angles, the base of which was a 400-meter baseline. Repetitive surveys over 10 range lines were conducted 46 times during the 13-month study period August 1971 through September 1972. Since the position of the inlet throat changed with time, three ranges, 2, 22, and 22A were established to accommodate the shifts in position. The echo sounder was calibrated for each survey using a bar check, and all soundings were corrected to mean tide level.

The precision of the profiling technique was tested by running ten consecutive profiles within a one-hour time span at Range 22. The mean area was 4,596 m^2 with standard deviation of 62 m^2. During the subsequent survey program area differences between successive profiling dates were considered significant if the difference, either positive or negative, exceeded 120 m^2.

Additional input data to the study consisted of continuous current velocities at one point in the inlet throat (0.6 depth) and tide-gage recordings at the inlet and at the town of Wachapreague. Daily visual wave observations were obtained from the Coastal Engineering Research Center, U. S. Army Corps observers on Assateague Island, some 45 km to the north.

Discharge gaging was performed at Ranges 22 (inlet throat) and 8. Six buoyed

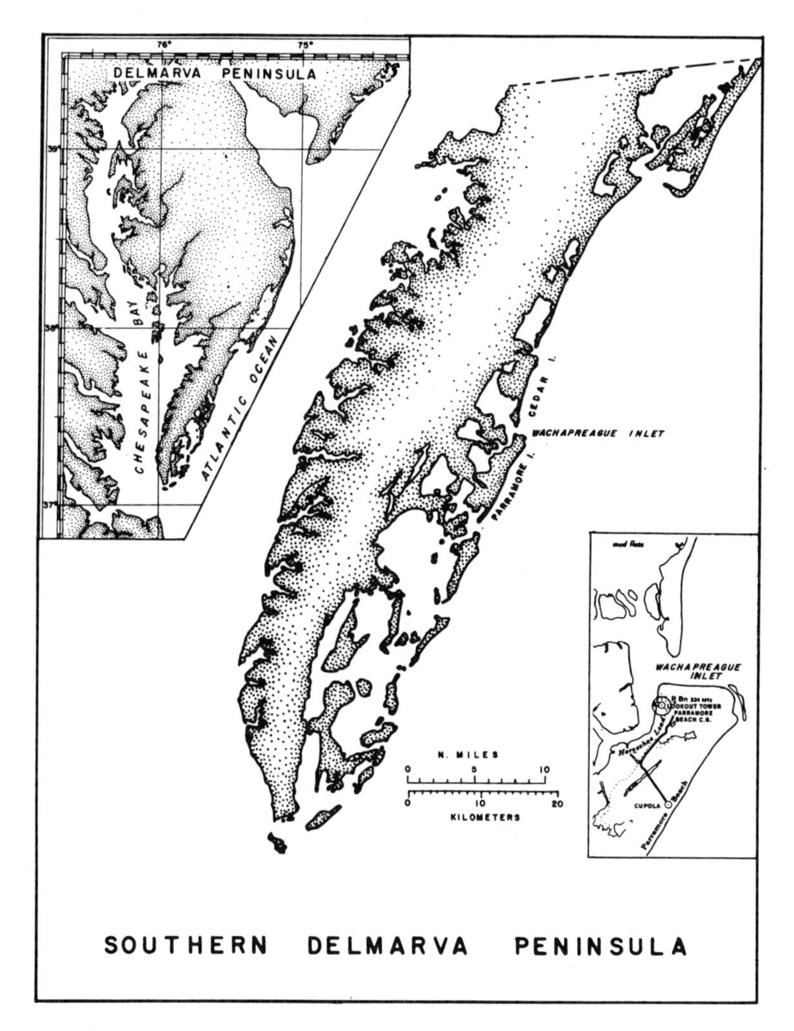

Figure 1. Southern Delmarva Peninsula. Extreme upper in lower Assateague Island and Fishing Point, a recurved spit formed since 1852. Chincoteague Inlet is directly west of Fishing Point.

stations were occupied at each range throughout the tidal cycle and the mean velocity over the vertical was sampled every 0.5 hr.

CHARACTERISTICS OF THE SYSTEM

Inlet Morphology and Sediments

A complete description of the geomorphology and recent history of the inlet

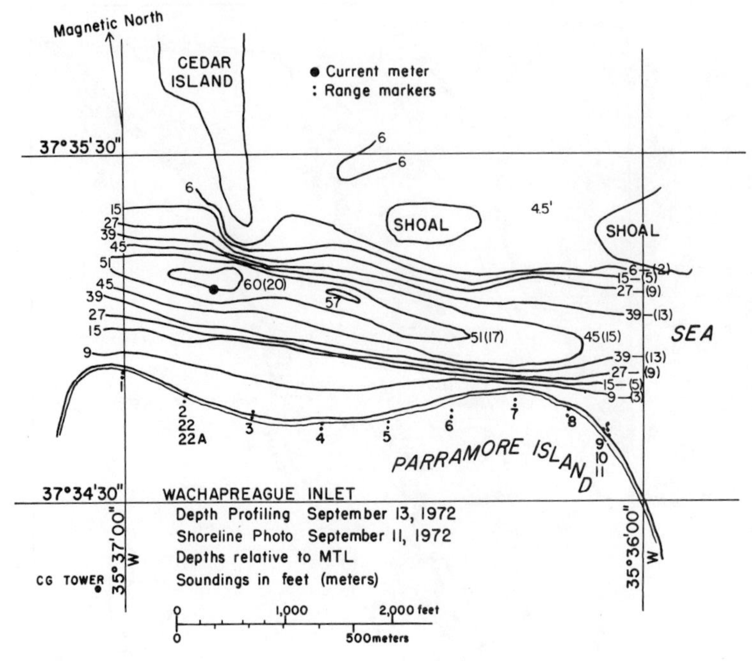

Figure 2. Wachapreague Inlet channel. Note shelf on south and asymmetry of channel cross-section. Numbers on south shoreline are survey range line positions, 1 thru 11.

is presented elsewhere in this volume (6).

Tide and Storage Characteristics

The regional tides are semi-diurnal with a mean ocean range of 1.16 m. Limited drogue studies indicate the coastal tidal flow is northerly during ebb flow in the inlet. The tide range within the lagoon-marsh complex is the same as the ocean range, thus the full potential tidal volume flows through the inlet to the interior system. The phase lags between high and low waters between the inlet and the town of Wachapregue, approximately 12 km distant via channels, are -0.6 hr and -0.7 hrs, respectively. Very little fresh water flows into the system.

The storage system with which the inlet interacts is complex, because there are channels connecting with the adjacent inlets. Even more important is the fact that the storage area does not remain constant with increase in tide stage, since channels, tidal flats and lagoons, and then extensive marshes become sequentially flooded. Gaging of the tidal discharge in the interior channels indicates that about 40 percent of the inlet prism passes into Swash and

Bradford bays (Fig. 3) via Horseshoe Lead. Assuming the lagoon depths and marsh elevations to be uniform throughout the system, the true storage function was approximated by three linear segments (Fig. 4) relative to the Wachapreague tide gage. The tide prism for any given tide could then be estimated as the difference in tidal volume stored at high and low water.

The existing storage characteristics apparently generate shallow-water, tidal components in the interior system that results in a difference between the duration of ebb current and flood current in the inlet, the flood duration being longer by 0.4 hr. This was verified by comparison between a 9-month record at the Wachapreague reference gage and the ocean gage at Wallops Island (30 km north), where the mean duration difference was only .05 hr. Further verification was obtained by comparing one month's data between the inlet currents and the tides at the reference gage, where the duration differences were 0.35 hr and 0.45, respectively. Given the longer average flood duration over ebb, the mean ebb discharge will be somewhat larger than the flood and the tendency for net outward sediment transport can be expected. This is discussed later.

Due to the storage characteristics of the interior system the discharge maxima in the inlet are shifted toward the time of high water. Limited gaging information at the inlet indicates that slack water follows the extremes of the tide by about 0.4 hr.

Flow Conditions in the Inlet Channel

The distribution of ebb and flood flow along the inlet channel is controlled by the degree of development of the lateral shoals flanking the north side of the channel and by the character of the entrances. In a very generalized sense, the flood flow behaves as a radial inflow to a point sink, whereas the ebb flow is more channelized and then issues as a plane jet over the ebb delta system. In order to discern the flow patterns in the channel flow, gaging was performed simultaneously for a 26-hour period at Ranges 22 and 8 (Fig. 2) in September 1972 when the flanking shoals on the north were at 1 m (MLW) depth (Fig. 5). During flood about 35 percent of the prism passing Range 22 occurred as lateral inflow over the north flank of the channel, while during ebb about 15 percent of the prism exited as lateral outflow. At Range 22 itself the ebb and flood flows are similarly distributed, but the ebb flow is more concentrated in the channel. The shelf area on the southern flank (Fig. 2) represents about 10 percent of the total cross-sectional area and about 10 percent of flood prism passes over the shelf. On ebb, however, the flow partially bypasses the shelf, with only 4 percent of the prism passing. Thus, higher velocities are experienced in the channel proper on ebb flow.

At Range 8, some 1200 m to the east, the flood discharge is less than the ebb, the difference being a function of the magnitude in lateral inflow. The ebb-flow

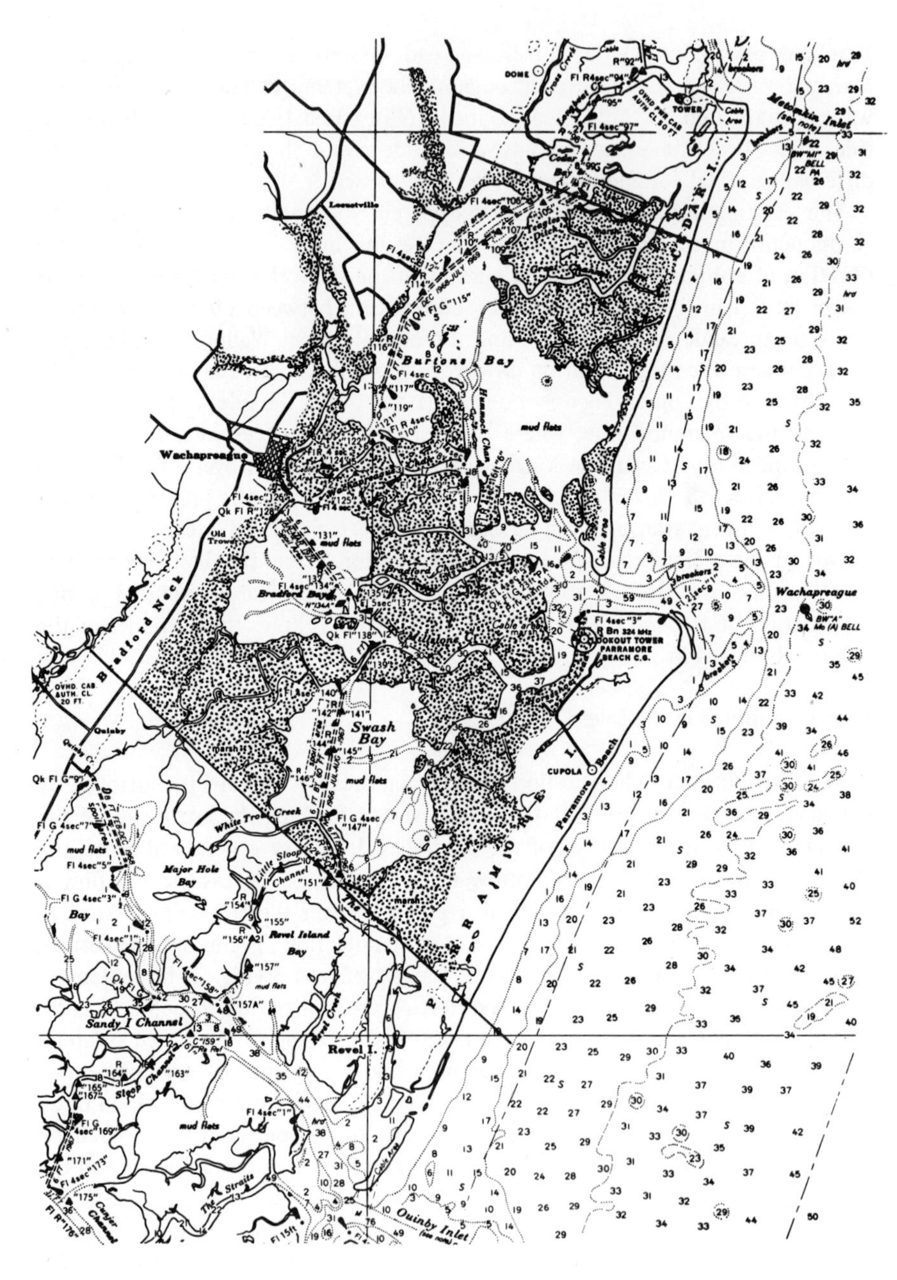

Figure 3. Wachapreague Inlet storage system. From U.S.C. & G. S. chart 1221. Distance between horizontal coordinates equals 18.3 km.

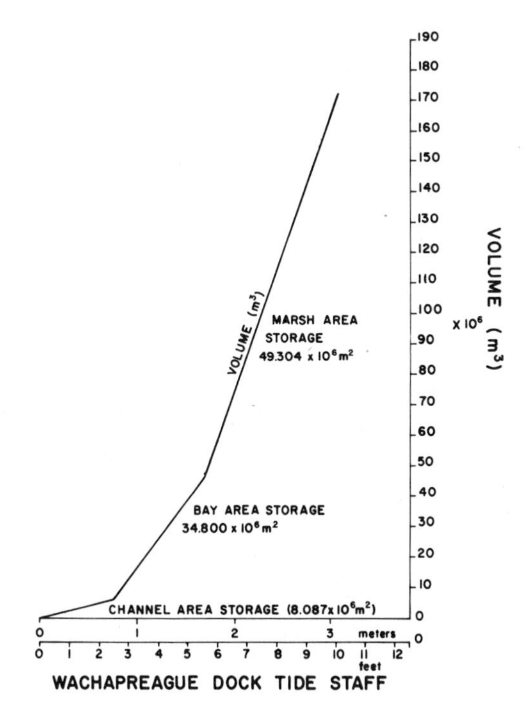

Figure 4. Estimated volume storage relative to tide gage at town of Wachapreague.

distribution is strongly skewed with higher speeds on the south side, whereas during flood the flow is slightly skewed with the higher velocities on the north side. Thus during times when the north flanking shoals are well developed and lateral inflow is reduced, there is probably a net inward sand transport on the north flank.

RESULTS AND DISCUSSION

Results

Virtually all the area modifications were the result of change in the volume of sand on the north flank of the inlet. The 8-m contour on the steep south flank remained with ± 7 m of the mean position in 91 percent of the cases; these were not real shifts but represent the range of positioning errors on the steep slope. Variation of maximum depth at each range line was small; 83 percent of the maximum depths fell within ± 0.5 m of their means. Range 1 showed the greatest depth variation with a decrease of 2 m between mid-January and mid-February, 1972. The horizontal position of maximum depth for each range remained stable; for all ranges and cases the position of maximum depth fell

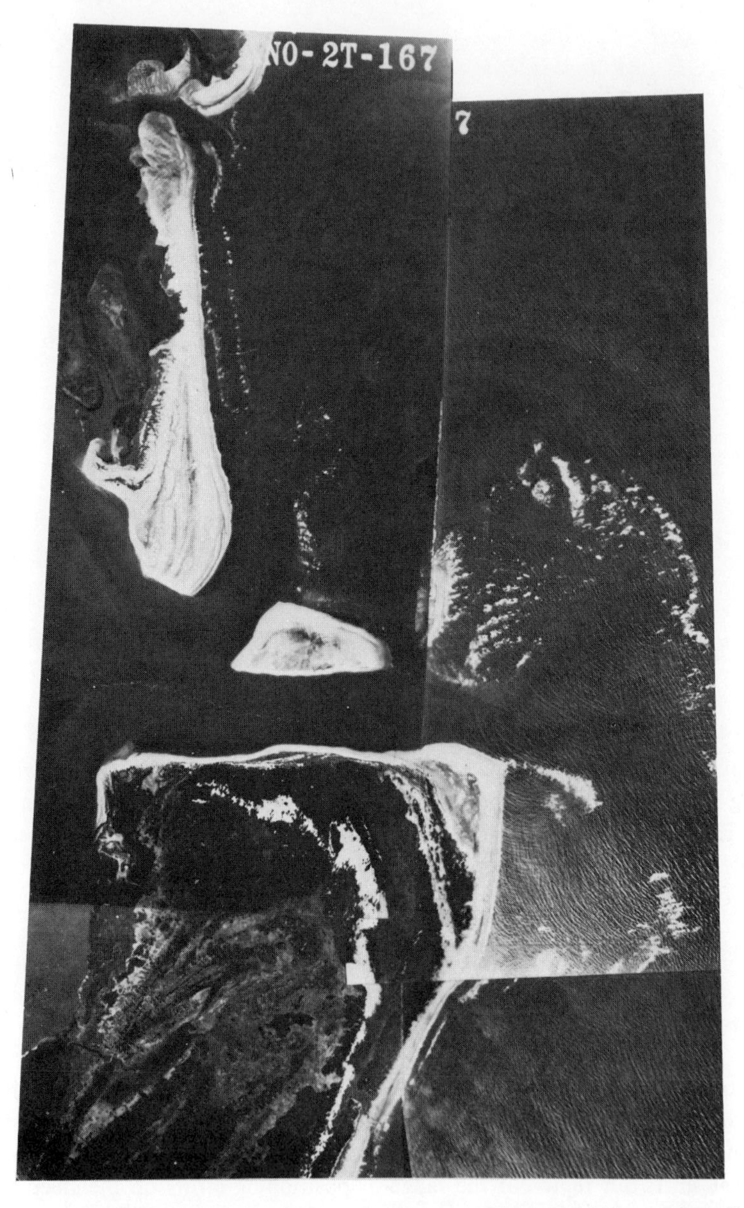

Figure 5. Diminution of north flanking shoals controlling lateral inflow and outflow. A. September 1971, B. February 1972, C. September 1972. NASA Wallops photography.

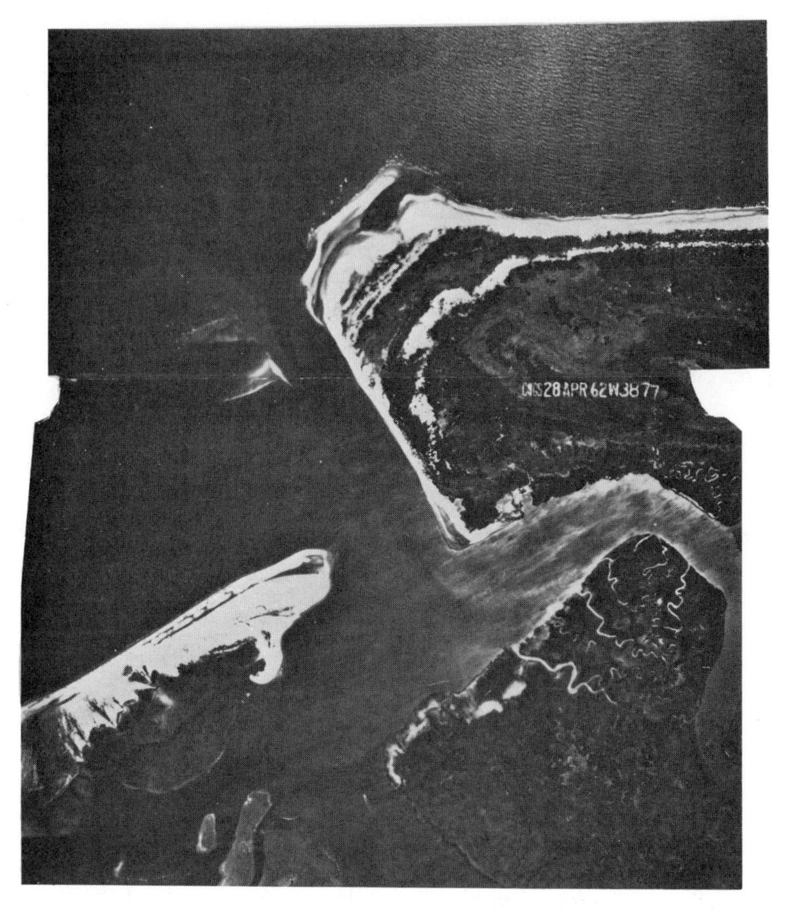

Figure 6. Area changes at Ranges 1 through 8. Time ticks within months are 5th, 15th, 25th of month. Survey starts on July 31, 1971, and ends on September 13, 1972.

with ± 15 m of the means 83 percent of the time.

The results of the repetitive cross-sectional area measurements are shown in Fig. 6. They indicate that adjustments in inlet cross-section can take place very rapidly. A case of rapid response is illustrated by the surveys of 28 September, 1 October, and 6 October, 1971. Between the first two dates Tropical Storm Ginger stagnated off the Virginia coast during the waning of neap tides. The heavy northeast seas resulted in large longshore sand transport and a consequent reduction in area throughout most of the channel. Between 1 October, and 6 October spring tides and residual storm surge resulted in large tidal prisms which expanded the cross-sections beyond the prestorm condition. Coherence, in the sense of response between ranges, was generally high for large storms or large prisms. At other times adjacent ranges frequently exhibited changes of opposite sense.

It is particularly interesting to note the behavior of Range 7, which exhibited a dramatic (~ 17%) reduction in area by February 1972 that persisted with modifications through September 1972. This reduction occurred as a result of the formation of a flood delta on the north which was time coincident with the diminution of the large lateral shoal (Fig. 5). It is interesting to note that the other ranges did not reflect this dramatic reduction in area.

The largest average cross-sectional area change occurred at the throat and at Ranges 7 and 8, while the least response was evidenced at Range 1. The throat (Ranges 22-22A) and Ranges 7 and 8 also exhibited the highest percentage of large area changes ($> 93\ m^2$).

Discussion

The tidal characteristics of the system result in such a duration difference between ebb and flood tide phases that the mean ebb discharge is somewhat greater than the flood. To assess qualitatively the net transport tendency during the study, the sediment-transport rate was assumed to be proportional to the cube of the mean discharge (13), which was estimated by using the prism calculated from the storage function (Fig. 4). The net sediment transport in the inlet channel for a given period is given by:

$$\text{Net sediment transport} \quad \alpha \ \Sigma \left(\frac{P_F}{\Delta t_F}\right)^3 \Delta t_F - \Sigma \left(\frac{P_E}{\Delta t_E}\right)^3 \Delta t_E$$

where P_F and P_E are flood and ebb prism and Δt_F and Δt_E are flood and ebb durations. The cumulative transport for the year is shown in Figure 7, as is the average daily net transport within survey periods. Finally, the daily maximum tide range is plotted with the monthly mean tide levels (MTL). Storm surges are evident in the plot. Although there were periods of net inward transport, the cumulative tendency over the long term is a net outward transport. These results agree with the analysis of Mota Oliveira (9), which predicts an ebb dominance for such storage systems. This characteristic of the system offers an explanation for the absence of flood-delta growth in recent times (120 years) (6), and the maintenance of the highly developed ebb-tidal delta system. This evidence and an examination of the morphology of the other deep inlets to the south along this reach of coast indicate that relatively small volumes of sand are trapped on the interior of the inlets. Caldwell (4), in contrast, finds that the flood deltas of the inlets of the New Jersey coast trap about 25 percent of the sand in the littoral drift system.

Many students of tidal inlets consider the cross-sectional stability to be

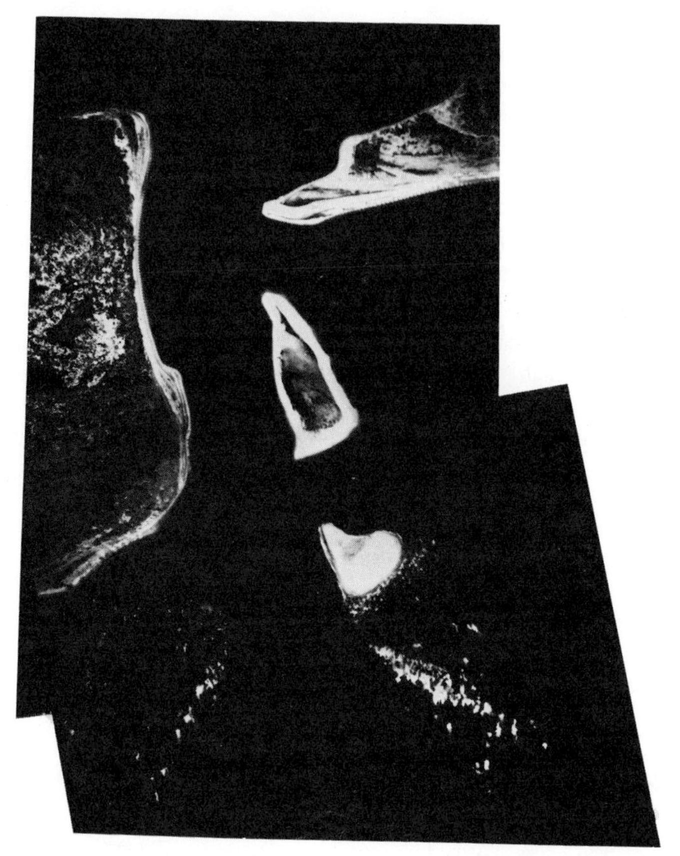

Figure 7. Transport tendency during survey period and daily tidal extremes. Cumulative and average daily sediment transport tending on relative scale.

controlled by the balance between the magnitude of littoral drift and the flushing power of the inlet currents. In order to examine the channel response relative to these parameters, the ebb-tidal power and the wave power were cast as a ratio (10) using the calculated prism and the visual wave observations from Assateague. Since the wave-observation program does not discriminate wave direction for small wave angles, the ratio was weighted using the U. S. Coast Guard observations at Chesapeake Light. The resulting ratio is proportional to the ebb-tidal power and the shallow water wave power:

$$\text{Channel maintenance ratio} \propto \overline{\frac{Q_E R_E}{H^{5/2} F}}$$

where Q_E = mean ebb discharge per day = prism $\div$ duration of ebb
R_E = ebb tide range

H = wave height
F = wave duration weighing factor
F = 3 waves approach 0 to 70° true
F = 2 waves approach 80° to 110°
F = 1 wave approaches 110° t0 180°

The wave-direction weighting factor, although arbitrary in its limit, was designed to increase the weight given to waves from the northeast, the dominant direction of storm conditions (15). Since the sediment transport relationships for the tidal flow and littoral drift are imperfectly known, the ratio has meaning only in a qualitative sense; that is, when the tidal power dominates, an increase in cross-section might be expected relative to those times when wave power dominates.

A comparison between the channel-maintenance ratio and the throat and average channel response is shown in Figure 8. There was general *qualitative*

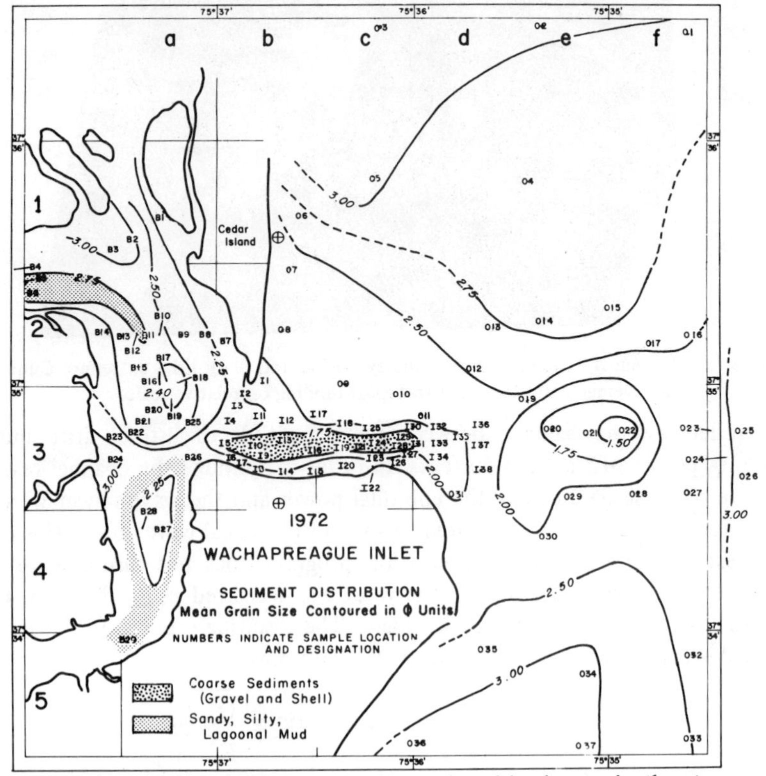

Figure 8. Comparison of channel-maintenance ratio with changes in throat area and averaged area for all ranges.

agreement between the sense of area change in the throat and the sense of the change in the maintenance ratio in 23 of the 31 cases compared. The hiatus in the calculated values for the maintenance ratio between December through March is due to the absence of wave information. In those 22 cases where an area change greater than 93 m^2 occurred, 19 agreed with the sense of change in the ratio. However, it is of interest to note that the same ratio unweighted for wave direction agreed with the sense of area changes in 20 of the 31 cases and in 15 of the 22 cases where large (> 93 m^2) changes occurred. Thus only relatively small improvements in the correlation resulted when the weighting scheme was used for wave direction. The averaged response of cross-section for the entire channel agreed with the ratio in 22 of the 31 cases. This was not surprising, as a poor coherence in response throughout the channel was observed in many cases.

It may be concluded that the ratio of ebb-tidal power to wave power is a potentially useful parameer for characterizing inlet channel response. Since most of the dramatic area reduction occurred during wave activity from the northeast or east, it is appealing to interpret the general correlation between the channel response and the maintenance ratio as indicating that channel closure is largely due to longshore drift from the north. However, several factors indicate other parameters are also controlling, in particular, the intensity and direction of net sediment transport in the channel:

(a) Addition of the incremental sand volumes deposited in the channel over the 13-month period total to a minimum of 2 x 10^6 m^3. Considerations of what is known of longshore drift rates in the region preclude the conclusion that the sand deposited in the inlet comes solely from input via longshore drift. For example, the Corps of Engineers (5) estimates that .46 x 10^6 m^3/yr drifts to the south along northern Assateague Island and that .3 x 10^6 m^3/yr is trapped in the growth of Fishing Point at the southern terminus of Assateague. Consideration of the recession rates from 1852-1962 of the island chain between Wachapreague Inlet and Chincoteague Inlet indicates a sand volume loss of .33 x 10^6 m^3/yr, if the eroding marsh is 25 percent sand (probably an overestimate). Thus a reasonable estimate for southerly drift to the inlet is .5 x 10^6 m^3/yr. The results of computed wave refraction (8) and field observations indicate that wave refraction patterns allow only small volumes of northerly drift. Recognizing the considerable risk in comparing events over a one-year period with averages based on decades, the estimate of drift versus the observed volumes deposited strongly suggests that a large fraction of the sand-volume change in the inlet channel is due to adjustments between the channel and the delta system.

(b) Particular cases during the survey period indicate that southeast wave activity also can result in channel area reduction, particularly during low or moderate tidal prisms and either on inward or low outward net transport conditions (14-19-26 July, 1972). In contrast, for a case (26 July -10 Aug. 1972) with similar wave conditions and a somewhat larger prism but with a calculated strong net outward transport, the channel widened dramatically (ratio predicted decrease in area). Finally, it is noteworthy that Range 1 exhibited a depth decrease during mid-January to mid-February 1972, a time of sustained low net-outward transport.

In summary, it appears that channel response, although reasonably correlated with the maintenance ratio, is also strongly controlled by the net sand transport characteristics during the given period. Moreover, the sand volume changes in the channel are, in part, due to shifts between the ebb delta and the channel which is enhanced by wave activity on the delta complex, regardless of wave direction.

The above interpretation is compatible with the fact that the area changes occur on the north flank of the channel as the influence of wave refraction, the regional tidal flow, and the flow distribution in the channel can be combined to form a conceptual model for a sediment flow loop on the north side of the inlet complex:

(a) Wave refraction tends to drive sand to the west from the northeast flank of the ebb delta. Evidence of this is shown in Figure 5, where the lateral shoals accrete by a succession of spits (6).

(b) The regional tidal flow is northerly during ebb in the channel which tends to drive material carried over the delta to the north.

(c) As previously discussed, the ebb flow is concentrated on the south side of the channel and, during periods of low lateral inflow, the flood currents on the north flank at Ranges 6, 7, and 8 are greater than ebb currents, so that a sediment-transport loop within the channel itself is possible. During times of high lateral inflow the flood currents cascade sand into the channel (i.e., Range 7 response) part of which is derived in the loop via steps a. and b. During periods of large prism the ebb velocities in the channel are sufficient to scour the north flank of the channel.

During northeast storms the wind-induced circulation probably overrides the regional tidal flow with resultant sediment flow to the south flank of the delta. Whatever sand is introduced into the channel from the south is scoured by the concentrated ebb flow.

Mean tide level shows significant variations in absolute level during the year as a result of steric fluctuations (14). An analysis of Wachapreague tides for a three-year period (1) showed mean tide levels are lowest in January and February, while the highest occur in September, October, and November. Mean tide levels for the survey period are shown in Figure 7, wherein it is to be noted that the October level is .3 m higher than the January level. The importance of this phenomenon in complex storage systems is evident, if a spring tide range (1.2 m) is considered at these times. Calculations using the storage relationship (Fig. 4) indicate that the October prism is 40 percent larger than the January prism. Thus, the period of enhanced prisms coincides with the advent of the northeast storm season on the east coast of the United States. During these months the largest longshore drift may be expected as the seasonal reduction in beach volumes occurs. Were it not for the enhanced prisms occurring simultaneously, inlet shoaling would be more severe.

ACKNOWLEDGMENTS

This study was supported by Office of Naval Research, Geography Programs, contract no. N000 14-71-C-0334, task 388-103. We wish to thank the Coastal Engineering Research Center, U. S. Army Corps of Engineers, for supplying the Assateague wave observations. We are grateful to J. D. Boon, III, of the Virginia Institute of Marine Science, for tidal computations and for very useful discussions on tides, and to J. T. DeAlteris, R. O'Quinn, and G. Anderson for much arduous field work. Finally, we must acknowledge the work on inlets by M. P. O'Brien which stimulated our interest.

REFERENCES

1. Boon, J. D., III
1974 Sediment transport processes in a salt-marsh drainage system. Doctoral dissertation, School of Marine Science, College of William and Mary, Williamsburg, Virginia. (Unpublished data.)
2. Brown, E. I.
1928 Inlets on sandy coasts. In **Proceedings ASCE,** 54(2): 505-553.
3. Bruun, P.
1967 Tidal inlets housekeeping. In **Jour. Hydraulics Div., ASCE,** 93(HY5): 167-184.
4. Caldwell, J. M.
1966 Coastal processes and beach erosion. **Jour. Boston Soc. Civil Engr.,** 53(2): 142-157.
5. Corps of Engineers
1973 Survey report, Atlantic Coast of Maryland and Assateague Island, Va.. **U. S. Corps of Engineers, Baltimore District.** (In preparation.)
6. DeAlteris, J. T. and Byrne, R. J.
1973 The recent history of Wachapreague Inlet, Virginia. In **The Second International Estuarine Research Conf.,** Myrtle Beach, South Carolina.
7. Escoffier, F. F.
1940 The stability of tidal inlets. **Shore and Beach,** 8(4): 114-115.
8. Goldsmith, V., Byrne, R. J., Sallenger, A. H., and Drucker, D. H.
1973 The influence of waves on the origin and development of the offset coastal inlets of the southern Delmarva Peninsula. **The Second International Estuarine Res. Conf.,** Myrtle Beach, South Carolina.
9. Mota Oliveira, I. B.
1970 Natural flushing ability of tidal inlets. In **Proceedings of 12th Conf. Coastal Engineering,** 1827-1845.

10. Nayak, I. V.
1971 Tidal prism-area relationship in a model inlet. **Univ. California, Hydr. Engr. Lab., Berkeley,** Tech. Rept. No. HEL-24-1, 72 p.

11. O'Brien, M. P.
1969 Dynamics of tidal inlets. In **Coastal Lagoons, a symposium,** p. 397-407 (eds. Ayola-Castanares, A., and Phleger, F. B.), (UNAM-UNESCO) Univ. Autonoma Mexico, Mexico, D. F.

12. O'Brien, M. P.
1970 Notes on tidal inlets on sandy shores. **Univ. California, Hydr. Engr. Lab., Berkeley,** Report HEL-24-5.

13. O'Brien, M. P. and Dean, R. G.
1972 Hydraulics and sedimentary stability of coastal inlets. In **Proceedings of 13th Conf. Coastal Engr.,** 761-780.

14. Pattulo, J., Munk, W., Revelle, R., and Strong, E.
1955 The seasonal oscillation in sea level. **Jour. Mar. Res.,** 14(1): 88-156.

15. Saville, T., Jr.
1954 North Atlantic coast wave statistics hindcast by Bretschneider-revised Sverdrup-Munk method. **Beach Erosion Board, U. S. Army Corps of Engineers,** Tech. Memo No. 55.

GENESIS OF BEDFORMS

IN MESOTIDAL ESTUARIES

by

Jon C. Boothroyd and Dennis K. Hubbard[1]

ABSTRACT

Velocity, depth, and temperature recorded over complete tidal cycles at 50 stations in two northern New England estuaries, combined with bottom profiles, grain-size analysis, 700 bedform scale and orientation readings, and SCUBA observations of bedform migration, indicate that bedform type is governed by: (a) maximum flood and ebb velocities (U_{max}) attained at a given locality; (b) velocity asymmetry (difference of maximum flood and ebb velocity); (c) velocity duration (time span above a given velocity).

The following sequence of intertidal bedforms based on increasing flow velocity has been recognized:

linear ripples → cuspate ripples ↘ linear megaripples ↗ cuspate megaripples → planed-off megaripples → rhomboid megaripples → flat beds

linear megaripples ↓ sand waves

In the intertidal and shallow subtidal (< 2m MLW) zones of tidal deltas, sand waves are characterized by $U_{max} < 80$ cm/sec, large velocity asymmetry, and dominant flood orientation. Cuspate megaripples are characterized by $U_{max} > 80$ cm/sec, slight velocity asymmetry, and no dominant orientation. In

1. Coastal Research Division, Department of Geology, University of South Carolina, Columbia, South Carolina.

deep subtidal (> 2m MLW) areas, sand waves are the principal bedform. Those on flood ramps are similar in morphology and flow characteristics but larger in scale than intertidal forms. Sand waves in ebb channels are characterized by large height and spacing, $U_{max} > 80$ cm/sec, superimposed megaripples, and a dominant ebb orientation.

Important topographic forms of a larger scale than bedforms are transverse bars and spill-over lobes. Both occur mainly in intertidal or shallow subtidal areas.

INTRODUCTION

Bedforms, primary sedimentary structures, and geometry of sandbodies in New England and Long Island estuaries have been studied by Coastal Research Group (3), DaBoll (4), Hartwell (9), Farrell (6), Rhodes (17), Hine (13), Kaczorowski (15), and Hubbard (14). The present study continues that work and goes on to document in some detail the development and migration of estuarine bedforms and larger-scale features in response to complex flow patterns and to differences in intertidal and subtidal topography.

The study area is located in the southern part of the Merrimack Embayment on the northeastern coast of Massachusetts, specifically in the Parker and Essex estuaries (Fig. 1, A). Mean tidal range is 2.6 meters, or mesotidal in the

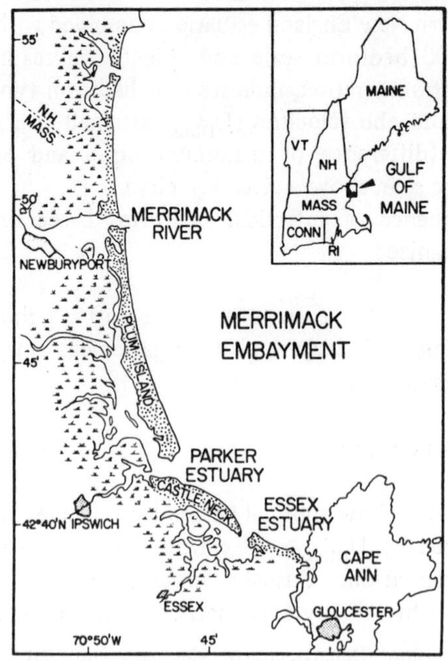

Figure 1a. The Merrimack Embayment. The study area includes the Parker and Essex estuaries.

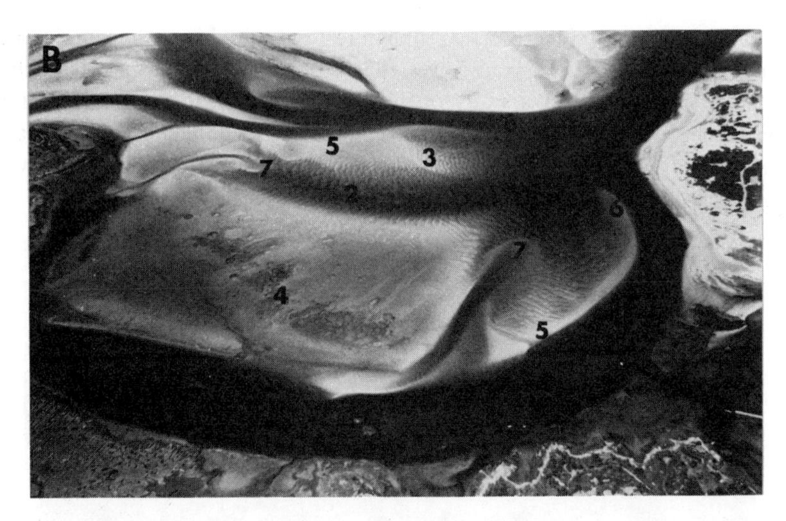

Figure 1b. Aerial view of the flood-tidal delta of the Essex estuary taken at spring low water. Various components of the estuary are numbered as follows: 1, flood ramp; 2, flood channels; 3, transverse bar; 4, high area (clamflat-marsh); 5, ebb shields; 6, ebb spits; 7, spill-over lobes; 8, ebb channels; 9, inlet throat.

classification system of Davies (5). There is a large spring-to-neap variation however; spring tidal range is 3.75 meters, while neap tidal range is only 1.65 meters. This factor is important in that it influences velocity-discharge relationships and, hence, bedform development.

Discussion will be limited to subtidal and intertidal portions of flood-tidal deltas, deep channels, and the inlet throat, with some comment on bedforms associated with ebb-tidal deltas. Emphasis is placed on the fact that bedform morphology and migration habit developed under varying hydraulic regimes in these estuaries are repetitive from estuary to estuary in the study areas of the above-mentioned authors.

SAND BODY GEOMETRY

Description of various portions of flood-tidal deltas and adjacent channels will follow the classification scheme of Hayes et al. (11) and Hayes (10). Figures 1, B, 4, A, and 9, A, illustrate this terminology as applied to flood-tidal delta complexes and associated channels in the Parker and Essex estuaries.

BEDFORM CLASSIFICATION

Bedforms, for the purpose of this discussion, are classified on the basis of spacing, not height. Ripples are bedforms with spacings less than 60 cm, megaripples have spacings from 60 cm to 6 m, and sand waves are bedforms over

6 meters in spacing. Figure 2, A, illustrates megaripples with superimposed ripples, Figure 2, B, shows sand waves. The sand waves have a spacing of about 14 meters. Klein (16), in a study of intertidal bedforms in the Bay of Fundy, grouped megaripples and small sand waves together as dunes and had a separate category for large sand waves. Allen (1) makes no distinction between megaripples and sand waves, classifying both as large-scale ripples.

Figure 2a. Megaripples on an intertidal sand body in the Parker estuary. Ripples are superimposed on the megaripple form. Bedform spacing is about 2 meters.

Figure 2b. Sand waves on the flood-tidal delta of the Parker estuary. Bedform spacing is about 14 meters.

As well as exhibiting a spacing difference, megaripples are morphologically distinct from sand waves. Megaripples are characterized by a sinuous to highly cuspate crestline, usually with well-developed scour pits downstream of the crestline. Some megaripples may have straight crests, but scour pits still occur at intervals along the crest line. Megaripples are also characterized by a small spacing-to-height ratio (ripple index) in comparison to sand waves. These bedforms are identical to dunes formed in a large flume described by Simons and Richardson (18) and illustrated by Guy et al. (7).

Sand waves have a straight to sinuous crestline; scour pits are absent or at best poorly developed. Sand waves, which have a very large spacing-to-height ratio, may be termed two-dimensional bedforms, whereas megaripples are three-dimensional bedforms.

Heights and spacings of bedforms were measured at the intersections of a 50-by-50 meter grid system on most of the intertidal positions of flood- and ebb-tidal deltas. Additional readings of height and spacing of subtidal bedforms were obtained from bottom profiles run parallel to current-flow direction. Some of these measurements are plotted on height-spacing diagrams shown in Figure 3. Megaripples (solid circles) show a concentration at spacings from 1.5 to 4 meters and heights of 8 to 40 cm. Intertidal and waves (solid squares) spacings tend to concentrate at 8 to 20 meters and heights at 15 to 40 cm. Subtidal sand waves (solid hexagons and triangles) are larger in height and spacing than are intertidal sand waves.

There is a pronounced gap in spacings at 6 meters, the afore-mentioned megaripple-sand wave division line (Fig. 3). The gap occurs at 6-9 meters for bedforms measured in the Parker River (Fig. 3, A) and at 4-6 meters for bedforms measured in Essex Bay (Fig. 3, B). The reason for this difference is not clearly understood but may be related to grain size and to current-velocity asymmetry. Mean grain size for intertidal megaripples in the Parker (those shown in Fig. 4) is .44 mm; mean grain size for intertidal sand waves (Fig. 4) is .38 mm. Mean grain size for both intertidal megaripples and sand waves in Essex Bay (Fig. 1, B) is .31 mm. Fine grain size may contribute to smaller bedform spacing, other conditions held constant, and thus displace the spacing gap downward, as is the case of Essex Bay. Velocity asymmetry will be discussed in a later section.

Bedforms shown by open triangles (Fig. 3) are concentrated around the 6-meter spacing area, in the gap discussed above. These are transition bedforms that occur in intertidal areas around the margins of sand-wave fields and at junctures between sand-wave and megaripple zones. They have the appearance of dwarfed sand waves, with straight crests, that could not fully develop in the hydrologic regime in which they are found.

Thus there is a distinct spacing difference between megaripples and sand waves as well as the morphological difference. The spacing differences of all three types can be explained by differences in hydrodynamic conditions

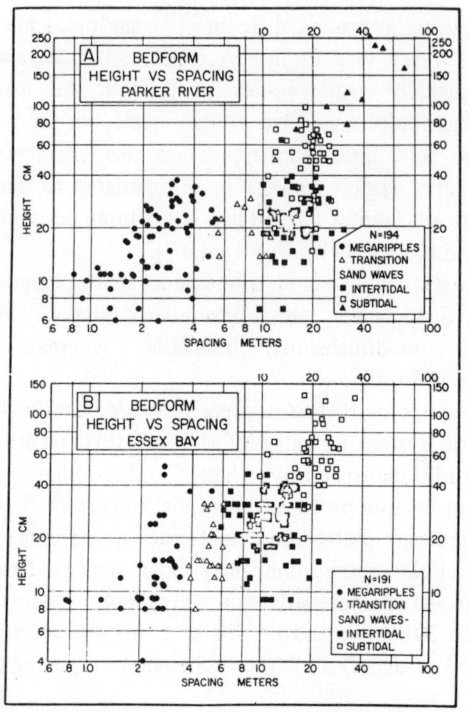

Figure 3. Height-spacing diagrams of large-scale bedforms for: A. the Parker estuary, and B. the Essex estuary.

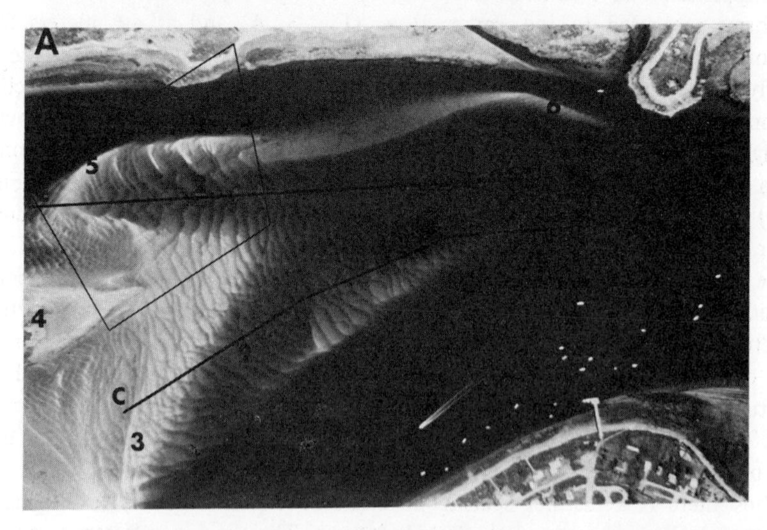

Figure 4a. Aerial view of a portion of Middle Ground, the flood-tidal delta of the Parker estuary. Inlet throat is to the right. Numbering system of components is the same as for Figure 1, B. Lines indicate the location of some bottom profiles illustrated in Figure 8. The area inside the box is shown in more detail in Figure 4, B.

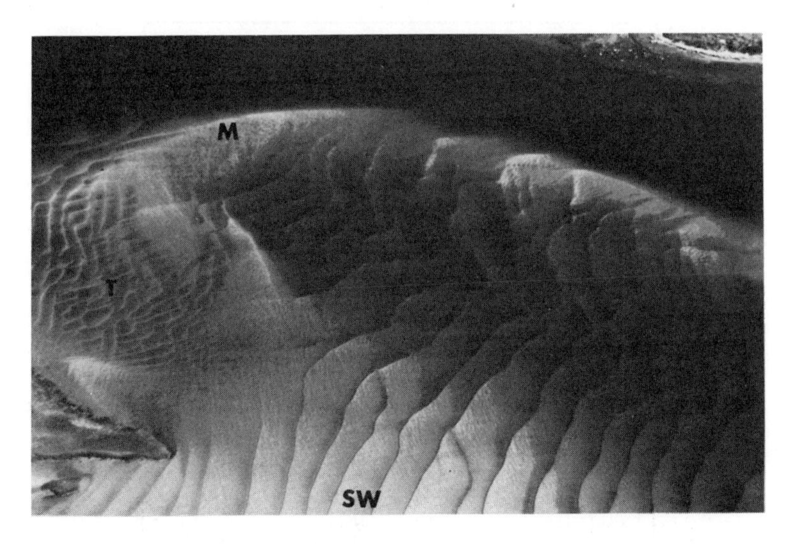

Figure 4b. Detail of a portion of Figure 4, A. Location of intertidal diving stations is shown by letters; M, megaripples; SW, sand waves; and T, transition bedforms.

governing the formation and migration of the bedform.

FLOW CONDITIONS

Current stations were occupied for a complete tidal cycle at 60 intertidal and subtidal locations in both estuaries. Data from these stations were compared with those obtained from approximately 250 stations in 9 other estuaries on the New England and Long Island coasts (see references cited in Introduction). Information presented in the following discussion is representative of, and may serve as a model for, bedform formation and migration in the mesotidal environment.

Three detailed intertidal diving stations, over megaripples, sand waves, and a transition form (small ebb-oriented sand waves), were continuously monitored by divers, with depth and velocity measurements recorded at intervals of 15 minutes or less. Locations of the stations are shown in Figure 4, B. Velocity measurements were made over the crest of the bedform 30 cm below the water surface, except in very shallow depths where surface or float velocities were obtained.

Velocity curves obtained over a complete tidal cycle (Fig. 5, A) illustrate that megaripples are characterized by a high maximum flow velocity (here, 106 cm/sec) and little or no velocity asymmetry. The velocity curve for sand waves shows a lower maximum flow velocity (80 cm/sec) and large velocity asymmetry. Transition forms fall between these two curves; they are

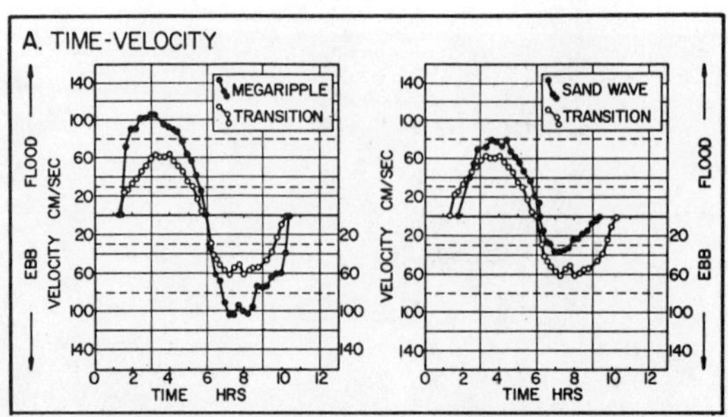

Figure 5a. Time-velocity curves obtained by measurements over a complete tidal cycle at each of the intertidal diving stations. Megaripples show high maximum velocity and little velocity asymmetry; sand waves, a lower maximum velocity and a large velocity asymmetry; and transition bedforms, a lower maximum velocity and little velocity asymmetry.

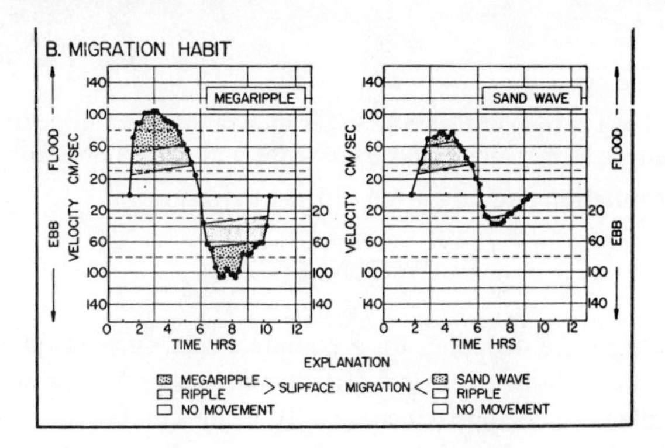

Figure 5b. Time and type of bedform migration observed over a complete tidal cycle at the megaripple and sand-wave diving stations. Sand-wave slip-face migration occurs only during the flood part of the cycle, whereas megaripple slip-face migration occurs during both the flood and ebb parts of the tidal cycle.

characterized by a lower maximum velocity (here, 64 cm/sec) and little velocity asymmetry.

Figure 5, B illustrates bedform migration differences between megaripples and sand waves. There, differences are due to the varying configuration of the velocity curves. Ripple migration begins at 30 cm/sec; megaripple slip-face migration begins at about 60 cm/sec. These estimates were obtained by direct diver observation of bedform morphology. There is some confusion in picking precise times because, of course, the bed must readjust from ebb-oriented bedforms to flood-oriented bedforms, or vice versa. Megaripples migrate during flood and then reverse and migrate in an ebb direction for an approximately equal time span.

Sand-wave migration occurs only during a small part of the flood-tidal cycle and not at all during the ebb cycle (Fig. 5, B). Ebb-oriented ripples form and the flood-oriented slip face is only slightly modified during ebb flow. Slip-face migration begins at approximately the same velocity as megaripple-slip-face migration, but flow over megaripples reaches a higher maximum velocity for a longer time span.

There is a reversal of slip-face orientation with change of tidal-flow direction for the transition bedforms, but at a much slower rate than that for megaripples. Fully developed bedforms, that is, those close to equilibrium at maximum flow velocity, do not occur in either the flood or ebb direction. These flow conditions impart a "peaked", or heavily modified, appearance to the bedform that is especially identifiable from the air.

BEDFORM MIGRATION

Bedform migration rates were obtained by placing approximately 40 stakes at crests of bedforms, at four localities, and monitoring slip-face migration over varying time spans. Figure 6 represents results at selected locations. Slip-faces of megaripples measured through time, during flood and ebb parts of the tidal cycle (Fig. 6, B), migrated a distance of 450 cm during ebb flow at an average rate of about 120 cm/hour. Slip-face migration rates of 0.7 up to 8 cm/min were recorded with a cluster of points falling around 1-3 cm/min (Fig. 6, B).

Slip-face migration of intertidal sand waves was measured for a three-month period; measurements were made once a day to three times a week to establish a pattern, and thereafter measured about every ten days (Fig. 6, A). The sand waves migrated a total distance of 16-18 meters during the three-month span, or about 1.5 times the spacing of the bedforms. Migration rates were 20-40 cm per tidal cycle on spring tides but only 1-5 cm during neap tides (Fig. 6, A). Diver measurements obtained through a complete cycle during a spring tide also indicate a migration rate of 40 cm/tidal cycle (Fig. 6, B).

The differences in magnitude of intertidal megaripple versus sand-wave migration rates are readily apparent, 450 cm of slip-face migration per half-tidal cycle for megaripples versus 40 cm for sand waves. Also, it is important to remember that megaripple migration is in both flood and ebb directions, whereas sand-wave migration is in the flood direction only.

BEDFORM SEQUENCE

Detailed study of flow conditions and bedform morphology has led to the recognition of a sequence of intertidal estuarine bedforms based on increasing flow velocity. This sequence (Fig. 7, A) is a modification of an earlier version

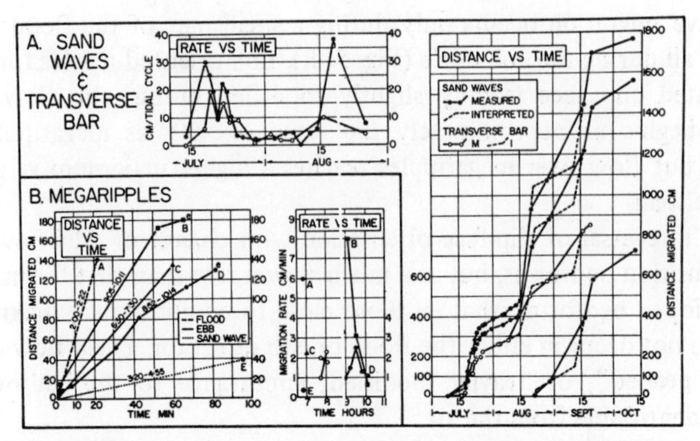

Figure 6a. Sand-wave and transverse-bar migration. Distances and rates of slip-face migration of selected sand waves were measured over a period of three and a half months. There was a substantial increase in the rate of migration during spring tides. Rate of migration of the transverse bar slip face is about a half that of surrounding sand waves. These bedforms are illustrated in Figure 4, A.

Figure 6b. Megaripple migration. Measurements of slip-face migration of selected megaripples were obtained over a complete tidal cycle at the same time as the velocity measurements illustrated in Figure 5. Migration rate for sand waves is shown for comparison. Note the extremely high rate of migration of megaripples when compared with that of sand waves.

(2), differing mainly in terminology. Note the comparison of this sequence to that of Simons et al. (19) (Fig. 7, A).

The sequence begins with linear ripples and goes to cuspate ripples, termed low- and high-energy ripples, respectively, by Harms (8). Height and spacing of these bedforms in relation to larger-scale forms can be seen in a plot by Hine (12). The megaripple sequence begins with linear forms that are transformed by increasing flow velocity to cuspate forms, with well-developed scour pits, and then the sequence continues to planed-off megaripples. Planed-off megaripples are of two types, those with short spacings analogous to washed-out dunes of Simons et al. (19), and those with long spacings which plot near the 6-meter boundary (Fig. 3). Rhomboid megaripples, the last form in the sequence, show almost no slip-face development and are essentially a flat-bed form. These bedforms were discussed by Smith (20).

A log-log plot of depth versus velocity (Fig. 7, B), similar to those of Southard (21), delineates fields where each member of the foregoing bedforms occurs. Since unsteady flow conditions occur throughout the tidal cycle, bedform morphology is constantly changing and a given bedform type may be stable only during a part of the tidal cycle. Diver observation of change of bedform morphology was used to establish field boundaries.

Sand waves plot in the linear megaripple field even though flood versus ebb velocity asymmetry varies markedly between the two forms. Sand waves may be

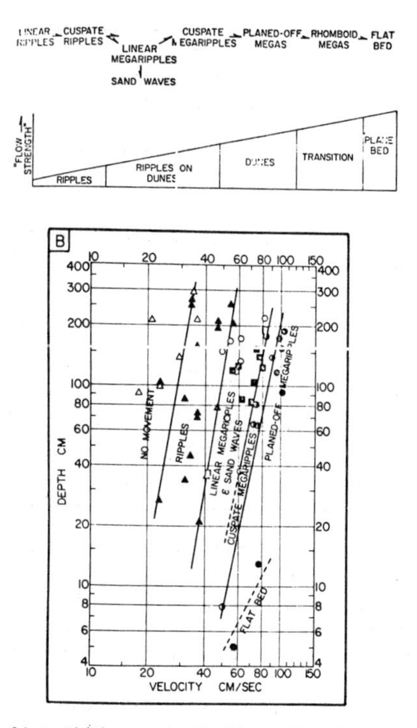

Figure 7a. Sequence of intertidal estuarine bedforms based on increasing flow velocity. The sequence is compared to that of Simons et al. (19).

Figure 7b. Log-log plot of depth versus velocity for the occurrence of intertidal bedforms. Diver observation of change in bedform morphology was used to establish field boundaries.

treated as linear megaripples with no ebb velocity component, and hence, no reversal of migration direction. This fact, continued migration in one direction only at relatively low velocities, may allow intertidal sand waves to develop into forms with large spacings.

SUBTIDAL FLOW CONDITIONS AND BEDFORM MORPHOLOGY

In subtidal channels, bedform type and orientation are also controlled by complex intertidal and subtidal topography. Bottom profiles, run parallel to migration direction, illustrate the variety in scale of bedforms in the subtidal environment (Fig. 8). The profile lines may be located by referring to Figures 4, A; 10, B; and 11, B. Most profiles shown were run at high tide. Heights and spacings of subtidal sand waves from lines A-B-C and D-E are plotted as hexagons in Figure 3, A; those from lines F-G and J-K are plotted as solid triangles.

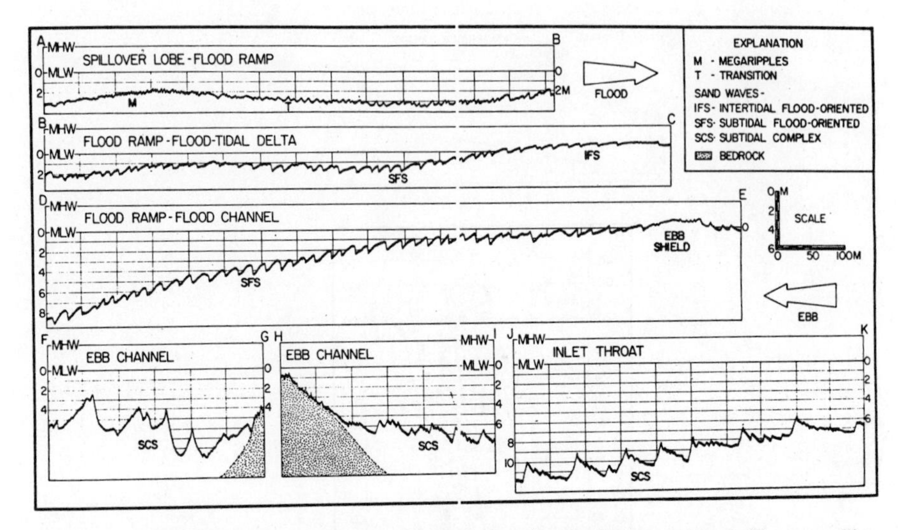

Figure 8 Bottom profiles run along selected lines in the Parker estuary. These profiles illustrate the variety in large-scale subtidal bedforms. Profile lines may be located by referring to Figures 4, A, 10, B, and 11, B. Heights and spacings of subtidal sand waves from lines A-B-C and D-E are plotted as hexagons on Figure 3, A; those from lines F-G and J-K are plotted as solid triangles.

Tidal-current velocity curves for two subtidal stations on profile D-E, flood ramp-flood channel (Fig. 8), are shown in Figure 9, A. These two velocity curves are similar in maximum flow velocity and amount of flood-current asymmetry. Note also their similarity to the velocity curve of the flood-oriented intertidal sand waves (Fig. 5, A). The conclusion is that flood-oriented sand waves, shielded from ebb flow, are similar in morphology and migration habit, differing only in scale; whether in 7 meters of water (at MLW), less than 2 meters of water, or intertidal.

In deep-ebb channels (carrying a large volume of ebb-tidal flow) flow velocities are very high, up to 200 cm/sec, but the dominant bedforms are sand waves (Fig. 8, lines F-G, H-I, and J-K). Where velocity asymmetry is low (station A-3, Fig. 9, B), the sand waves are nearly symmetrical; but where velocity asymmetry is high, the sand waves are oriented (e.g., ebb-oriented in profile on right, Fig. 9, B). Diver observation confirms that when surface velocity exceeds 80 cm/sec, megaripples are superimposed on the sand-wave form. Planed-off megaripples are a common occurrence at these high velocities, as are regressive ripples, and ripples migrating transversely down troughs and across sand-wave slip faces.

Major sand movement in the deep-ebb channels is by migration of megaripples, even though the principal bedforms are large-scale sand waves. Sketches made by divers, the sequence in Figure 9, B, summarize bedform

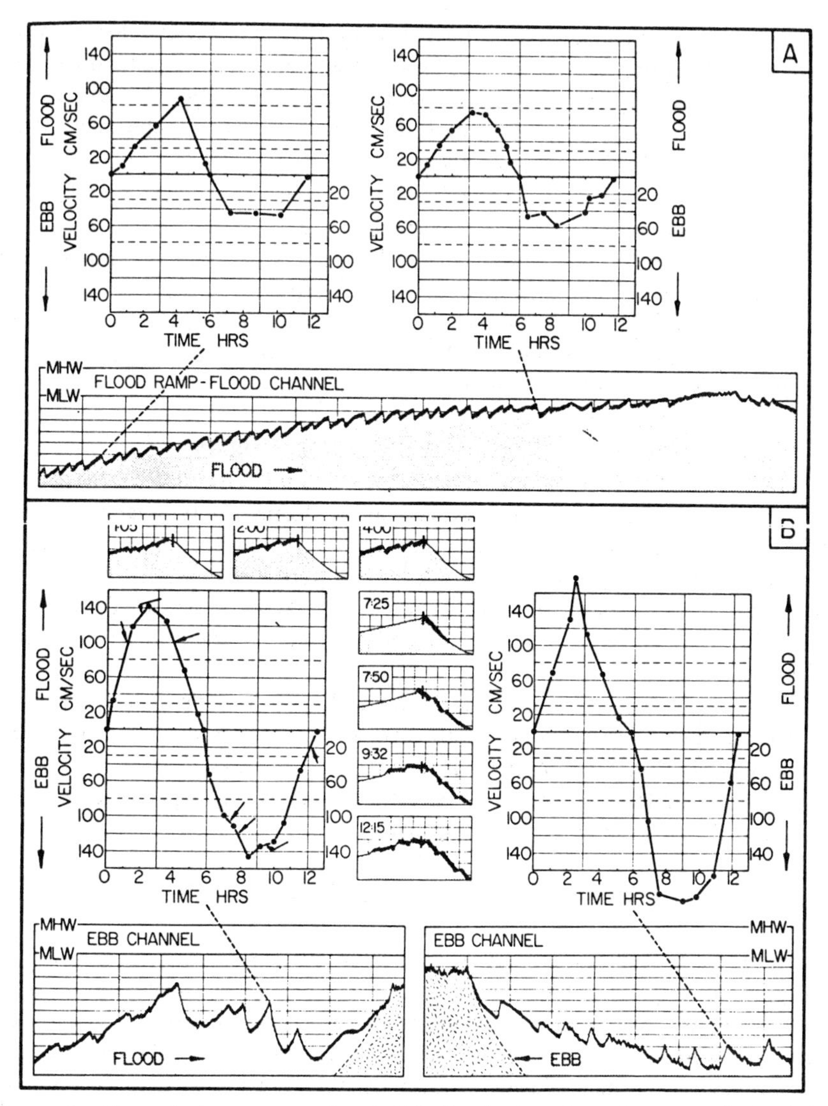

Figure 9a. Time-velocity curves of subtidal flood-oriented sand waves obtained by measurement over a complete tidal cycle. These two stations are on profile line D-E (Fig. 8). The curves are similar to that obtained over the intertidal flood-oriented sand waves (Fig. 5, A).

Figure 9b. Time-velocity curves obtained from measurements over complex sand waves in deep-ebb channels in the Parker estuary. The curve on the left shows high velocity but little asymmetry. The accompanying sketches illustrate change in bedform morphology over the tidal cycle. Arrows on the curve correspond to times indicated on each sketch. The curve on the right shows high velocity but a marked ebb asymmetry, hence the sand wave is ebb-oriented.

movement at the symmetrical sand-wave station. During the flood-tidal cycle, megaripples migrate up to the sand-wave crest and deposit sand down the avalanche slip face of the sand wave. During the ebb-tidal cycle, the flood-oriented slip face of the sand wave is modified by ebb megaripple migration up the slip face and over the crest.

There are two distinct classes of estuarine sand waves; (a) intertidal and subtidal flood-oriented forms with superimposed ripples, found in flood channels on flood ramps and flood-tidal deltas; (b) flood- or ebb-oriented subtidal forms with superimposed megaripples found in ebb channels. Intertidal to shallow subtidal ebb-oriented sand waves with superimposed ripples do occur but are not common. Intertidal forms with superimposed megaripples are extremely rare in mesotidal estuaries.

LARGE ACCUMULATION FORMS

Topographic forms of a larger scale than bedforms are transverse bars and spill-over lobes. Both occur mainly in intertidal to shallow subtidal areas (Figs. 1, B and 4, A). Spill-over lobes may be either flood- or ebb-oriented, but are dominantly ebb-oriented in the Parker and Essex estuaries. They occur predominantly in areas subject to flow-velocity conditions producing megaripples.

Transverse bars occur in sand-wave areas and are flood-oriented. They appear as large sediment forms mantled with sand waves, moving in a flood direction, but at a slower rate than individual sand waves. Rate and distance of slip-face migration at the leading edge of a transverse bar shown in Figure 6, B, is about half that of sand waves in the same vicinity.

SUMMARY OF BEDFORM DISTRIBUTION

The bedform distribution pattern shown in Figures 10 and 11 is based on mapping of intertidal flats, and 75 km of bottom profiles plus bottom observation and interpretation by divers of forms recorded on those profiles. Bedforms observed can be tied directly to tidal-current velocity patterns and direction and magnitude of sand migration. Compare the two diagrams (Figs. 10, B and 11, B) with the accompanying photographs (Figs. 10, A and 11, A) and with the profile lines (Fig. 8).

Megaripples occur on low-intertidal ebb shields and spill-over lobes subject to high-velocity flood and ebb flow. Flood-oriented intertidal and subtidal sand waves occur in areas shielded from ebb flow; on flood ramps and in flood channels. Transition bedforms occur in partially shielded areas high on ebb shields or in shallow channels. Subtidal sand waves, flood- or ebb-oriented, with superimposed megaripples, occur in deep channels around the tidal delta wedge.

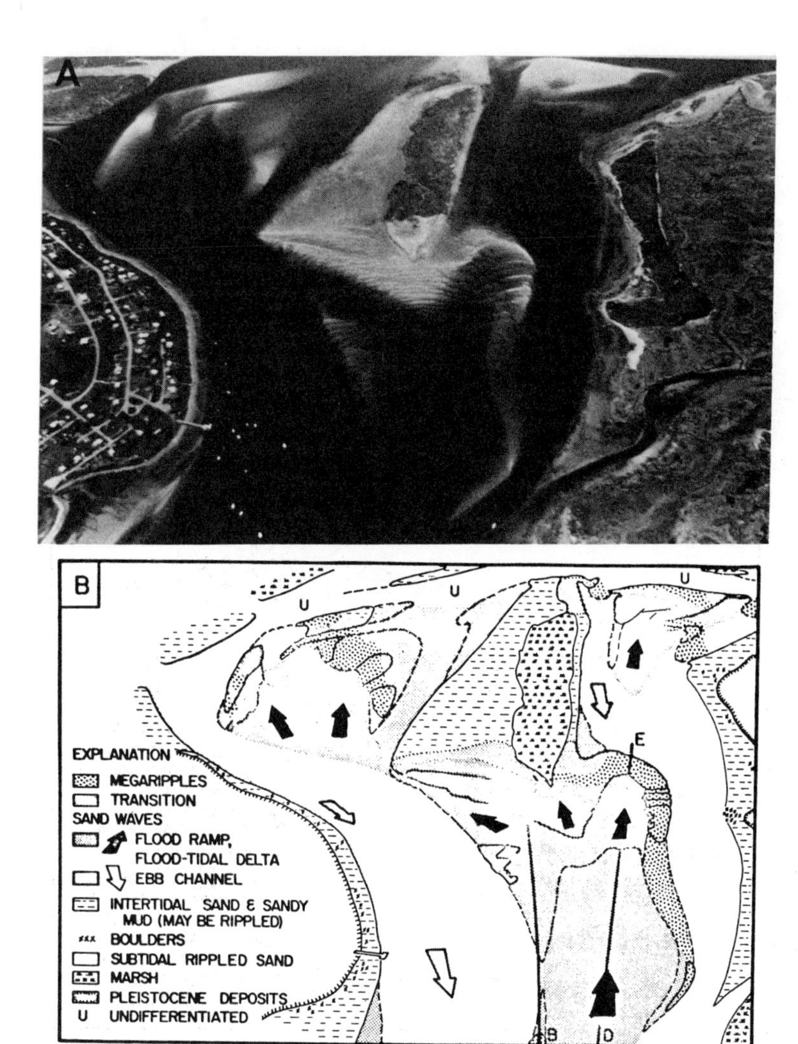

Figure 10a. Aerial view of the Middle Ground, the flood-tidal of the Parker estuary, and surrounding portions of the estuary. This photograph, taken at spring low water, illustrates the relationship among intertidal, shallow subtidal, and deep subtidal areas.

Figure 10b. Sketch made from the above photograph illustrating bedform distribution patterns and other sub-environments of deposition. Bottom profile lines are also shown. Arrows indicate orientation of major bedforms and net direction of sediment transport.

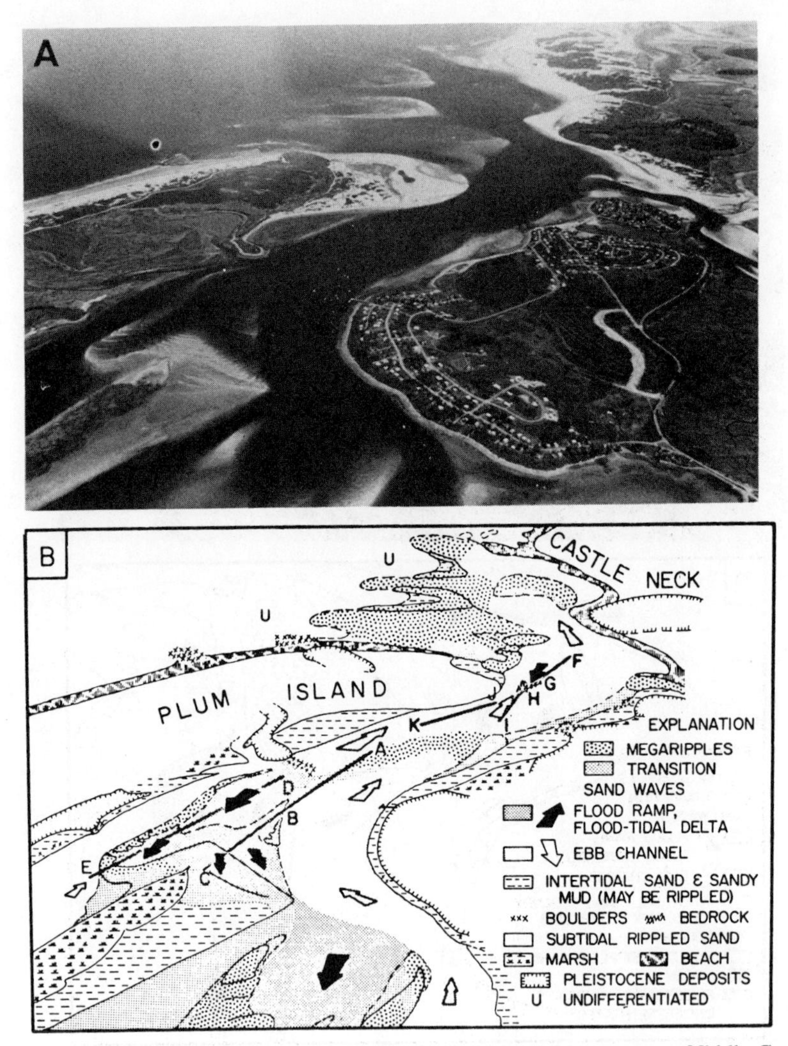

Figure 11a. Aerial view of the lower portion of the Parker River estuary. Middle Ground (flood-tidal delta) is at the lower left; the ebb-tidal delta is at the top center between the two barrier islands (Plum Island, left center, and Castle Neck, top right); and a Pleistocene drumlin complex is at the lower right. This photograph also taken at spring low water.

Figure 11b. Sketch made from the above photograph illustrating the bedform-distribution pattern and other sub-environments of deposition. Bottom profile lines are also shown. Arrows indicate orientation of major bedforms and net sediment transport.

REFERENCES

1. Allen, J. R. L.
   1968 **Current ripples: their relation to patterns of water and sediment motion.** North-Holland Publishing Co., Amsterdam, 433 p.
2. Boothroyd, J. C.
   1969 Hydraulic conditions controlling the formation of estuarine bedforms. In **Coastal environments: N. E. Massachusetts and New Hampshire,** Cont. No. 1-CRG, p. 417-427. Univ. Massachusetts, Dept. of Geology Publication Series, 462 p.
3. Coastal Research Group
   1969 Coastal environments: N. E. Massachusetts and New Hampshire, Cont. No. 1-CRG, Univ. of Massachusetts, Dept. of Geology Publication Series, 462 p.
4. DaBoll, J. M.
   1969 Holocene sediments of the Parker River estuary, Massachusetts: Cont. No. 3-CRG, Dept. of Geology, Univ. Massachusetts, 138 p.
5. Davies, J. L.
   1964 A morphogenic approach to world shorelines. **Zeit. Für Geomorphologie, Bd.** 8: 27-42.
6. Farrell, S. C.
   1970 Sediment distribution and hydrodynamics, Saco River and Scarboro estuaries, Maine: Cont. No. 6-CRG, Dept. of Geology, Univ. Massachusetts, 129 p.
7. Guy, H. P., Simons, D. B., and Richardson, E. V.
   1966 Summary of alluvial channel data from flume experiments, 1956-61: U. S. Geol. Survey Prof. Paper 462-I, 96 p.
8. Harms, J. C.
   1969 Hydraulic significance of some sand ripples. **Geol. Soc. America Bull.,** 80: 363-396.
9. Hartwell, A.
   1970 Hydrography and holocene sedimentation of the Merrimack River estuary, Massachusetts: Cont. No. 5-CTG, Dept. of Geology, Univ. Massachusetts, 166 p.
10. Hayes, M. O.
    Morphology of sand accumulation in estuaries. In Proceedings of 2nd International Estuarine Research Federation Symposium, introduction to geology sessions on **Coarse-Grained Sediment Transport and Accumulation in Estuaries,** Myrtle Beach, South Carolina, October 15-18, 1973. (In press.)
11. Hayes, M. O., Owens, E. H., Hubbard, D. K., and Abele, R. W.
    1973 Investigation of form and processes in the coastal zone. In **Coastal geomorphology,** p. 11-41. (ed. Coates, D. R.) Proceedings of 3rd Annual Geomorphology Symposium Series, Binghamton, New York, Sept. 1972, 404 p.
12. Hine, A. C.
    Bedform distribution and migration patterns on tidal deltas in the Chatham Harbor estuary, Cape Cod, Massachusetts. In Proceedings of 2nd International Estuarine Research Federation Symposium, Myrtle Beach, South Carolina, October 15-18, 1973. (In press.)

13. Hine, A. C.
1972 Sand deposition in the Chatham Harbor estuary and on the neighboring beaches, Cape Cod, Massachusetts. M.S. thesis (unpublished), Dept. of Geology, Univ. Massachusetts, 154 p.

14. Hubbard, D. K.
1973 Morphology and hydrodynamics of Merrimack Inlet, Massachusetts, Part I, p. 1-162. In **Tidal inlet morphology and hydrodynamics of Merrimack Inlet, Massachusetts and North Inlet, South Carolina.** Final Report for Contract DACW 72-72-C-0032, Coastal Engineering Research Center, 260 p.

15. Kaczorowski, R. T.
1972 Offset tidal inlets, Long Island, New York. M.S. thesis (unpublished), Dept. of Geology, Univ. Massachusetts, 150 p.

16. Klein, G. DeV.
1970 Depositional and dispersal dynamics of intertidal sand bars. **Jour. Sed. Pet.,** 40(4): 1095-1127.

17. Rhodes, E. G.
1971 Three dimensional analysis and interpretation of Pleistocene-Holocene deposits by the seismic reflection-refraction method, Merrimack Embayment, Gulf of Maine. M.S. thesis (unpublished), Dept. of Geology, Univ. Massachusetts, 134 p.

18. Simons, D. B. and Richardson, E. V.
1962 Resistance to flow in alluvial channels: **Am. Soc. Civil Engineers Trans.,** 127: 927-954.

19. Simons, D. B., Richardson, E. V., and Nordin, C. F., Jr.
1965 Sedimentary structures generated by flow in alluvial channels. In **Primary sedimentary structures and their hydrodynamic interpretation,** p. 34-52. (ed. Middleton, G. V.) Soc. Econ. Paleo. and Min., Spec. Pub. No. 12, 265 p.

20. Smith, N. B.
1971 Pseudo-planar stratification produced by very low amplitude sand waves. **Jour. Sed. Pet.,** 41(1): 69-73.

21. Southard, J. B.
1971 Representation of bed configurations in depth-velocity-size diagrams. **Jour. Sed. Pet.,** 41(4): 903-915.

BEDFORM DISTRIBUTION AND MIGRATION PATTERNS ON TIDAL DELTAS IN THE CHATHAM HARBOR ESTUARY, CAPE COD, MASSACHUSETTS

Albert C. Hine[1]

ABSTRACT

Well-developed, multi-component flood- and ebb-tidal deltas exist in the lower portion of the Chatham Harbor estuary, Cape Cod, Massachusetts. Each tidal delta contains several smaller sand bodies, a dominant bedform type, and a preferred bedform orientation, which indicates a distinct sediment transport pattern. The hydrography and overall geometry of the estuary are the critical factors in controlling these features on the tidal deltas. The estuary has two inlets, each facing a separate large body of water whose tidal ranges are significantly different. The tidal-range difference develops a steep hydraulic slope that occurs during flood and results in pronounced time asymmetry and tidal-current segregation.

The ebb-tidal delta conforms to the model proposed by Hayes et al. (7). Tidal-current segregation has developed a deep, main ebb channel, which is flanked by two shallow marginal flood channels. These three channels are floored with unidirectionally-oriented sand waves. Net sand transport by tidal currents occurs in the proximal portion of the ebb-tidal delta, while net transport by wave-generated currents occurs in the distal portion.

Nearly complete flood-tidal current dominance exists on the flood-tidal delta. The margins of this multi-lobate sand body have migrated approximately 900 m during a ten-year period, indicating rapid sand transport. Three bedform

1. Coastal Research Division, Department of Geology, University of South Carolina, Columbia, South Carolina 29208.

orientations develop during flood because of two changes in the direction of flow. The resulting dominant bedform feature is an intersecting pattern of two sand-wave orientations.

INTRODUCTION

General Setting

The Chatham Harbor estuary is in the extreme southeastern section of Cape Cod, Massachusetts (Figs. 1 and 2). This linear body of water is approximately 6 km long, 1 km wide, and has a maximum depth of 7 m. Farther to the north, the estuary becomes wider and shallower and is called Pleasant Bay. Nauset Beach, a Holocene barrier beach system prograding south by recurved spit growth, is the eastern boundary, and the Pleistocene mainland of Cape Cod is the western boundary. Pleistocene sediments in this area are composed of glaciofluvial and glaciolacustrine gravel, sand, and silt (5).

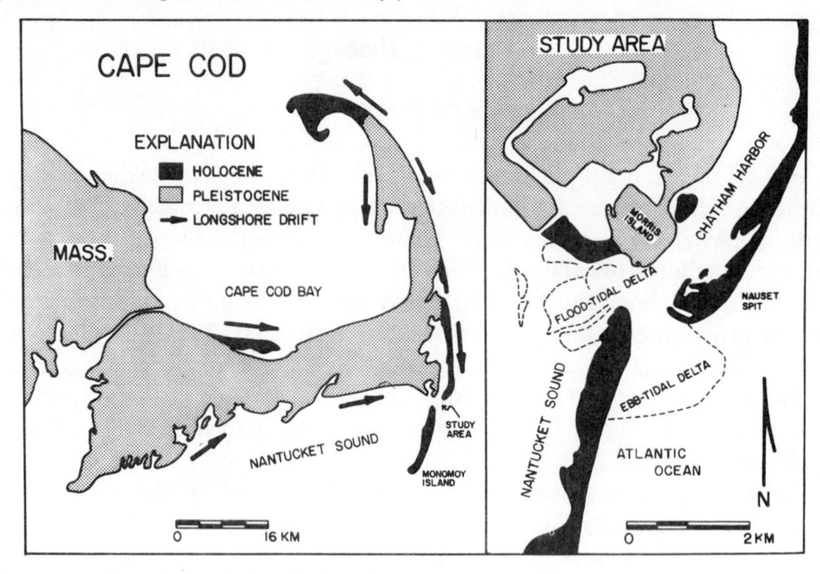

Figure 1. Cape Cod and the Chatham Harbor estuary. Note the regional longshore drift system around Cape Cod.

The main inlet to the estuary is enclosed by the active, recurved spit complex of Nauset Beach on the north and the northern tip of Monomoy Island on the south. However, the estuary is unique because a second, it has a shallower inlet facing Nantucket Sound, between Monomoy and Morris islands. This second inlet was formed in 1958 when waves breached the sand barrier connecting these two islands.

Figure 2. Aerial photograph of Chatham Harbor estuary and Nauset Spit. Monomoy Island is in the lower left corner of photograph.

Within the estuary, the most prominent zones of sand deposition are two well-developed tidal deltas (ebb and flood). The purpose of this study was to analyze how these sand bodies have responded to the regional longshore transport system, wave regime, and to the tidal flow regime. The basic approach used was to measure the significant parameters of tidal flow, study the sand-body geomorphology and its individual components, map bedform types, and measure bedform orientation, height, and spacing. From the data and observations, net sand-transport patterns could be determined.

HYDROGRAPHY

This estuary is a type-D estuary under Pritchard's classification (9). Since little fresh water is being introduced by run-off, sea water is not significantly diluted, nor is there any suspended-sediment load.

The tides in the estuary are semidiurnal with little observed diurnal inequality. The mean tidal range in the Atlantic Ocean, approximately 2 km off

Monomoy Island, is 200 cm. In a westerly direction across the lower portion of the estuary, it decreases quite rapidly. At the inlet, the mean tidal range is 130 cm, and approximately 3 km to the west of the inlet in Nantucket Sound, it drops to 90 cm. Along this 5-km distance, the mean tidal range decreases by 110 cm.

This tidal range difference produces hydraulic slopes that are steepest at high water and at low water (Fig. 3). Because the slopes are steepest at these times, maximum flood and ebb tidal-current velocities occur at high water and low water, respectively. This relationship develops because the tidal regime is responding to a progressive wave and not to a stationary wave.

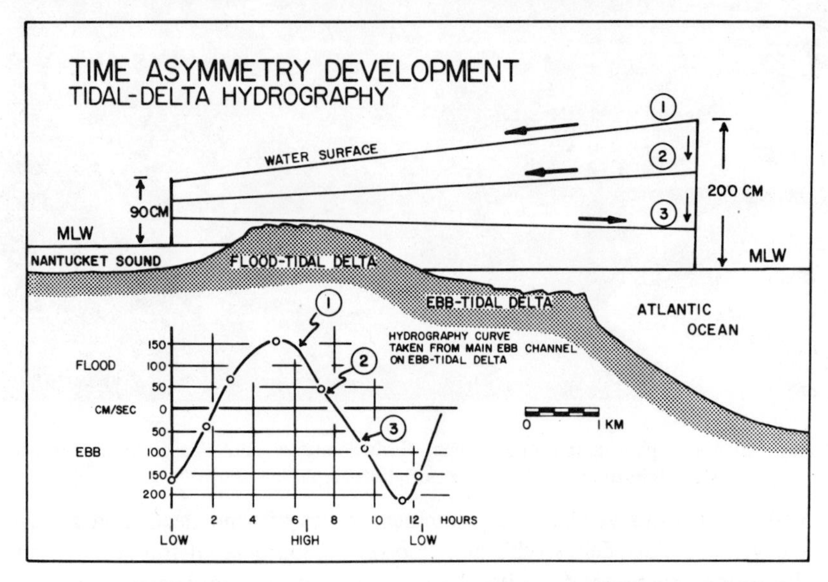

Figure 3. Time-asymmetry development caused by tidal range differences. (1) Maximum flood velocity occurs at high water. (2) Water level has dropped, but flow still is in flood direction. (3) Hydraulic slope is reversed, developing ebb current, but flood-tidal delta is nearly exposed, preventing a significant volume from passing into the Atlantic.

This high water-maximum flood velocity relationship is called time asymmetry in contrast to velocity asymmetry, where there is little current flow at high water and low water and the maximum flood or ebb velocities occur either early or late in their portion of the tidal cycle. This time asymmetry develops pronounced tidal-current segregation, which means that the shallower intertidal portions within the estuary are flood-dominated and the deeper intertidal and subtidal portions are ebb-dominated. The shallower zones will be flood-dominated because, while they are influenced by the maximum flood velocities at high water, they are subaerially exposed at low water, when maximum ebb velocities occur.

The magnitude of the time asymmetry in the Chatham Harbor estuary is increased because Nantucket Sound reaches high water approximately thirty minutes after the time high water occurs in the Atlantic Ocean. This increases the steepness of the westerly dipping hydraulic slope and allows even greater flood velocities to occur. The progressive wave-induced time asymmetry also develops a situation where strong flood currents persist even after the water level has been dropping for more than two hours after high water (Fig. 3). Ebb currents are initially produced approximately three hours after high water, at a time when the hydraulic slope direction is reversed (Fig. 3). At this time, the flood-tidal delta is nearly exposed and blocks most of the ebb flow from Nantucket Sound to the Atlantic Ocean. Water flows out of Nantucket Sound farther to the south, in the large opening between Nantucket and Monomoy islands. Because there is practically no ebb flow across the flood-tidal delta, this sand body is completely flood-current dominated.

EBB-TIDAL DELTA AND INLET MORPHOLOGY

The ebb-tidal delta is a complex, multi-component, asymmetrically shaped sand body located seaward of the major inlet of the estuary (Figs. 4 and 11). Most of the sand accumulated in this area has been derived from the southerly,

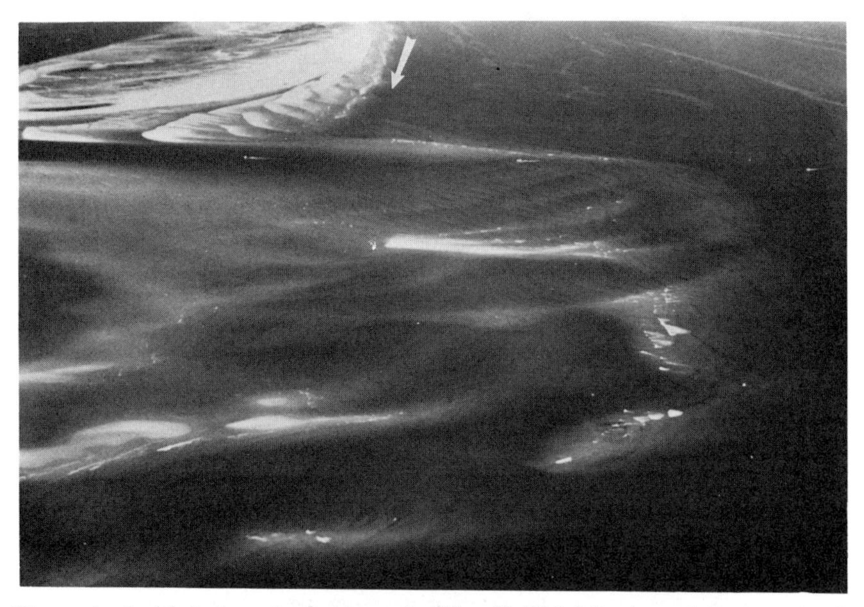

Figure 4. Aerial photograph of outer part of the ebb-tidal delta. Arrow indicates regional longshore drift direction. Note terminal lobe and distal swash bars.

longshore transport system. Sand is carried onto the subtidal platform from the main ebb channel through bed-load transport. The subtidal platform is the broad, flat, featureless base of the ebb-tidal delta on which the smaller sand bodies (spill-over lobes, swash bars) and bedforms lie. This tidal delta contains most of the components proposed in the ebb-tidal delta model (7) and is typical of most ebb-tidal deltas.

The inlet is morphologically updrift-offset (6). As Nauset Spit migrated south (Fig. 5), Monomy Island moved to the west (4), probably in response to the approaching recurved spit. By the time Nauset Spit had nearly reached its present location, Monomoy Island had migrated westerly far enough to produce the updrift-offset morphology.

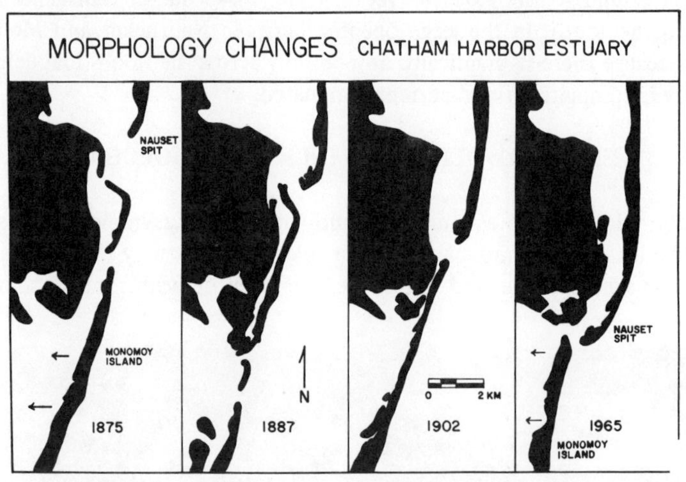

Figure 5. Historical morphological changes of southeastern Cape Cod. Note that as Nauset Spit migrated down from the north, Monomoy Island moved to the west.

In many inlets, downdrift-offset is developed by a localized reversal in the longshore drift system, transporting sand back up towards the inlet (8). The drift reversal is generated by wave refraction across the ebb-tidal delta. However, exposed peat layers, a narrow beach, and an active dune scarp indicate that little sand is being furnished to the beach along the north end of Monomoy Island. Presently, this portion of the Monomoy barrier system is still migrating to the west.

Current Segregation

Tidal-current segregation plays a dominant role on the ebb-tidal delta. With maximum ebb currents occurring at or near low water, the deeper main ebb channel (Fig. 6) becomes the main conduit for water leaving the estuary.

Figure 6. Aerial photograph of Nauset Spit and the main ebb channel (arrow), which is the primary conduit for ebb flow out of the estuary. The intertidal part of Nauset Spit is the northern marginal flood channel.

Surrounding the main ebb channel are two shallower marginal flood channels (Figs. 7 and 8) which are relatively unmodified by the maximum ebb current. During early flood, these marginal flood channels are active and act as conduits for water entering the inlet peripherally. Water cannot enter the inlet during early flood through the main ebb channel because a strong ebb lag is still flowing seaward. However, at maximum flood, water enters the inlet in a broad, unconfined manner, with water flooding through both the marginal flood channels and the main ebb channel.

Because much of the surface of the ebb-tidal delta is topographically high and very shallow, it is not significantly modified by ebb currents. The only visible ebb influence, other than in the major ebb channel, are ebb-oriented spill-over lobes (1) moving seaward along the major ebb channel margins (Fig. 9). By the time maximum ebb velocities are attained, the water level has dropped so that the banks of the major ebb channel confine the flow and prevent the higher portions of the ebb-tidal delta from being modified.

Channel-Bottom Bedforms

On the floor of the channels associated with the ebb-tidal delta are sand waves

Figure 7. Detail of intertidal part of Nauset Spit. This is the northern marginal flood channel. Note sand waves in runnel (arrow).

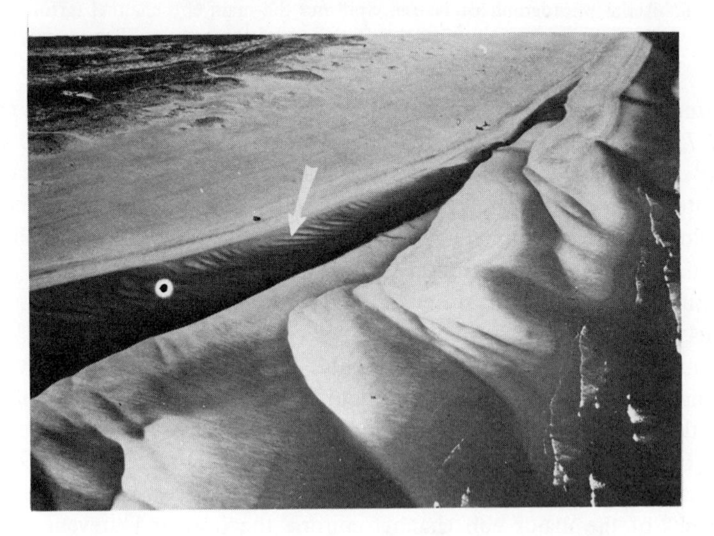

Figure 8. Southern marginal flood channel. Sand waves and megaripples are oriented toward the estuary (left).

Figure 9. Aerial photograph of inlet. Arrow indicates ebb-oriented spill-over lobes.

averaging approximately one meter in height. These sand waves are asymmetrical and remain uniformly flood- or ebb-oriented throughout a tidal cycle. From observations and other work (3), one can conclude that uniformly oriented sand waves, particularly in the intertidal or shallow subtidal environment, indicate a net sand-transport direction. Since there are well-developed, ebb-oriented sand waves migrating towards the terminal lobe in the main ebb channel, net sand transport must be seaward. In the northern marginal flood channel, which actually consists of the intertidal and subtidal portions of the recurved spit complex, small-scale sand waves as well as obliquely oriented ridges indicate net transport into the inlet. In the southern marginal flood channel, which runs parallel to the north end of Monomoy Island, well-developed sand waves with linear megaripples superimposed on top also indicate net sand transport into the inlet. Profiles A-A' and D-D' (Fig. 11) illustrate this bedform orientation and morphology.

Distal Swash Bars

Three large, linear, intertidal swash bars are exposed on the distal portions of the ebb-tidal delta at low water (Figs. 10, 11, and 12). These are part of a continuous line of swash bars that lie along the outer margin of this tidal delta. Bedform morphology and orientation measurements were taken on the northern and middle intertidal swash bars. (Fig. 12).

The bedform orientations on the swash bars indicate the effectiveness of the ebb-tidal delta in refracting waves. The ebb-tidal delta refracts the dominant northeast waves so that their approach along the southern portion of the tidal

Figure 10. Aerial photo graph of middle and southern swash bars on ebb-tidal delta. Arrow indicates large landward-oriented slipface.

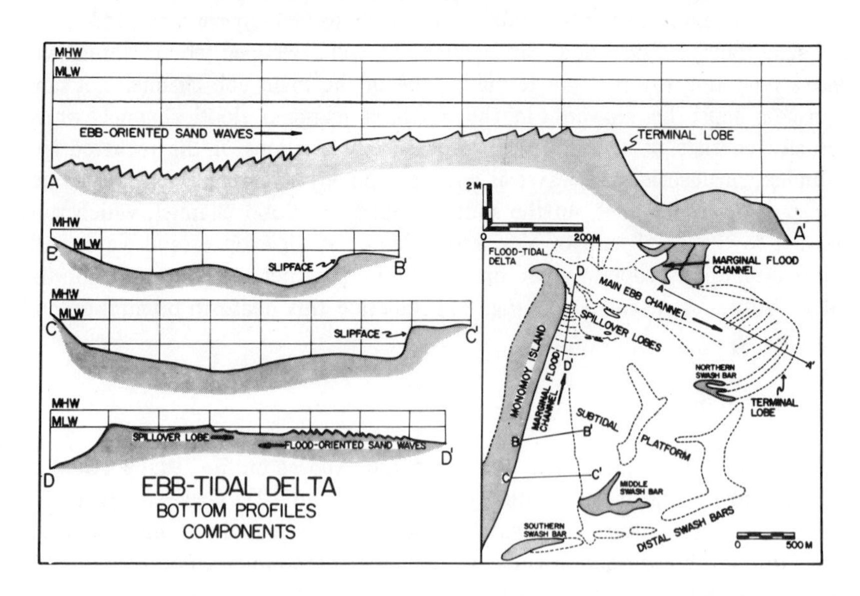

Figure 11. Ebb-tidal delta components and bottom profiles. Note that the main ebb channel has ebb-oriented sand waves, while the marginal flood channels have flood-oriented sand waves. Note also the distribution of swash bars.

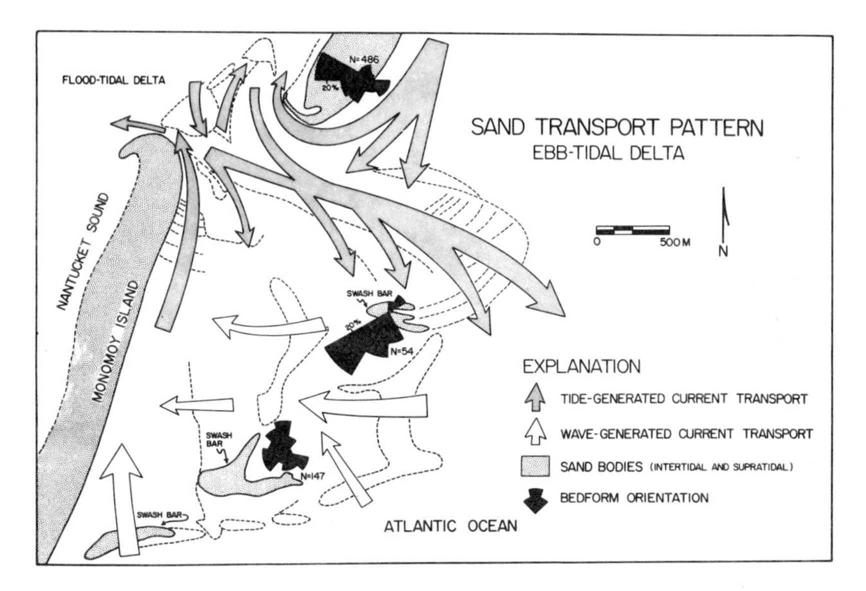

Figure 12. Net sand transport pattern of ebb-tidal delta based upon presence of sand waves and sand-wave orientation. Proximal part of ebb-tidal delta is dominated by tidal currents, while the distal part is dominated by wave (swash)-generated currents. Note cumulative bedform orientation of the intertidal sand bodies.

delta is more from the east and southeast. A comparison of the bedform orientation rose diagrams of the northern and middle swash bars illustrates this wave refraction (Fig. 12). The northern swash bar receives waves from the northeast that have not yet been significantly refracted. Consequently, bedform orientation is towards the southwest. The middle swash bar receives refracted waves that now approach from the southeast and the resulting cumulative bedform orientation is toward the northwest. The presence of large, active slip faces (up to 6 m in height) on these swash bars indicate that sand transport is generally landward, away from the distal margin of the ebb-tidal delta. The exact transport direction is dependent upon the degree of wave refraction and the swash-bar location on the ebb-tidal delta.

Net Sand Transport Pattern

From the bedform orientations and field observations, a net sand-transport pattern for the ebb-tidal delta can be determined (Fig. 12). Sand enters the system from the longshore transport system that carries sand down from the north. Some sand enters the inlet through the northern marginal flood channel, but most is carried into the main ebb channel which transports it seaward to the distal subtidal platform and off the terminal lobe. Along the distal portions of the ebb-tidal delta, the sand is transported landward in large, actively migrating swash bars. When the sand is carried far enough landward by the wave-generated currents, it is transported back towards the inlet by flood currents in the southern marginal-flood channel. Most of the sand in this channel probably is carried around the north end of Monomoy Island and ultimately deposited on the flood-tidal delta. A significant quantity of sand also is probably transported to the beaches along Monomoy Island, where it is incorporated into the regional southern longshore transport and is carried south down Monomoy Island.

FLOOD-TIDAL DELTA

The flood-tidal delta complex in the Chatham Harbor estuary is located between Monomoy and Morris islands and consists of two distinct spill-over lobe complexes. The flood-spill-over lobe complex, composed of several large, coalescing spill-over lobes and dominated by flood-tidal currents, has been formed as a result of the break between Monomoy and Morris islands that occurred in 1958 (Fig. 13). The ebb-spill-over lobe complex is located in the center of the inlet and is composed of five coalescing spill-over lobes (Fig. 14).

Figure 13. Areial photograph of the flood-spill-over lobe complex of the flood-tidal delta. Note several large coalescing lobes that form this sand body. Arrows indicate break between Morris Island (top) and Monomoy Island (right).

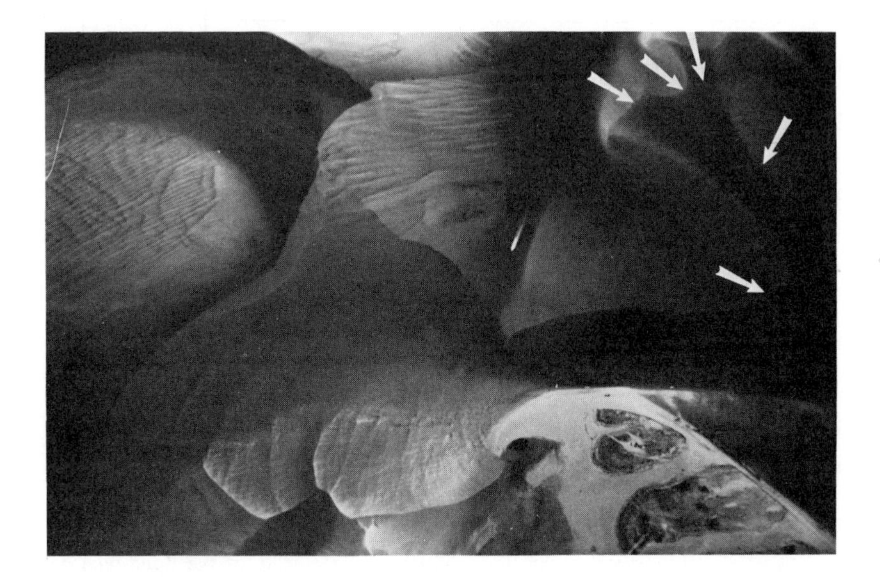

Figure 14. Aerial photograph of ebb-spill-over lobe complex of the flood-tidal delta. Arrows indicate the individual ebb-oriented spill-over lobes. Channel in center of this sand body is flood dominant.

Sand is being transported into the area of the flood-tidal delta complex by two different longshore drift systems (Fig. 1). Most of the sand is carried south along Nauset Beach, transported into the inlet via the southern marginal flood channel (Fig. 12), and deposited on the flood-tidal delta complex. However, a small amount of sand is being carried easterly, along the south shore of Cape Cod, and is being deposited on the distal margin of the flood-spill-over lobe complex.

The flood-spill-over lobe complex is a relatively new feature that has grown very rapidly since 1958, when storm waves broke through the barrier system and separated Monomoy Island from the mainland. Because of the tidal range differences in the Atlantic Ocean and Nantucket Sound, the resulting time asymmetry created nearly complete flood-tidal current dominance in the breached zone. As a result, several rapidly migrating flood-oriented spill-over lobes developed. The largest of these spill-over lobes has migrated approximately 900 m in ten years and now threatens to fill in an artificially dredged channel that allows access to an adjacent harbor.

The flood-tidal current dominance is illustrated by velocity curves plotted from data taken at several hydrographic stations (Fig. 15). Velocities were measured under mean tidal range conditions. Stations 4, 5, 6, and 7 were located on the intertidal portion of the sand body and were exposed for nearly four

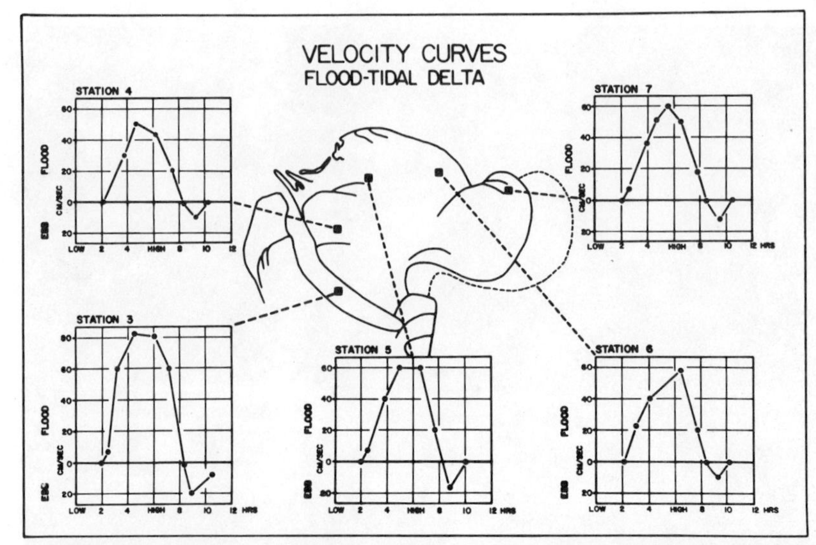

Figure 15. Velocity curves for the flood-spill-over lobe portion of the flood-tidal delta. Note the pronounced time asymmetry. During a tidal cycle, the sand body is exposed for approximately 4 hours, influenced by flood for 6 hours, and influenced by ebb for 2 hours.

hours during the tidal cycle. They were under the influence of ebb currents for less than two hours, with ebb velocities approaching only 20 cm/sec. For the remaining six hours in the tidal cycle, the stations were dominated by the flood-tidal current, which reached a maximum velocity of 85 cm/sec.

During flood, the current flow changes direction twice resulting in three distinct directions of flow (Fig. 16). As the early flood current enters the estuary and begins to spill over into Nantucket Sound, it is directed to the north across the proximal portions of the flood-spill-over lobe complex by the topographically higher distal portions which are still exposed. As the distal portions are covered by the rising water, the flow is no longer directed to the north and now swings through an arc of approximately 70° to the west.

As the water level drops during late flood, the distal margin again becomes an effective current-directing agent and the late flood-tidal current swings back more to the north. Note that the velocity curves in Figure 15 all show that maximum flood velocity is reached at approximately high water, and that the flood velocity continues but declines in strength as the water level drops. By the time the ebb commences, much of the flood-spill-over lobe complex is exposed and is not influenced by the ebb-tidal current.

Bedform Orientation

Bedform orientation resulting from the flood-tidal current flow pattern is

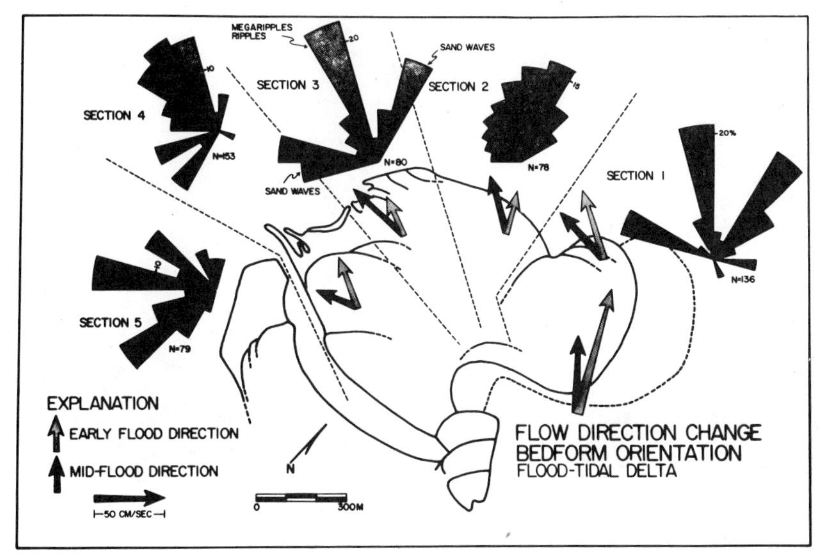

Figure 16. Flow direction change during flood and cumulative bedform orientation on the flood-spill-over lobe complex. Two different directions of flood have developed an intersecting sand-wave pattern. Late flood (not shown) develops the ripple/megaripple mode in the trimodal rose diagrams.

illustrated in Figure 16. The sand body is divided into sections with the cumulative orientation of bedforms in each section plotted separately. Sections 1, 3 and 5 show a distinct trimodal trend in the rose diagrams. The trimodality in section 4 is not as distinct and is disguised by the proximity of two modes. The trimodality does not exist in section 2 because the area is protected somewhat by the topographically higher flood-spill-over lobe adjacent to it in section 1.

These three modes reflect distinct bedform responses to the three different directions of flow. The dominant bedform pattern is an intersecting array of two sand-wave fields (Fig. 17). The right main mode in the rose diagrams represents orientation measurements of sand waves that were developed during early flood. The left main mode represents sand waves that were generated furing mid-flood after the current had changed direction once. Field observations show that these sand waves are topographically higher and actually migrate over the earlier formed sand waves.

During late flood, after the current has changed direction a second time and has switched back more to the north, many minor bedforms (linear and cuspate ripples) are produced. Megaripples are also produced at this time in some areas. These bedforms are represented in the center mode of the trimodal rose diagrams. They also are nearly completely unmodified by the small ebb flow. Practically all sand transported on the flood-spill-over lobe complex is in the

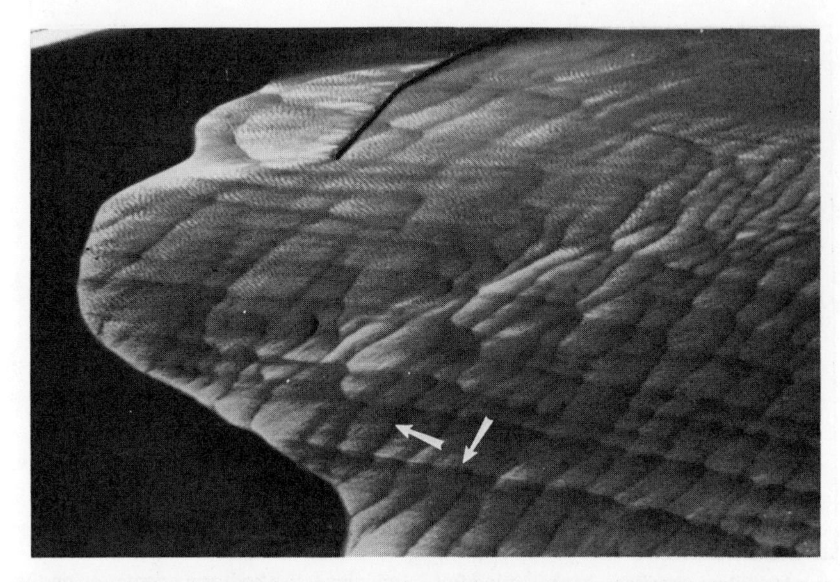

Figure 17. Aerial photograph of intersecting sand-wave pattern on one of the sand lobes of the flood-spill-over lobe complex. Arrows indicate each sand-wave orientation.

flood direction.

Bedform morphology measurements (height and spacing) were taken and plotted to determine whether population densities existed and to further verify that bedforms could be adequately classified by spacing measurements (2). The measurements were also taken to determine the height/spacing relationships within each bedform type.

The height/spacing plots (Fig. 18) indicate a paucity of bedforms in the 40-50 cm area and the 400-500 cm area along the spacing axis. No such areas exist along the height axis, as there is a continuum of data points. These two data-point gaps on the spacing axis define the three bedform population areas: ripples, megaripples, and sand waves. Within the ripple field (5-50 cm spacing) and within the megaripple field (50-500 cm spacing), height increases with spacing. There is no such relationship in the sand-wave field (500 cm spacing).

It should be emphasized that these sand waves exist in the intertidal environment and that possibly subtidal sand waves do show an increase in height with an increase in spacing.

The megaripple bedforms have been classified in the field as either linear or cuspate megaripples, depending upon the degree of linearity of the bedform crest. The cuspate megaripple data points are more closely spaced together, indicate smaller spacing, and show a greater height increase with the same spacing increase as linear megaripples.

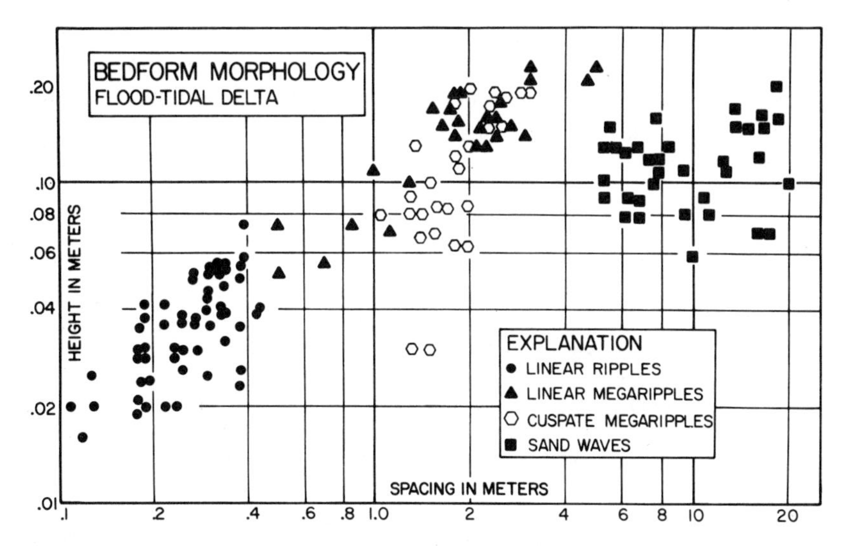

Figure 18. Bedform height-spacing diagram based on data taken from flood-spill-over lobe complex. Note the three distinct population densities, with breaks along the 60-cm and 4-m spacing axis.

CONCLUSIONS

1. Significant tidal range differences in two connected adjacent bodies of water can develop pronounced time-asymmetry and tidal-current segregation. With maximum flood and ebb velocities occurring at or near high and low water, respectively, the shallower, shielded zones in an estuary will be flood-dominated and the deeper zones will be ebb-dominated.
2. The ebb-tidal delta displays a morphology that closely conforms with the established ebb-tidal delta model. Current segregation plays a dominant role in establishing the main ebb channel and surrounding marginal flood channels. Sand enters the estuary through the marginal flood channels and is carried to the terminal lobe through the main ebb channel. Wave-generated currents are dominant along the middle and distal portions of the ebb-tidal delta and develop large landward-migrating swash bars. Tidal currents are dominant on the proximal portion of the ebb-tidal delta and produce uniformly oriented sand waves.
3. The flood-tidal delta consists of two distinct spill-over lobe complexes. The flood-spill-over lobe complex is the more important and has developed rapidly as a result of nearly complete flood-current dominance. During flood, the current changes direction twice in response to topographic

control. Three distinct bedform orientations result from these three directions of flow. The dominant feature is an intersecting array of two sand-wave fields that form from the early and mid-flood flow. Bedform morphology measurements show that bedforms can be classified on the basis of spacing.

REFERENCES

1. Ball, M. M.
   1967 Carbonate sand bodies of Florida and the Bahamas. **Jour. Sed. Pet.,** 37: 556-591.
2. Boothroyd, J. C.
   1969 Hydraulic conditions controlling the formation of estuarine bedforms. In Coastal environments: **NE Mass. and N. H.** p. 417-427. (ed. Hayes, M. O.), Cont. No. 1, Coastal Research Group, Dept. of Geology, Univ. Massachusetts.
3. Boothroyd, J. C. and Hubbard, D. K.
   Genesis of mesotidal bedforms. In **Proc., 2nd International Estuarine Research Federation Symposium,** Myrtle Beach, South Carolina, October 16-18, 1973. (In press.)
4. Goldsmith, V.
   1972 Coastal processes of a barrier island and adjacent ocean floor: Monomoy Island, Nauset Spit, Cape Cod, Massachusetts. Doctoral dissertation, Univ. Massachusetts. 469 p. (Unpublished data.)
5. Hartshorn, J. H., Oldale, R. N., Kotoff, C.
   1967 Preliminary report on the geology of Cape Cod National Seashore. In **Economic Geol. in Massachusetts.** p. 49-58. (ed. Farquhar, O. C.) Graduate School, Univ. Massachusetts.
6. Hayes, M. O., Goldsmith, V., Hobbs, C. H., III
   1970 Offset coastal inlets. **Am. Soc. Civil Engineers Proc. Twelfth Coastal Eng. Conf.,** p. 1187-1200.
7. Hayes, M. O., Owens, E. H., Hubbard, D. K., Abele, R. W.
   1973 The investigation of form and processes in the coastal zone. In **Coastal Geomorphology,** p. 11-41. (ed. Coates, D. R.) State Univ. New York, Binghamton, New York.
8. Hubbard, D. K.
   Tidal inlet morphology and hydrodynamics of Merrimack Inlet, Massachusetts. **U. S. Coastal Engineering Research Center Technical Memorandum.** (In press.)
9. Pritchard, D. W.
   1955 Estuarine circulation patterns. **Am. Soc. Civil Engineers Proc.,** 81: 717/1-717/11.

MORPHOLOGY AND HYDRODYNAMICS OF THE MERRIMACK RIVER EBB-TIDAL DELTA

by

Dennis K. Hubbard[1]

ABSTRACT

A well-developed ebb-tidal delta occurs at the mouth of the Merrimack River, Massachusetts. The inlet is the site of artificial structures (groins, jetties, sea walls) that affect the ebb-tidal delta and adjacent beaches. Old maps and historical records show that, before stabilization, the ebb-tidal delta displayed all the features of the models of Oertel (5) and Hayes et al. (2) and that the inlet behaved as do natural mesotidal inlets observed today. Changes in the present ebb-tidal delta are the amplification of adjacent nearshore bars and the elimination of marginal flood channels and channel margin linear bars.

Hydrography stations were monitored for 13 hours to determine tidal-current velocity and direction, salinity, and water temperature in and around the inlet. Bottom samples were collected by scuba divers at over 200 locations on the ebb-tidal delta. Semi-daily wave readings (breaker height and type, approach angle, wave period and longshore drift) were taken along the beach and were related to changes in beach and inlet morphology.

Wave-refraction diagrams simulating storm conditions (9-second waves approaching from the northeast) show a complex relationship between the morphology of the ebb-tidal delta, wave refraction, and man's intervention in the form of jetties and groins. The refraction of northeast waves around the ebb-tidal delta creates a reversal in the direction of longshore transport south of the inlet causing serious beach erosion. Groins have proved ineffective in

1. Department of Geology, Coastal Research Division, University of South Carolina, Columbia, South Carolina.

maintaining the present beach and continue to cause local erosion.

INTRODUCTION

General Setting

The area studied (Fig. 1) consists primarily of two active barrier beaches, Salisbury Beach to the north and Plum Island to the south, separated by a tidal inlet at the mouth of the Merrimack River. The barriers are Holocene in age and are backed by Holocene marsh. The origin of the barrier islands was considered by McIntyre and Morgan (4) who concluded they were transgressive features that formed in response to changes in post-glacial sea level.

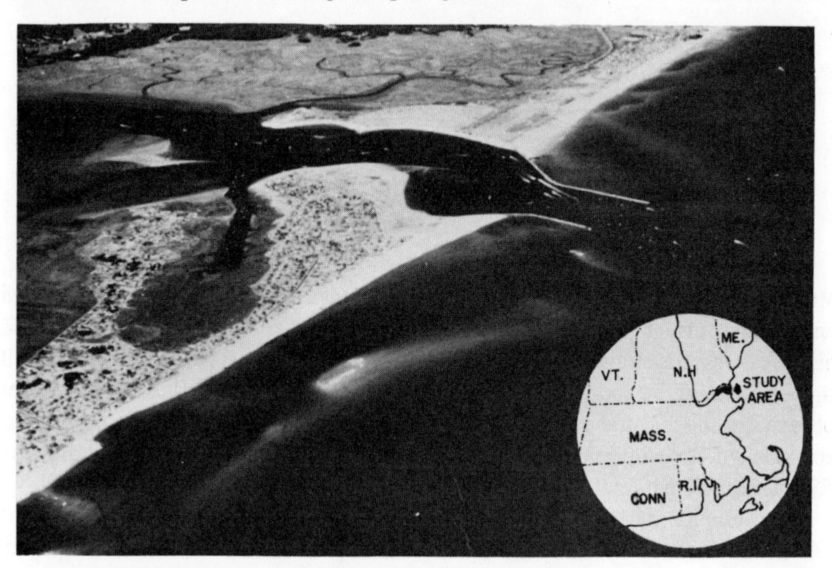

Figure 1. Low-tide aerial photograph looking northwest across the Merrimack River Inlet. The location of the area ia shown in the inset.

The Merrimack River, the fourth largest river in New England, has an average discharge in excess of 250 m^3/sec. Its semi-diurnal tide averages 2.53 m. Spring and neap tides are 3.20 m and 1.82 m, respectively.

The upper reaches of the Merrimack River channel are, for the most part, deep and narrow. In the main estuary, however, vast intertidal areas are exposed. One of these areas, Joppa Flats, supported extensive clamming until pollution made commercial digging impossible, except in a few selected areas.

The near-shore zone is characterized by a well-developed ebb-tidal delta flanked on either side by submarine bars (3) and swash bars (Fig. 2). The southern bar system trends parallel to the shore and is broken periodically by

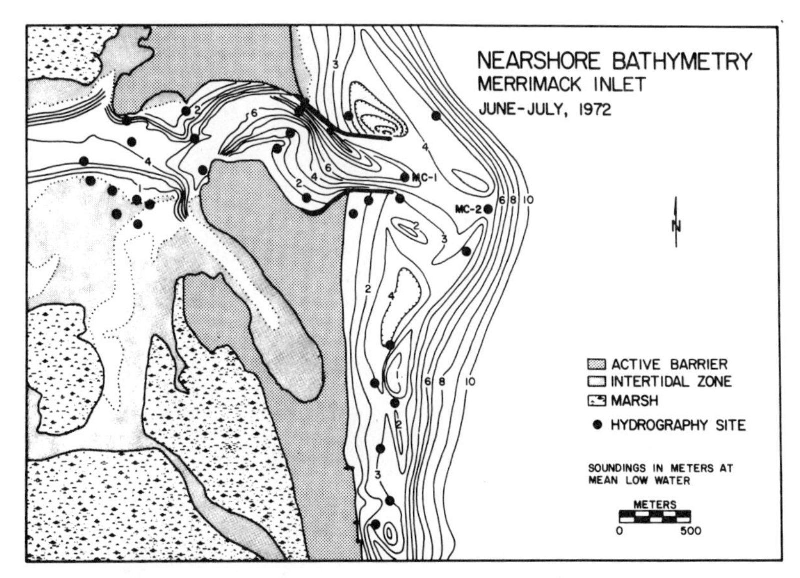

Figure 2. Bathymetric map of the Merrimack Inlet. Contours are based on over 100 controlled offshore profiles.

storm channels. The bar to the north is somewhat more continuous and is oriented into the predominant northeast waves.

Modifications in ebb-tidal delta morphology

The Merrimack ebb-tidal delta has been modified from the models of Oertel (5) and Hayes, et al. (2) by inlet stabilization and channelization since 1881. The jetties on Plum Island and Salisbury Beach have replaced the channel margin linear bars and block the formation of marginal flood channels. Also, stabilization has caused the system of swash bars to become morphologically more significant than those observed on natural ebb-tidal deltas. These modified swash bars have combined with the submarine bars (3) normally occurring opposite the barriers to form a system of bars that can be divided into four segments (Fig. 3):

1) **Updrift Leading Bar**: a continuous bar extending north from the ebb-tidal delta. This modified swash bar is oriented into the dominant northeast waves and joins a system of rhythmic topography approximately .5 km updrift of the inlet.
2) **Downdrift Trailing Bar**: a somewhat more extensive series of bars trending parallel to the downdrift shorefront of the inlet and broken by periodic storm channels.
3) **Downdrift Nearshore Bar**: An arcuate nearshore bar continuing south as an

Figure 3. Nearshore bar types in the Merrimack embayment. Numbers correspond to those appearing in the text.

extension of the downdrift trailing bar. This submarine bar is not directly associated with the ebb-tidal delta.

4) **Rhythmic Topography**: a variable system of oblique bars occurring updrift of the next inlet. This system may or may not be reflected on the beach in the form of cusps.

This assemblage of bar types has been observed near other tidal inlets. In the Merrimack embayment the bars are best developed where the inlets have been stabilized and the beaches are oriented obliquely to the dominant northeast waves.

Historic inlet changes

Charts dating from 1741 document changes in the shoreline configuration around the Merrimack River inlet. Before stabilization in 1881, the Merrimack ebb-tidal delta had all the components of the models of Oertel (5) and Hayes, et al. (2) and behaved like those associated with natural mesotidal inlets observed today in New England and Alaska. Figure 4 shows the components of the Merrimack ebb-tidal delta in 1741, 1809, and 1826.

PRESENT PROCESSES

The morphology of the present ebb-tidal delta represents a complex interrelationship between waves, tidal currents, and the man-made jetties. In order to determine the effect of tidal currents on the area, a series of thirty hydrography stations were established (Fig. 2). Each station was monitored at one-hour intervals through a 13-hour tidal cycle. Measurements of tidal-current velocity and direction, salinity, and water temperature were taken at one-meter intervals throughout the water column.

The Merrimack River estuary is a type B, tilted-boundary estuary of Pritchard (6). Early flood currents enter the inlet from the south by passing up over the southern margin of the terminal lobe and flowing into the estuary as a salt wedge beneath the residual ebb-flow of the river. The lower-density river water on the surface is separated from the more saline ocean water by a sharp boundary inclined to the south.

Tidal-current velocity curves for the main ebb channel (Fig. 5) show some degree of time-velocity asymmetry. In the inlet throat (station MEC-1) maximum ebb-current velocities of 125 cm/sec were measured, while the highest flood velocity attained was 80 cm/sec. Also, ebb currents continued to flow for almost two hours into the flood stage of the tidal cycle, while currents on the bottom turned to flood 45 minutes into the tidal cycle. Similar results were recorded at station MEC-2. The smaller velocities measured there (maximum flood = 35 cm/sec; maximum ebb = 105 cm/sec) were a function of the distance

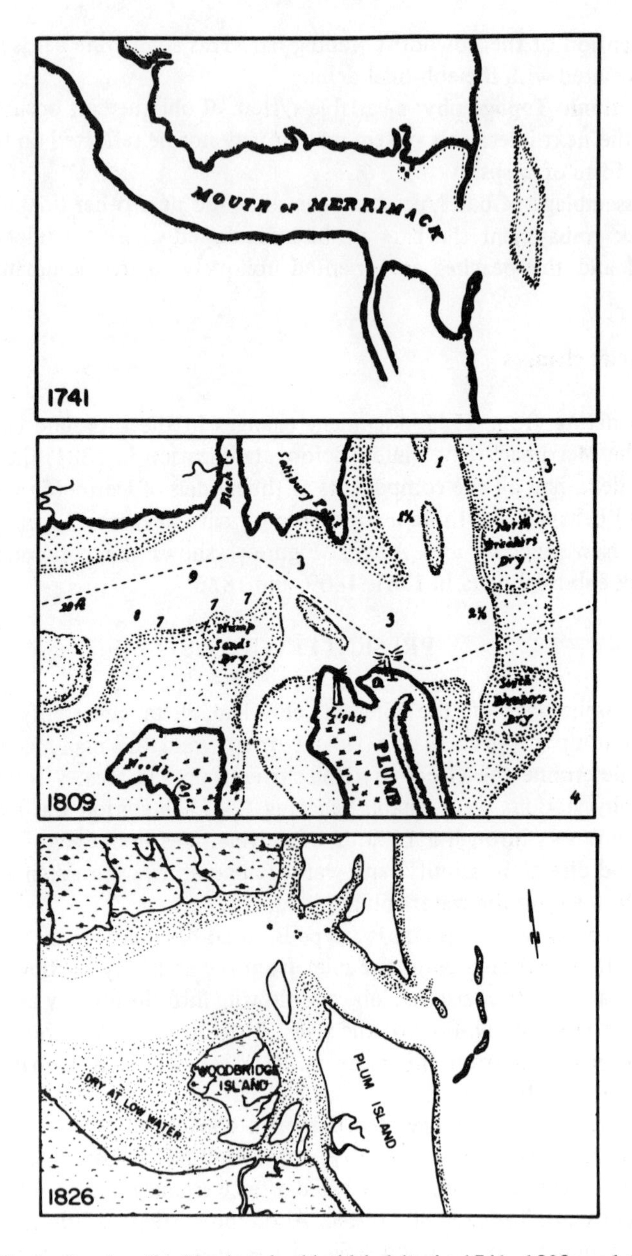

Figure 4. Charts showing the Merrimack ebb-tidal delta in 1741, 1809, and 1826. The 1809 and 1826 charts show a well-developed ebb-tidal delta. The 1741 chart, although not as detailed, still shows a shoal at the mouth of the inlet.

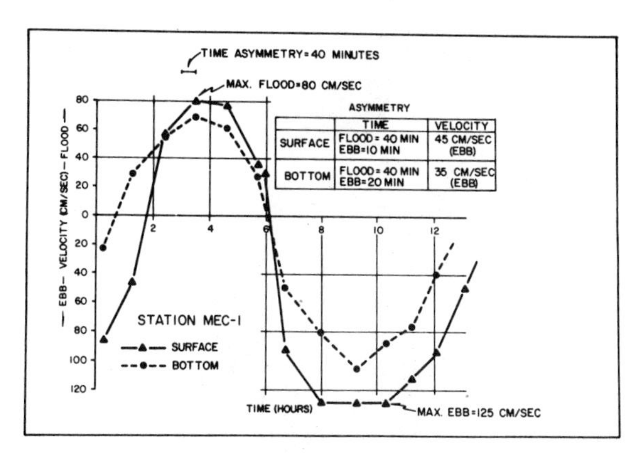

Figure 5. Tidal-current velocity curves for station MC-1 (Fig. 2). Note the velocity and time-velocity asymmetry in both surface and bottom curves. Also, note the opposing surface and bottom currents occurring until almost one hour into the flood stage.

from the inlet mouth.

Sediment sampling

Hydrography serves as a valuable means of directly determining the distribution of tidal currents and their effect on the morphology of an area. The evaluation of grain-size distribution combined with an analysis of bedform type, scale, and orientation throughout the area can provide valuable insight into the pattern of sediment dispersal in the near-shore zone. Bottom samples were collected by scuba divers at over 200 locations on the ebb-tidal delta and the associated system of near-shore bars. At each site a 30-cm core was taken through the crest of the existing bedform. This method is thought to be superior to grab sampling because it eliminates both loss of a portion of the sample in transport to the surface, and blind sampling without regard for bottom morphology. Also, the investigator gains a much clearer idea of the agents responsible for trends in the data. Grain size was estimated in the lab by comparing the samples to grains of known size.

The coarsest material (-0.75ϕ) occurs in the main ebb channel (Fig. 6). This is a result of the rapid tidal currents (> 150 cm/sec at spring tides) occurring there. Other coarse material (-0.2 to 0.4ϕ) is associated with the system of near-shore bars downdrift of the inlet. The finest grain sizes (> 1.0ϕ) occur in the deep offshore areas seaward of the near-shore bar system. Generally, grain size becomes finer to the south and seaward. The fining with depth is a function of the reduction in wave energy along the bottom. The fining southward is a result of the selective transport of finer material by the waves.

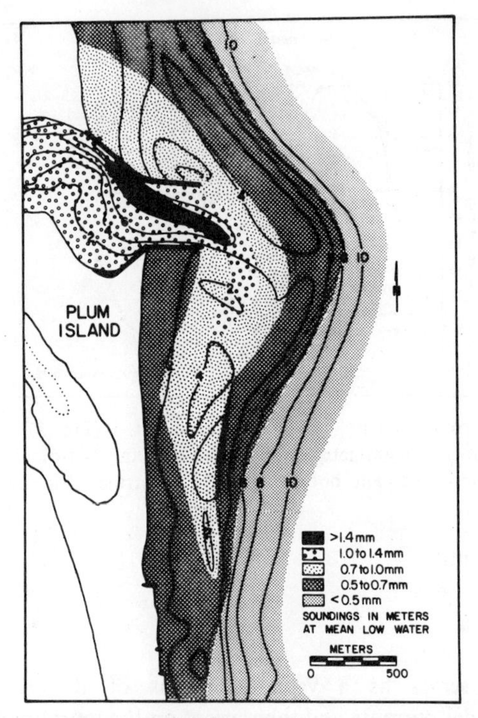

Figure 6. Grain-size distribution on the ebb-tidal delta and associated nearshore bars.

The distribution of bedform scale and orientation follows the same pattern as that for grain size. Bedform height and spacing generally varies inversely with water depth (Fig. 7). Bedform orientations (Fig. 8) reflect the adjustment of waves to local topography.

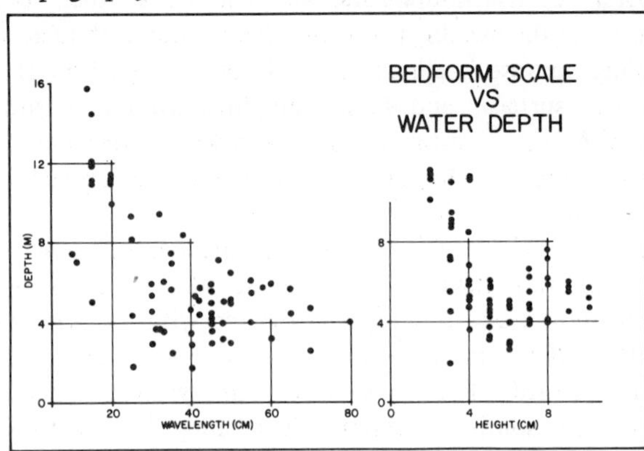

Figure 7. Bedform scale vs water depth. The anomaly occurring near 5 meters is associated with the downdrift trailing bar. Depths are in meters at mean high water.

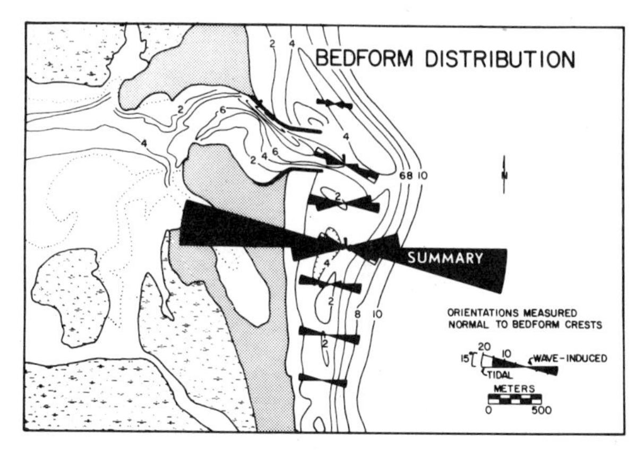

Figure 8. Bedform orientations in the nearshore zone. Orientations are measured normal to ripple strike.

Waves and wave refraction

The movement of sediments on the ebb-tidal delta and adjacent beaches is related primarily to the refraction of northeasterly waves around the ebb-tidal delta. Waves approaching from the northeast refract around it, causing a reversal in longshore current direction immediately downdrift of the inlet (Fig. 9). The resulting divergence generally causes a break in the downdrift trailing bar due to the movement of sediment away from that point in all directions.

The sediment-dispersal pattern for the Merrimack River inlet

Combination of the hydrodynamic parameters and environmental indicators collected during the study period has led to the delineation of the sediment-dispersal pattern around the Merrimack River inlet (Fig. 10). The gross morphology of the Merrimack inlet is a function of the tidal currents within it. The cycle of erosion and deposition along the adjacent beaches is related to the wave climate. The shape of the ebb-tidal delta and the associated near-shore bars is a function of the complex interrelationship between the two parameters. The pattern of sediment dispersal in and around the inlet is controlled by the nature of the hydrodynamic parameters and their response to the morphology of the area.

The movement of sand in the Merrimack embayment is a result of competition between northeast waves during storms and southeast waves during calm periods. The dominant direction of longshore transport is to the south. During storms sediment moves southward along Salisbury Beach until it is interrupted by the north jetty at the Merrimack inlet. The effectiveness of the

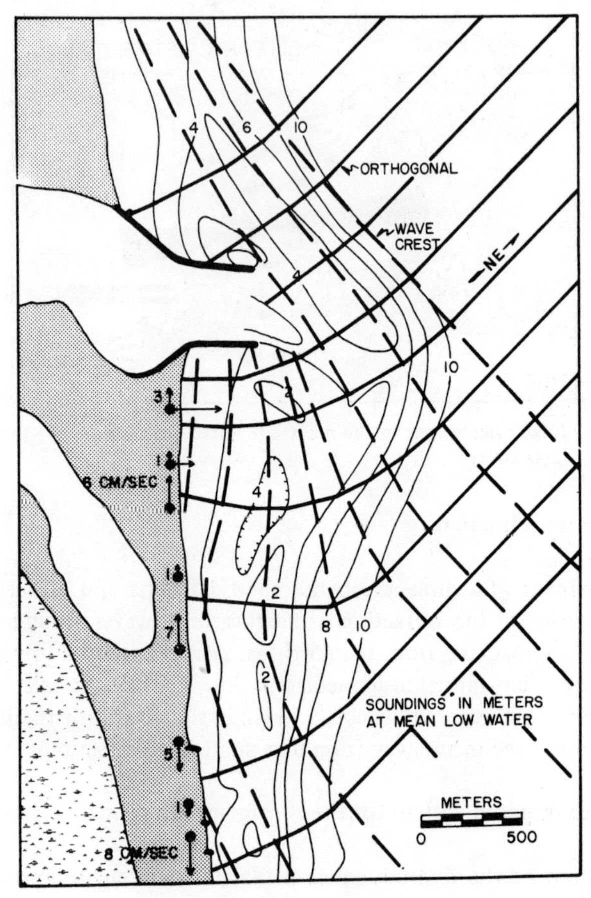

Figure 9. Wave refraction around the ebb-tidal delta. Note the agreement between the reversal predicted by the diagram and measured under similar conditions. (T + 10 sec.)

jetty as a sediment trap is evidenced by the large accumulation of sand along the adjacent beach. The sediment that bypasses the north jetty is joined by a small amount of sediment contributed by the Merrimack River. The sand is then transported southward along the near-shore bar system.

During calm periods the sediment that has not moved southward or offshore out of the area is transported onto the beach by the landward migration of swash bars and transverse bars. The sediment in the littoral zone is moved northward by southeast waves until it either accumulates on the south side of the Plum Island jetty or moves offshore and back into the southward drift system along the downdrift bar.

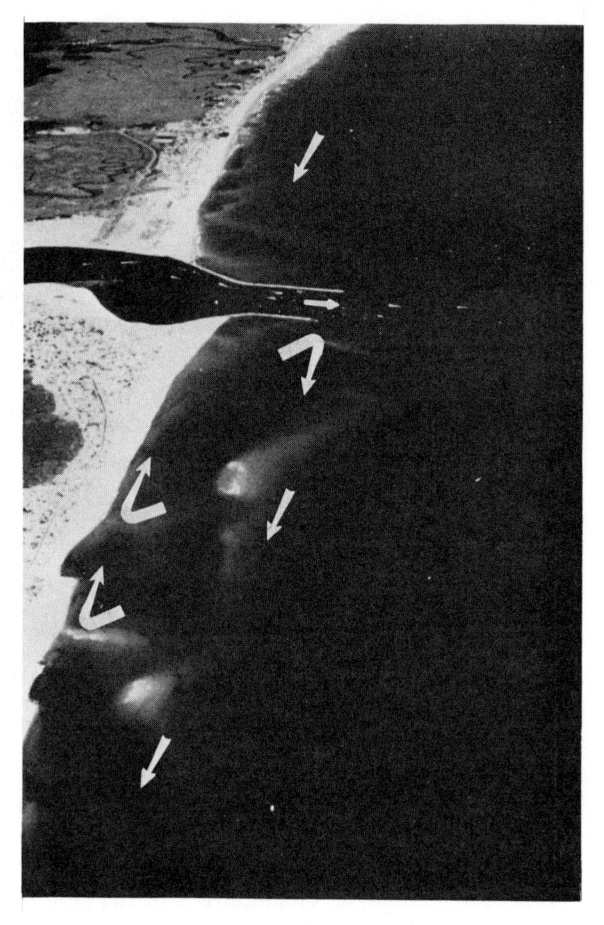

Figure 10. Proposed sediment-dispersal pattern along the ebb-tidal delta and downdrift trailing bar.

Problems related to man-induced change

Several critical erosion areas exist near the Merrimack inlet. The problems occurring there can be directly linked to attempts by man to control the inlet-beach complex. Critical erosion of an area 2.5 km downdrift of the inlet occurs on the north side of a large rubblestone groin (Fig. 11). Erosion can be related to three factors: (a) the occurrence of a reversal in longshore-current direction at this point during northeast storms; (b) the resulting break in the near-shore bar opposite the eroding area, allowing storm waves to pass through the bar and break on the beach; and (c) sediment entrapment and modification of wave parameters by the groin. The most serious erosion occurring during

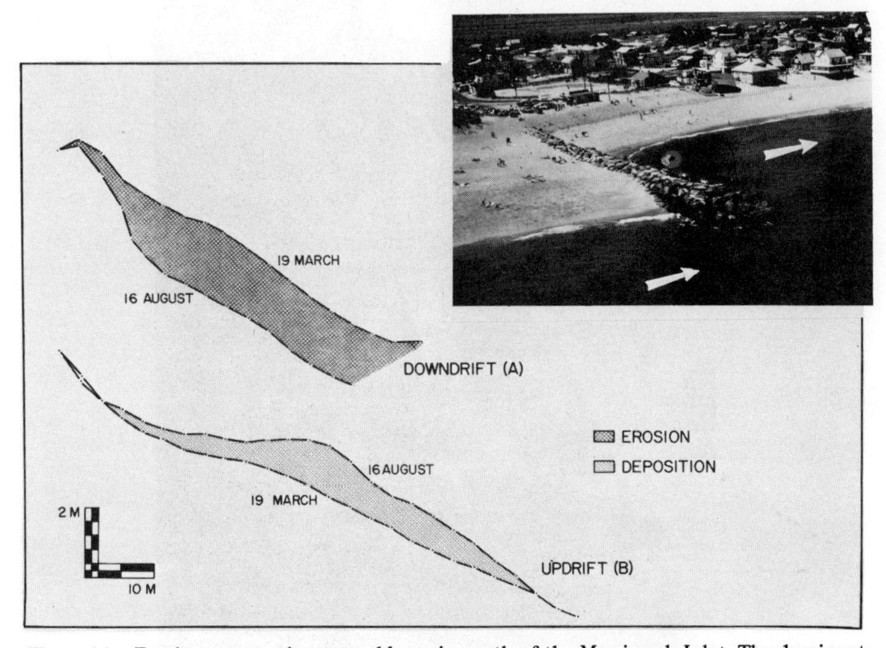

Figure 11. Erosion near an impermeable groin south of the Merrimack Inlet. The dominant longshore current is northward (arrow).

northeast storms is related to the first two factors. During calm periods, however, the waves approach from the southeast and the effects of sediment entrapment and modification of wave parameters become the dominant factors. Between 19 March and 18 August, 1972, the beach updrift of the groin showed accretion, while the beach directly behind the large groin was lowered by as much as 2 m (Fig. 11). The updrift, or south, side of the groin is characterized by a gentle, uniform, near-shore slope leading to a well-developed berm. A gentle slope normally induces spilling or spilling-plunging waves (1) which dissipate their energy over a wide area, inducing longshore rather than offshore currents. On the downdrift side of the groin, the beach face is much steeper, resulting in surging or plunging waves. Although waves on this side of the groin are on the average slightly smaller than those on the updrift side (25.5 cm and 30.7 cm, respectively), their energy is expended directly on the beach. Current measurements downdrift of the groin showed a strong offshore component to beach-face and near-shore currents related to breaker type and beach-face gradient. Beach-face slope, wave type, and degree of sediment entrapment by the groin are therefore more important than wave height during calm periods in determining losses from the beach.

Erosion inside the Merrimack inlet can also be related, at least in part, to the jetties placed there. Between 19 and 27 February, 1969, three major low-pressure systems entered the Merrimack embayment. These slow-moving storms remained in the area for a period spanning 18 high tides (C. K. Wentworth, personal communication) causing severe erosion of the river bank near the Merrimack Coast Guard Station inside the Plum Island jetty (Fig. 12). Waves overtopping the north jetty removed 826,000 cubic meters of sand, lowering fore-shore and near-shore areas by nearly 60 cm and back-shore areas by as much as 300 cm.

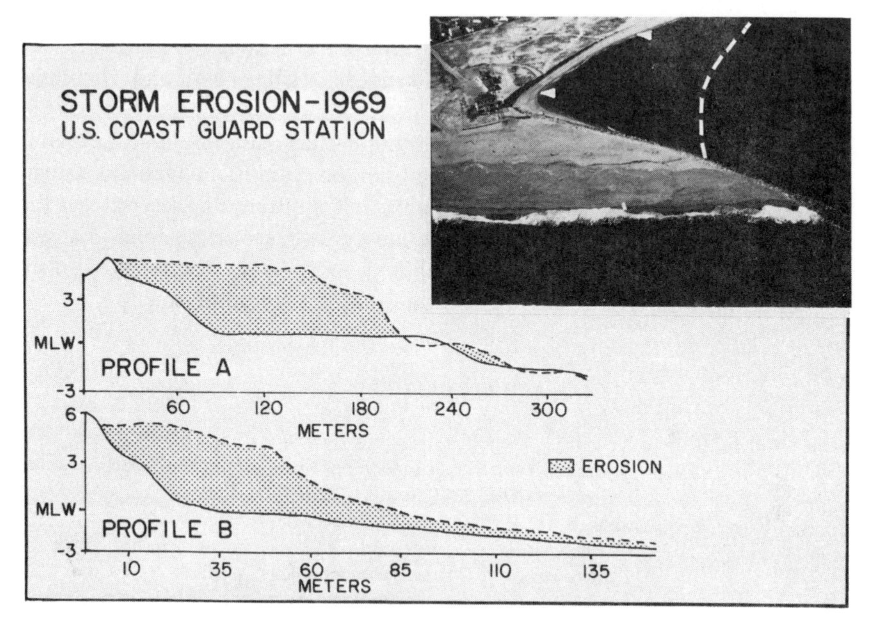

Figure 12. Storm erosion at the Merrimack Coast Guard Station. Dashed line represents the location of the pre-storm beach.

The damage from these storms can be related to three causes: (a) the overtopping of the north jetty by northeast waves; (b) the slow passage of the low-pressure system through the Merrimack embayment; and (c) rehabilitation of the south jetty in 1968. Although the first two factors are probably the most significant, the latter is important because it reflects the response of the inlet to stabilization by a jetty complex. Sand which previously overtopped the south jetty during nor'easters (due to refraction of waves around the ebb-tidal delta) and nourished the river bank fronting the Coast Guard Station was trapped on the adjacent beach.

CONCLUSIONS

(a) Historical studies have shown that in its natural state the Merrimack River inlet displayed all the features of the ebb-tidal delta model and behaved much like natural inlets observed today in Alaska, Baja California, and New England.

(b) The sediment-dispersal pattern in and around the Merrimack River inlet is a result of the complex relationship between tidal currents and the refraction of northeast waves around the ebb-tidal delta, both of which have been modified by the presence of man.

(c) Serious erosion has occurred in populated areas downdrift of the inlet. This erosion can be related to the divergence of longshore currents, the location of the break in the downdrift trailing bar, and the large rubblestone groin built in 1952.

(d) Wave refraction around the Merrimack ebb-tidal delta would occur even if the jetties were not there; but the erosion on Plum Island would not be as severe. If the inlet had been allowed to migrate, the drift reversal and the resulting erosion zone would have moved in response to inlet changes. Also, the stabilization of ebb-tidal-delta morphology has resulted in an unchanged pattern of wave refraction for a long period of time.

REFERENCES

1. Galvin, C. J.
1968 Breaker type classification on three laboratory beaches. **Jour. Geophys. Res.,** 73(12): 3651-3659.

2. Hayes, M. O., Owens, E. H., Hubbard, D. K., and Abele, R. W.
1973 Investigation of form and processes in the coastal zone. In **Proceedings of Coastal geomorphology,** p. 11-41. (ed. Coates, D. R.) 3rd Ann. Geomorphology Symposia Series, Binghamton, New York, September 1972.

3. King, C. A. M.
1972 Beaches and Coasts. 2nd Ed., Edward Arnold Ltd., London, 570 p.

4. McIntire, W. G. and Morgan, J. P.
1963 Recent geomorphic history of Plum Island, Massachusetts, and adjacent coasts. Louisiana State Univ., Coastal Studies, No. 8, 44 p.

5. Oertel, George F.
Ebb-tidal of Georgia estuaries. In **Proceedings of 2nd International Estuarine Research Federation Symposium,** Myrtle Beach, S. C., October 15-18, 1973, (In press.)

6. Pritchard, D. W.
1955 Estuarine circulation patterns: Am. Soc. Civil Engineers, Proc. 81, 717/1-71/11.

EBB-TIDAL DELTAS OF GEORGIA ESTUARIES

by

George F. Oertel[1]

ABSTRACT

Configurations of tidal deltas are determined by the interactions between inlet tidal drainage and longshore currents in the near-shore zone. The tidal range along the Georgia shoreline is 2-3 meters and produces moderate to strong tidal currents adjacent to estuary inlets. Wave energy dissipates over a wide (11 kilometers), shallow (less than - 10 meters) zone and attenuates significantly before it impinges on the shoreline. Therefore, along the Georgia coast, the zones adjacent to estuary inlets experience moderate-velocity to high-velocity tidal currents, whereas, the wind-and wave-induced longshore currents are generally buffered before they reach the shoreline. Tidal deltas produced by the current interactions at tidal inlets, illustrate the predominant influence of inlet tidal drainage. The peripheral-shoal complexes of the tidal deltas have well-developed marginal shoals that are approximately 2 kilometers apart and parallel each other for approximately 5.5 kilometers seaward of the inlet. Distal shoals are poorly developed or absent.

The marginal shoals on the south sides of these channels are elongate, triangular sand bodies that are broadest at their point of attachment to the barrier. Ephemeral spill-over channels and runnels transect these sand bodies at oblique angles. These channels help distribute the centrifugal flow of inlet drainage away from the entrance.

On the north sides of the main axial channels, spill-over channels are deeply incised into the marginal shoals. One of these channels separates the marginal shoals from the barrier.

1. Skidaway Institute of Oceanography, P. O. Box 13687, Savannah, Georgia 31406.

There are eleven major tidal inlets along the coast of Georgia that exhibit well-developed tidal deltas. Sea islands along the coast of South Carolina and the coast of northern Florida also have tidal inlets with well-developed tidal deltas. These deltas all have geomorphic similarities that are related to the inlet flow dynamics. Several of the inlets have large fluvial sources of water (St. Mary's Entrance, Altamaha Entrance, Ogeechee River Entrance, and Savannah River Entrance). However, the majority of the entrances of the southeast coast of the United States have small fluvial sources of water. The tidal deltas of the river-dominated estuaries exhibit different forms than the tidal deltas of the tide-dominated estuaries. The tidal deltas described in this report refer to the inlets that are associated with tide-dominated estuaries.

Tidal inlets are primarily influenced by reversing currents that are issued into the inlet during the flooding tide and are drained out of the inlet during the ebbing tide (10). The forces of the reversing currents at the tidal inlets are determined by the water-storage capacities of the lagoonal-marsh complexes. The lagoonal-marsh complexes behind the Georgia barrier islands are broad and extensive, and store large volumes of water during high tide. The thickness of the tidal prism is approximately 2 to 3 meters during the respective neap and spring tides, and the width of the marshes behind the barriers averages approximately 6.5 to 7.5 kilometers. Water flushes in and out of these marshes as the tides rise and fall twice a day. The velocity of the tidal currents is determined by the volume of the lagoonal-marsh drainage basin and the width of the inlet. Small changes in the thickness of the water prism are amplified as they are spread over the broad, lagoonal-marsh complexes.

The synodic variations of the tides are continually changing the thickness of the tidal prism and changing the forces of water being issued through the inlets. Wind and precipitation also play an important role in determining the forces of water masses that are issued through or are drained out of tidal inlets. Lagoonal-marsh basins vary in area from 120 to 170 square kilometers. When a relatively small quantity of precipitation falls on this area, the volume of the draining-water mass is increased significantly. Winds work in a similar way. Offshore winds inhibit the flow of the issuing-water masses and decrease the volume of water entering the lagoonal-marsh basin. Onshore winds enhance the flow of the issuing-water masses and increase the volume of water entering the lagoonal-marsh basin. The cross-sectional area at an inlet also determines the velocities of the flooding currents being issued through the inlet and the ebbing currents draining out of it (1 and 2). These reversing currents generally accelerate as they approach the inlets. However, the magnitude of this acceleration is greater at constricted inlets than it is at broad-mouthed inlets. While the velocities of currents vary in magnitude, the directions and patterns of water flow remain relatively consistent for the different inlets.

Near-shore currents in the coastal waters of Georgia are generally transient.

During the spring and summer months, wind-drift currents and wave-induced currents generally flow in a southerly direction. During the winter, patterns of near-shore currents are more variable. High-energy storms are generally from the northeast, but short durations of offshore winds also produce periods of near-shore calm.

The near-shore zone of Georgia's coast is relatively shallow (less than -10 meters) for approximately 11 kilometers offshore. This relatively broad shallow zone buffers the force of many of the larger storm waves, and the force of the wind-induced currents is spread over an 11-kilometer zone, rather than being concentrated directly adjacent to the shoreline.

At the tidal inlets; the interactions between longshore currents and tidal drains result in the diversion of both currents (12). However, in the shallow, near-shore zone (less than -10 meters), the tidal drains generally overwhelm the force of the buffered coastal currents. As the tidal currents centrifugally drain away from the inlet they decrease in force until coastal currents become the dominant force.

The interactions of the coastal currents with the inlet currents produce an accumulation of sand seaward and adjacent to tidal inlets. Characteristics of these interactions influence the patterns of sedimentation (1, 2, and 11). Sand accumulations (tidal deltas) at Georgia's tidal inlets are geomorphically and sedimentologically very similar. The pronounced offshore elongation of the marginal portions of peripheral shoals is one of the unique and diagnostic characteristics of the tidal deltas along the mesotidal shoreline of the coast of Georgia. In other coastal areas, the interactions between coastal currents and inlet currents may have different resultant forces and the peripheral shoals of tidal deltas may be semicircular or elongate to the left or right of the inlet (Fig. 1).

The channels of tide-dominated inlets are generally deepest at inlet troughs that are located just seaward of the inlet (Fig. 2). The troughs may be centrally located in the inlet channel or may occur on the south side of the channel. Inlet troughs vary in depth from approximately 15.0 meters to 30.0 meters below mean low water and represent the deepest areas of the coastal waters before the shelf slopes to -15 meters approximately 28 kilometers offshore, and -30 meters approximately 93 kilometers offshore. Landward of the inlet trough, the channel bottom gets shallower over a ramp-to-the-sound. This ramp often merges with a "flood" tidal delta (3) that may be centrally located, but is generally located nearer to one of the margins of the sound. Seaward of the inlet trough, the bottom gets shallower over a channel-bottom ramp (11) or a ramp-to-the-sea (6). Ramps-to-the-sea slope seaward for 3 to 7 kilometers, where they merge with the upper shoreface or with a poorly developed distal shoal. Three to 7 kilometers seaward of the inlet, the channels above the ramps-to-the-sea may bifurcate or curve right or left.

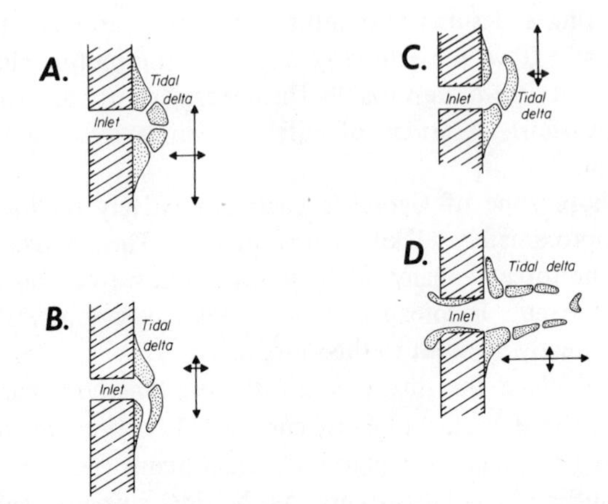

Figure 1. Ebb-tidal deltas of non-stratified estuaries. Arrows illustrate the relative forces of the onshore, long-shore, and offshore currents. A, the prevailing force of longshore and onshore currents is greater than the offshore force. B, the prevailing force of the southward longshore current is greater than the three remaining component forces. C, the prevailing force of the northward longshore current is greater than the three remaining component forces. D, the prevailing forces of the inlet currents are greater than the forces of northward and southward longshore currents.

The ebb-tidal deltas (3) are composed of three morphologically distinct shoals that are peripherally located around the ramp-to-the-sea. Ramp-margin shoals are present on both the north and south sides of the ramp-to-the-sea and portions of these shoals are generally exposed at low water. Distal shoals are located at the seaward end of the ramp-to-the-sea, and they are often poorly developed or may be absent. Although wave crests are sometimes amplified as they cross distal shoals, shoals are not exposed even during the lowest spring tides. Ramp-margin shoals have one of two configurations: they may be attached to the shoreline ("attached" ramp-margin shoals) or they may be separated from the shoreline by tidal channels ("segmented" ramp-margin shoals). See Figure 2.

Shoals that are separated from the beach are generally segmented into several sections by tidal channels that appear to "spill" away from the main axial channel of the inlet (Fig. 3). These spill-over channels are long-term structures that are deeply incised in the shoals at regular spacings. Spill-over channels help distribute the centrifugal flow of the inlet drain away from the entrance. Longshore bars are often developed adjacent to the seaward sides of these channels. Thus, although segmented ramp-margin shoals are elongated normal to the shoreline, they may be composed of a number of sub-parallel bars that are oriented parallel to the shoreline. During periods of low water, these bars act as shields that restrict the surge of waves and flooding currents from the spill-over channels. During the half tides, waves break across these bars and transport

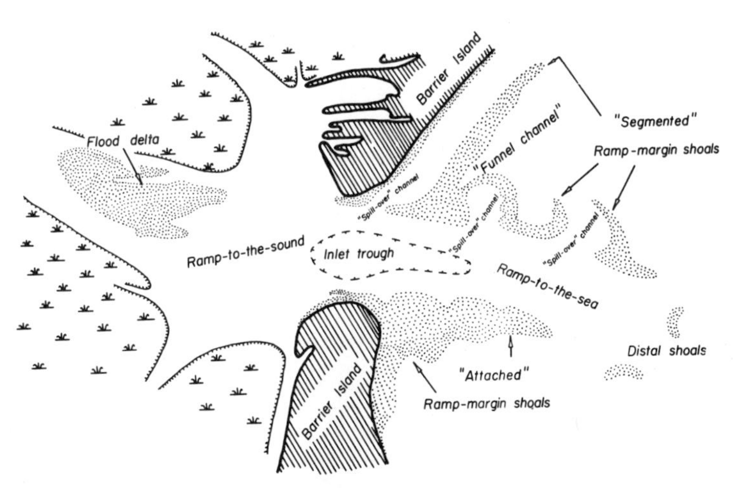

Figure 2. A Georgia tidal inlet with suggested terminology for prominent geomorphic features. The sketch of the inlet was modified after the inlet at St. Catherines Sound, Georgia. Flood delta is the name given to relatively large shoals that are located in the center of a sound, and landward of the inlet (no genetic implications are intended by the use of "flood" in this name). The ramp-to-be-sound is a wedge-shaped accumulation of sediment that originates at a scoured inlet trough and slopes upward to the floor of the sound or a flood delta. The ramp-to-the-sea is a wedge-haped accumulation of sediment that originates at a scoured inlet trough and slopes upward to the floor of the sound or a flood delta. The ramp-to-the-sea is a wedge-shaped accumulation of sediment that originates at a scoured inlet trough and slopes upward to the shoreface or some distal shoals. The "ebb" delta is composed of a peripheral shoal complex that is formed on the margins of the ramp-to-the-sea. Three different types of shoals may be present in this shoal complex, viz. segmented ramp-margin shoals, attached ramp-margin shoals, and distal shoals. Spill-over channels are tidal channels that issue draining water through segmented ramp-margin shoals. A "funnel channel" located between the ramp-margin shoals and the barrier beach (or a longshore bar) issues flooding water across the shoals toward the inlet. Funnel channels are considerably shallower than the channels above the ramp-to-the-sea.

sediment landward. Bars have gentle seaward slopes and relatively steep slip faces adjacent to the spill-over channels. During low water, the planar surfaces of these alongshore trending bars are often exposed. Numerous similarities exist between these offshore, intertidal bars and swash bars (4 and 5) that are found along many coastal beaches. Along the Georgia coast, segmented ramp-margin shoals have preferred locations on the north sides of ramps-to-the-sea. However, they may also occur on the south sides of the channel or along both margins of the channel.

Attached ramp-margin shoals, are triangular in plain view. These triangular

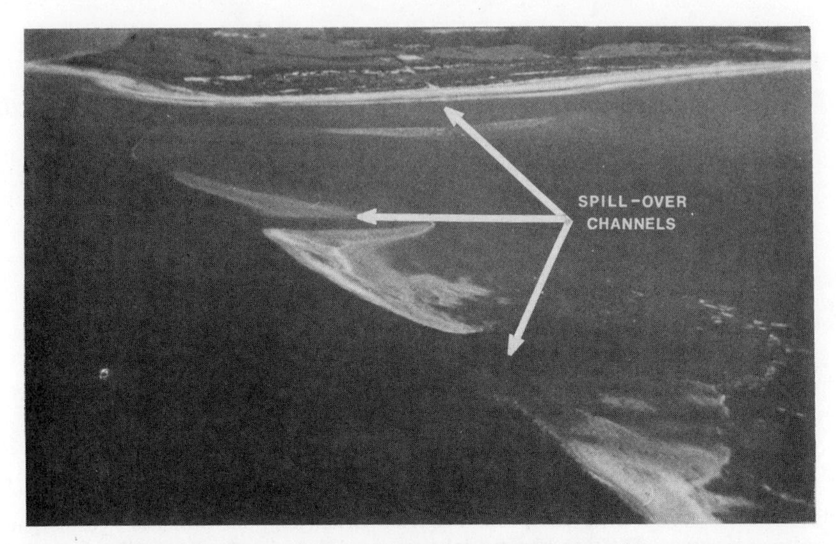

Figure 3. Oblique air photo of the southern tip of Sapelo Island illustrating the spill-over channels that transect the segmented ramp-margin shoals.

sand bodies are elongated in the offshore direction and are broadest at their point of attachment to the barrier. Small spill-over channels that transect these sand bodies at oblique angles ephermeral features, continually changing in size, shape, and orientation. The surface of an attached ramp-margin shoal has gentle undulations with a planar crest (swash platform) (6) that is oriented perpendicular to the shoreline. Observations have shown that planar crests are located below areas experiencing wave interference (6). Internally, swash platforms are generally composed of horizontal laminations rich in heavy minerals. Large cuspate and linguoidal megaripples form an undulating surface along the margins to these platforms. At low water, the slipfaces of the megaripples are generally oriented in a centrifugally distributed pattern away from the inlet, whereas, bedforms in the ephemeral channels and runnels have crests normal to the axis of their respective channels. Sedimentary structures within bedforms generally indicate only one flow direction, and bedforms appear to be completely reworked during high-velocity currents following the reversing of the tide. Cores that penetrated the amplitude of some of these bedforms sometimes illustrated "herring-bone" structures that are commonly produced by reversing tidal currents.

Along the margins of the ramp-to-the-sea, shoals generally absorb most of the wave energy, and wave processes play an important role in sedimentation. A high percentage of clean, horizontally-laminated sand that is rich in heavy minerals is present along the crests of the ramp-margin shoals. Many of the bedforms along the shoal crests and along the seaward side of the shoal also exhibit flood orientations throughout the tidal cycle. These flood orientations are often affected by shoaling waves that transport sand across shoal surfaces in a turbulent flow. Residual currents in the funnel channels adjacent to the ramp-margin shoals also produce on shore transfers of sediment (Fig. 4) (9). Drift-bottle monitoring, continuous current data and bedform analysis has documented the characteristics of residual onshore currents in these "funnel channels" (9).

Landward of the ramp-margin shoals, residual ebb currents build large sand waves that have seaward orientations. Echo soundings along the axis of the ramp-to-the-sea illustrated ebb-oriented sand waves with amplitudes as high as 5 meters. The water depths over the ramp-to-the-sea were generally much greater than the shoreface surrounding the shoals, and waves apparently do not "touch" at the sand bed of the ramp-to-the-sea. In this area, wave-induced currents were not important agents of sedimentation. The amount of clay in the sediment of the ramp-to-the-sea was also greater than the amount of clay observed in the shoals. The textures and structures of the interbedded layers of mud and sand illustrated characteristic changes with depth and location (Fig. 5) (7, 8).

It appears that the "ebb-tidal" deltas have ebb- and flood-dominated zones that are separated by the crest of a peripheral shoal complex (Fig. 5). Landward of these crests, major bedforms illustrate ebb orientations, whereas seaward of these crests, swash bars, swash platforms, and sand waves generally illustrate flood orientations.

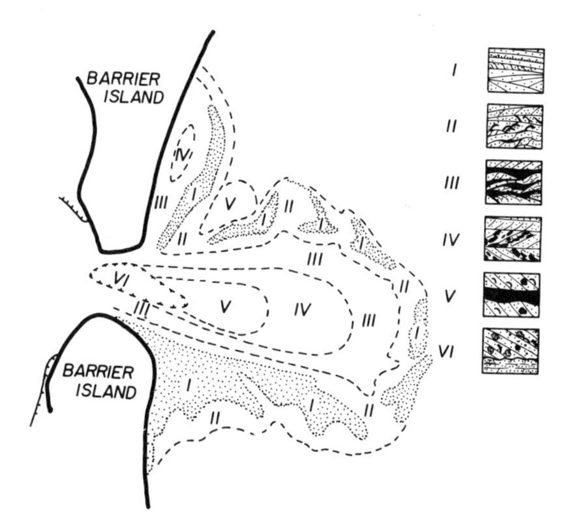

Figure 4. Patterns of residual water flow through tidal inlets. The flow above the ramp-to-the-sea is ebb-dominated, the flow in the funnel channel is flood-dominated. Flow assymetry is interpreted from drift-bottle monitoring, continuous current data, and bedform analysis.

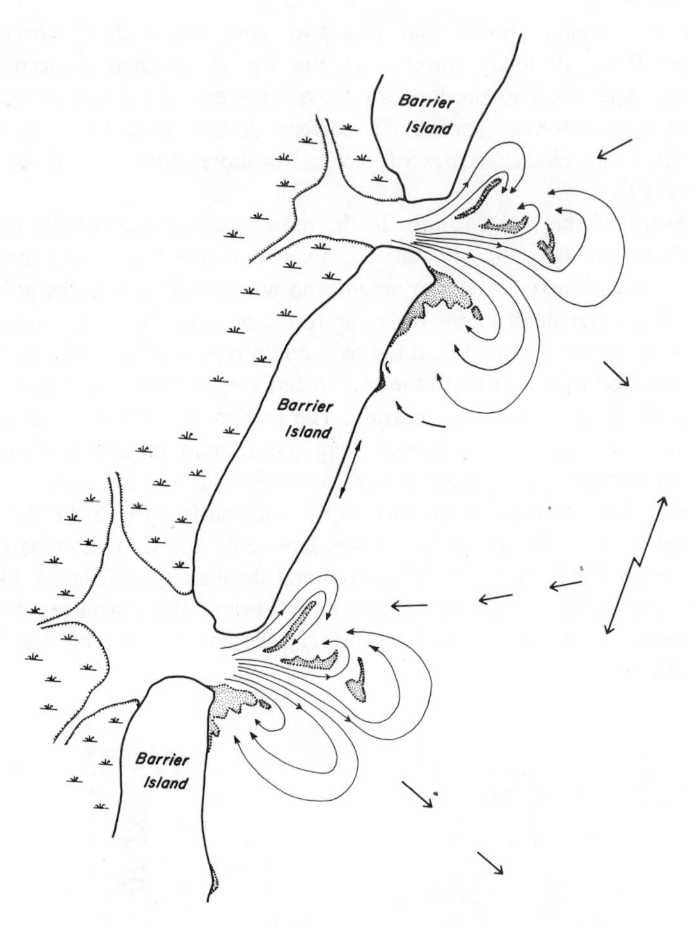

Figure 5. Textural and structural characteristics of the sea bed adjacent to Georgia estuary entrances (Adopted from Oertel (9)). Zone I - clean sand; Zone II - clean sand with mud lenses; Zone III - interbedded sand and mud; Zone IV - mud pebbles in clean sand; Zone V - interbedded mud and poorly sorted sand; Zone VI - coarse sediment "lag" over Miocene sandstone outcrop.

In conclusion, tidal deltas landward of Georgia inlets (flood deltas) generally are centrally located, and tidal deltas seaward of Georgia inlets (ebb deltas) are composed of a peripheral shoal complex that is more extensive in its offshore direction than it is in its alongshore direction (Fig. 1, D). The peripheral shoal complex may be described further as having two ramp-margin shoals and a distal shoal. The overall shape and orientation of these shoals are determined by the interactions between inlet tidal currents and coastal currents. Tidal-current and wave-current processes combine to produce a variety of textures, structures, and bedforms. Tidal-current processes are generally most important in the "shielded" area above the ramp-to-the-sea. Wave-current processes are generally most important at the "unshielded" portions of the tidal deltas around the peripheral shoal complex.

The term *ebb tidal delta* is somewhat misleading, in that sand accumulations are produced by the interactions of ebb- and flood-tidal currents with other currents, and both ebb and flood currents contribute to patterns of deposition. I feel that continued use of this term should refer only to sediment accumulations seaward and adjacent to tidal inlets, and there should be no genetic implications associated with the term.

ACKNOWLEDGMENTS

My initial work on Georgia inlets began with Dr. J. D. Howard in 1970. At that time, work was supported by the U. S. Army Corps of Engineers, contract DACW 72-68-C-0030 and Oceanographic Section, National Science Foundation, NSF grant GA-30565 (J. D. Howard, Principal Investigator). Subsequent work leading to the descriptions in this report have been supported by the Skidaway Institute of Oceanography.

The author is indebted to many people who assisted in various studies related in this paper; Stephen Arpadi, Eric Knutdson (Antioch College, Yellow Springs, Ohio), and Brian Edwards (University of Southern California, Los Angeles, California.)

The air photo, Figure 3, was taken from a private plane owned by Dr. J. D. Howard while the author was observing coastal geomorphology with Dr. Howard.

Thanks are also extended to Ms. Janette Davis and Ms. Barbara McNair who assisted in the preparation of the manuscript.

REFERENCES

1. Bruun, P. F. and Gerritsen, F.
   1959 Natural by-passing of sand at coastal inlets. Reprinted from: Amer. Soc. of Civil Engineers. Waterways and Harbors Division Journal 85 WW4 Paper 2301: 75-107.
2. Bruun, P. F. and Gerritsen, F.
   1961 Stability of coastal inlets. In **Proceedings of Seventh Conference on Coastal Engineering,** p. 386-417. Berkeley, California.
3. Coastal Research Group, Department of Geology, Univ. Massachusetts.
   1969 Coastal environments of Northeastern Massachusetts and New Hampshire. Field Trip Eastern Section Society of Economic Paleontologists and Mineralogists. May 9-11, 458-462.

4. Hoyt, J. H.
1962 High angle beach stratification, Sapelo Island, Georgia. **Jour. Sed. Pet.,** 32: 309-311.

5. King, C. A. M.
1959 **Beaches and Coasts.** E. Arnold (Publ.), London.

6. Oertel, G. F.
1972 Sediment transport on estuary entrance shoals and the formation of swash platforms. **Jour. Sed. Pet.,** 42(4): 858-863.

7. Oertel, G. F.
1973 Examination of textures and structures of mud in layered sediments at the entrance of a Georgia tidal inlet. **Jour. Sed. Pet.** 43(1): 33-41.

8. Oertel, G. F.
1973 A sedimentary framework of the substrate adjacent to Georgia tidal inlets. In **The Neogene of Georgia,** Guidebook for the Eighth Annual Field Trip of the Georgia Geological Society.

9. Oertel, G. F.
1973 Residual currents and sediment exchange between estuarine margins and the inner shelf, southeast coast of the United States. In **Symposium Volume for the International Symposium on interrelationships of estuarine and continental shelf sedimentation,** Bordeaux, France.

10. Price, W. A.
1963 Patterns of flow and channeling in tidal inlets. Abstract **Amer. Assoc. Pet. Geo. Bull.,** 47(2): 366-367.

11. Scruton, P. C.
1956 Oceanography of Mississippi Delta sedimentary environments. **Bull. of the Amer. Assoc. of Pet. Geol.,** 40(12): 2864-2952.

12. Todd, T. W.
1968 Dynamic diversion - influence of longshore current tidal flow interaction on Chenier and Barrier Island Plains. **Jour. Sed. Pet.,** 38: 734-746.

HYDRODYNAMICS AND TIDAL DELTAS

OF NORTH INLET, SOUTH CAROLINA

Robert J. Finley[1]

ABSTRACT

A one-year study at North Inlet, South Carolina, has documented a pattern of changing bathymetry, eroding shorelines, and generally westward migration of the entire system. The ebb-tidal delta is large with prominent and changeable channel-margin linear bars. Tidal currents have maintained an equilibrium cross-section despite infilling of the main ebb channel with sediment eroded from updrift beaches. The only beaches not eroding are those protected from northeast waves by the ebb-tidal delta.

The flood-tidal delta is much smaller than those of New England estuaries and is strongly influenced by wave swash at high tide. Only a few poorly developed sand waves are present, and the area of bedforms that remains flood-oriented at low tide is small. The seaward projecting ebb spit divides the flow between two main tidal creeks. The delta and flanking arcuate sand ridges are migrating over the adjacent *Spartina* marsh. Measurements of the beach profile provide additional evidence of a transgressive shoreline.

Hydrographic measurements over fourteen complete tidal cycles reveal generally ebb-dominated channels and flood-dominated subtidal flats. Maximum velocities measured are near 80 cm/sec, and time-velocity and current-velocity asymmetry are present.

INTRODUCTION

North Inlet is a natural tidal inlet in Georgetown County, South Carolina, on

1. Coastal Research Division, Department of Geology, University of South Carolina, Columbia, South Carolina 29208.

property held by the Belle W. Baruch Foundation. The former plantation is being preserved in a relatively natural state and, as such, the inlet, adjacent beaches, and intertidal areas provide an undisturbed area for geologic investigation (Fig. 1). The tidal regime at North Inlet is semidiurnal with a mean range of 1.37 m, or 4.5 feet. The mean spring range is 1.62 m, and a diurnal inequality is present which averages 37 cm (6).

There is no permanent fresh-water influx into the North Inlet system and summer salinities range from 32.5°/oo to 34°/oo. Summer vertical variation from hourly readings taken at one-meter intervals of depth rarely exceeded 0.8°/oo at any one station, except where surface water was diluted by heavy

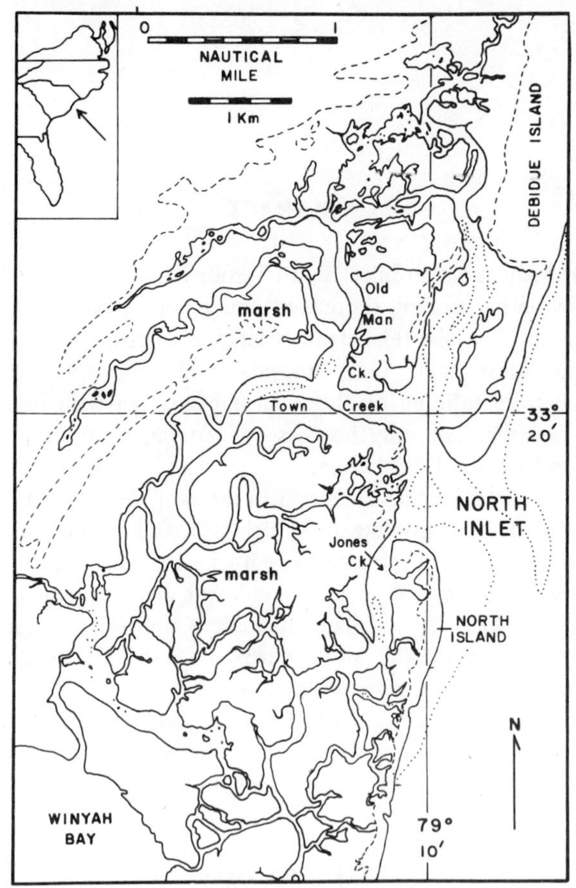

Figure 1. North Inlet is midway between Charleston, South Carolina, and the South Carolina-North Carolina border. The dashed line on the northwest side of the area represents the marshforest boundary and is the approximate Holocene-Pleistocene contact (3). Enclosed arcuate area east of the contact represents a probable Late Holocene beach ridge.

rainfall. Fresh water, however, does enter Winyah Bay (Fig. 1) from the Pee Dee and Waccamaw rivers, whose combined average discharge is 286.4 c.m.s. (5). Under conditions of high spring run-off, as occurred in March and April, 1973, brackish water does reach North Inlet through the tidal creek system. A salinity minimum of 10.2°/oo was recorded in mid-April, 1973. Unusually strong southeast winds also result in the ebb flow of Bay water through North Inlet.

Prevailing wind directions (Fig. 2) are from the southwest (241°-270°) and the southeast (151°-180°), associated with the flow of warm-weather, high-pressure systems. The dominant wind is an onshore northeast wind (31°-60°) associated with the passage of cyclonic low-pressure systems. The oblique approach of northeast wind-generated waves results in active drift of littoral materials to the south, where much of the sand is incorporated into the ebb-tidal delta.

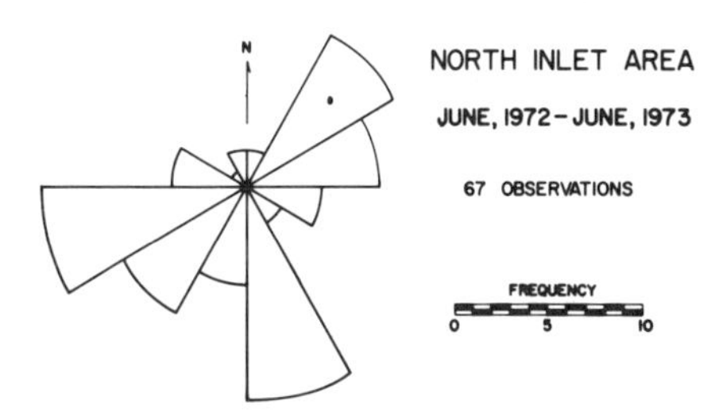

Figure 2. Wind observations were taken at irregular intervals during field work, using hand-held instruments. Southwest and southeast winds prevail while the northeast wind is dominant.

The average mean grain-size of seventeen mid-tide level beach samples updrift of the inlet in August, 1972, is 2.05ϕ, just barely into the fine-sand range. A similar set, collected in December, 1972, shows a mean of 1.93ϕ, an insignificant difference compared to the summer samples. Standard deviations of 0.35ϕ and 0.37ϕ, respectively, indicate a very-well-sorted sediment. Analysis was by settling tube (1), and statistics calculated according to the methods of Folk and Ward (4).

RECENT MORPHOLOGICAL HISTORY

Original U. S. Coast and Geodetic Survey hydrographic sheets dated 1878, 1925, and 1964, are available. The earliest survey (H-1419) shows little of the exact configuration of the inlet mouth, since the offshore bathymetry is emphasized. The survey of 1925 (H-4521) forms the basis for the first edition of Chart 787, which was not published until 1938 (Fig. 3). When surveyed, the

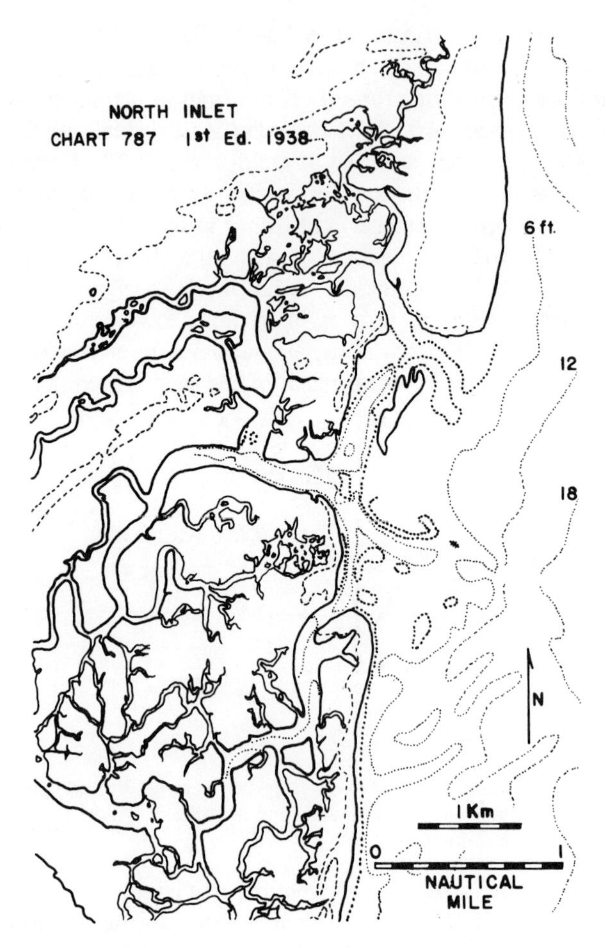

Figure 3. Note the absence of the southward projecting spit on Debidue Island when surveyed in 1925.

inlet was about 2.8 km wide and generally less than 2 m deep. By 1939, however, spit growth had commenced, as shown by aerial photographs, and the area began to assume the configuration charted in 1964, which chart forms the basis for the third edition (Fig. 4). Note that, in the interim between surveys, Debidue Island and North Island (Fig. 1) had narrowed and creek drainage had changed as a result of westward migration of the shoreline. The Debidue Island spit had extended some 1600 meters to the south (2).

The most recent (1970) edition of Chart 787 shows minor revisions of the 1964 survey and forms the basis of Figure 1, which approximates the present configuration of the inlet.

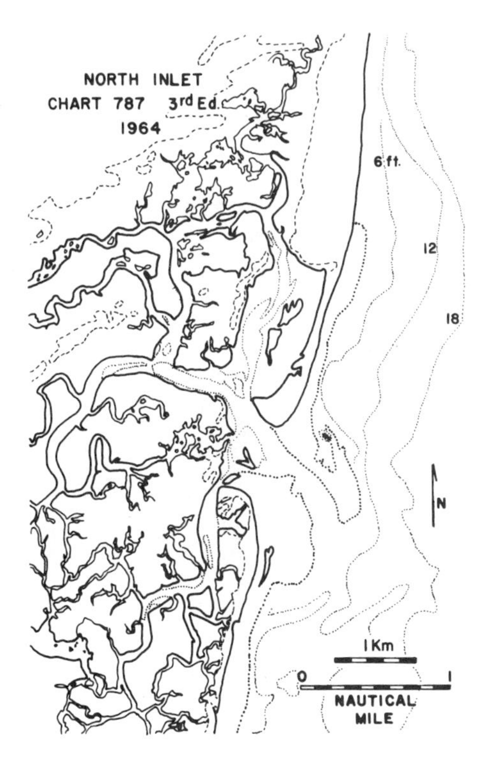

Figure 4. Compare the changes in tidal-creek drainage with the previous survey, especially directly opposite the inlet.

EBB-TIDAL DELTA

The ebb-tidal delta at North Inlet (Fig. 5, A) includes channel-margin linear bars on either side of the main ebb channel, marginal flood channels, swash bars, and a terminal lobe. These components are consistent with those comprising the ebb-tidal delta model developed by members of the Coastal Research Division, Geology Department, University of South Carolina (Hayes, 1973, this volume), largely as a result of field work conducted in New England.

The main ebb channel at North Inlet is 3.5 m below MLW through most of its length and drops to 6.5 m below MLW at the margin of the terminal lobe. Comparison of aerial photographs taken in June, 1972, and April, 1973, shows that the northern channel-margin bar has increased in sand volume and consequently areal extent. It has migrated to the south and infilled part of the main ebb channel, as is shown on the inlet-throat profiles (Fig. 5, B) and discussed below. Much of the increased sand volume was derived from the

Figure 5, A. The ebb-tidal delta components shown here are the channel-margin linear bars (A) on either side of the main ebb channel (large arrow), marginal flood channels (small arrows) and swash bars (B).

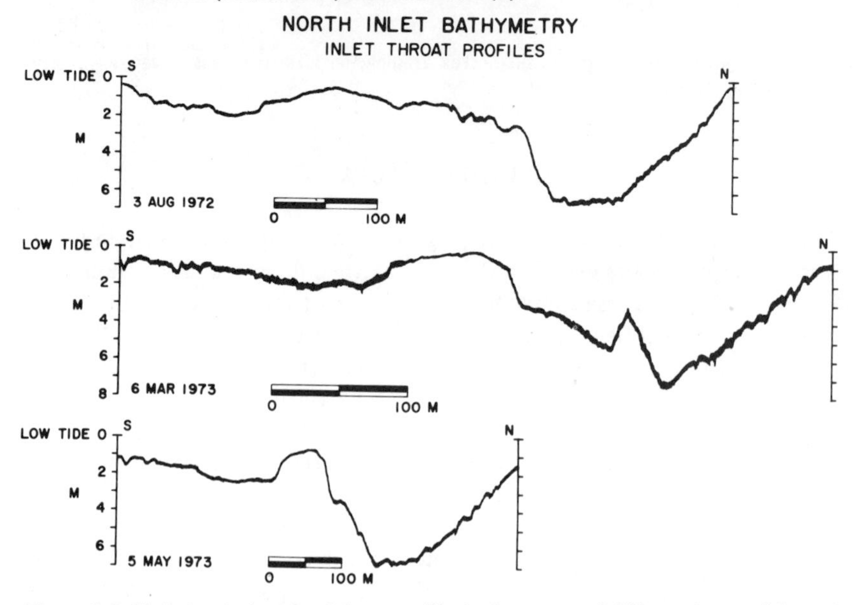

Figure 5, B. Variation in length of these profiles is due to use of different boats with varying minimum speeds. Profile location is shown in Figure 6. Central high on the 5 May 1973 profile is the ebb spit projecting from the flood-tidal delta.

extensive erosion of the Debidue Island foredune ridge. At beach profiling locality DBI-10, located north of the inlet (Fig. 6), 7.3 m of foredune ridge was lost in a two-month period. Most of this loss was caused by the severe northeaster of 9-11 February 1973, which resulted in dune scarps of up to 2 m in height.

As a result of this extra sediment moving south, the inlet-throat profiles (Fig. 5, B), located between the northern channel-margin bar and the beach on the south side of the inlet (Fig. 6), have changed through time. The changes between the profile of 3 August 1972 and 6 March 1973 are probably due to tidal-current scour leaving the ridge shown on the latter profile as an erosional remnant. This erosion represents the natural system's tendency to maintain a balance between

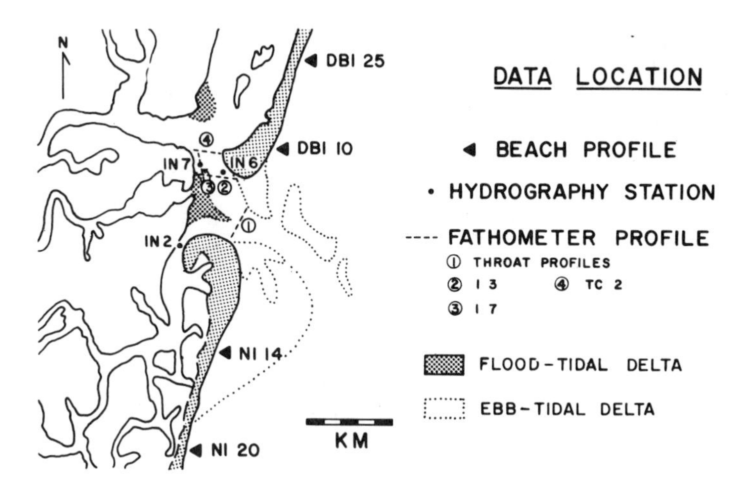

Figure 6. The less dense stippled pattern shows the areas of beach and dune. All data referred to in this paper are shown here.

the cross-sectional flow area and the volume of the tidal prism. The throat section of 5 May 1973 does not show the ridge, indicating further erosion and return to a more stable configuration.

FLOOD-TIDAL DELTA

The main flood-tidal delta which faces North Inlet is small compared to those of New England estuaries. Its major components (Fig. 7) include ebb shields, a flood ramp, and a seaward projecting ebb spit, which is evident on the inlet-throat profiles (Fig. 5, B). The landward margin of the flood-tidal delta has migrated steadily to the west during field observations, the migration distance being on the order of 42 m in the period July, 1972, through March, 1973.

The flood-tidal delta is limited in the variety and size of bedforms present, with ripples of 2 to 5 cm in height and 17 to 50 cm in wavelength predominating. Mapped at a spring low tide (Fig. 8), only four poorly developed sand waves were present. The orientation of these bedforms and the megaripples at the base of the ebb spit is the result of Jones Creek ebb flow prior to subaerial exposure. A plot of slip-face azimuths (Fig. 8) shows the dominant ebb orientation of all bedforms, with the few flood-oriented ripples preserved in the central higher part of the delta, protected by the ebb shield.

A small additional flood-tidal delta is located at the junction of Town and Debidue creeks. Morphologically it consists of a flood ramp, an ebb spill-over lobe, an ebb shield, and a small trailing ebb-spit. Much of the late ebb flow over the spill-over lobe is a result of drainage from a small tidal creek that has to cross

Figure 7. The components of the main flood-tidal delta include ebb shields (A), an ebb spit (B) and a flood ramp (large arrow). An additional flood-tidal delta is located in the upper left (C) at the junction of Town and Debidue creeks. Wash-over berms are indicated by small arrows, which also show the generally westward migration direction. View is toward the north.

the delta to reach Debidue Creek.

The flood-tidal delta is flanked to the west (Fig. 7) by two arcuate sand ridges, or wash-over berms, which are also migrating westward and steadily

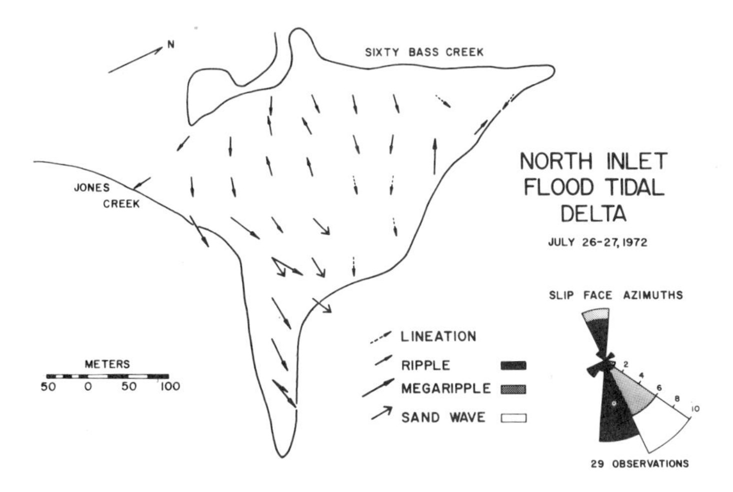

Figure 8. Main flood-tidal delta map with bedform orientation sampled on a 50 by 50 m grid. Lineations consisted of aligned shell debris; bedform name is based on wavelength: ripple, up to 60 cm; megaripple, 60 cm - 6 m; and sand wave, greater than 6 m.

inundating *Spartina alterniflora* marsh. The northern wash-over berm is the lower of the two and has an active slip-face on its western margin. Spring tides in combination with an onshore wind are sufficient to cause wash-over at high tide; a major storm is not required. *Spartina* peat is gradually being uncovered on the seaward side of each ridge and some recolonization by the grass is taking place.

CHANNEL CONFIGURATION AND TIDAL CURRENTS

The main flood-tidal delta (Fig. 7) which faces North Inlet divides the tidal flow between Jones Creek to the south and the channel leading to Debidue and Town creeks on the north. Hydrographic measurements have been made over fourteen full tidal cycles at eleven stations within the inlet area. These include readings of current velocity and direction at least hourly and at one-meter intervals of depth, utilizing either a Savonius rotor or ducted impellor type of meter. The fathometer profiles are run with a Bludworth Marine Depth Recorder.

Velocity station JC-2 (Fig. 9) is typical of those run in Jones Creek (location shown on Fig. 6) and indicates that, for the neap tidal cycle measured, current-velocity asymmetry was negligible. Time-velocity asymmetry is present with flood peak velocities occurring five or more hours after the time of low water. This may be attributed to some residual ebb flow remaining after the time of low water.

The main tidal channel is located northeast of the flood-tidal delta and is

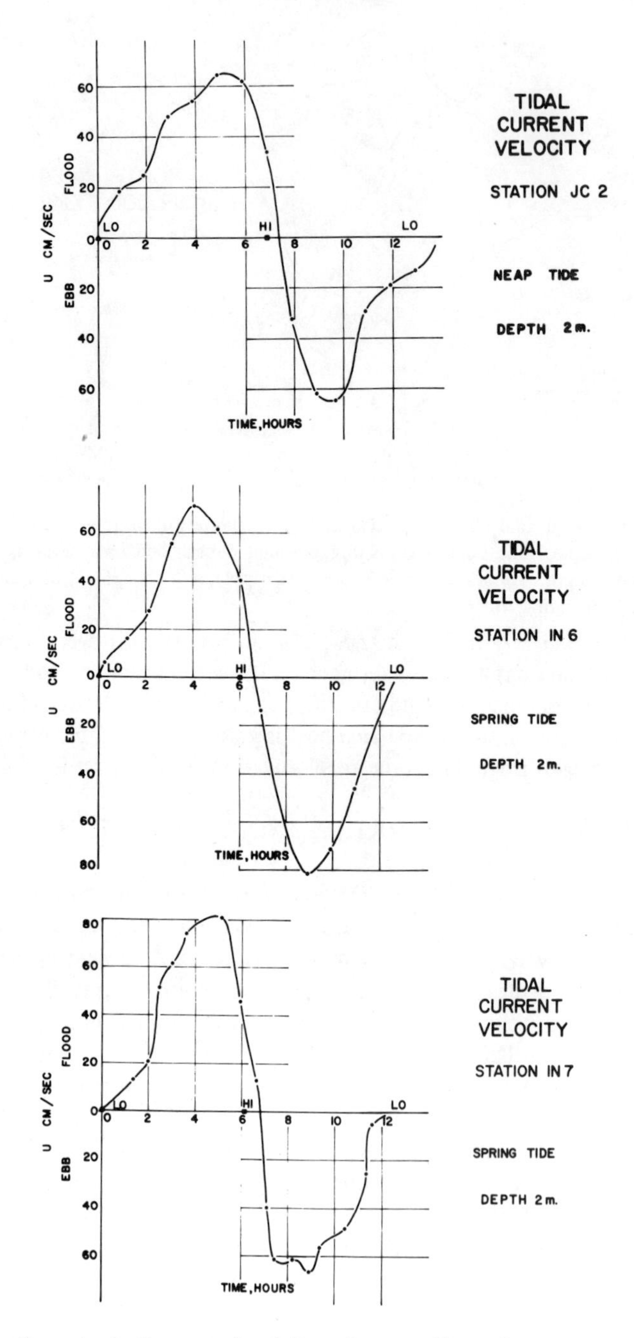

Figure 9. Current velocity curves for stations shown on Figure 6.

shown in cross-section on profiles I-3 and TC-2 (Fig. 10; location on Fig. 6). Closer to the inlet, the channel is broad (I-3) and the north end of the flood-tidal delta shows up as two distinct ridges, which may be an ebb-oriented subtidal spit. At profile TC-2 the main channel has narrowed and deepened slightly with a flatter subtidal area present to the west. A fathometer profile over this location, I-7 (Fig. 10), shows a series of sand waves and megaripples whose crests are aligned in a northeast direction and whose orientation changes with tidal flow. These bedforms are symmetrical to slightly flood-oriented at high water and ebb-oriented at low water; they are large enough to cause flow-separation zones which are evident at the water surface.

Measurements of current velocity in the main ebb channel at IN-6 and over the shallow subtidal area at IN-7 (Fig. 9; location on Fig. 6) show a small amount of asymmetry characteristic of the ebb-dominated deeper chanel vs. the flood-dominated shallow subtidal flat. The ebb-dominance at IN-6 for the spring tide measured is about 12 cm/sec. Note that peak flood velocities are not reached until two-thirds flood, probably the result of a residual ebb current in the channel which the flood must first overcome as the tide turns. At IN-7, over the shallow subtidal area, flood velocities exceed ebb by about 15 cm/sec; however, the late flood peak indicates some ebb-inhibiting effect in this area also.

IN-7 is located on fathometer profile I-7, which shows (Fig. 10) symmetrical to slightly flood-oriented bedforms at high water and ebb-oriented forms at low water. With the flood currents slightly dominant, it was expected that the flood bedform orientation would be the best developed. The fact that some ebb orientation seems to be retained even at high water is not fully understood as yet, but may be related to the detailed flow configuration immediately over the bedforms.

BEACH PROFILES

Additional evidence for the transgressive nature of the shoreline comes from successive observations of beach profiles, taken utilizing the Emery method (3). Significant erosion prior to the start of field studies was indicated by the steep foredune ridge shown on summer profiles at locality DBI-10 (Fig. 11). Note the landward migration of a ridge during the period 25 June - 22 July 1972. At locality DBI-25 the changes resulting from the severe northeaster of 9-11 February 1973 are evident on the post-storm profile. Much of this eroded beach sediment was incorporated into the ebb-tidal delta and contributed to the changes in the throat-profile sections discussed above.

Beaches immediately south of the inlet, as at NI-14 (Fig. 12), showed little overall change because they are protected against storm waves approaching from the northeast by the ebb-tidal delta. In addition, sand accumulation has taken

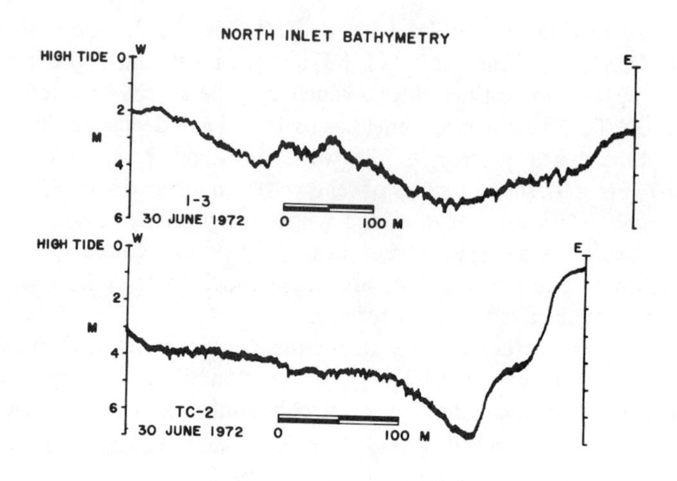

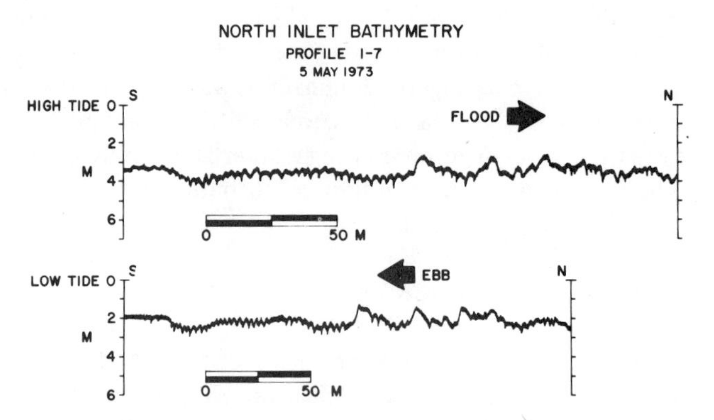

Figure 10. Bathymetry of the main tidal channel between Debidue Island and the main flood-tidal delta. Variation in length at the two I-7 profiles is a result of varying boat speed.

place as a result of a drift reversal toward the inlet caused by wave refraction around the ebb-tidal delta. Comparisons of Figures 3 and 4 show the increase in sand volume on the south side of the inlet, once North Inlet reached the position charted in 1964. Only after the Debidue Island spit grew to approximately its present position did the ebb-tidal delta begin to influence the wave-refraction pattern.

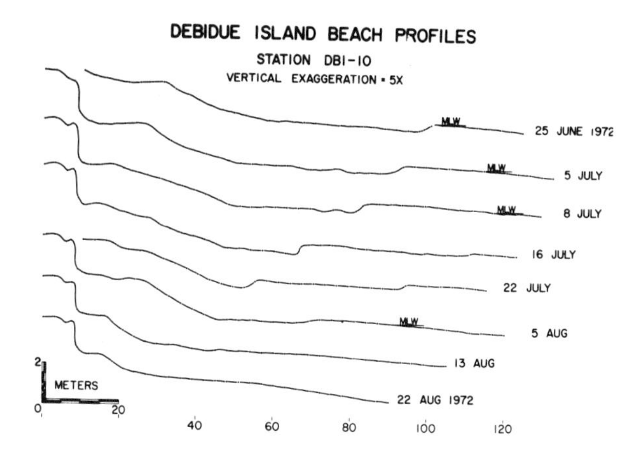

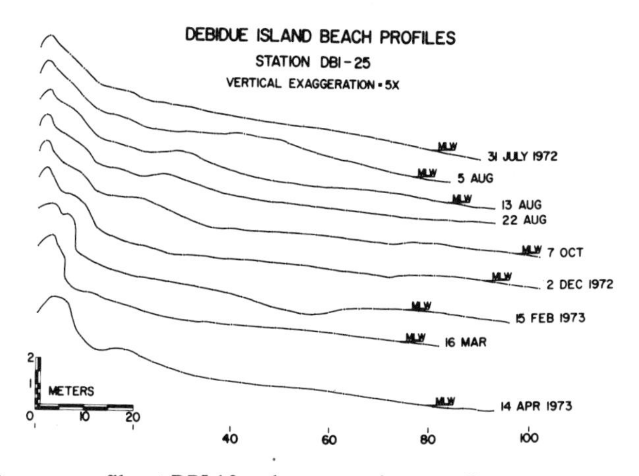

Figure 11. Summer profiles at DBI-10 and summer-winter profiles at DBI-25.

South of NI-14, and beyond the zone of protection and of wave refraction influenced by the ebb-tidal delta, the beach narrows and returns to an erosional morphology. At profile locality NI-20 (Fig. 12) the winter northeast storms have affected the seaward margin of the foredunes ridge and destroyed the berm which was present the previous summer. About 100 m south of NI-20, a *Spartina alterniflora* peat is exposed on the foreshore which has been radiocarbon-dated at 920 $\pm$ 120 years B. P. This exposure is the result of westward migration over the active marsh of the sand prism comprising the beach. Measurement of the distance from the peat margin to active salt-marsh

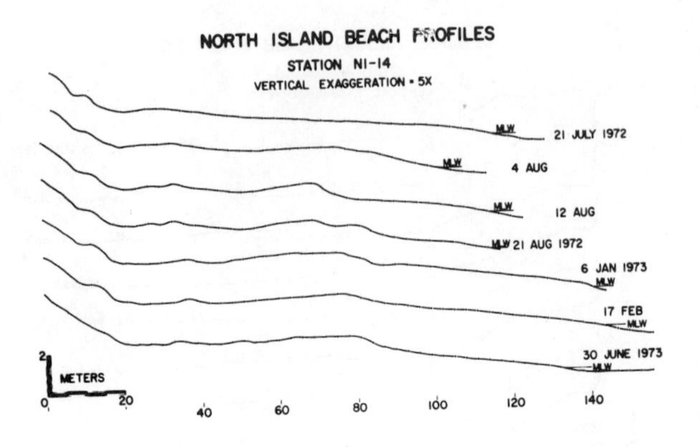

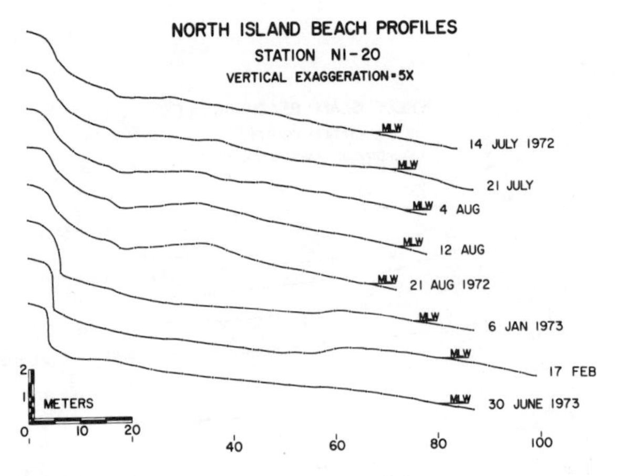

Figure 12. 1972-73 profiles at two North Island localities.

cordgrass gives a minimum recession rate on the order of 400 m/920 years, or roughly 0.5 m per year. Comparison of the 1925 and 1964 hydrographic surveys gives a recession rate on the order of 6m per year.

SUMMARY

The highly transgressive nature of the shoreline at North Inlet, together with the extensive areas of actively growing *Spartina alterniflora* low marsh, are factors tending to minimize the size of the flood-tidal delta. The main flood-tidal delta is limited to an area between the two main channels of tidal flow and the active marsh, where a large proportion of the delta's migration is due to incoming wave swash at high tides. In a sense, then, this tidal delta is a combination of a swash and tidal-current-generated feature. It is smaller than those in New England estuaries and contains a much more limited set of bedforms. The ebb-tidal delta, however, is large and contains a greater volume of sand than, for example, New England shoals for an inlet of comparable size.

ACKNOWLEDGMENTS

This study was sponsored by the Coastal Engineering Research Center, U. S. Army Corps of Engineers, under contract DACW72-72-C-0032. Dr. Miles O. Hayes provided overall direction for the project. The writer is thankful to his associates at the Coastal Research Division, who gave generously of their time in field assistance, discussion of problems associated with the project, and preparation of the manuscript.

REFERENCES

1. Anan, F. S.
   1972 Hydraulic Equivalent Sediment Analyzer (HESA), Tech. Rept. No. 3-CRD, Coastal Research Center, Univ. Massachusetts.
2. Buffington, R. M. and Randall, J. P.
   1964 Descriptive report to accompany Hydrographic Survey H-8838, U. S. Dept. of Commerce, E.S.S.A.
3. Emery, K. O.
   1961 A simple method of measuring beach profiles. **Limnol. Oceanogr.**, 6: 90-93.
4. Folk, R. L. and Ward, W. C.
   1957 Brazos River Bar: a study in the significance of grain size Parameters. **Jour. Sed. Pet.**, 27(1): 3-27.
5. Thom, B. G.
   1967 Coastal and fluvial landforms: Horry and Marion counties, South Carolina. Louisiana State Univ. Press, Coastal Studies Series No. 19, 75 p.
6. U. S. Department of Commerce
   1972 Tide Tables, 1973, NOAA, 288 p.

INTERTIDAL SAND BARS IN COBEQUID BAY (BAY OF FUNDY)

R. W. Dalrymple, R. J. Knight, and G. V. Middleton[1]

ABSTRACT

The largest tides in the world, with an average perigee spring range of 15.4 m and a maximum measured range of 16.3 m, occur at Burntcoat Head, on the south shore of Cobequid Bay, The tides generate currents with speeds of 1 to 1.5 m/sec in the centre of Cobequid Bay. These currents, and waves approaching mainly from the west, have reworked sand derived from glaciofluvial outwash and cliff erosion of bedrock into a major sand body which occupies the eastern part of the Bay. The sand body is 30 km long and 6 to 25 m thick. Most of it is subtidal, but close to shore and at the east end of the Bay there are several intertidal sand bars, which reach 6 m or more in relief and thickness, and have dimensions up to 4 kms in length. Maximum speeds of tidal currents over the bar surfaces generally range from 0.5 to 1 m/sec; the strength and direction are largely determined by shore and bar topography. In many areas either flood or ebb currents dominate, producing strongly asymmetrical patterns of sand dispersal. Bars are covered by sand waves (with wavelengths of the order of 30 m; these are not found on all bars), megaripples (with wavelengths in the range 1.5 to 5 m) and ripples. Many megaripples can be observed on depth recordings to reverse their orientation during each ebb or flood. On Selmah Bar, sand waves are flood-oriented and covered at low tide by ebb-oriented megaripples. The sand waves are found only on the south side of the bar in a low area (not affected by wave action) that is strongly dominated by flood currents, and they migrate about 20 cms per tidal cycle.

INTRODUCTION

The Bay of Fundy is famous for its tides, which are reputed to have the

1. Department of Geology, McMaster University, Hamilton, Ontario, L8S 4M1, Canada.

largest range of any in the world. Until 1970, the only permanent tide gauge was at St. John, New Brunswick, where a continuous record is available dating back to 1896. Temporary gauges have been set up for short periods in other parts of the Bay, and it was established by Dawson (8) that the greatest tides were to be found in Cobequid Bay (the eastern arm of the Bay). The highest, calm-weather tidal range actually measured is 16.3 m (8), from the western side of Burntcoat Head. Data from more recent surveys are summarized in the report prepared by the Atlantic Tidal Power Engineering and Management Committee (2, Appendix I).

The mean tidal range increases from about 3.5 m at the mouth of the Bay to 11.7 m at Burntcoat Head (Fig. 1). The tide is strongly semidiurnal and of the anomalistic type. Dawson (8) recorded a mean spring range of 12.3 m at lunar apogee and 15.4 m at lunar perigee. The diurnal inequality in relatively small, with a maximum of about 0.8 m at Burntcoat Head.

The cause of the very large tidal range is generally stated to be resonant amplification of the semidiurnal tidal component. However, studies by Redfield (22), Harleman (13), Rao (21) and Yuen (29), suggest that the resonant period of the Bay of Fundy itself is by no means coincident with the semidiurnal period of 12.42 hours. These studies suggest that the large tidal range is due to

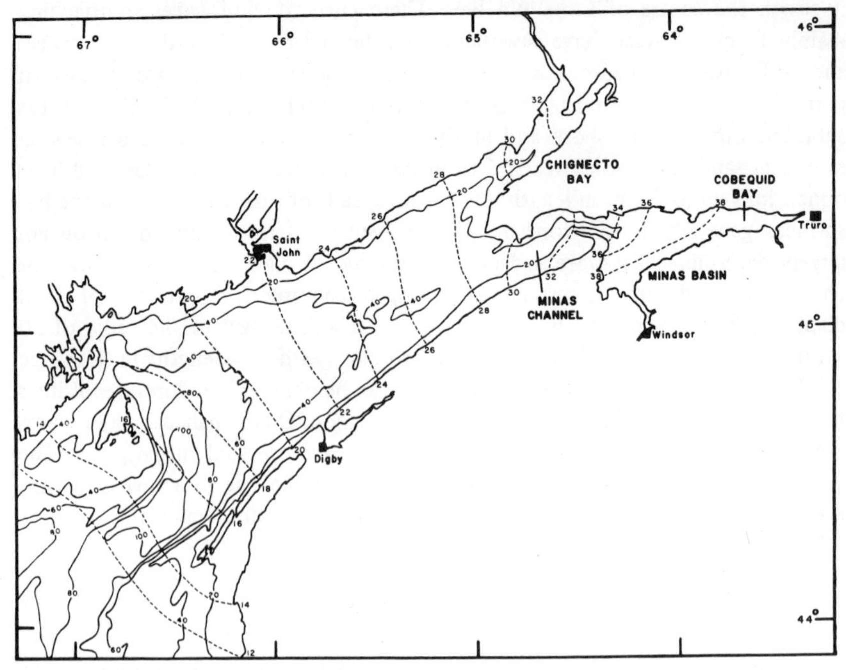

Figure 1: Bay of Fundy, showing major place names, bathymetry (solid lines) in fathoms (from Pelletier and McMullen, 20), and mean tidal range (dashed line) in feet (from Canadian Hydrographic Survey).

two-stage amplification. A relatively large tide is present at the mouth of the Bay (Fig. 1) caused by amplification of the oceanic tide as it moves across the broad continental shelf southeast of Nova Scotia. This tide is further amplified (by about two and a half times) within the Bay itself to give the very large tidal range. Duff (9) and Garrett (10) have suggested that the large tides may be considered to be due to resonance if the system is enlarged to include not only the Bay but also adjacent parts of the continental shelf.

The cause of the large tides is not without geological significance. Because it has been argued that, if the large tides are caused by resonance, they must be sensitive to changes in the present dimensions of the Bay. These dimensions have resulted from a combination of rising sea level and isostatic rebound, and have been achieved only during the last 4000 years. Independent stratigraphic and geomorphic evidence presented by Swift and Borns (24) and by Grant (12) appears to confirm that the tides have grown very substantially in amplitude during this same period of time.

The general distribution of sediments in the Bay of Fundy has been discussed by Swift et al. (26, 27) and by Pelletier and McMullen (20). Most of the detailed studies of intertidal sediments have been made in the Minas Basin and Cobequid Bay, a clearly defined subdivision of the Bay of Fundy, and only this area will be described in the present paper.

MINAS BASIN AND COBEQUID BAY

Tidal Currents

The Minas Basin is the eastern arm at the head of the Bay of Fundy (Fig. 1). Cobequid Bay refers to the shallower, eastern part of the Minas Basin. Depths in the Minas Passage, which connects the Minas Basin with the main Bay, reach 120 m, but depths are generally not more than 10 m below low water in the Basin itself (Fig. 2).

The major pattern of tidal currents in the Minas Basin was established on the basis of high-altitude air photography by Cameron (5) and on the basis of current and ice surveys by ATPEMC (2), but it is still not clear whether or not there is a predominant tidal drift in Cobequid Bay. Swift and McMullen (25) and Pelletier and McMullen (20) suggested an anticlockwise drift due to coriolis forces.

Speeds of tidal currents are as high as 5.5 m/sec in the Minas Passage, but do not generally exceed 2 m/sec in the Basin and decrease to 0.5 to 1.5 m/sec over the intertidal sand bars (2, 11, 16, 18, 25). For a large sand bar in the centre of Cobequid Bay, Swift and McMullen (25) found that the bar surface was flood-dominated and the channels between bars were ebb-dominated. Large

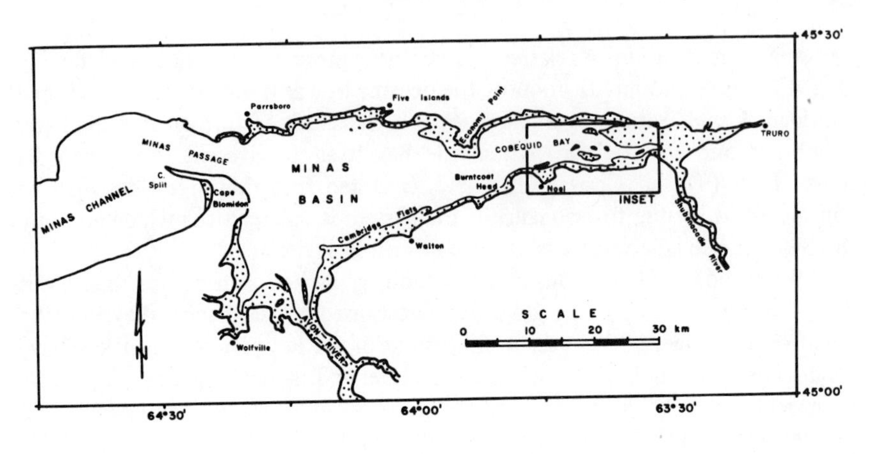

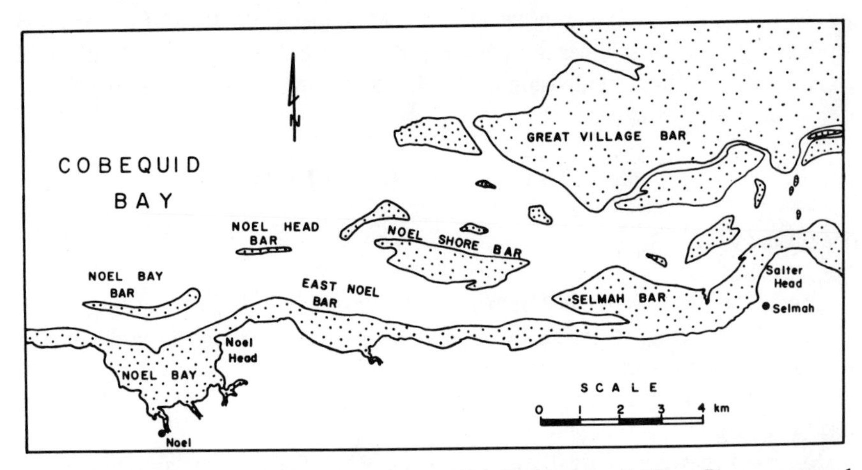

Figure 2: Minas Basin – Cobequid Bay, Bay of Fundy, Nova Scotia; Top: Place names and main intertidal areas. Bottom: Noel – Selmah area, Cobequid Bay, showing the distribution of intertidal zone (from air photos taken on July 22, 1972).

areas of Selmah Bar (for locations, see Fig. 2) are also flood-dominated, but Klein (16) has presented data that suggest that the sand bars south of Economy Point are mainly ebb-dominated. Locally however, bar topography is the major control on the relative strengths of flood and ebb currents.

Ice and Waves

The Minas Basin experiences a temperate maritime climate. Mean annual precipitation is about 110 cms, and mean annual snowfall is about 180 cms. July is the warmest month, with a mean daily average temperature of 20°C. January is the coldest month, with a mean daily average temperature of -6°C. Westerly winds prevail throughout most of the seasons, but major shifts in wind direction during a single tidal cycle are not uncommon. The most severe storms occur from June to November.

ATPEMC (2) has recorded in detail the results of ice surveys made during the winters of 1967 and 1968. From January to March or early April, much of the Minas Basin is covered by floating ice, and there are accumulations of shore-fast ice along the high-tide level. Ice grounding along the shore or on intertidal areas picks up great quantities of sediment, including large boulders (14), and the ice and sediment are redistributed by tidal currents and wind. Thus even muddy tidal flats may contain considerable amounts of gravel and sand. Large boulders or fragments of tidal marsh on the surface of the sand bars were presumably also moved by floating ice.

Observations along the south shore of Cobequid Bay were made by Dalrymple and Knight from February 19 to 24, 1973. They observed that the surface of Selmah Bar was frozen to a depth of several centimeters. In some restricted areas, the frozen "crust" had been breached and rapid local erosion by undercutting was taking place. Over most of the bar surface, however, the sand was essentially immobilized, preventing bedform migration or sediment movement. This phenomenon, together with the damping of wave action by floating ice and protection of the shore by shore-fast ice, may considerably reduce the sedimentological and geomorphological effectiveness of waves during the winter months.

Wave action within the Minas Basin is restricted by the relatively small size of the basin and by its isolation from the main Bay of Fundy. ATPEMC estimated that wave heights in the Basin reach 6 m once every 20 years. Qualitative observations of summer wave conditions indicate that wave heights do not generally exceed 2 m.

Wave conditions in Cobequid Bay may be expected to be most effective for waves moving from the west, because this direction corresponds both to the predominant wind direction and to the direction of maximum fetch. Cliff erosion by waves is obviously effective (see discussion below) and some of the major sand "ridges" observed on sand bars (especially the bars at Economy Point) may be formed or strongly modified by waves. Wave action may also contribute significantly to the accumulation of sand in the eastern part of Cobequid Bay.

Sediment Sources

The rivers entering the Minas Basin are not presently contributing large volumes of water or sediment.

Fresh water contributed by the rivers is sufficient to lower the salinity in the Minas Basin and Cobequid Bay to 26 to 30 parts per thousand. However, the very large tidal range results in the flushing of large volumes of water into and out of the Bay, so that the fresh and saline waters are well mixed. Profiles in the central parts of Cobequid Bay and over the major sand bars show no evidence for any major temperature or salinity stratification (2, Table A3-10; 16).

Suspended-sediment concentrations in the rivers are about 0.02 mg/1, generally less than in the estuaries and in open water. In the Minas Basin concentrations are generally in the range of 20 to 200 mg/1. The rivers undoubtedly contribute some sand and gravel to the Bay, as shown, for example, by the small gravelly deltas that have built out into the intertidal areas at the mouths of some of the steeper creeks, especially on the north shore (19, 23). Most of the major rivers, however, do not seem to be supplying much bed material at present.

Probably most of the sand in Minas Basin-Cobequid Bay has been derived from the erosion of Pleistocene, Triassic, and Carboniferous deposits in and around the Basin. The Pleistocene is composed of unconsolidated till and outwash, generally less than 10 m thick. The Triassic consists of basalt, sandstone, siltstone, and shale. The basalts are resistant and form the Cape Blomidon-Cape Split peninsula and some of the islands and promontories on the north shore. The Triassic sediments, which are widely exposed throughout the Basin, are friable and easily eroded. Carboniferous shales and fine-grained sandstones are well indurated but only moderately resistant to erosion because of extensive joints and fractures. They appear to yield mainly platy or blade-shaped gravel, rather than sand.

These source deposits are eroded and redistributed by tidal currents and wave action from the Basin floor and sea cliffs. The importance of sea-cliff erosion as a source of sand is confirmed by (a) the mineralogical composition (and reddish colour) of the sand, (b) the broad wavecut platform present along both the north and south shore of Cobequid Bay, and (c) the observed rapid rate of erosion of the present cliffs, which may reach 2 m per year (6).

Recent Sediment Distribution

Pelletier and McMullen (20) have published maps showing the general distribution of bottom sediments in the Minas Basin. Most of the sediment is sand or sand and gravel. Continuous refraction profiles (2, 25; observations of Knight, 1973) indicate that the recent sediments occur as local accumulations over a roughly flat surface underlain by Pleistocene deposits and bedrock. For example, in Cobequid Bay there is a major subtidal sand body 6 to 25 m thick, and some 10 by 30 kms in area, parts of which become exposed as intertidal sand bars.

Most of the bars that have been studied in detail are close to the shore and to some extent shielded by headlands, islands, or rock ledges from the main force of the tidal currents. Examples included the megarippled sand bars at Five Islands and Economy Point (16), Selmah Bar, East Noel Bar, Noel Bay Bar (17, 18), and Walton Bar and Cambridge Flats (25; see Fig. 2 for locations). Other bars are found within the mouths of estuaries, such as the Avon River estuary. All these bars appear to be relatively stable in position, based on five years observation in the field and the record of air photographs taken in 1963, 1947 and 1938 (for examples, see 16).

There are a number of other bars exposed some distance from the shore. These include Noel Head Bar and several others in the inner parts of Cobequid Bay, east of Noel Head. One of these, "Betsy Bob" Bar, was described in some detail by Swift and McMullen (25). Sand bars exist in approximately the same area today, but the outlines of the "Betsy Bob" Bar are no longer recognizable in recent air photographs. Thus it appears that the offshore bars may shift their position much more rapidly than those protected by shore topography.

Shoreward of the intertidal sand-bar complex, along most cliffed shore margins, a harrow-marked, sandy lag-gravel or a megarippled coarse sand occurs (Fig. 3). These grade shoreward into a thin (15 cm) mud veneer underlain by lag-gravel. This changes abruptly into a gravel beach at the high-tide line. In some places, always immediately adjacent to Triassic sandstones or Pleistocene outwash sands, small sand beaches replace the gravel beach. Bedrock and glacial till outcrop in places throughout these zones, indicating that the recent sediments are only a thin mantle. Wave action has built out gravel spits from some headlands.

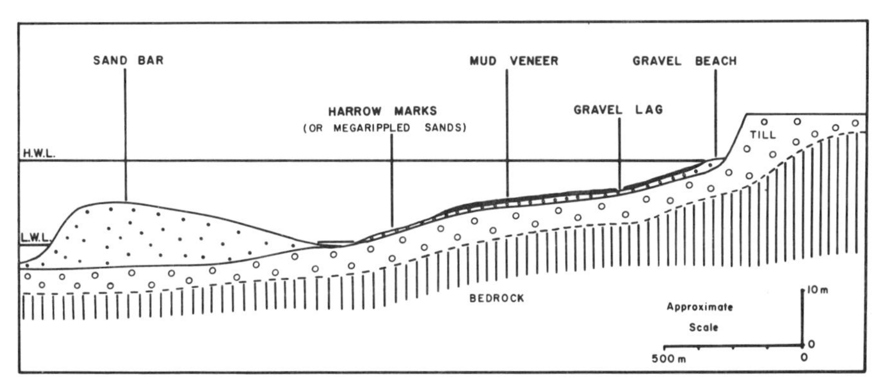

Figure 3: Schematic cross-section of the intertidal zone showing sediment facies distribution outward from a cliffed shoreline, across Selmah Bar. Vertical exaggeration approximately 20X.

In bays and other areas protected from wave action and strong tidal currents, mudflats occur in the middle and upper intertidal zone, in place of the above sequence. The mudflats terminate abruptly at their seaward edge with a break in slope. The flats display many of the features described from other areas (e.g., seaward zonation of sedimentary structures and textures, meandering tidal channels) but no detailed studies have yet been made in the Minas Basin. Behind the shore margin there may be areas of supratidal marsh, which are extensively flooded only by high spring tides.

SELMAH BAR

The remainder of the paper will be devoted to observations made on one sand bar, as typical of the type of studies carried out along the south shore of Cobequid Bay during the summers of 1971 to 1973.

The field work has consisted of topographic mapping of the bar and the collection of samples on a regular grid for textural and mineralogical analysis. Tidal current profiles have been measured using a Kelvin Hughes direct-reading current meter for 13-hour periods at each of forty-six stations twenty of these on or near Selmah Bar). In addition, tidal currents have been recorded at a number of stations by the use of continuous recording current meters. Water samples were also collected for suspended-sediment distribution. The behaviour of megaripples was monitored by the use of stakes and a precision depth recorder. Trenching and epoxy peels were used to study sedimentary structures, and dyed sand to determine sand dispersal. Vertical air photos of the intertidal area of Cobequid Bay were taken at low tide on July 22, 1972, and June 5, 1973.

The topographic survey of Selmah Bar, and the textural and sand-dispersal studies have been carried out mainly by Dalrymple. The studies of tidal currents, bedforms, and sedimentary structures are mainly the work of Knight. Data collection and analysis are still not completed and this paper gives only a short preliminary report and tentative conclusions.

Topography and Currents

Selmah Bar is a large sand bar, about 4½ kms long by 1½ kms wide, and with a total relief of about 6 m, lying just offshore on the south side of Cobequid Bay (Figs. 3 and 4). It is elongate in an east-west direction, parallel to the shore and to the direction of the tidal currents. The bar is partly shielded from ebb currents by Salter Head on the east. Between Salter Head and the sand bar there is an extensive area of mudflats. The transition between the sand bar and the mudflats is abruptly marked by a major tidal channel. The bar is separated from the shore along the entire south side by a low area, most of which emerges only

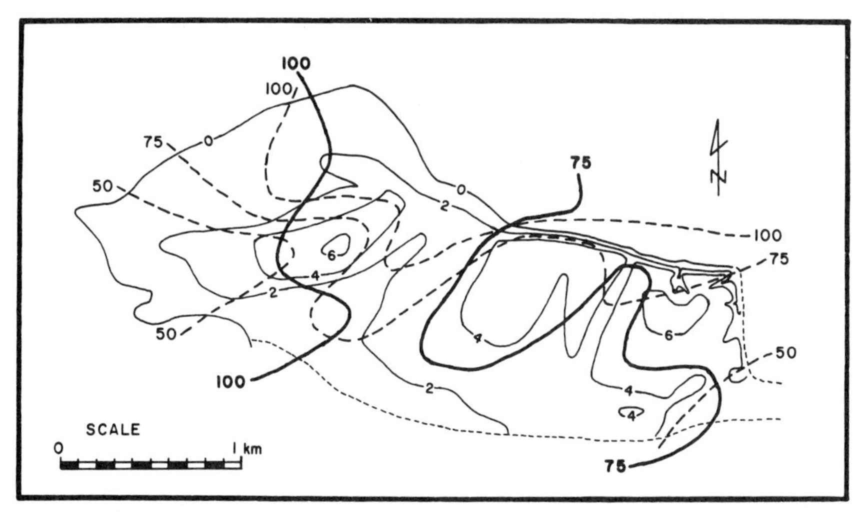

Figure 4: Selmah Bar: Topography (thin solid lines) and distribution of maximum bottom current speeds for flood (heavy solid lines) and ebb (dashed lines) tides. Topography in meters; currents in centimeters per second.

at low water. The sediment zonation shoreward of the low area is as shown in Figure 3 and described above. The topography of the bar and the distribution of maximum bottom currents for flood and ebb tide are shown in Figure 4.

The strongest flood currents are found at the exposed west end and along the north side of the bar (Fig. 4). Strong flood currents are also found in the low area between the bar and the shore, and extend northeastwards across the bar along broad "swales" at three points. During the ebb tide, however, most of the south side of the bar is shielded from strong ebb currents by the high central part of the bar, resulting in a strong flood dominance for this area. Strong ebb currents are found only along the north side of the bar, which slopes steeply (at an angle of about 5 degrees) to the north.

Figure 5 shows the distribution of mean sediment size for Selmah Bar. Almost all the sand in the bar falls into the medium sand grade. The coarsest sediment is found along the south side, presumably because this is the area closest to the major sources of sediment (in the cliffs south and west of the bar). A coarse tail is prominent in the sediment-size distributions of this part of the bar and these sands show a negative skewness. Those on the north side show a slight positive skewness. The sands are well, to very well sorted (Fig. 5).

Bedforms

The major bedforms are, in order of increasing size: (a) ripples, with wavelengths less than 15 cm and amplitudes less than 5 cm; (b) megaripples,

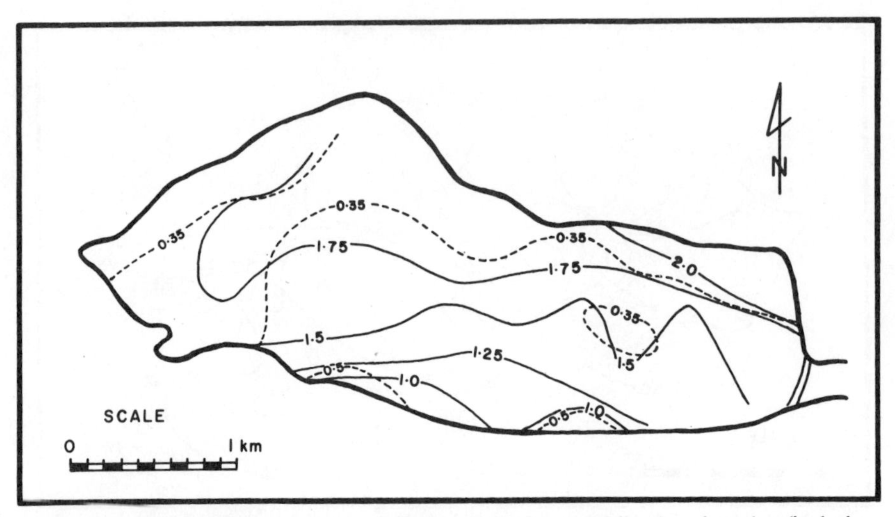

Figure 5: Selmah Bar: Distribution of mean grain size (solid lines) and sorting (inclusive graphic standard deviation – dashed lines).

with wavelengths from 1 to 12 m, and amplitudes from 10 to 70 cm; (c) sand waves, with wavelengths of 15 to 30 m and amplitudes of 40 to 150 cm; and (d) the bar itself.

Megaripples are the commonest and most prominent bedforms. The megaripples directly observed are, of course, seen only at low water following an ebb tide. Therefore, they are either ebb-oriented megaripples or flood megaripples that have been strongly modified by ebb flow. All medium- to small-scale bedforms are modified to some extent by wave action, and by shallow sheet or channel flow during the last stages of emergence. Echo soundings made during high slack water, however, indicate that many of the bedforms that were ebb-oriented at low water reverse their asymmetry and become flood-oriented during the flood current. The entire flood (or ebb) megaripple may not be reworked by the succeeding ebb (or flood) current. In some cases such partially reworked megaripples can be recognized by the presence of two distinct angles of slope on the stoss and/or lee sides (18). Varying degrees of megaripple reorientation and reworking result in variable migration rates, but in many areas of the bars studied, net rates of movement are only a few decimeters per tidal cycle.

Most megaripples observed have crest lines that are more or less linear to moderately sinuous in plan form. Knight (18) has distinguished three main morphological types: "unmodified" megaripples, scoured megaripples, and planed-off megaripples. "Unmodified" megaripples occur as bedforms little altered by late-state ebb flows. Scoured megaripples are found in areas whre the mean current flow is deflected during a later stage of flow by local bar topography. At this stage, therefore, flow is oblique to the crests of megaripples formed earlier, and a series of vortices is shed at more or less regular intervals from the old megaripple crest (see Allen, (1), "swept catenary ripples", p. 78-80, 271-275). The vortices erode deep scours into the surface of the earlier megaripple field. Megaripples with a prominent flat crestal platform have been called "planed-off megaripples" because they are formed by modification of the basic ripple shape by relatively high speed and shallow flows during emergence (or submergence) of the bar. They are most prominent in areas that are locally high topographically (18), where currents are constricted during the falling (or rising) tide, causing local steep water slopes, accelerating currents, high Froude Numbers and, consequently, erosion of the megaripple crest.

Sand waves are composite features and lack long avalanche faces. They are flood oriented and occur only on the south side of Selmah Bar. This part of Selmah Bar is not exposed to wave attack and it is clear that the flood orientation and large size of the bedforms must be due to some combination of the strong flood currents, weak ebb currents, relatively low position on the bar (hence greater duration of flood flow as a result of large depth), and coarse grain size.

At low tide, the sand waves are covered by small, ebb-oriented megaripples. Echo soundings show that at high water, the sand waves have superimposed flood-oriented megaripples. The migration rate for the sand waves on Selmah Bar is about 15 to 30 cm per tidal cycle.

Sedimentary Structures

Migration of megaripples and ripples produces most of the sedimentary structures preserved in the bar sands. Unfortunately most of the fine- to medium-grained sand does not reveal the details of cross-stratification in grenches, due to a combination of the good sorting, short exposure time, and low permeability, which prevent the sand from drying at low tide. Therefore, information about sedimentary structures depends upon peel techniques that can be used in wet sand (4), or upon samples taken using box cores. In studying structures, it is important to observe the structures below the surface layer. Because of the reversal of bedforms mentioned above, large trenches are needed in order to see the structures in the deeper parts of the bedform and thus obtain a complete picture. Examples of structures from one such trench (dug in the coarse part of East Noel Bar) are shown in Figure 6.

Thicknesses of the preserved sets of cross-bedding are usually considerably less than the height of the active megaripples in the area. This is a result of stoss-side erosion during megaripple migration and of erosion by the subordinate tidal current. The large sand waves, in particular, rarely show thick, unbroken cross-bedding sets. In the process of bedform reversal, the upper portion of the

Figure 6: Trench through sand wave on East Noel Bar showing successive cross-sets and reactivation surfaces. Sand wave on East Noel Bar with superimposed megaripples and current ripples. Location of trench to right of man.

previously deposited slip-face laminae is eroded, producing prominent "reactivation surfaces" (7, 16), that might, in a poor exposure, be mistaken for large-scale, low-angle cross-bedding.

Sand Movement

The pattern of sand movement and thus the origin of the bar itself and of the major textural gradients in it, have been studied by the use of fluorescent tracers. Sand was collected from the area where dispersal was to be monitored. It was dyed with fluorescent paint (28), returned to the collection area, and placed in a trench, 5-10 cm deep and 1 by 1.5 m in size. The trench was located about half-way up the stoss side of a megaripple or sand wave and oriented perpendicular to the crest. About 50 samples were collected along lines radiating from the injection point, one day (2 tidal cycles) and again 8 days after injection. Data were obtained for five stations on Selmah Bar (Fig. 7).

The results cannot be reported in detail here, but they support the general conclusions about the importance of shielding and the distribution of flood- or ebb-dominated areas that have been derived from the study of bedforms and tidal currents. The strongly flood-dominated transport seen at D3 clearly indicates the extent to which the bedforms can reverse, as only ebb-oriented megaripples are present here at low tide.

The overall pattern (Fig. 7) suggests an anticlockwise movement of sand around Selmah Bar. Flood currents transport sand eastwards along the south side

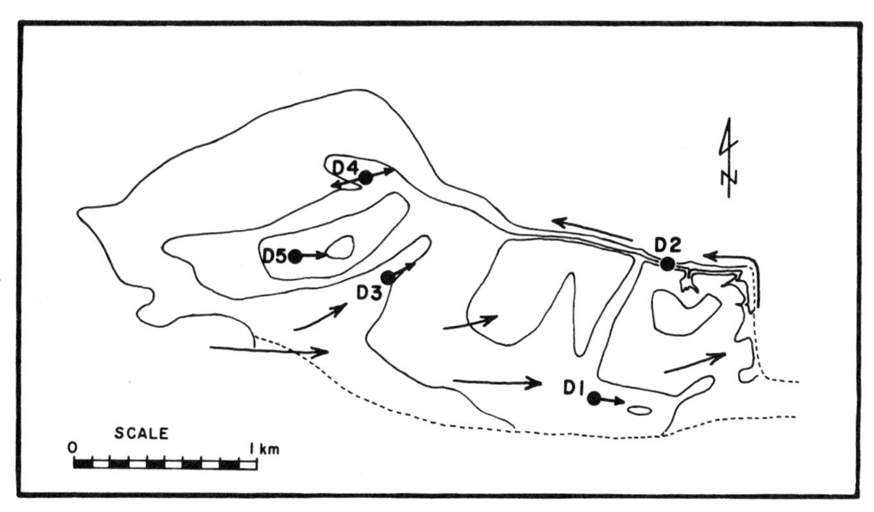

Figure 7: Selmah Bar: Location of tracer dispersion studies with resultant vectors. Longer arrows show inferred net sediment circulation pattern (arrow length not significant).

of the bar and ultimately deposit it in the tidal channel separating the bar from the mudflats. Here, the sand is moved northwards, and a return westerly transport is inferred from current measurements as the tracer data are at present incomplete. This supports the "racetrack" model proposed for subtidal bars in the North Sea by Houbolt (15) and for intertidal bars on the north shore of the Minas Basin by Klein (16). Our data suggest strongly, however, that the average distance migrated by sand grains during a single tidal cycle is on the order of 1 meter, rather than the 100 m suggested by Balazs and Klein (3).

ACKNOWLEDGMENTS

George Klein introduced one of us (G. V. Middleton) to the Minas Basin. Financial support has been provided by the National Research Council of Canada and the Department of Energy, Mines and Resources.

REFERENCES

1. Allen, J. R. L.
1968 **Current Ripples.** North-Holland Publ. Co., Amsterdam, 433 p.

2. Atlantic Tidal Power Engineering and Management Committee
1969 Report to Atlantic Tidal Power Programming Board on Feasibility of Tidal Power Development in the Bay of Fundy. Halifax, Nova Scotia.

3. Balazs, R. J. and Klein, G. deV.
1972 Roundness-mineralogical relations of some intertidal sands. **Jour.**

Sed. Pet., 42: 425-433.

4. Barr, J. L., Dinkelman, M. G., and Sandusky, C. L.
1970 Large epoxy peels. **Jour. Sed. Pet.,** 40: 445-449.

5. Cameron, H. L.
1961 Interpretation of high-altitude small-scale photography. **Can. Surveyor,** 15(10): 567-573.

6. Churchill, F. J.
1924 Recent changes in the coastline in the county of Kings. **Proc. Trans. Nova Scotian Inst. Sci.,** 16: 84-86.

7. Collinson, J. D.
1970 Bedforms of the Tana River, Norway. **Geografiska Annaler,** 52A: 31-56.

8. Dawson, W.
1917 Tides at the head of the Bay of Fundy. Department of Naval Service, Ottawa, 34 p.

9. Duff, G. F. D.
1970 Tidal resonance and tidal barriers in the Bay of Fundy system. **Jour. Fish. Res. Bd. Canada,** 27(10): 1701-1728.

10. Garrett, C.
1972 Tidal resonance in the Bay of Fundy Gulf of Maine. **Nature,** 238: 441-443.

11. Godin, G.
1968 The 1965 current survey of the Bay of Fundy – a new analysis of the data and an interpretation of the results. Bedford Institute of Oceanography, Dept. Energy, Mines and Resources, Marine Sci. Branch, Manuscript Rept. Ser. No. 8, 97 p.

12. Grant, D. R.
1970 Recent coastal submergence of the Maritime Provinces, Canada. **Can. Jour. Earth Sci.,** 7: 676-689.

13. Harleman, D. R. L.
1966 Real Estuaries. In **Estuary and Coastline Hydrodynamics.** p. 522-545. (ed. Ippen, A. T.) McGraw-Hill Book Co. Inc.

14. Hind, H. Y.
1875 The ice phenomena and the tides of the Bay of Fundy, considered in connection with the construction of the Baie Verte canal. **Can. Monthly and National Review,** 8(3): 189-203.

15. Houbolt, J. J. H. C.
1968 Recent sediments in the southern bight of the North Sea. **Geol. en Mijnbouw,** 47: 245-273.

16. Klein, G. deV.
1970 Depositional and dispersal dynamics of intertidal sand bars. **Jour. Sed. Pet.,** 40: 1095-1127.

17. Knight, R. J.
1971 Cobequid Bay sedimentology project – a progress report. **Maritime Sediments,** 7: 33-37.

18. Knight, R. J.
1972 Cobequid Bay sedimentology project – a progress report. **Maritime Sediments,** 8: 45-60.

19. Laub, M. G.
1968 The origin and movement of gravel bars in the intertidal zone of

Parrsboro Harbour, Nova Scotia. M.S. thesis, Univ. Pennsylvania, 41 p. (Unpublished data.)

20. Pelletier, B. R. and McMullen, R. M.
1972 Sedimentary patterns in the Bay of Fundy and Minas Basin. In **Tidal Power.** p. 153-187. (eds. Gray, T. J., and Gashus, O. K.) Plenum Press, New York.

21. Rao, D. B.
1968 Natural oscillations of the Bay of Fundy. **Jour. Fish. Res. Bd. Canada,** 25: 1097-1114.

22. Redfield, A. C.
1950 The analysis of tidal phenomena in narrow embayments. Papers Phys. Oceanog. Meteorol., M.I.T., and Woods Hole Oceanog. Inst., 11(4), 36 p.

23. Smith, R. E.
1969 Sedimentology of the deltas of the Moose and Diligent Rivers, Minas Basin, Bay of Fundy, Nova Scotia. M.S. thesis, Univ. Pennsylvania, 112 p. (Unpublished data.)

24. Swift, D. J. P. and Borns, W. W., Jr.
1967 A raised fluviomarine outwash terrace, north shore of the Minas Basin, Nova Scotia. **Jour. Geol.,** 75: 673-711.

25. Swift, D. J. P. and McMullen, R. M.
1968 Preliminary studies of intertidal sand bodies in the Minas Basin, Bay of Fundy, Nova Scotia. **Can. Jour. Earth Sci.,** 5: 175-183.

26. Swift, D. J. P., Pelletier, B. R., Lyall, A. K., and Miller, J. A.
1969 Sediments of the Bay of Fundy – a preliminary report. **Maritime Sediments,** 5: 95-100.

27. Swift, D. J. P., Pelletier, B. R., Lyall, A. K., and Miller, J. A.
1973 Quaternary sedimentation in the Bay of Fundy. In **Earth Science Symposium on Offshore Eastern Canada.** p. 113-151. Geol. Survey Can. Paper 71-23.

28. Yasso, W. E.
1966 Formulation and use of fluorescent tracer coatings in sediment transport studies. **Sedimentology,** 6: 287-301.

29. Yuen, K. B.
1969 Effect of tidal barriers upon the M_2 tide in the Bay of Fundy. **Jour. Fish. Res. Bd. Canada,** 26: 2477-2492.

SEDIMENT TRANSPORT AND DEPOSITION IN A MACROTIDAL RIVER CHANNEL: ORD RIVER, WESTERN AUSTRALIA

by

L. D. Wright, J. M. Coleman,[1] and B. G. Thom[2]

ABSTRACT

The funnel-shaped channel of the lower Ord River in Western Australia, experiences a semidiurnal spring-tide range of 5.9 meters. The tidal prism substantially exceeds river discharge throughout the lower 65 km of the channel, and most of the sediments contributed to the system by the river are transported and deposited by tidal currents.

The tide wave is symmetrical at the mouth but becomes deformed upstream owing to a high amplitude/depth ratio. Accordingly, the velocity of flood currents increasingly exceeds ebb velocities upstream, whereas ebb flow increases in duration.

In and seaward of the mouth, bed-load transports by flood- and ebb tide currents are approximately equal. Linear subaqueous sand ridges parallel to tidal currents separate mutually evasive zones of flood- and ebb-dominated sediment transport and appear to be related to convergence of flood- and ebb-oriented bedforms. In response to the upstream increase in tide-wave asymmetry, the largest bedforms within the channel migrate upstream under the influence of flood currents. The upstream increase in asymmetry of the channel cross section and in channel sinuosity. This results in concentrating ebb flows in the decreased

1. Coastal Studies Institute, Louisiana State University, Baton Rouge, Louisiana 70803.

2. Department of Biogeography and Geomorphology, Australian National University, Canberra, Australia.

cross section of channels, where bed-load transport is ebb-dominated, thereby balancing the sediment budget.

INTRODUCTION

Many of the world's major rivers debouch along coasts with extreme tidal ranges. These rivers differ from those that experience small tidal ranges in that for much of the year flow in the lower river course is subject to reversals over a tidal cycle; tidal currents play a significant role in determining the ultimate fate of the riverborne sediments. Within the lower 65 km of the Ord River of Western Australia, tides account for an appreciably greater fraction of the sediment-transporting energy than does the river flow. The accumulation forms and sediment-transport patterns that characterize the lower Ord appear to be common among many macrotidal rivers. The purpose of the present paper is to examine some of the more conspicuous of these patterns and their relationships to tidal processes.

THE STUDY SITE AND FIELD METHODS

The Ord River has built a recent deltaic surface consisting largely of tidal flat deposits into the structurally dominated Cambridge Gulf, in which depths average about 20 meters (Fig. 1). The lower Ord River channel has a funnel-shaped form in which channel width and depth decrease upstream at exponential rates; the channel narrows progressively upstream from a total width of 9 km at the mouths to 90 meters at 65 km above the mouth. This form has been attributed previously to mutual interactions between the channel and a standing tide wave (7).

The Ord River rises from a tropical drainage basin 78,000 km^2 in area within the West Australian Shield. The entire river system experiences a warm, dry, tropical-monsoon climate (Bsh by Köppen classification) and has a summer wet season and a winter dry season. There is a water balance deficit during every month except January and February. Discharge ranges on the average from a low of 0.07 $m^3 sec^{-1}$ in September to a maximum of 730.4 $m^3 sec^{-1}$ in January; the annual mean is 163 $m^3 sec^{-1}$ [these figures are based on data for 1955-1968 compiled by Sadler (6)]. The total sediment load of the Ord during an average hydrologic year has been estimated to be 22 x 10^9 kg (6).

Field observations in the lower Ord River were conducted during June and July 1971. Simultaneous tide observations were made by means of two pressure-type wave-tide recorders (Hydro-Products Model 521). These instruments record continuously, unattended, for a period of seven days. Tide recorders were time-synchronized with each other and with a permanently fixed recording tide gage at Wyndham, which served as a third source of tide data.

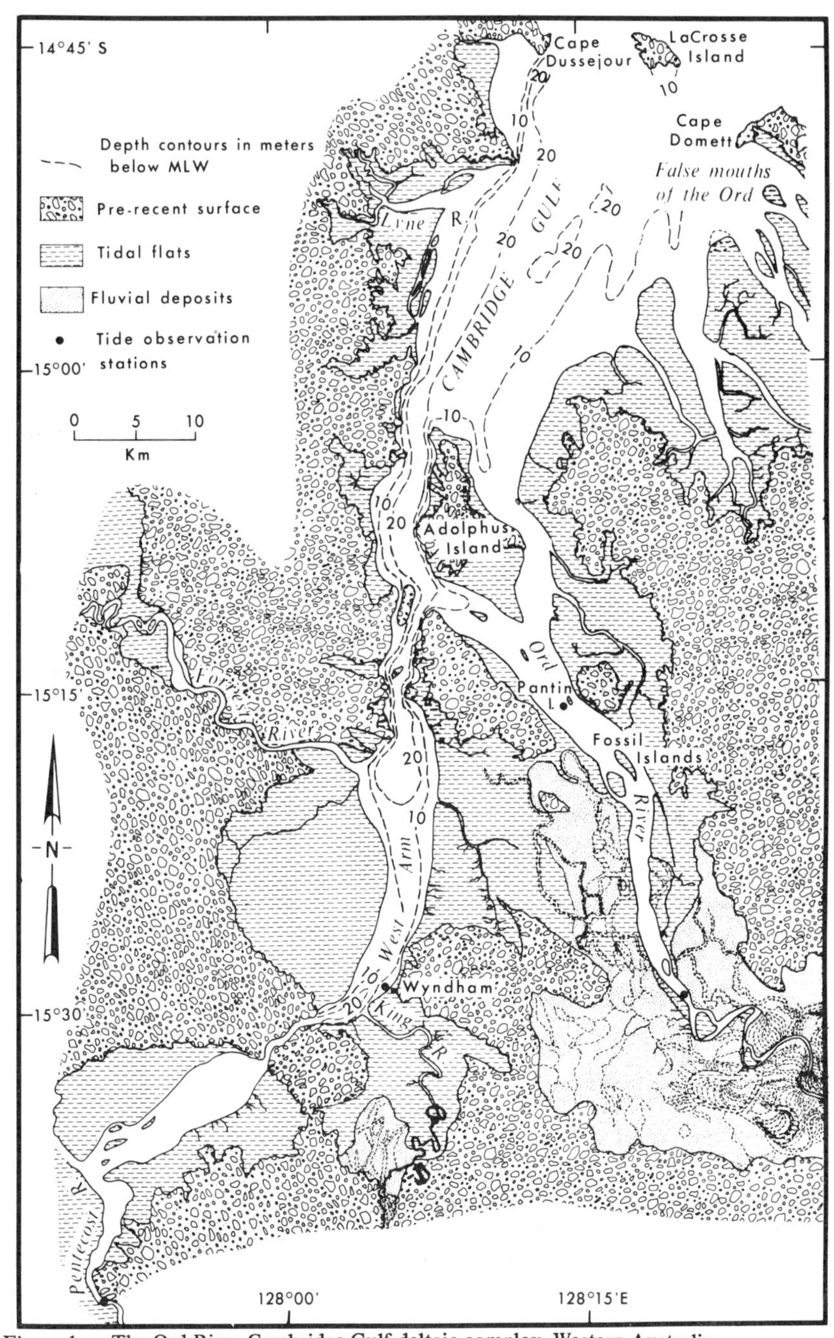

Figure 1. The Ord River-Cambridge Gulf deltaic complex, Western Australia.

Current velocities were measured with a vane-type current drag (5) and with free-drifting drogues. Water salinity and temperatures were determined with a Beckman Model RS 14 induction salinometer. Depths, cross-sectional areas, and bottom profiles were measured with an electronic depth recorder (Raytheon Model 721). Positions were determined by horizontal sextant angles.

TIDAL PROCESSES

The mean and spring ranges of semidiurnal tides at the mouth of the Ord are 4.3 meters and 5.9 meters, respectively (based on interpolations from U. S. Department of Commerce Tide Tables). In contrast with tides within the deep Cambridge Gulf, which undergo appreciable landward amplification, tidal amplitudes in the lower Ord decrease progressively upstream. Figure 2 shows simultaneous tide curves for the mouth, for Pantin Island (kilometer 15.6), and at thc downstream limit of the meandering zone (kilometer 42). From Figure 2 the upstream decrease in amplitude is seen to be accompanied by a rapid increase in tide-wave asymmetry. The simultaneity of high water at the three stations is in conformity with the theory for a standing tide wave. Current observations in the lower Ord (Fig. 3) are also consistent with standing-wave theory. In the Ord, tidal currents are slack at high and low water, the times at which tidal currents in a classic progressive wave attain maximum velocity. Maximum velocities of over 3 meters per second were recorded near times of mid-tide (the time of progressive-wave slack water). The nature of the standing-wave properties and their relationships to the morphology of the lower Ord channel have been discussed in detail by Wright et al. (7).

As in most natural estuaries, however, the tide possesses both standing-wave and progressive-wave properties. The asymmetry in the duration of the flooding and ebbing phases is a manifestation of the finite-amplitude progressive-wave characteristics of the tide. The lower Ord is relatively shallow, overall average depth being only 4 meters, and consequently the ratio of tidal amplitude to depth in high: the mean value is 1.3. The celerity, c, of the tide wave is a function of channel depth, h, approximately according to the relationship

$$c \approx \sqrt{g\,h}$$

(where g is the acceleration of gravity); therefore, the celerity of the flood phase substantially exceeds that of the ebb phase. Because of this tendency, flooding tide at a position 42 km above the mouth takes place over a period slightly greater than 2.5 hours, whereas ebb requires nearly 9.5 hours. Owing to the duration inequality between flood and ebb, average flood velocities exceed average ebb velocities by a significant amount (Fig. 3).

Tidal prism values and flood and ebb discharge rates averaged for observations

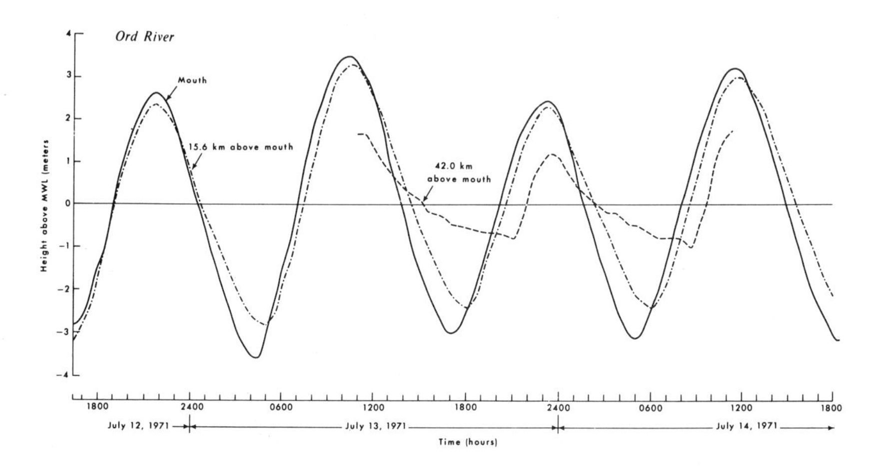

Figure 2. Simultaneous tidal curves for three stations in the lower Ord River.

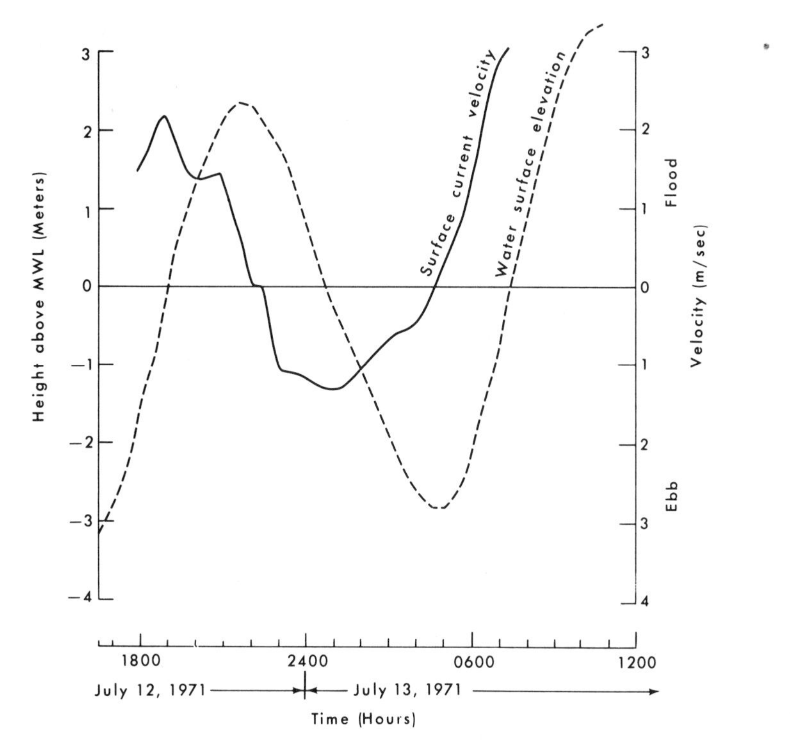

Figure 3. Relationships between tidal-current and water-surface elevations at Pantin Island (15.6 km above the mouth).

over the period July 12-14, 1971, are listed in Table 1, together with associated tidal ranges for stations at the mouth and at 15.6 km and 42 km upstream from the mouth.

Tidal prism decreases upstream at an exponential rate; only 2 percent as much water is exchanged past the section at kilometer 42 as at the mouth. Nevertheless, even the lowest tidal discharge rates are appreciably greater than the average riverine discharge rate of 163 m^3 sec^{-1}, indicating tidal domination of the lower Ord channel. From Table 1 it is seen that the observed flood discharge rate at kilometer 42 was 3.78 times as great as the ebb rate. During the observation period, marine salinities averaging 33 $^{0}/oo$ were observed throughout the estuary, indicating insignificant fresh-water contributions. However, during the summer wet season (January-February), fresh-water run-off should add considerably to the total volume of ebb discharge and reduce the inequality between flood and ebb discharge rates.

TABLE 1

Tidal Prism and Discharge

	Tidal Range (m)	Tidal Prism (m^3)	Flood Discharge Rate (Q_f) (m^3/sec)	Ebb Discharge Rate (Q_e) (m^3/sec)	Q_f/Q_e
Mouth	6.20	1.02×10^9	4.72×10^4	4.72×10^4	1.0
Km 15.6	5.60	2.56×10^8	1.19×10^4	1.01×10^4	1.18
Km 42	2.54	2.09×10^7	2.32×10^3	6.11×10^2	3.78

SEDIMENT TRANSPORT AND DEPOSITIONAL PATTERNS

Tidal and riverine currents in the lower Ord River transport large quantities of sediment both in suspension and as bed load. Suspended sediments consisting of fine silts and clays are deposited primarily on intertidal surfaces such as tidal flats and where they are trapped by mangroves. Bed-load deposits consist of quartzose and feldspathic medium-grained sands in the active channel. The patterns of bed-load transport, as indicated by the configurations of associated bedforms, change progressively upstream in response to the upstream increase in the asymmetry of the tide wave. Depositional forms within the lower Ord reflect these changes: near the mouth and in the downstream part of the estuary the channel is relatively straight and has a symmetrical cross-sectional profile; with increasing distance upstream, the channel sinuosity and cross-sectional asymmetry increase radically. These progressive changes in form are characteristic of most funnel-shaped river estuaries, including the Victoria and Daly rivers to the north of the Ord, and are directly related to variations in the direction of net bed-load transport.

In the mouths of the Ord and immediately beyond the mouths in the Cambridge Gulf where flood and ebb currents are of approximately equal magnitude, net sediment transport was roughly bidirectional during the lower-river-stage observation period. (It is probable that net transport is seaward during times of high river stage; however, this situation occurs during only a small fraction of the year.)

Bidirectional transport of bed load is accomplished partially by reversal in the migration direction of intermediate- and small-scale bedforms: bathymetric profiles in the southern mouth revealed bedforms with heights of about 30 cm and chords of 5-10 meters which reversed their orientation with tidal phase. Of probably greater significance, however, is the tendency for flood- and ebb-dominated bedform migrations to follow proximate but mutually evasive courses, a commonly observed characteristic of tidal inlets and estuaries, e.g., Ludwick (2) and Price (4).

Linear, elongate depositional shoals aligned parallel to each other and to the direction of tidal flow are the most prominent channel accumulation forms within and seaward of the mouth. They appear to be directly related to the bidirectional sediment transport patterns, high tidal amplitudes, and tidal current symmetry. The planar configurations and distribution of these features are shown in Figure 4. The shoals are abundant within the extreme lower reaches of the Ord channel, adjacent to and seaward of the northern mouth of the Ord in the Cambridge Gulf, and seaward of the entrance to the Cambridge Gulf. They range in relief from 10 to 22 meters and compositely account for over 5×10^6 m^3 of total sand accumulation. This type of accumulation form has been described from tide-dominated bays, coasts, and estuaries in various parts of the world by Off (3), who refers to them as "tidal current ridges."

In the lower reaches of the Ord River channel these shoals average roughly 2 km in length and 300 meters in width; they present hazards to navigation. Their crests are emergent or near the surface at low tide, and a few (notably Barnes Island, in the southern mouth) are permanently emergent and surmounted by mangrove swamp. More extensive shoals lie just beyond the northern mouth within the confines of the Cambridge Gulf and constitute the Australind Bank, Guthrie Banks, and East Banks (Fig. 4). The linear, mangrove-fringed islands which separate the so-called "False Mouths of the Ord," immediately to the east of the true mouth, parallel East Banks and appear to be of the same general origin as the subaqueous shoals. The King Shoals, located off the entrance to the Cambridge Gulf, are narrower, more elongate, and more widely spaced than

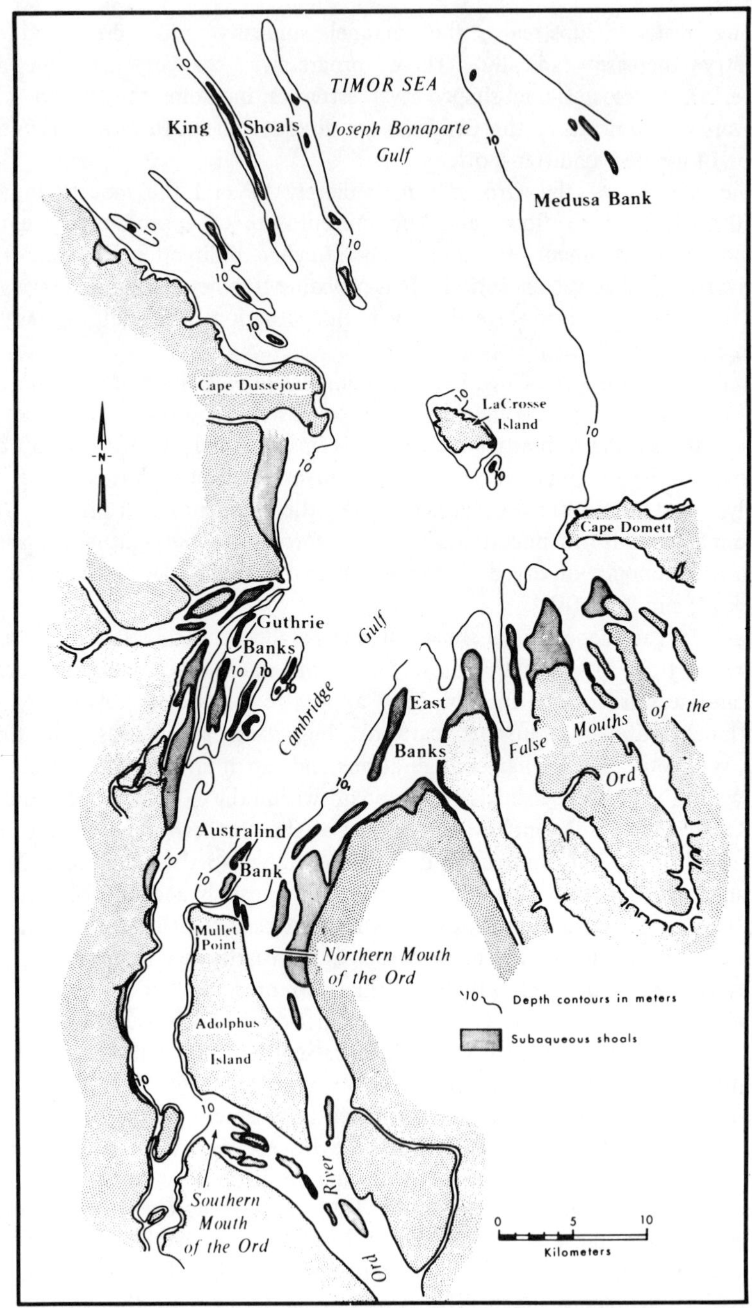

Figure 4. Linear tidal shoals in the mouths of the Ord River and in the entrance to the Cambridge Gulf.

those within the Cambridge Gulf and Ord channel. The King Shoals rise from water depths of 21 meters and are awash at low tide.

All the shoals appear to be remarkably stable: their present positions are almost identical to those charted in 1891. Stability of analogous shoals in other parts of the world is reported by Off (3).

Unfortunately, logistic difficulties precluded accurate surveys of the shoals; however, a ground reconnaissance was made of one of the smaller shoals, in the northern mouth of the Ord near Mullet Point, at low tide (Fig. 5A). The surface of this shoal exhibited bedforms with heights of 20-30 cm and chords of 5-10 meters (Fig. 5B). On the western flank of the shoal, bedforms were flood-oriented, whereas the eastern side of the shoal exhibited ebb-oriented bedforms of similar dimensions. Bedforms along the center line of the shoal exhibited no preferential orientations. These patterns suggest that these linear sand accumulations may result from the convergence of flood- and ebb-dominated bed-load transport.

As tidal current asymmetry increases with increasing distance upstream, bedforms on the shallower portions of the channel bed increase in size and migrate upstream with the swifter flood-tide currents. The bedforms are strongly asymmetrical, and their steep lee slopes face upstream; they retain this orientation throughout the tidal cycle. Figure 6 shows flood-oriented megaripples at low tide on a shallow portion of the channel in the vicinity of the

Figure 5A. Linear sand shoals in the mouth of the Ord River. Arrows indicate directions of bedform migration.

Figure 5B. Flood-oriented bedforms on linear sand shoal near Mullet Point.

Figure 6. Flood-oriented megaripples exposed at low tide in Fossil Islands region, 27 km upstream from the mouth.

Fossil Islands, 27 km upstream from the mouth. These megaripples were composed of medium sand and had heights of 30 cm to 1 meter and chords of 15-20 meters. Small-scale ripples with amplitudes of 3-5 cm and chords of 15-20 cm covered the stoss slopes of the larger bedforms and were ebb-oriented, reflecting the immediately preceding ebb phase.

Smaller, less conspicuous bedforms in the adjacent deeper portions of the channel retained ebb orientations throughout the tidal cycle. Downstream migration of bed load in the deeper channels can be attributed to the concentration of ebb flow in the reduced channel cross section, where ebb velocity and duration attain maxima.

Farther upstream, in the region between 45 and 65 km above the mouth, the channel meanders intensely and has a sinuosity ratio of 1.55 (Fig. 1). Abundant meander scars indicate that this high sinuosity is not unique to the present day. Within this meandering zone the tide-wave asymmetry becomes extremely pronounced (Fig. 2), and bedform migrations over the shallower point-bar sides of the channel are strongly flood-dominated during low-river stage. This upstream migration of bed load augments point-bar accretion; its importance is apparent from Figure 7, which shows stabilized, inactive, flood-oriented bedforms on the downstream side of the seawardmost point bar. These bedforms had heights of 1.5 meters and chords of about 20 meters. A narrow channel and intertidal mud flat separated these forms from an active ridge covered by smaller flood-oriented bedforms. Farther upstream within the meandering zone, point-bar surfaces exhibited alternations between flood- and ebb-dominated transport owing to the increased riverine influence. Convergence between upstream and downstream sediment transport undoubtedly plays a major role in accentuating the sinuosity of the channel.

In a channel of uniform cross section the upstream increase in tidal asymmetry and accompanying flood-dominated bed-load transport would, in the absence of significant riverine flow, lead to an upstream accumulation of sediment to clog the channel; only during river flood would this sediment be flushed. However, as in normal meandering rivers, cross-section asymmetry is associated with sinuosity. In the lower Ord channel asymmetry increases upstream, so that a small portion of the channel (adjacent to cut banks in meandering zone) becomes appreciably deeper than the mean depth of a given section. At the same time, as Figure 2 shows, the duration of ebb flow increases as the absolute water level decreases. Hence, with increasing distance upstream, an increasing fraction of the ebb flow becomes concentrated in a decreasing fraction of the total channel cross section. The overall effect is to produce an increase in ebb velocities within the deepest parts of the channel. This tendency for ebb currents to exceed flood currents in channel bottoms has been observed by Klein (1) in Nova Scotia. It causes ebb-dominated bed-load transport in channels, and a net balance of the sediment budget is maintained.

Figure 7. Stabilized, flood-oriented megaripples on downstream side of point-bar in meandering zone 44 km upstream from mouth.

Flood-dominated bed-load transport augments point-bar growth and necessitates lateral channel migration in order that a channel of sufficient depth be maintained. As downstream riverine flow increases in relative strength within the meandering zone, increasing asymmetry is required for flood and ebb transports to be balanced. Significantly, it is within the meandering zone that the ratio of tidal discharge to riverine discharge reaches unity and ultimately approaches zero. It is also within the meandering zone that the unequal flood- and ebb-discharge rates must converge in order to reach zero at a common point.

CONCLUSIONS

In the lower courses of nontidal rivers, sand accumulations in active channels are relatively unimportant as compared with river-mouth bar sands deposited just beyond the mouth. In macrotidal river channels such as the lower Ord, however, the river supplies sediments but riverine forces play the lesser role in their transport and deposition: tidal processes dominate. Significant upstream transport of bed-load favors accumulation of sand within the channel. The geometry of the resulting sand bodies can be expected to reflect the degree of deformation of the tide wave, which, in turn, depends on the ratio of tidal amplitude to water depth. Salient diagnostic characteristics of macrotidal channel sands include large-scale linear sand ridges formed in and seaward of the river mouth and flood-oriented megaripples within the confines of the channel.

ACKNOWLEDGMENTS

This study was supported by the Geography Programs, Office of Naval Research, under contract N00014-69-A-0211-0003, project NR 388 002, with Coastal Studies Institute, Louisiana State University.

REFERENCES

1. Klein, G. de V.
1970 Depositional and dispersal dynamics of intertidal sand bars. **Jour. Sed. Pet.,** 40: 1095-1127.
2. Ludwick, J. C.
1970 Sand waves and tidal channels in the entrance to Chesapeake Bay. Old Dominion Univ., Norfolk, Va., Inst. Oceanography, Tech. Rept. 1, 79 pp.
3. Off, T.
1963 Rhythmic linear sand bodies caused by tidal currents. **Bull. Am. Assoc. Petrol. Geol.,** 47: 324-341.
4. Price, W. A.
1963 Patterns of flow and channeling in tidal inlets. **Jour. Sed. Pet.,** 33: 279-290.
5. Pritchard, D. W. and Burt, W. V.
1951 An inexpensive and rapid technique for obtaining current profiles in estuarine waters. **Jour. Mar. Res.,** 10: 180-189.
6. Sadler, B. G.
1970 The hydrologic investigations for the Ord River Dam. Public Works Dept., Kununnura, Western Australia. (Unpublished report.)
7. Wright, L. D., Coleman, J. M., and Thom, B. G.
1973 Processes of channel development in a high-tide-range environment: Cambridge Gulf-Ord River Delta, Western Australia. **Jour. Geol.,** 81: 15-41.

A STUDY OF HYDRAULICS AND BEDFORMS AT THE MOUTH OF THE TAY ESTUARY, SCOTLAND.

by

Christopher D. Green[1]

ABSTRACT

An attempt has been made to link directly the bedforms and sequence of bedform changes in a tidal channel to the tidal currents and associated critical erosion velocities. No satisfactory relationship has been established.

Reynolds numbers show the tidal currents to be in fully developed turbulent flow. Froude numbers, whilst characterizing the transitional and lower flow stages, do not allow the delimitation of the stage of initiation of motion, neither do they permit the characterization of the sequences of bedform change.

Shear stress and shear velocity values based on bed-slope calculations are inappropriate in tidal waters and those based on water-surface slope are of limited value. Shear stresses derived from velocity gradients enable a critical shear stress value for the initiation of particle motion to be suggested. Sequences of bedform change can be correlated more directly with von-Karman-Prandtl values than with any other single parameter.

In attempting to define the ranges of flow conditions associated with various bedform stages it is necessary to question the validity of extrapolating from steady-state experiments to the natural tidal system. The study illustrates this point and suggests a basic four-fold division of bedform stages.

1. Tay Estuary Research Centre, The University of Dundee, Newport-on-Tay, Fife, Scotland, United Kingdom.

INTRODUCTION

The existence of bedforms and their relationship with conditions of flow have long interested both geologist and hydraulician. Many fundamental investigations have been conducted in flumes under conditions of controlled steady-state flow. Little systematic work has been reported using field conditions. This study conducted in a tidal channel near the mouth of the Tay estuary, Scotland, forms a part of continuing work attempting to understand and establish controls of sediment motion in the field situation of reversing tidal currents. Pool channel provides an ideal field area in which the suitability of conventional bedform description and prediction criteria can be examined. The author defines his bedforms according to the nomenclature proposed by the Task Force on "Bedforms in Alluvial Channels", the American Society of Civil Engineers, May 1966.

THE STUDY CHANNEL AND INSTRUMENTATION

Pool channel stretches on a NW-SE line joining the River Tay in the north to the northern part of St. Andrews Bay (Fig. 1). From the Pool Bar in the northwest the channel falls with mean bedslope gradient 0.0026 so that at the Salmon Nets (the southeastern extremity) the bed elevation is 2.56 metres lower. The channel length is some 1000 metres, with cross-sectional water surface length of approximately 70 metres. Measurements and observations were made at three sites indicated in Figure 1. The first is on the Pool Bar, whilst the second, Mid-Point, and third, Salmon Nets, are situated within the well-defined channel.

The study relied on both direct and indirect observational methods. Currents were monitored by Dentan CM2 and Bray stoke direct reading current meters. The bedform changes were plotted on Kelvin Hughes MS 26 and Seascribe echo-sounders. Underwater measurements of non-emergent structures and observations of the sequences of bedform changes were made by divers. All measurements were undertaken on spring tides of equal tidal ranges and similar fluvial discharge.

BEDFORMS AND HYDRAULICS OF SEDIMENT MOTION

The existence of bed structures has long been recognized and these features subjected to extensive study in flume and field. Most field studies of bedforms have been essentially descriptive. Large-scale ripples have been described by Cornish (5), Von Straaten (22), and non-emergent trains of bedforms beneath tidally flowing waters by Cloet (4) and Stride (15, 16, 17). Sonar and echo-sounding observations on the continental shelf of the British Isles have

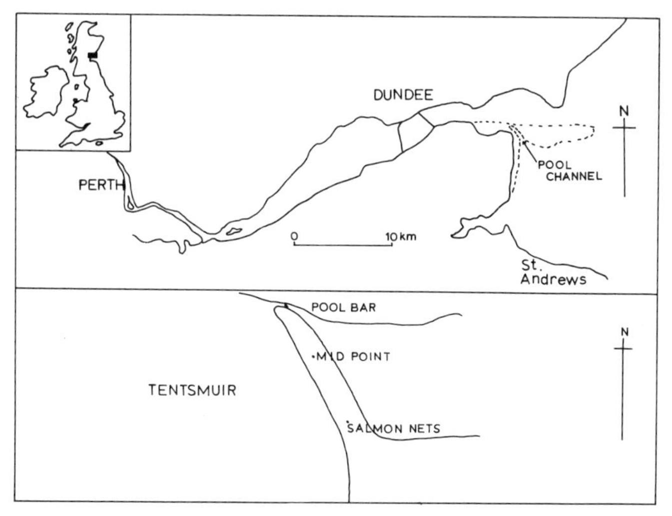

Figure 1. The Tay estuary, Scotland.

revealed the occurrence of large dunes in many areas (12, 21). McCave (13) examined the tidal currents and associated sand-wave field in the southern North Sea to estimate the rates of migration of major sand waves.

The interaction of moving water and sediment load have been examined by many workers since Du Buat (6), including Gilbert (8), whose experiments provided valuable data for many later analyses. Hjulstrom (11) introduced the concept of a critical erosion velocity for the entrainment of particles of any given size. The most extensive studies were provided by Simons and Richardson (18), who worked with recirculatory flumes and recorded a sequence of bedforms permitting a primary relationship with hydraulic conditions to be suggested. Empirical approaches attempting to define the ranges of flow conditions associated with the bedform stages have been made by Znamenskaya (25), Yalin (23), Bogardi (3), Grade and Rangaraju (7), and the extensive studies of Simons and Richardson (20).

Bogardi (3) related the bedform type to the mean diameter D of the sediment and to the parameter

$$\frac{gd}{\tau\rho}$$

where g is the acceleration due to gravity, d is the depth of flow, τ is the shear stress and ρ is the density of the fluid.

Garde and Rangaraju (7) produced a graph showing the ordinate consisting of

$$S \quad (\gamma_s - \gamma_F / \gamma_F)$$

where S is the water surface slope, γ_s is the specific weight of the sediment and γ_F the specific weight of the fluid. The abscissa gives values of the ratio wherein the hydraulic radius of the channel and the depth of flow.

The plots of Simons and Richardson (20) show the relationship of the bedform to the "streampower" and median diameter of the bed material. The steampower is the product of the mean velocity $\overline{u}$ and the shear stress τ.

More recently shear stress and shear velocity have been recognized as important, especially in the initiation of particle motion (1, 24). Halliwell (10) provided a method for the calculation of shear stress and shear velocity from field observations.

THE PHYSICAL SCALE OF THE BEDFORMS

Allen (1) recognized two ripple populations based on considerations of bedform chord (B) and height (H) and the vertical form index expressed by $^B/H$. The morphological distinction may be defined

Small scale ripples 60 cm chord
V.F.I. 5 - 20
Large scale ripples 60 cm chord 4 cm height
V.F.I. 10-100

The field measurements in the Pool area (Fig. 2) for ripples and dunes exceeding 12 cms in height and 1 metre chord reveal a vertical form index of 10-40, the mean of which is below the 50 recorded by Allen (1) but close to the mean value of 20 obtained by Guy et al. (9) probably because the latter structures were strongly three-dimensional in the field study area. Allen (1) did not include three-dimensional ripples in his natural scale data.

THE SEQUENCES OF CHANGE: BEDFORMS AND CURRENT

Before presenting detailed descriptions, the general setting of the field area and its water flow characteristics will be considered. Pool Channel is in an unusual natural situation from two viewpoints. First, the tidal channel is protected by an offshore bar system which filters incoming waves. Calm days were selected for field investigations to take advantage of the natural filtering process. Second, the falling tide uncovers a bar at the northwestern end of the channel so that for three hours of the semi-diurnal tide the water flow virtually ceases. The flood tide flows from southeast to northwest through the channel, the ebb the reverse, from northwest to southeast.

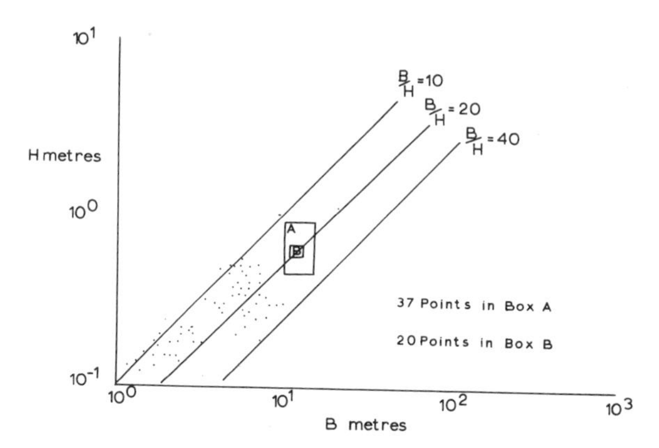

Figure 2. Mean ripple height (H) as a function of mean ripple chord (B). Based on data from Pool Channel.

THE SALMON NETS

The current velocities measured at the Salmon Nets are shown in Figures 3 and 4. These represent the semi-diurnal characteristics at the surface and bottom (Fig. 3) and current profiles plotted against tidal height and state (Fig. 4). In both cases the dominance of the ebb currents is plain. The velocity profiles indicate that the ebb velocities 0.25m from the bed reach a maximum of 0.77 m/sec. compared with 0.30 m/sec. on the flood.

The bedforms at the Salmon Nets consist of ebb-orientated asymmetrical dunes with associated backing structures. They are 60-90 cms in height and 10-15 metres in chord. These dunes do not reverse with the flood tide. No supracrestal ripples are formed and underwater observations have shown that no rounding of the dune crest is seen during the flood tidal period. The situation is thus of ebb dunes maintaining their dimensional characteristics throughout the entire tidal cycle.

MID-POINT

The water flow and bedform sequences are more varied and complex than at the Salmon nets. Surface and bottom currents are shown in Figure 5 and the current profile distribution in Figure 6. There is a more balanced distribution to the flow, with a slight indication of flood dominance. Tidal velocities 0.25 m from the bed reach 0.85 m/sec. on both flood and ebb tides.

At low water ebb dunes 20 cms in height and 2-3 metres in chord cover the bed. The turn of the tide and rise in water level culminate in the bar becoming breached 1½ hours after low water. Two hours after low water, small asymmetrical flood ripples can be observed backing the ebb dunes. Two and a

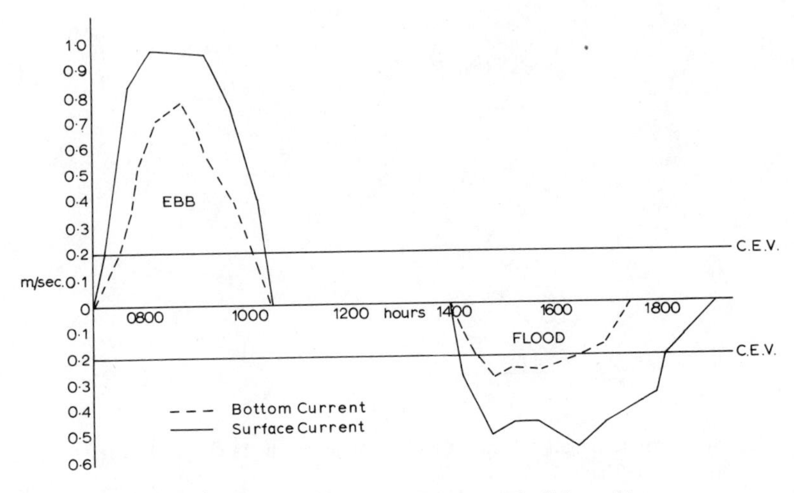

Figure 3. Semi-diurnal surface and bottom current speeds, Salmon Nets.

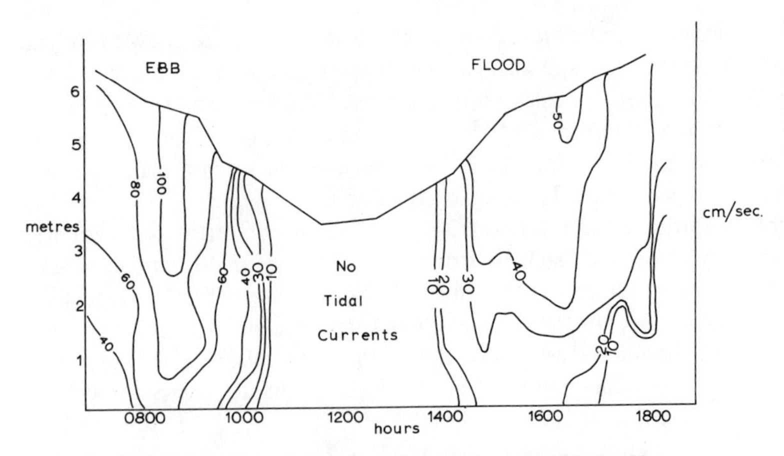

Figure 4. Vertical distribution of semi-diurnal current speeds, Salmon Nets.

half to three hours after low water, "rounding" of the ebb dunes is seen until, at 4-5 hours after low water, a destructive plane bed (lower flowstage) is reached. The bed is then relatively flat. The last hour of the flood tide results in the growth of small flood-orientated ripples. From high water these small bedforms are quickly destroyed and ebb structures are generated approximately two hours after high water.

POOL BAR

A very dynamic situation exists at Pool Bar, the flood tide showing greater complexity than the ebb.

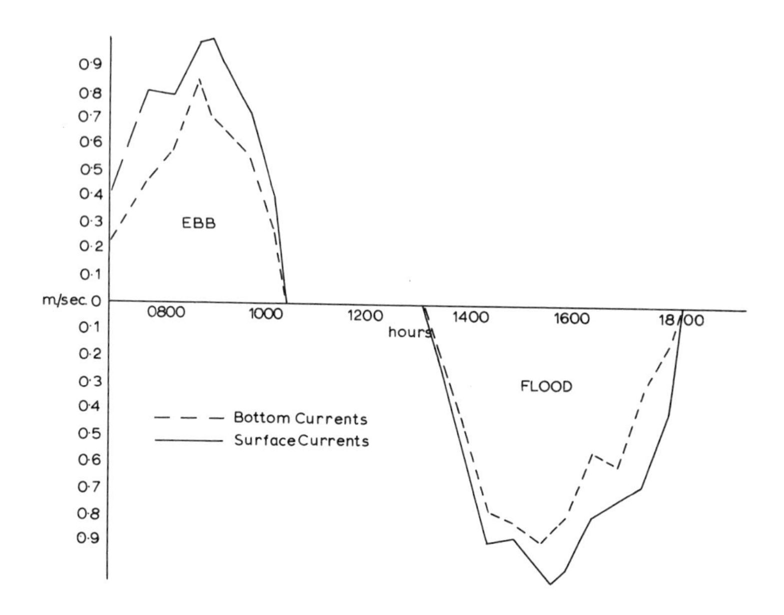

Figure 5. Semi-diurnal surface and bottom current speeds, Mid-Point.

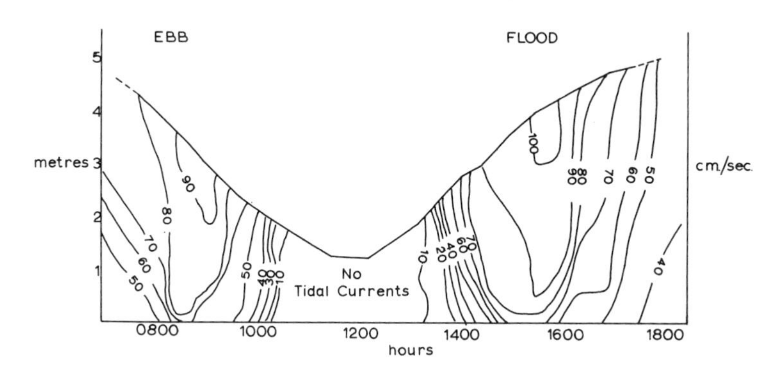

Figure 6. Vertical distribution of semi-diurnal current speeds, Mid-Point.

The Pool Bar is covered 1¼-1½ hours after low water. A rapid sequence of bedform changes occurs. Remnant ebb-orientated dunes are rapidly destroyed by the flooding tide. The bedforms enter the transition stage of Simons and Richardson (18) and washed-out dunes alternate with plane bed stages. A standing wave field is established at the northern edge of the bar. The water returns to the lower flow regime after 15-20 minutes and flood-orientated ripples are created. The ebb tide destroys the flood-orientated dunes and generates ebb dunes which dry out at low water. The flood current curve (mean) shows the very rapid buildup of water flow following the covering of the bar. The high rate of flow is maintained for much of the flood tidal period. This can be compared with the more typical rise of ebb curve (Fig. 7).

The sequence of bedform change is varied both with tidal state and position within the channel.

According to Hyjulstrom (11), critical erosion velocities of 18-25 cm/sec. are required to transport sediment of 0.20 mm diameter. Thus, the observed velocities are competent to transport sediment on both flood and ebb tides at all the study sites. However, no transportation of sediment occurs at the Salmon Nets on the flood tide. Also the sequences of bedform change cannot be explained simply from a description of the current velocities or by using cirtical erosion velocities.

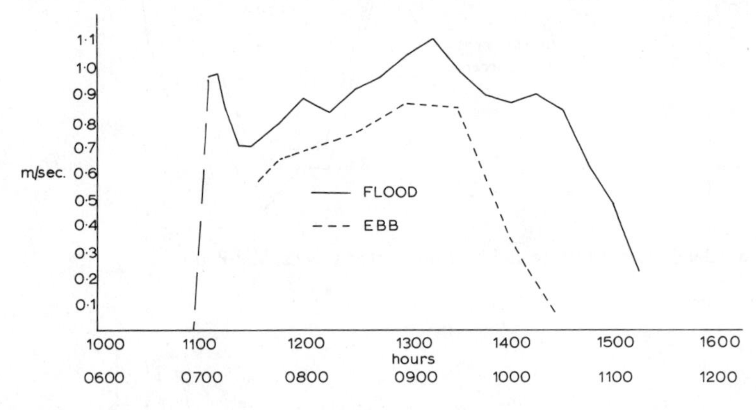

Figure 7. Semi-diurnal mean current speeds, Pool Bar.

HYDRAULIC CONSIDERATIONS OF CHANGE

It is important to examine the hydraulic parameters that describe the flow and establish whether they relate directly to observed bedforms and sequences of bedforms.

The Reynolds numbers for the study sites (Table 1) show the waters throughout the pool area to be in a state of turbulent flow. Near Laminar conditions may occur for about 15 minutes when the bar is first covered. Reynolds number values on the ebb are highest at the Salmon Nets ($>$50,000), followed by the Mid-Point (25,000), and at the Pool Bar (23,000). On the flood tide the highest values occur at the Mid-Point (34,000) at the Salmon Nets (20,000) and at Pool Bar (17,000).

Froude numbers N_F may be used to distinguish between sub- and super-critical flow. The Simons and Richardson classification (18) is based largely on flow regimes. In Figures 8, 9, and 10 the Froude numbers are shown for each study site plotted against time and the known sequence of bedform

TABLE 1.

Reynolds Numbers

Ebb Tide

Pool Bar		Mid-Point		Salmon Nets	
0700	22,631	0745	23,653	0715	15,400
0745	22,574	0815	22,845	0740	30,427
0800	22,469	0845	25,195	0815	38,467
0830	21,238	0900	21,840	0845	40,341
0900	15,004	0915	18,844	0852	52,039
0930	6,234	0930	18,259	0915	36,081
1015	2,769	0945	12,914	0945	21,848
		1000	6,178	1015	9,567
		1015	2,637	1045	1,203

FLood Tide

Pool Bar		Mid-Point		Salmon Nets	
1105	1,220	1315	864	1415	8,522
1110	2,407	1330	3,085	1450	17,410
1200	9,298	1408	12,507	1515	16,185
1300	16,621	1430	16,985	1545	17,535
1400	16,554	1500	24,049	1630	19,333
1515	5,344	1535	32,149	1700	14,388
		1600	33,839	1730	12,789
		1630	21,215	1800	13,738
		1700	22,647	1815	6,044
		1725	20,482		
		1745	17,141		
		1800	10,616		
		1805	7,160		

change.

At the Salmon Nets (Fig. 8) the values of N_F do not exceed $N_F = 0.2$; for much of the semi-diurnal period they are below $N_F = 0.1$. Similarly at the Mid-Point (Fig. 9) the values of N_F reach a maximum of $N_F = 0.25$ on the flood and $N_F = 0.17$ on the ebb. In both cases the absolute values observed under field conditions are much lower than those recorded by Simons and Richardson (18) as indicative of bedform generation from flume studies.

At the Pool Bar (Fig. 10) the N_F values are greatest; a N_F value of 0.77 is recorded soon after the bar is covered by the flood tide. This places the bedform

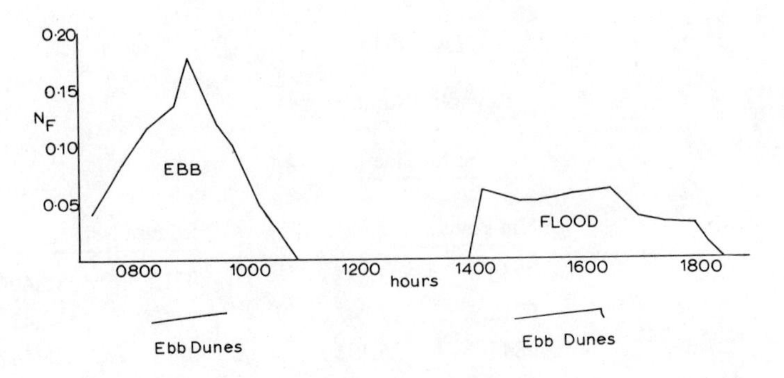

Figure 8. Froude numbers and the sequence of bedform change, Salmon Nets.

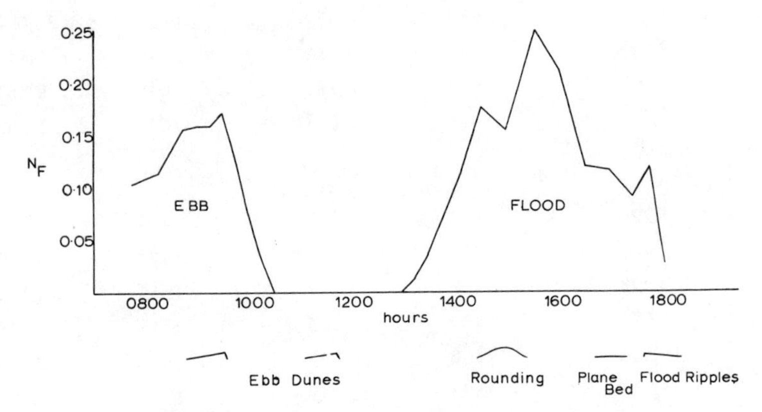

Figure 9. Froude numbers and the sequence of bedform change, Mid-Point.

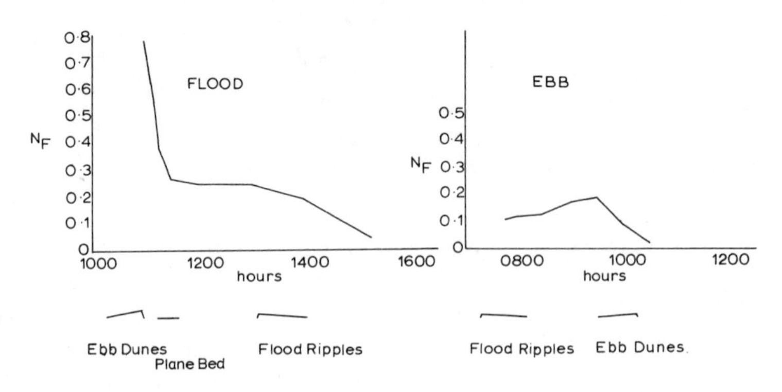

Figure 10. Froude numbers and the sequence of bedform change, Pool Bar.

stage within the transition phase of Simons and Richardson (18). As the tide rises the N_F values decrease from 0.2 - 0.25.

The N_F values on the ebb tide range from 0.02 - 0.18.

At each study site the absolute value of N_F is only qualitatively significant. It helps to describe the sequences at any one locality but it is impossible to relate the growth and destruction of a particular bedform to a single value of N_F. It is, however, useful in characterizing the change from the transition stage to the lower flow stage. As the great majority of bedforms studied on the natural scale form in the lower flow stage the usefulness of the Froude number is limited, particularly when comparisons are made between different localities.

SHEAR STRESS AND SHEAR VELOCITY

For uniform flow, the shear stress can be calculated by the formula

$$\tau = \gamma d S'$$

where γ is the specific weight of water and S' is the slope of the bed. Similarly the shear velocity can be computed by

$$u_* = \sqrt{g d S'}$$

Shear stress and shear velocity are basic parameters that govern the initiation and transport of sediment in uni-directional flow. They are equally important in the reversing current conditions of tidal flow.

In this field study tidal currents transport sand with and against the bed gradient. A derivation invoking motion exclusively down the bed gradient is therefore not appropriate to the tidal condition and the calculation using

$$\tau = \gamma d S'$$

is not suitable.

Two alternatives will be examined. Shear stress and shear velocity values can be calculated by using values of the water-surface slope which may be computed by using the basic energy equation. An appropriate modification introduced by Halliwell (10) for estuaries allows for density, inertial and kinetic energy terms.

Secondly, Prandtl (14) introduced the expression

$$\rho l^2 \left(\frac{du}{dy}\right)^2$$

to determine the shear stress assuming uniform but fully turbulent flow. In the expression ι is a quantity having length known as the mixing length, $\frac{du}{dy}$ the velocity difference at a distance above the bed.

RESULTS

An important step in deriving the values of u and Υ yields information on the water-surface slope. Table 2 shows the results obtained on flood and ebb tides.

From high tide when there is no measurable water-surface slope along the channel the tide commences to ebb. The constriction of flow at the Pool Bar on the Ebbing tide creates an energy gradient which fluctuates about a mean value of 0.00132. At 1½ hours before low water the bar is exposed, water ceases to flow, and the energy gradient is destroyed.

TABLE 2.

The water surface slopes, Pool Channel

Floodtide

Low Water	Water surface slope
+ 1 hr. 43 m.	.0009
+2 hr. 33m.	.0006
+ 3 hr. 03 m.	.0005
+ 3 hr. 33 m.	.0004
+ 4 hr. 03 m.	.00026
+ 4 hr. 33 m.	.0001
+ 5 hr. 48 m.	0

Ebb Tide

High Water	Water surface slope
+ 1 hr. 45 m.	.0010
+ 2 hr. 45 m.	.0014
+ 3 hr. 45 m.	.0011
+ 4 hr. 45 m.	.0017

After the tide turns the water level builds up in Pool Channel and northern St. Andrews Bay. Once water overtops the Pool Bar it flows into the main tidal stream of the River Tay with the result that an energy gradient is established in a reverse (flood) direction. This gradient decreases with time until at high water there is no measurable gradient.

The shear stress and shear velocity measurements based on water-surface slope (Table 3) allow comparison with reversing tidal flows. Their use is however

TABLE 3.

Shear stress and shear velocity based on the water surface slope.

Ebb Tide

Pool Bar Time			Mid-Point Time			Salmon Nets Time		
0745	23.37	.559	0745	23.22	.55	0715	27.14	.751
0800	22.85	.533	0815	22.32	.509	0740	26.96	.742
0830	21.7	.481	0845	20.81	.442	0815	26.56	.72
0900	19.7	.396	0900	19.7	.396	0830	25.49	.663
0930	16.73	.286	0915	18.68	.357	0852	25.24	.65
1000	16.79	.288	0930	18.12	.335	0915	25.63	.671
1015	15.56	.247	0945	17.68	.319	0945	22.85	.533
			1000	16.74	.286	1015	22.09	.498
			1015	15.87	.257	1045	19.18	.376

Flood Tide

Pool Bar			Mid-Point			Salmon Nets		
1105	5.77	.014	1315			1415	17.94	.329
1110	5.14	.027	1330	13.86	.190	1450	16.85	.29
1200	8.33	.071	1408	14.39	.207	1515	15.59	.248
1300	8.74	.078	1430	12.84	.168	1545	13.02	.173
1400	6.78	.047	1500	12.76	.166	1630	7.605	.059
1500	5.42	.03	1535	12.40	.157	1700	—	—
			1600	10.34	.109	1730	—	—
			1630	6.261	.04	1800	—	—
			1700	—	—	1815	—	—
			1730	—	—			
			1800	—	—			
			1805	—	—			

limited, the influence of d upon the values is fundamental in this tidal situation of decreasing and increasing water depth.

The calculation of shear stresses using the Prandtl (14) derivations allows comparison of shear stress values through time and at different stations. The vertical distribution of shear stress through time at the Salmon Nets and Mid-Point is shown in Figures 11 and 12 and at Pool Bar in Table 4.

The rapid decrease in τ from near the bed is apparent. It is most useful to look at the lower set of τ values generally obtained 25 centimetres above the bed. At all localities a sequence of bed response to prevailing shear stress was observed (Fig. 13).

(a) $\tau < 300$ gm cm^{-1}sec^{-2} grain motion is absent

(b) $\tau \geq 300$ gm cm^{-1}sec^{-2} grain motion commences building or destroying

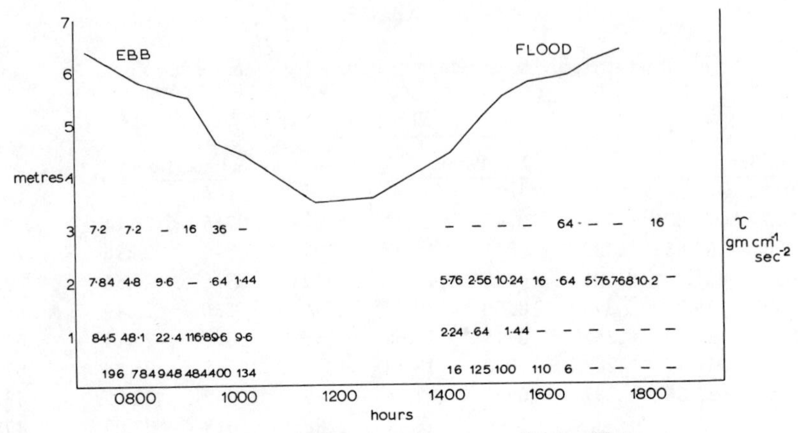

Figure 11. Semi-diurnal vertical distribution of shear stress, Salmon Nets.

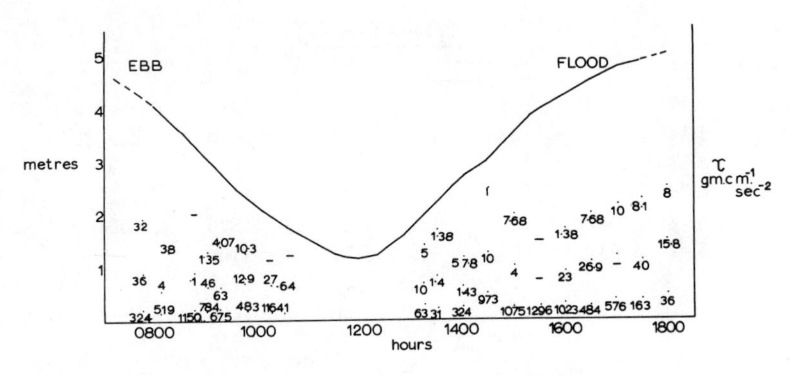

Figure 12. Semi-diurnal vertical distribution of shear stress, Mid-Point.

TABLE 4.

Shear stress based on the velocity gradient, Pool Bar, gm cm^{-1} sec^{-2}.

Ebb Tide		Flood Tide	
0700	1845	1105	1273
0800	1822	1110	1521
0830	1769	1200	1238
0900	1535	1300	1762
0930	973	1400	1634
1000	414	1515	97.2
1015	165.5		

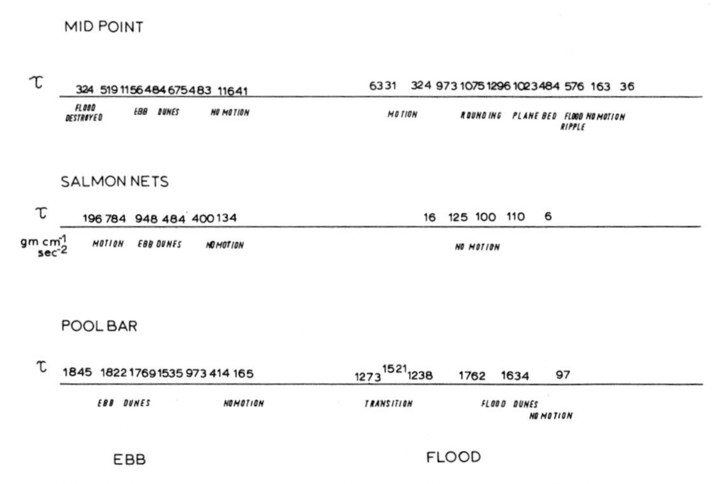

Figure 13. Near bed shear stress as a function of bedform type and change.

bedforms.

(c) At τ values in excess of 500 gm $cm^{-1}sec^{-2}$ ripple growth and destruction is normal. Dunes are common where the values of τ exceed 600-700 gm $cm^{-1}sec^{-2}$ for part of the tidal cycle.

At the Salmon Nets the maximum calculated value of τ on the flood tide is 165 gm $cm^{-1}sec^{-2}$ and this is insufficient to permit entrainment of sediment (confirmed by diver observation). All bedforms are generated during the ebb tide when τ rises to 950 gm $cm^{-1}sec^{-2}$.

At the Mid-Point the values of τ reach 1,296 gm $cm^{-1}sec^{-2}$ on the flood tide and the progressive increase of τ towards this maximum is matched by a well-defined sequence of bedform generation and destruction. During the ebb tide the shear stresses increase to over 1000 gm $cm^{-1}sec^{-2}$ with the steady destruction of flood-dominant bedforms and their replacement by ebb-dominant bedforms. This situation shows a range of changing bed structures from ebb-dominant dunes to flood-orientated ripples in response to the local dynamics.

At Pool Bar the value of τ reaches over 1,750 gm $cm^{-1}sec^{-2}$ on both flood and ebb tides and the resultant bedforms range from ebb to flood dunes with all intervening stages present.

Dunes form at each locality investigated. Their dimensions are greatest at the Salmon Nets (1m height, 15m chord) where the shear stress is appreciably less than at the other two sites. This situation may be rationalized in relation to the reversing tidal currents and the ranges of shear stress to which the bed sediments are subjected during an normal tidal cycle.

At the Salmon Nets the flood currents are incompetent to transport material in terms of a critical shear stress value. The ebb dunes are the result of successive ebb tidal periods and must represent a maximum growth stage under the physical constraints of the area.

At the Mid-Point and Pool Bar sites the shear stresses are sufficiently high on both flood and ebb tides to move sediment, destroy pre-existing structures, and generate fresh bedforms. During the early parts of each tidal phase the bedforms of the previous tidal half-cycle must be destroyed before the initiation of growth of bottom structures consonant with the acting currents. Much of the energy of the flow in the early part of each tidal half-cycle is expended in destruction and the resultant time under generative conditions of flow may be inadequate to construct fully mature sets of dunes. Thus, although the calculated values of shear stress are greater, the bedforms resulting are of diminished order due to the required destruction in the early flow stages.

THE EMPIRICAL APPROACH

Many largely empirical attempts have been made to define the ranges of flow conditions associated with various bedforms.

In many cases the parameters used include a shear stress value derived from the bed slope, e.g. Bogardi (3) and Simons and Richardson (20).

The plot of Bogardi's parameter based on the Pool Channel data is shown in Figure 14. The data are plotted for a grain size of 0.2 mm and thus represent only one abscissal condition with respect to Bogardi's original graph. The transitional form stage is not delineated.

Similarly a plot of the streampower and the median diameter of the bed material (20) using the Pool Channel data (Fig. 15) shows no delineation between the three dynamic states of interest. Indeed the transitional stage occurs at some of the lowest values of streampower. In this case the streampower was calculated using the water-surface slope to derive the shear stresses. The limitation is in this method of calculation; the water-surface slope varies with tidal state and concomitantly the hydraulic head has different values.

In Figure 16 the streampower values are based on Prandtl's (14) determination of shear stress. These are independant of water-surface slope, bed slope, and hydraulic depth. There is a clear separation of the sediment motion and no motion stages. Although there is some overlap the transitional form stage lies in the upper part of the sediment motion plot.

The transitional form stage is associated with high Froude numbers from N_F = 0.7. A plot of streampower against the Froude number separates the bedform stages (Fig. 17). In order to allow comparison of the three stages with varying median diameter of bedform material, a useful parameter can be derived from the product of the shear stress

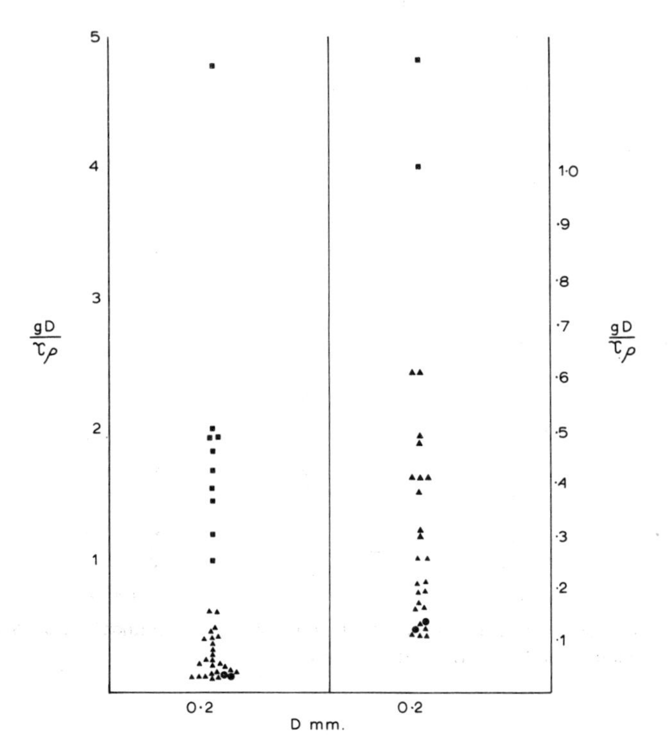

Figure 14. Occurrence of bedform using the criteria of Bogardi (3). – no motion, –motion with the generation and destruction of bedforms, – transition stage.

$$\rho \iota^2 \left(\frac{du}{dy}\right)^2$$

and the Froude number and plotting this against the median diameter.

In Figure 18 each bedform stage is clearly delineated. Whilst Figure 18 shows the separation on the basis of bed material size of 0.2 mm, studies in different field systems and areas could build the picture into an effective system for delimiting the basic bedform stages exhibited in the natural tidal situation.

CONCLUSIONS

The bedforms and sequences of bedform changes have been related to the tidal currents, associated critical erosion velocities, and various hydraulic parameters.

The most important comment on any study of bedforms is that results based on steady-state experiments are not suitable for the discussion and interpretation

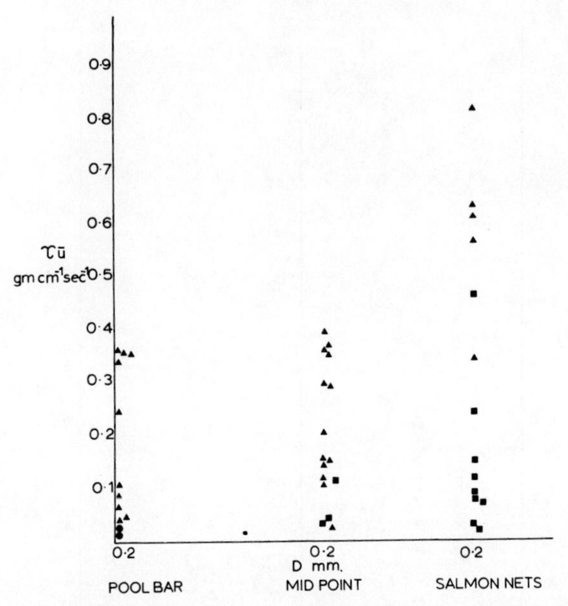

Figure 15. Bedform as a function of streampower and sediment size. Shear stress determined from the water-surface slope.

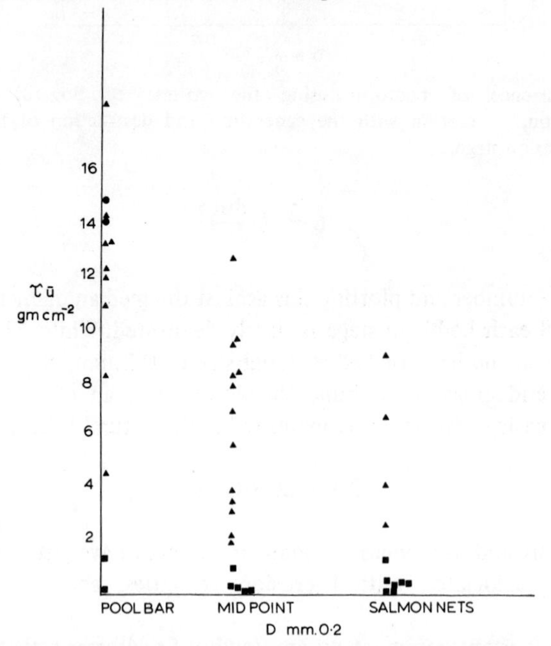

Figure 16. Bedform as a function of streampower and sediment size. Shear stress determined from velocity gradient.

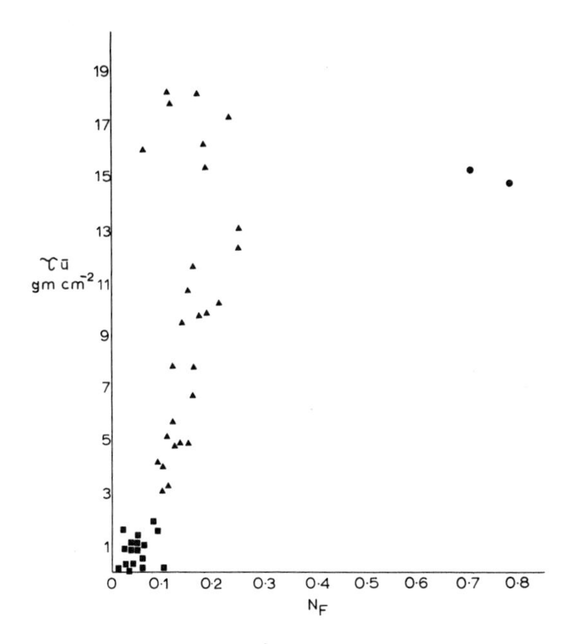

Figure 17. Streampower as a function of the Froude number.

of bedforms in the natural situation. This point has been well expressed by Allen (2).

The field worker cannot rely on traditional steady-state derivations for the parameters that characterize the conditions associated with water-sediment interactions. Reynolds and Froude numbers in particular are of qualitative importance but the steady-state derivations of shear stress and shear velocity are inappropriate to the tidal situation of reversing currents and constantly changing hydraulic head.

The shear stress values based on Prandtl (14) are useful in attempting to characterize the sequences of events in this study. The values have been based mainly on a determination of shear stress at 25 cms from the bed. This is at the limit of the present instrumentation. A higher degree of control and instrumentation both of currents and bedforms could yield promising results. It is to be remembered that this study has worked in only one dimension.

Shear stress is of paramount importance in the evaluation of factors controlling water sediment interactions. It appears that a critical shear stress of value 300 gm cm^{-1} sec^{-2} marks the condition at which sediment (0.2 mm) starts to move.

The use of the basic energy equation or its modified form (10) greatly aids in understanding the water transfer systems and energy gradients in tidal situations. It has been of particular relevance in Pool Channel.

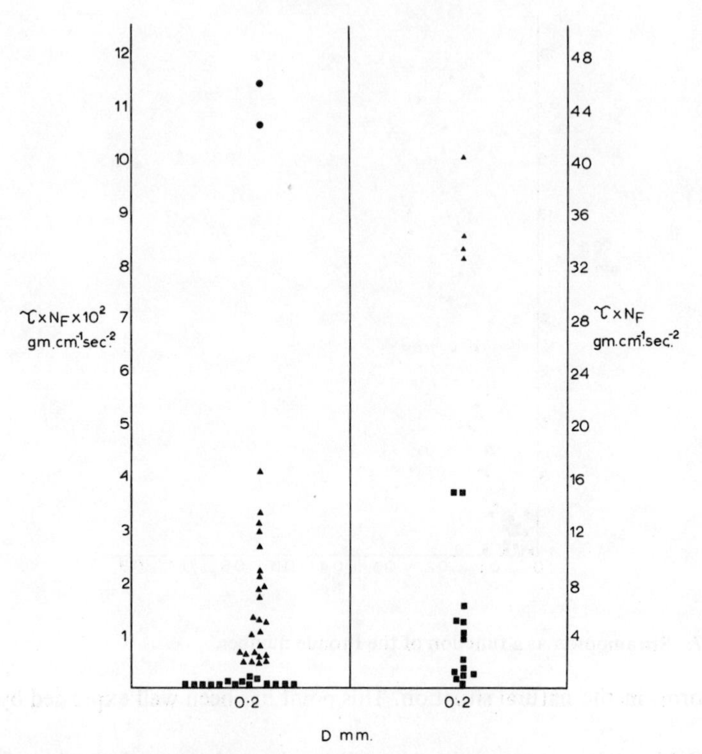

Figure 18. Bedform as of function of

$$\rho \iota^2 \left(\frac{du}{dy}\right)^2 \times \frac{\bar{u}}{\sqrt{gD}}$$

and the sediment size.

An effective separation of the basic division of bedform stages may be achieved from the use of the parameter

$$\rho \iota^2 \left(\frac{du}{dy}\right)^2 \times \frac{\bar{u}}{\sqrt{gd}}$$

plotted against the median diameter of the bed material.

Summarizing, the validity of extrapolating from steady-state experiments to natural systems must be questioned. Simons and Richardson (19) recognized that in unsteady flow there cease to be unique hydraulic limits between different bedforms. This point has been illustrated beyond reproach from Pool Channel.

Four basic bedform stages are recognizable in the field: -

(a) No motion

(b) Motion - generation and destruction of bedforms
(c) Transition stage
(d) Upper flow regime with plane beds and antidunes.

The first three have been recognized and delimited in this study, more detailed observations may elucidate upon the complexities associated within stage (b). A higher degree of control and instrumentation in the field may bring answers that have so far eluded both sedimentologist and hydraulician alike.

ACKNOWLEDGMENTS

The writer wishes to acknowledge Mr. C. Charley, Kingston Polytechnic, for his help with the field work. He is also indebted to Dr. J. McManus, Dundee University, for his constructive criticism and advice during the preparation of the manuscript.

The investigation was carried out during the tenure of a Dundee University Research Studentship.

REFERENCES

1. Allen, J. R. L.
1968 Current Ripples, their relation to patterns of water and sediment motion, North-Holland, Amsterdam.
2. Allen, J. R. L.
1973 Phase differences between bed configuration and flow in natural environments and their geological relevance. **Sedimentology,** 20(2): 323-329.
3. Bogardi, J. L.
1965 European concepts of sediment transportation. **Proc. Am. Soc. Civ. Eng.,** Vol. 91, H71.
4. Cloet, R. L.
1954 Sand waves in the southern North Sea and Persian Gulf. **Jour. Inst. Nav.,** Vol. 7, No. 3.
5. Cornish, V.
1901 On sand waves and tidal currents. **Geog. Jour.,** 8: 170-200.
6. Du Buat, P.
1786 **Principes d'Hydraulique,** 2nd Ed., 2 Books, De L'Imprimerie de Monsieur, Paris.
7. Garde R. and Rangaraju, K.
1966 Resistance relationships in alluvial channel flow. **Proc. Am. Soc. Civ. Eng.,** Vol. 94.
8. Gilbert, G. K.
1914 The transportation of debris by running water. U. S. Geol. Survey Prof., Paper 86.
9. Guy, H., Simons, D., and Richardson, E.
1966 Summary of Alluvial Channel data from flume experiments 1959-61. U. S. Geol. Survey Prof., Paper 462-5.

10. Halliwell, A. R.
1968 Shear velocity in a tidal estuary. **Coastal Eng.,** Vol. 2, Part 4.

11. Hjulstrom, F.
1935 In **The River Klaralven, A study of fluvial processes,** A. Sundborg Evette Aktreberg, Stockholm, 1956.

12. Kenyon, N. and Stride A.
1970 The tide-swept continental shelf between the Shetland Isles and France. **Sedimentology,** Vol. 14, No. 3: 159-173.

13. McCave, I.
1971 Sand waves in the N. Sea. **Mar. Geol.,** 10: 199-225.

14. Prandtl, L.
1926 In **Mechanics of Sediment Transport.** Yalin, M., 1972. Pergammon, Oxford and New York.

15. Stride, A.
1959 A linear pattern on the sea floor and its interpretation. **Jour. Mar. Biol. Ass. U. K.,** 38: 313-318.

16. Stride, A.
1961 Mapping the sea floor with sound. **New Scientist,** 10: 304-306.

17. Stride, A.
1963 Current swept sea floor near southern half of G. Britain. **Q. Jour. Geol. Soc. (London),** 119: 175-199.

18. Simons, D. and Richardson, E.
1961 Forms of bed roughness in alluvial channels. **Proc. Am. Soc. Civ. Eng.,** Vol. 87, H73.

19. Simons, D. and Richardson, E.
1962 The effect of bed roughness on depth-discharge relationships in alluvial channels. Wat. Supply Irrigation Paper. Washington 149 8E, 26.

20. Simons, D. and Richardson, E.
1965 U. S. Dept. Agric. Miscellaneous Pub. 970, 193-207.

21. Terwindt, J.
1971 Sand waves in Southern bight of the N. Sea. **Mar. Geol.,** 10: 51-67.

22. Von Straaten, L.
1950 **Tisdsch. Kon. Ned. Aardrijksk Geroot.,** 67: 76-81.

23. Yalin, M.
1964 Geometrical properties of sand waves. **Proc. Am. Soc. Civ. Eng.** Vol. 96, No. H75.

24. Yalin, M.
1972 Mechanics of sediment transport. Pergammon, Oxford and New York.

25. Znamenskaya, N.
1963 Calculation of dimensions and speed shifting of channel formations. Soviet Hydr. Sel. papers 1962, No. 3: 323-328.

CIRCULATION AND SALINITY DISTRIBUTION IN THE RIO GUAYAS ESTUARY, ECUADOR

Stephen Murray, Dennis Conlon[1], Absornsuda Siripong[2], and Jose Santoro[3]

ABSTRACT

Observations of velocity and salinity profiles over a tidal cycle were made throughout the Guayas estuarine complex. Bihourly maps of the tidal current field show a nearly 2-hour phase shift from the estuary mouth to Guayaquil and a slack-water/high-tide phase lag in accordance with a frictionally retarded long wave in a nonuniform channel. A zone of intense mixing of river water and sea water is identified south of Guayaquil, and the flushing time indicated by the longitudinal salinity distribution is calculated at about 21 days. Gravitational convection is poorly developed and present only locally, but a significant upstream salt flux does occur in the tidal prism. The Hansen and Rattray circulation-stratification diagram successfully classifies the Guayas system as a poorly developed, partially mixed (type 2B) estuary.

INTRODUCTION

As part of a larger study including the ecology, geology, and boundary-layer meteorology of the lower Rio Guayas basin, the physical oceanography of the

1. Coastal Studies Institute, Louisiana State University, Baton Rouge, Louisiana 70803.

2. Coastal Studies Institute, Louisiana State University, Baton Rouge, Louisiana 70803. Now, Department of Marine Science, Faculty of Science, Chulalongkorn University, Bangkok, Thailand.

3. Armada del Ecuador, Servicio Hidrografico y Oceanografico, Guayaquil, Ecuador.

interior portions of the Gulf of Guayaquil were studied in detail in October and November, 1970, in cooperation with the Instituto Nacional de Pesca of Ecuador (INSNAPES). Thirty-three anchor stations were occupied on two separate cruises, of which nineteen stations involved measurements of velocity and salinity depth profiles over at least one tidal cycle (see Fig. 1).

Perhaps the largest and certainly the most important estuary on the western coast of South America, the Guayas serves as a conduit for the dense mesh of smaller rivers that drain and irrigate the 36,000 square kilometers of fertile coastal plains lying north of Guayaquil, the principal seaport of Ecuador (4). The estuary proper has two principal openings from the Gulf of Guayaquil, the Canal del Morro and the Canal de Jambeli, on the northern and southern shores, respectively, of the 22-km-wide Isla de Puná (see Fig. 1). At present, the Estero Salado, its high salinity evidenced by its name, has no natural connection to the Rio Guayas at its northern extremity near Guayaquil and apparently carries most of the flow coming through the Canal del Morro. There is minimal flow of water into the lower reaches of the Estero Salado through the channel north of Puná connecting the Canal del Morro and the Canal de Jambeli, hereafter referred to as the Puná channel. The Guayas, the principal conduit of fresh water to the sea, is formed only 5 km north of Guayaquil by the merger of the Rio Daule, from the northwest, and the Rio Babahoyo, from the northeast.

At the latitudes of the study area the air temperature changes only about 2°C (10) annually (mean value about 25°C), more as a function of the advance and retreat of the cold Peru Current and changing cloud cover than of solar altitude. Winds during the study interval were southwesterly to westerly trades (8) and rarely exceeded 7 m/sec; a marked diurnal variation occurred (evening die-off). In contrast to air temperature, rainfall has a marked annual variation into wet and dry seasons. Our study was conducted during the dry season, but some orographic rainfall occurs nearly all year along the foot of the Andes, eventually draining into the Rio Guayas and maintaining a fresh-water supply for Guayaquil (4). River stages at Babahoyo, about 60 km upstream from Guayaquil, vary by some 5 m between wet and dry seasons (12).

Tides nearly double in range, from about 1.8 m at the entrances to the Canal del Morro and Canal de Jambeli to about 3.3 m at Guayaquil (data from British Admiralty Chart No. 586).

The diurnal inequality of tides, tidal currents, and salinity oscillations amounts usually to no more than 5% of range, as evidenced by the two semidiurnal cycles observed at the INSNAPES floating dock (see Fig. 2). While maximum diurnal inequality is expected at southerly tropic tides (November 3) and minimum diurnal inequality six days later (November 9; equatorial tides), this bimonthly cycle, a function of lunar declination, is unimportant in the study area.

The objectives of the present study are (a) to determine the velocity field

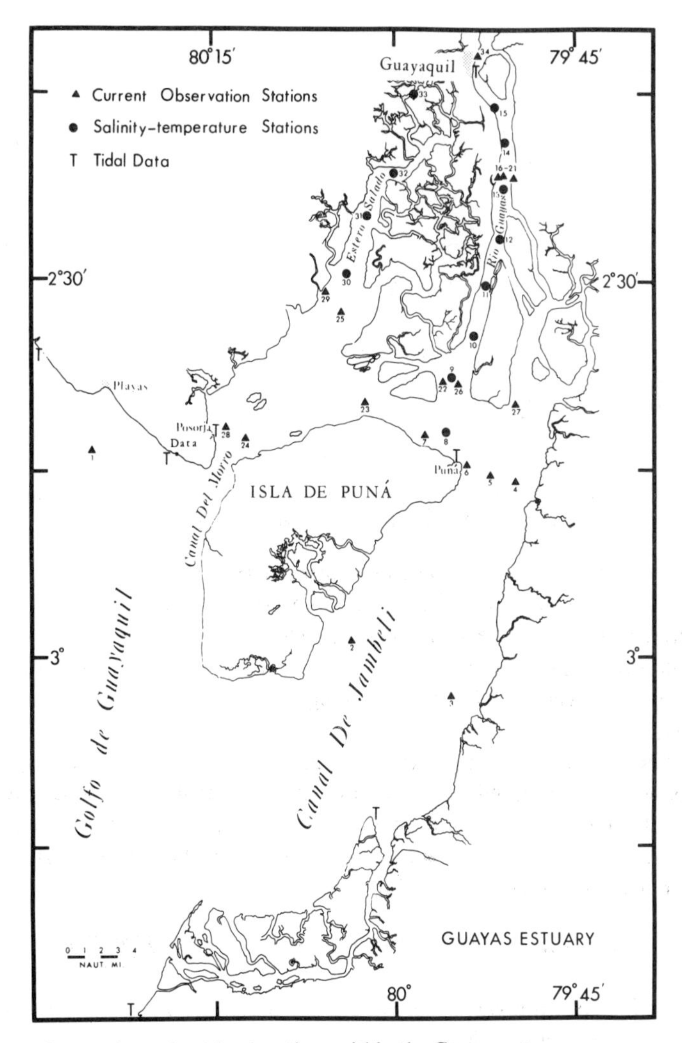

Figure 1. Geography and station locations within the Guayas estuary.

throughout the estuary complex as a function of time in the tidal cycle; (b) by detailed observations over a tidal cycle to ascertain the vertical and lateral variations of salinity and velocity at one channel cross section and thence their respective time-averaged distributions, (c) to determine the character of the longitudinal salinity gradient and thence the flushing time of the estuary below Guayaquil; and (d) to examine classification of the Guayas estuary according to the Hansen and Rattray (7) circulation-stratification scheme.

Current measurements were taken with both Marine Advisors ducted meters

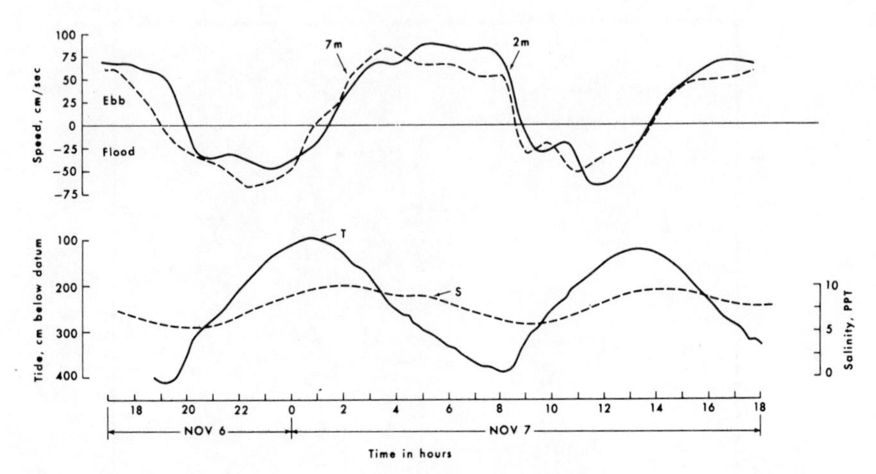

Figure 2. Time series of salinity, water-surface elevation, and longitudinal-current speed at two levels below the time-dependent water surface.

(Q15) and Hydro-Products Savonius rotor meters, which were cross calibrated periodically throughout the study interval. Salinity and temperature measurements were made with Beckman RS-5 induction salinometers. Although data on tidal heights is available for several locations in the estuary, direct measurements were made at Data (Fig. 1) and Guayaquil for comparative purposes.

Profiles of current were taken at least once each hour, and the observed vectors were decomposed into longitudinal and lateral components with respect to the channel and then plotted in a time series. Values of current speed at lunar-hour intervals were read from the smoothed time series. Salinity measurements were treated in a similar fashion.

STAGES IN THE TIDAL CURRENT VELOCITY CYCLE

Figure 3 shows the vertically averaged current speed at time intervals of two lunar hours throughout the estuary. At certain locations, one vector arrow may represent the cross-channel averages of two or three stations (cf. Fig. 1). Lunar hour zero (Fig. 3a) corresponds to the exact time of slack current after ebb flow at the 2-m-depth level at station 3. This is 2330 hours local time, October 29, 1970, or 7 hours and 36 minutes before predicted high tide at Guayaquil. Data at the other stations were, of course, taken in different tidal cycles, but they were easily referenced to the same lunar hour in a 12-lunar-hour cycle; diurnal inequalities were disregarded, as already discussed. Figure 3a shows that at lunar hour zero the current has barely begun to flood after passing through slack at the entrance to the Canal de Jambeli (stations 2 and 3). The rest of the

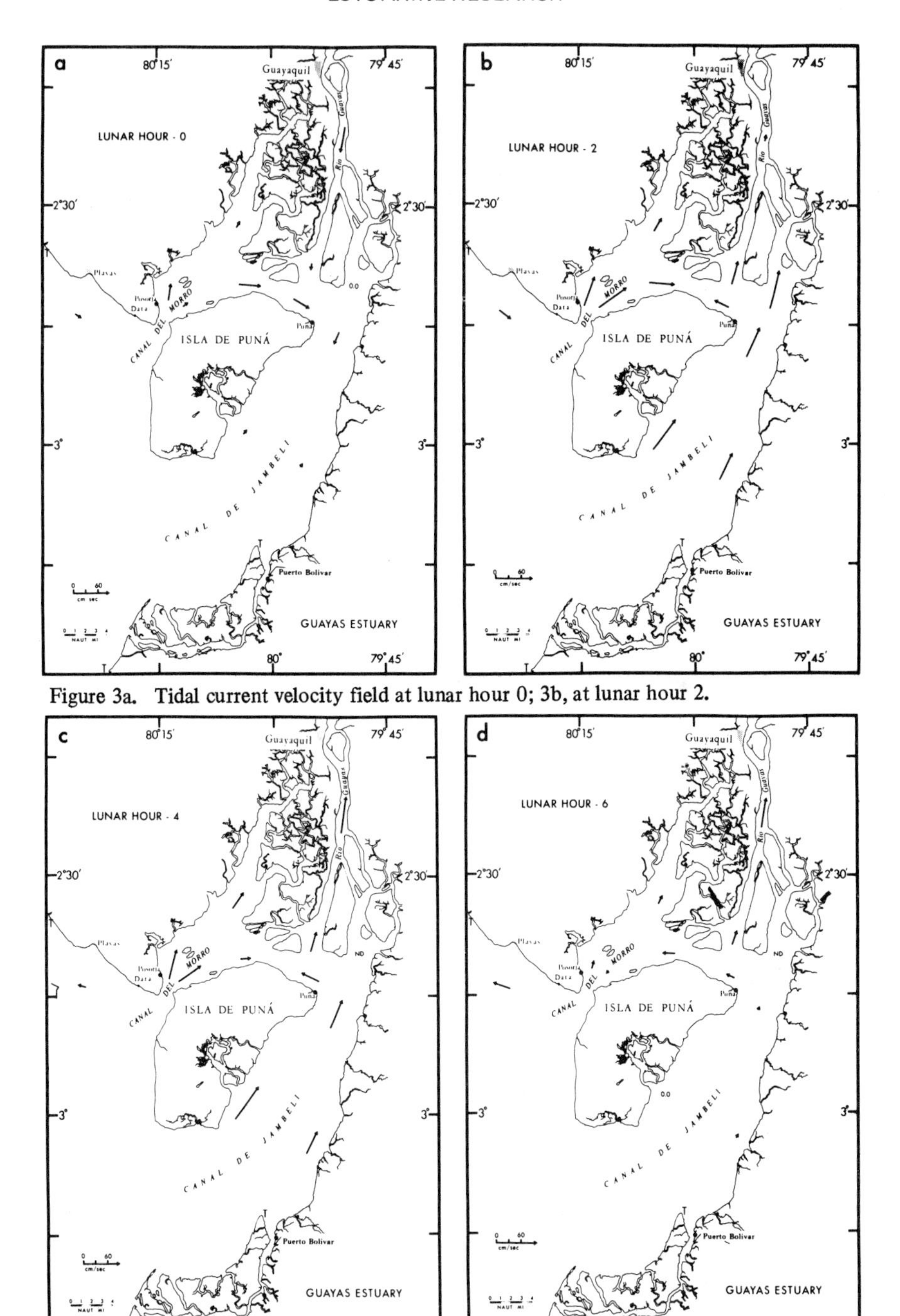

Figure 3a. Tidal current velocity field at lunar hour 0; 3b, at lunar hour 2.

Figure 3c. Tidal current velocity field at lunar hour 4; 3d, at lunar hour 6.

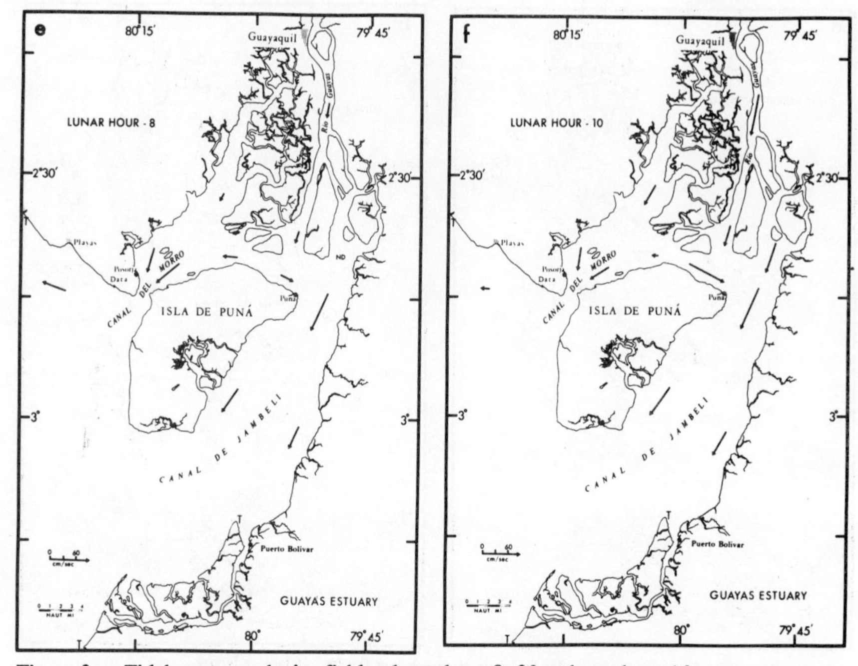

Figure 3e. Tidal current velocity field at lunar hour 8; 3f, at lunar hour 10.

Guayas-Jambeli channel, however, is still in strong ebb, seaward currents reaching 60 cm/sec in the Rio Guayas proper (stations 6-21) and decreasing to about 30 cm/sec in the northern end of the Jambeli channel (stations 4, 5, and 6). It is interesting to note that the current is already strongly flooding at this time (> 30 cm/sec) through the Canal del Morro and has even reversed upchannel in the Estero Salado (stations 25, 29). Owing to this flooding, current flow is eastward through the Puná channel.

Flood currents occur earlier at Canal del Morro despite the fact that both high and low tides occur earlier at Puerto Bolivar, near the entrance to the Canal de Jambeli, indicating that tidal dynamics at the entrances are probably strongly influenced by both friction and the nonlinear field accelerations. By one hour later (LH 1, not shown), all currents are upstream except at stations 16-21, which still show an ebb current of 60 cm/sec, and station 7, at the east end of the Puná channel, which is near slack as the effect of the flood current up the Canal de Jambeli is first felt.

By LH 2 (Fig. 3b) flood currents prevail throughout the estuary. The flow in the Puná channel is now convergent as the two major flood streams approach equal strength. At LH 3 (not shown), currents are similar to those of the previous hour, all stations showing larger magnitude except stations 7 and 23, in the Puná channel, where convergence is tending to reduce the opposing speeds.

At LH 4 (Fig. 3c) the flood-tide velocity field still prevails, but the earlier weakening of the flood current in the Canal del Morro is clearly shown by a marked drop in speed at station 23. One hour later, LH 5 (not shown), zero speed occurs in the west end of the Puná channel (station 23), while decelerating flood currents prevail through the rest of the channel.

At LH 6 (Fig. 3d) currents are already ebbing at the entrance to the Canal del Morro, and slack before ebb is imminent in the Canal de Jambeli. Note the westerly flow through the Puná channel. With the passage of one hour (LH 7, not shown), ebb currents occupy the lower reaches of the two major channels, while the upper reaches are still flooding, indicating about a 1.5-hour lag upchannel in the current stages. Divergent flow now exists (LH 7) in the Puná channel. Ebb currents predominate throughout at LH 8 (Fig. 3e), and divergent flow remains in the Puná channel. During LH 9, essentially the same velocity field occurs as in the previous hour.

Figure 3f (LH 10) shows the full strength of the ebb current in the Canal de Jambeli and a persistent but weakening divergent flow in the Puná channel. One hour before the completion of the tidal cycle (LH 11, not shown), an easterly through-flow exists in the Puná channel as a result of both the drawdown in the Canal de Jambeli and the early inception of flood currents through the Canal del Morro entrance.

The principal features emerging from these tidal velocity field maps (3a-3f) are the documentation of the phase shift of the tidal current stages upchannel in both major channels and the interesting current structure in the Puná channel, which shows the tidal cycle partitioned into four successive segments: (a) three hours of easterly flow, (b) four hours of convergent flow, (c) one hour of westerly flow, and then (d) four hours of divergent flow. Sedimentation studies (Rudolfo Cruz-Orosco, personal communication) clearly show this channel to be an area of rapid sedimentation, undoubtedly associated with a nodal point which should be present during eight hours of the tidal cycle.

The character of the tidal wave in the estuary is of interest both from the dynamical and geomorphological points of view (13). Table 1 gives the progression of high tide up the estuary to Guayaquil and would normally be interpreted as showing the "progressive wave" nature of the tide. We agree with Hunt (9), however, who discourages the use of "progressive" and "standing" wave terminology as being very misleading in the case of real estuaries, where friction must be considered in the momentum balance. Hunt's analytical solutions show that the effect of friction on the tide in a channel is to cause a progression of high water up the channel, as in a progressive wave, while the phase difference between high water and slack current remains small, close to the classical standing-wave value of zero. Table 2 shows that the Guayas data agree with Hunt on this point and support his contention that it is the effect of friction on tidal oscillations, and not any combination of "standing" and

TABLE 1.

Progression of Time of High Water

Location	Lead Time on Guayaquil	Distance Below Guayaquil
Guayaquil	0:0	0.0 km
Gaging Station	0:42	16.5
Puna	1:27	59.0
Puerto Bolivar	2:03	88.8

TABLE 2.

Phase Lag between Slack Water and High Tide

Location	Phase Lag
Guayaquil	0:35
Gaging Station	0:40
Stations 4, 5, 6	0:38
Stations 2, 3	0:45

"progressive" natures, that determines the character of the tide in nonuniform channels.

VELOCITY AND SALINITY SECTIONS IN THE RIO GUAYAS

From November 18 to 20, six anchor stations (16-21) across one section of the Rio Guayas were each occupied for one tidal cycle by two ships operating in tandem, thus requiring only three tidal cycles to complete the gaging. By treating the data in the same manner as discussed previously, maps of the cross-sectional distribution of speed and salinity were constructed for each lunar hour of the tidal cycle. Lunar hour zero in this series of figures (Figs. 4a-4f) was chosen as the time at this section of slack current at the 2-m depth after ebb flow, or 20 minutes before predicted low tide at Guayaquil. The view is upstream, toward the north, and so references to the right and left sides of the channel are reversed from the standard usage.

As seen in Figure 4a, only the middle section of the channel is at slack; ebb (positive) currents are still prevailing near the surface, and flood (negative) currents are already appearing near the bottom. This could be considered

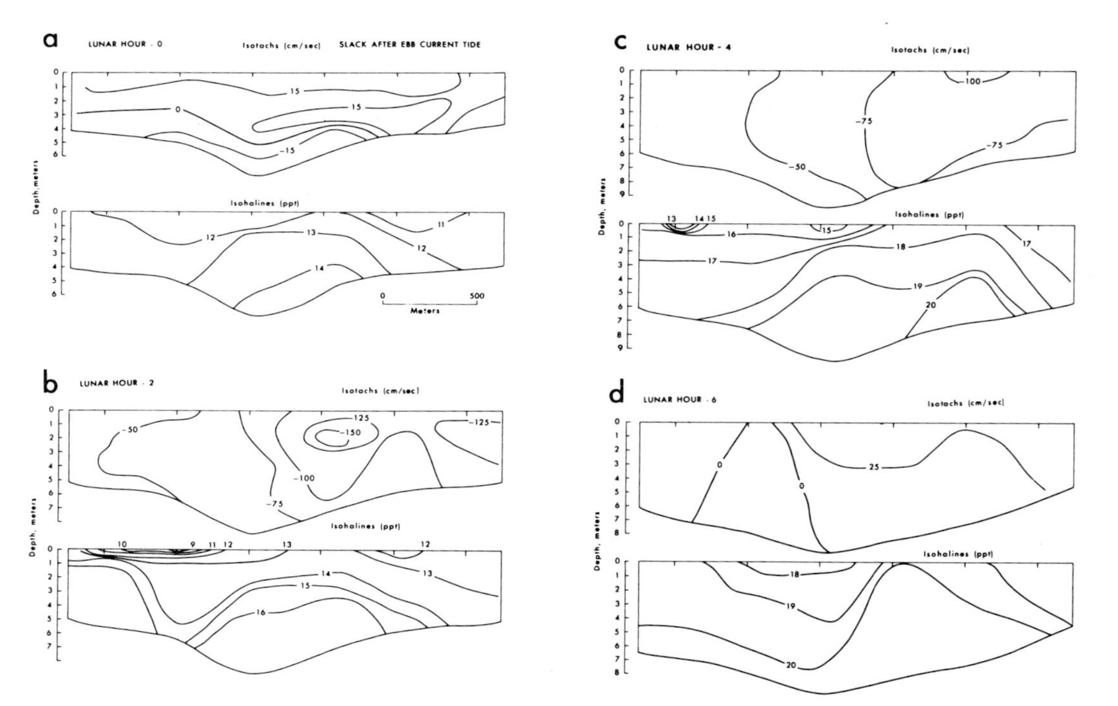

Figure 4. Velocity and salinity distributions at the gaging station through the tidal cycle.

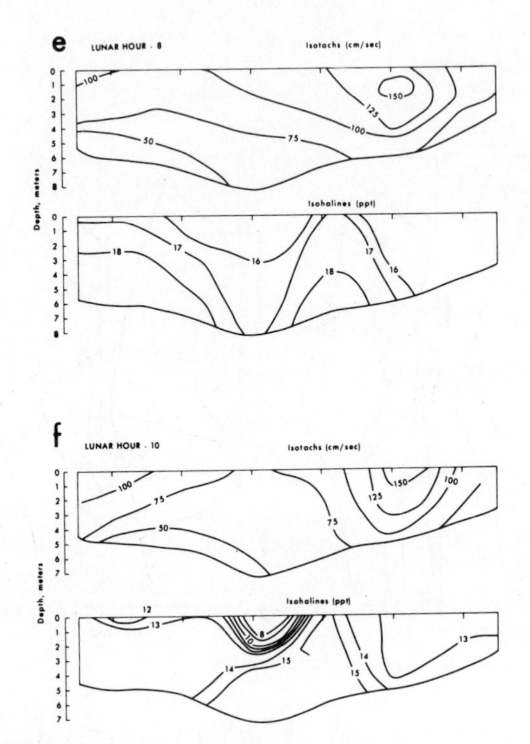

Figure 4 (continued.)

evidence for a weak density-driven flow, but it is equally likely that this advancement of the stage of tide closer to the bottom (priming) results from the role of the bottom frictional forces, as discussed by Proudman (11) and Cannon (3). Inasmuch as this is slack after ebb current, salinity is at its minimum for this location. Note the distinct stratification in the central, deep part of the channel but the well-mixed aspect along the channel walls. Within one lunar hour, flood speeds (negative) are well in excess of 50 cm/sec across the channel, and a pronounced subsurface jet occupies the east side of the channel, while the salinity is relatively unchanged.

At LH 2 (Fig. 4b), the flood velocity field is nearly at its maximum development; high velocities in excess of 100 cm/sec still occupy the east side of the channel, and the center channel pool of high-salinity water has expanded considerably as a result of the upstream movement of the densest water in the deepest part of the channel. Figure 4c (LH 4) shows that, even as the flood decelerates, maximum velocities occupy the east side of the channel; salinities have increased markedly (~5 ppt) throughout the channel in the previous two hours, maintaining a distinct stratification despite the considerable current speeds and shear. High water occurs at LH 5, while the decelerating current still exceeds 50 cm/sec upstream in more than half the channel; at this time, salinities are close to maximum, water of more than 21 ppt being present along the bottom in the east side of the channel.

The distributions at LH 6 (Fig. 4d) are very interesting in that currents have begun to ebb on the west side, while flood is still in progress on the east side. Note that the salinity distribution mirrors the velocity distribution inasmuch as the 20-ppt isohaline has bulged up to enclose most of the eastern half of the channel. An hour later (LH 7), the ebb is already well established, velocities of more than 75 cm/sec occurring in the eastern half of the channel. By LH 8 (Fig. 4e), the ebb is fully established; the speed distribution, except for direction reversal, is similar to that of the flood in that there is a subsurface jet again on the eastern side of the channel and speeds drop off slowly laterally across to the western bank. Clearly seen in Figure 4e and persistent, but more pronounced, throughout the remainder of the ebb cycle is a depressed lens of fresher water occupying the west-central part of the channel adjacent to the dome of saltier water in the east-central channel noted during the flood cycle. At LH 9 (not shown) and LH 10 (Fig. 4f), the velocity distributions differ only in minor details from those at LH 8, but the fresher water lens in the west-central channel has both deepened and strengthened considerably.

It appears from the hourly salinity maps that a major zone of fresh-water advection down the estuary in the shape of a tongue lies in the surface layer of the west-central channel. In contrast, a dome of high-salinity water persistently occupies the east-central channel bottom. Coriolis forces in the southern hemisphere should deflect the fresher waters to the eastern side of the channel and the denser saltier waters to the west side of the channel. Hansen (5) noted similar discrepancies in the Columbia River estuary and attributed them to accelerations related to channel shape. It is open to question whether a large bar growing out from the west channel wall about 750 m upstream of our section is deflecting the flow in such a manner as to cause these observed concentrations of high- and low-salinity waters. Waters near the western and eastern channel walls (roughly one-third the channel width on each side) are usually the best mixed in the channel. The role of these regions as mixing zones owing to channel-wall turbulence also can only be the subject of conjecture at this point.

TIME MEAN PROPERTIES

Tidal currents are generally considered to be the prime mixing agent in an estuary, and in this regard the root-mean-square tidal current speed over the tidal cycle is generally considered a critical parameter. Figure 5 shows the root-mean-square speed distribution across the channel at the gaging station. The dominant features are the subsurface jet on the right side, the general subsurface

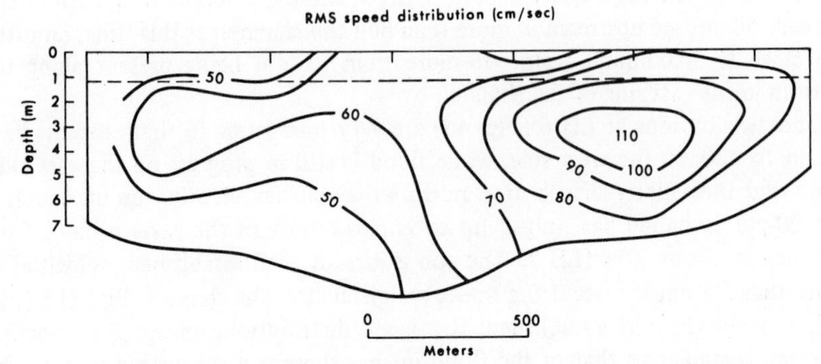

Figure 5. The root-mean-square speed distribution for the semidiurnal tidal cycle at the gaging station. The dashed line at 1.4 m is the mid-tide level.

maximum (frequently observed in rivers), and the significant lateral variation. Although such high root-mean-square speeds lead us intuitively to expect intense mixing, it is interesting to note that the 2-ppt salinity gradient over the channel depth was maintained through the tidal cycle.

Prime indicators of the type of circulation characteristic of a given stretch of estuary, i.e., the relative importance of gravitational convection and associated upstream advective salt flux on the salt balance, are the distributions of speed and salinity averaged over a tidal cycle. Figure 6 shows that net downstream (positive) flow occurs in two distinct zones (shaded) separated laterally across the channel. While some net upstream flow occurs in two small regions along the bottom, and in a vertical section at east-center channel, the major upstream flow is confined to the intertidal prism above the mean tide level (the dashed lines at 1.4 m in Figs. 5 and 6). There is a distinct correlation between the two centers of time/mean low salinity in the channel (locate the two 15-ppt isohalines in Fig. 6) and the zones of net downstream flow. Likewise, the 16-18 ppt water is associated with the zone of net upstream flow. Thus it appears that, although the gravitational convection mode of circulation is indeed present, it is only poorly developed and distributed nonuniformly across the channel. Very likely, these lateral inhomogeneities in root-mean-square speed and time-averaged speed and salinity are topographic effects related to the bar occupying the western side of the channel upstream of our section.

Owing to extreme shallowness of the estuary with respect to tide range, caution must be exercised in interpreting the time/mean quantities in the intertidal prism, i.e., 0-2.8 m in Figures 5 and 6. The averaging operation was performed through time at fixed levels above the bottom; thus above the low-tide level there is the disadvantage of having less than 12 lunar hour values in the average. On the other hand, this method clearly points out the importance of upstream advection of higher-salinity water in the intertidal prism and

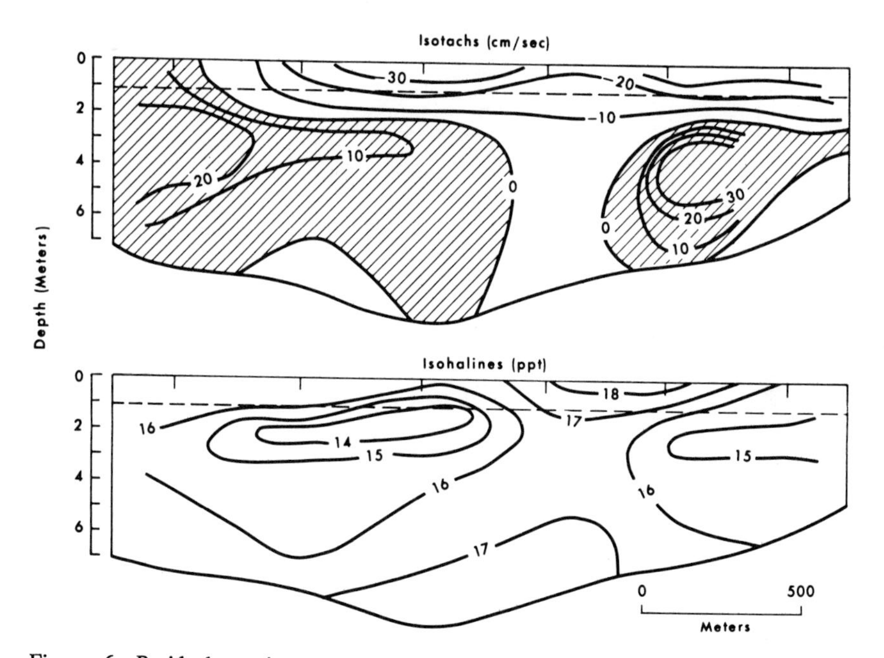

Figure 6. Residual or time-averaged velocity and salinity distributions at the gaging station. The dashed line at 1.4 m is the mid-tide level.

additionally meets the condition of conserving discharge of volume and salt over a tidal cycle. In other words, the discharge obtained by integrating the speed over the cross-sectional area in Figure 6 is equivalent to that obtained by averaging the time-step discharges through the tidal cycle. With an average discharge of 409 m^3/sec and an average area of 14,378 m^2, the fresh-water river discharge per unit area through this section is ~3 cm/sec.

LONGITUDINAL SALINITY DISTRIBUTION

The time/mean-salinity distribution in the Canal de Jambeli-Rio Guayas channel is shown in Figure 7. The sharp decrease in mean salinity between kilometer 75 and kilometer 120 identifies this stretch of channel as the zone where the river water and the sea water are most intensely mixed. The net flow of the river has been nearly fully diluted by sea water by the time it emerges from the Rio Guayas into the Canal de Jambeli.

Figure 8 shows the double amplitude or range of the salinity oscillations as a function of distance along the Jambeli channel. Above Guayaquil, the data are inferred from spot measurements of the longitudinal salinity gradient. The greatest temporal changes, exceeding 7 ppt at the gaging stations (stations 16-21), are also located in the zone of intense mixing. Examples of the vertical

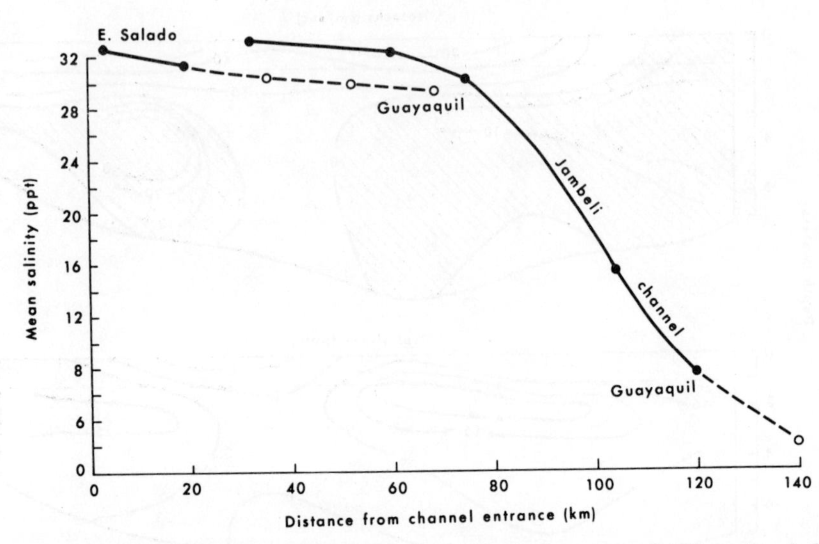

Figure 7. Distribution of time-averaged salinity along the two major channels of the Guayas estuarine complex.

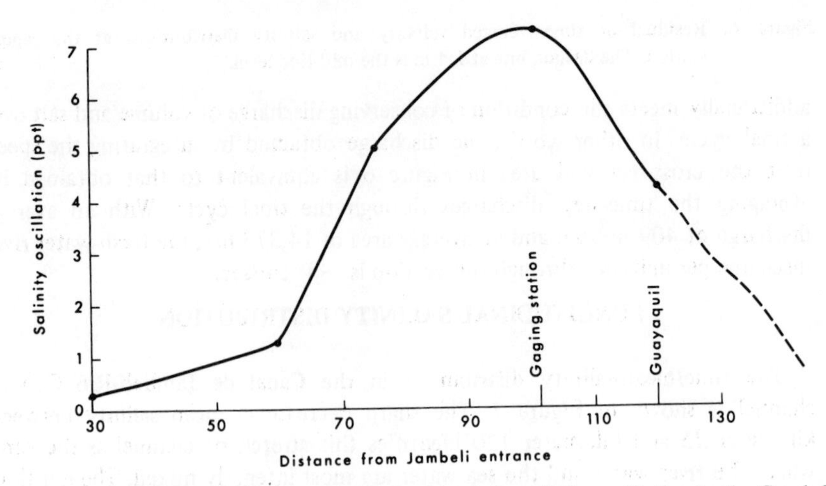

Figure 8. Maximum value (double amplitude) of the salinity oscillation in the Canel de Jambeli - Rio Guayas channel. Distance is in kilometers.

salinity gradients along the channel, shown in Figure 9, indicate vertical gradients $\partial S/\partial z$ also to be important only in the vicinity of the gaging station. Temporal changes in the vertical salinity gradients at all stations are negligible.

The data from the Estero Salado in Figure 7 indicate the waters near its mouth to be slightly fresher on the average than the Jambeli water. Comparison

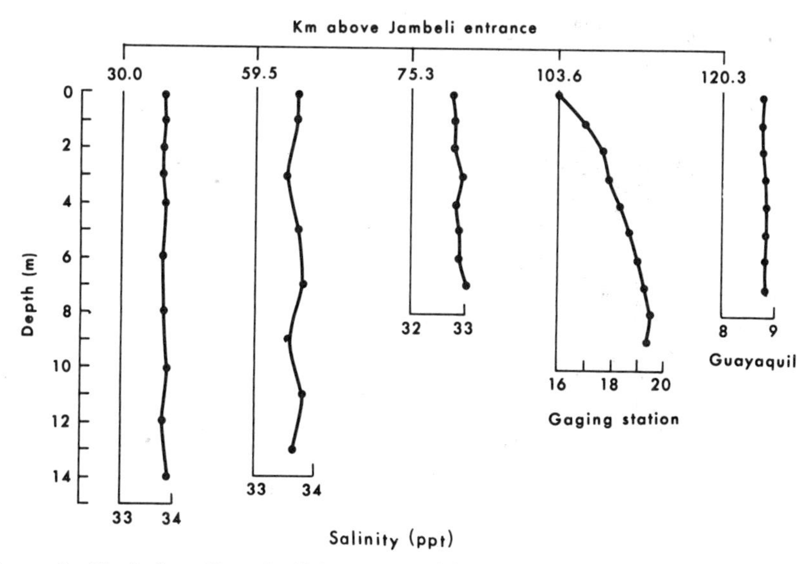

Figure 9. Vertical profiles of salinity at several locations in the Canal de Jambeli - Rio Guayas channel.

of ebb salinities from the Jambeli channel to flood salinities through the Canal del Morro support the idea that water flushed from the Canal de Jambeli is advected north for recycling into the Canal del Morro. Current-meter data from station 1, along the open coast off Playas (Fig. 1), are in agreement, indicating a net west-northwesterly drift along the coast of nearly 11 cm/sec. The high mean salinities in the Estero Salado all the way to Guayaquil (Fig. 9), together with the current observations in the Puná Channel, indicate a lack of circulation of fresher Rio Guayas water westward through the Puná channel.

With this knowledge of the salinity along the channel, a simple estimate of the flushing time t can be made (1) from t = F/R, where F is the accumulated volume of fresh water in the estuary and R is river discharge. Considering the estuary from Guayaquil to the seaward end of the Canal de Jambeli, the flushing time is calculated at 20.7 days. During high-river stage we can expect the flushing to decrease, but not in direct proportion to the increase in R, inasmuch as F also increases with increasing R.

CLASSIFICATION ON A CIRCULATION-STRATIFICATION DIAGRAM

Hansen and Rattray (7) presented a new classification scheme for estuaries based on their previous theoretical work (6) which was recently used (2) with good success in representing the dynamical condition in the Mersey estuary.A summary of this scheme is given in Figure 10, where the ratio of the

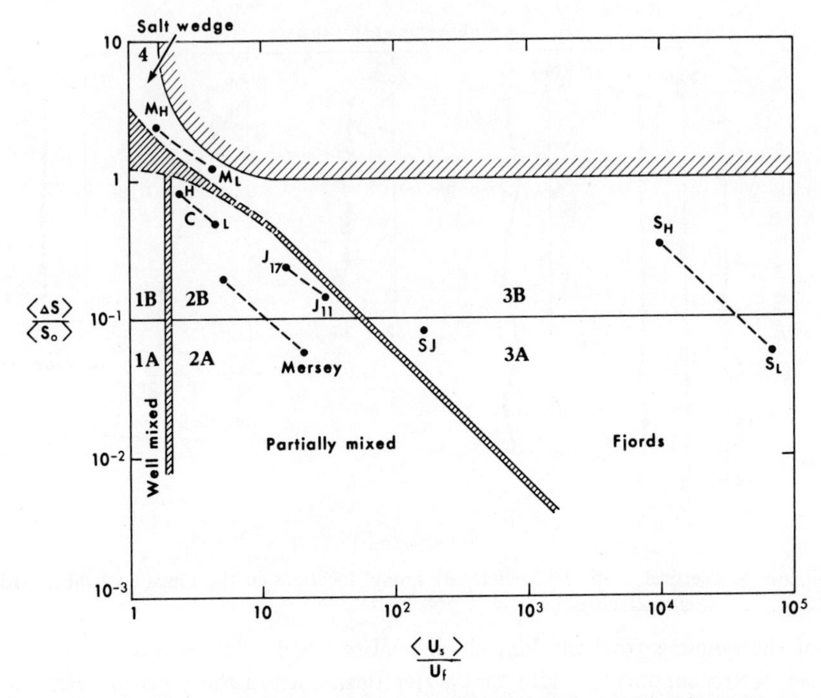

Figure 10. Stratification-circulation diagram showing the seven proposed estuary types. Examples are Mississippi River (M); Columbia River (C); James River (J); Silver Bay (S); the Strait of Juan de Fuca (JF); and the Mersey estuary. Subscripts H and L refer to high and low river discharge (after Hansen and Rattray, 1966).

surface-bottom salinity difference ΔS to depth mean salinity S_o is plotted as the vertical axis of a log-log scale, with U_s/U_f on the horizontal axis, where U_s is the surface velocity and U_f is the river discharge per unit area. All four quantities are residual or time-averaged over a tidal cycle.

On this type of diagram, seven types of estuaries are delineated. Type 1 is the well-mixed estuary in which the net flow does not reverse with depth and the upstream salt flux is all by diffusion. Type 1A is the archetype well-mixed estuary, whereas 1B shows appreciable stratification. For type 2 the net flow reverses with depth, and both advection and diffusion are important to the net upstream salt flux. Types 2A and 2B have the same stratification boundary as type 1. The type 3 estuaries are differentiated primarily by the clear dominance of advection in the upstream salt transfer. Type 4 is the salt wedge. Points from a given estuary are expected to determine one line on the C-S diagram analogous to the water-mass characteristics on a temperature-salinity diagram. The estuary characteristics should move along this line with seasonal changes in discharge and location changes in the estuary.

Data points from the present study of the Guayas estuary are plotted on a C-S diagram in Figure 11. Also drawn on this figure are three isopleths of ν, an important parameter in the Hansen-Rattray (6) theory, which is the fraction of the total salt flux (advective plus diffusive) accounted for by the diffusive upstream salt flux term. It is, in a sense, an inverse measure of the importance of the gravitational convection mode of circulation typified by a net upstream flow in the bottom layer.

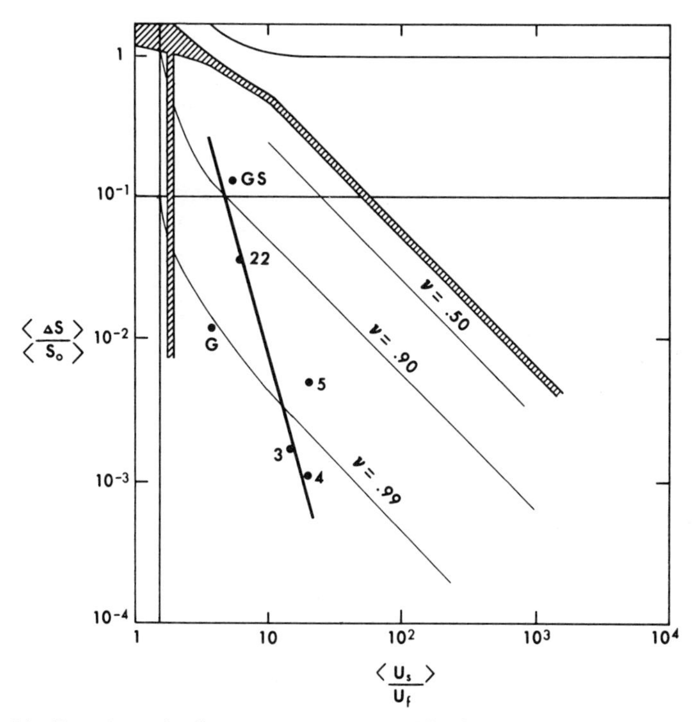

Figure 11. Data from the Guayas estuary on a stratification-circulation diagram. See Figure 1 for station locations; G is Guayaquil INSNAPES dock, GS is the gaging station (16-21). The ν isopleths representing the diffusive fraction of the total salt flux are replotted from Hansen and Rattray (1966).

The location of the least-squares line of best fit on Figure 11 ($r = 0.75$) is in good agreement with the theoretical expectations. The Guayas is indicated to be a partially mixed estuary but strongly increasing in ν values down-estuary as upstream diffusive flux increases and the gravitational mode of circulation decreases. Contrary to the restrictions of the theory behind the C-S diagram, U_f has been adjusted for change in cross-sectional area A along the channel using the continuity relation $A_1(U_f)_1 = A_2(U_f)_2$, where the numerical subscripts indicate different channel locations.

SUMMARY

Measurement of the velocity field throughout the Guayas estuarine complex documented a 1.5-2-hour phase shift in the dominant semidiurnal tidal current stages upchannel. The phase lag between slack water and high tide remains on the order of one-half hour along the channel, in agreement with Hunt's (9) analysis of frictionally retarded waves in nonuniform channels.

Measureable salinities extend 20 km north of Guayaquil, or 140 km upchannel from the entrance to the Canal de Jambeli, and indicate a flushing time south of Guayaquil of about 21 days.

Detailed measurements of velocity and salinity at one section over a tidal cycle show root-mean-square speeds as high as 100 cm/sec. Despite associated large vertical and lateral speed gradients, a vertical salinity gradient of more than 3-4 ppt over the 8-m channel depth is maintained through the tidal cycle. While highest salinity waters confine themselves to the deepest part of the channel, as expected, a major zone of fresh-water advection down-estuary lies in the surface layer in left (looking upstream) center channel and is probably related to the location of a large bar upstream.

Maps of the residual speed and salinity distributions show gravitational convection to be present but poorly developed and irregularly distributed across the channel. A major contribution to the upstream flux of mass and salt is seen to be located in the tidal prism owing to the relative shallowness of the estuary and the phase relationship between high water and current direction.

The estuary is successfully classified on the Hansen and Rattray circulation-stratification diagram. The C-S diagram, in agreement with our observations, indicates a poorly developed, partially mixed estuary in the zone of intense mixing near the gaging station, grading down-estuary on the C-S diagram into a region where the upstream salt flux is 99% accounted for by the diffusive flux. It seems doubtful that a completely well-mixed estuary would exist in nature in a confined channel.

Computational work is in progress on (a) partitions of the salt flux into contributions from vertical and lateral shear effects and time deviations of cross-sectional area, speed, and salinity; (b) vertical and longitudinal diffusion coefficients and vertical and lateral eddy stresses; and (c) finite-difference modeling of the propagation of the tide up the channel.

ACKNOWLEDGMENTS

We owe special thanks to the Instituto Nacional de Pesca and the Servicio Hidrografico y Oceanografico, Armada del Ecuador, both of Guayaquil, for outstanding logistical support of our research effort. The study was sponsored by the Office of Naval Research, Geography Programs, under contract N00014-69-A-0211-0003, project NR 388 002, with Coastal Studies Institute, Louisiana State University. Dr. William Gill Smith provided critical assistance in the program's formative stages. Mrs. G. Dunn provided cartographic services.

REFERENCES

1. Bowden, K. F.
1964 Circulation and diffusion. In **Estuaries,** p. 15-36. (ed. Lauff, G. H.) Am. Assoc. Adv. Sci. Publ. 83.

2. Bowden, K. F. and Gilligan, R. M.
1971 Characteristic features of estuarine circulation as represented in the Mersey estuary. **Limnol. Oceanogr.,** 16(3): 490-502.

3. Cannon, G. A.
1969 Observations of motion at intermediate and large scales in a coastal plain estuary. Johns Hopkins Univ., Chesapeake Bay Inst., Tech. Rept. 52, 114 pp.

4. Ferdon, E. N., Jr.
1950 Studies in Ecuadorian geography. Univ. Southern California, School of American Research, No. 5, 86 pp.

5. Hansen, D. V.
1965 Currents and mixing in the Columbia River estuary. **Ocean Science and Ocean Engineering,** Trans. Joint Conf. Marine Tech. Soc. and Amer. Soc. Limnology and Oceanography, Washington, D.C., p. 943-955.

6. Hansen, D. V. and Rattray, M., Jr.
1965 Gravitational circulation in straits and estuaries. **Jour. Mar. Res.,** 23: 104-122.

7. Hansen, D. V. and Rattray, M., Jr.
1966 New dimensions in estuary classification. **Limnol. Oceanogr.,** 11(3): 319-326.

8. Hsu, S. A.
1972 Boundary layer trade wind profile and stress on a tropical windward coast. **Boundary-Layer Meteorology,** 2: 284-289.

9. Hunt, J. N.
1963 Tidal oscillations in estuaries. **Geophys. Jour.,** 8(4): 440-455.

10. James, P. E.
1969 **Latin America.** Odyssey Press, New York, 947 pp.

11. Proudman, J.
1953 **Dynamical Oceanography.** Methuen, New York, 409 pp.

12. Servicio Nacional de Meteorologia e Hidrologia
1967 Annuario Hidrologico, 1967. No. 5, Quito, Ecuador.

13. Wright, L. D., Coleman, J. M., and Thom, B.
1973 Processes of channel development in a high-tide-range environment: Cambridge Gulf-Ord River Delta, Western Australia. **Jour. Geol.,** 81(1): 15-41.

TIDAL CURRENTS, SEDIMENT TRANSPORT, AND SAND BANKS IN CHESAPEAKE BAY ENTRANCE, VIRGINIA

by

John C. Ludwick[1]

ABSTRACT

Taking the mean over all stations in the entrance area, and with reference only to a level 100 cm above the bed, tidal currents at ebb and flood strength are nearly equal in speed and average 42 cm/sec. Taking individual stations, ebb maximum is 79 cm/sec; flood maximum is 66 cm/sec; however, at more than half the observation stations, maximum flood speed and duration exceed maximum ebb speed and duration.

If transport rate of bed sediment is proportional to stream power, $\tau_0 u_{100}$, where τ_0 is shear stress at the bed and u_{100} is current speed 100 cm above the bed, then at 19 of 24 stations, ebb-directed sediment transport exceeds flood-directed sediment transport. Much of the longer flood is incompetent and the greater ebb shear stress more than offsets larger flood speeds. Net sediment transport is directed landwards only in parts of channels and in dead-end sinuses re-entrant into shoals and open to flood currents.

In Chesapeake Bay entrance, a wide entrance of moderate tidal range, net sediment transport at the headlands is directed seaward. Individual shoals within the entrance are bounded on one side by a net sediment transport to seaward and on the other side by a net sediment transport to landward. Ebb deltas and flood deltas occur in alternate succession across the entrance with four to five of the former and three to four of the latter. There is also an interior flood delta found where ebb-directed sediment is interdicted and swept landwards in a flood-dominated channel.

1. Institute of Oceanography, Old Dominion University, Norfolk, Virginia 23508.

INTRODUCTION

In this paper a summary is given of sediment transport phenomena relating to shoals and sand banks in the tidal entrance to Chesapeake Bay, Virginia (Fig. 1). The entrance to this body of water is wide, the distance between the two entrance capes amounting to 17 km. Mean water depth is 11.3 m in the entrance section. Tidal range is 1 m, wave action is not intense; hence, the estuary is properly classified as one of moderate energy.

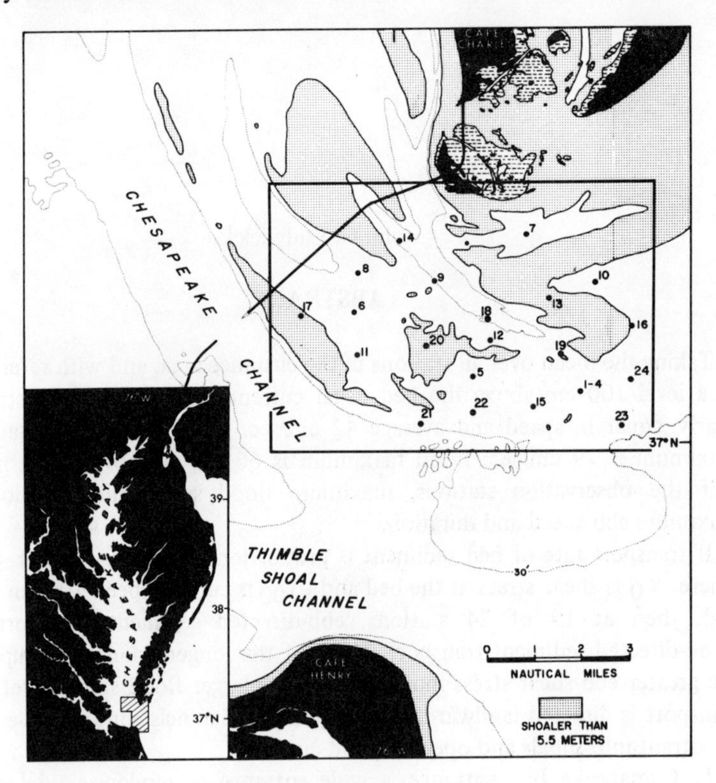

Figure 1. Index chart to the location of tidal current stations.

Sand banks, shoals, and tidal channels are prominent features of the entrance area, particularly in the northern part where sand-sized sediment is debouched into the estuary mouth by the littoral drift system along the ocean side of the Delmarva Peninsula. In contrast with smaller inlets, Cheseapeake Bay entrance lacks a typically developed single ebb-tidal delta and single flood-tidal delta. Instead there is a complex arrangement of sand banks separated by interdigitating tidal channels. It is this pattern of banks and the associated tidal currents that is the subject of this investigation.

Previous work on estuarine shoals in other areas is extensive and has been summarized elsewhere (6, 12, 13). Study of the entrance area to Chesapeake Bay has been conducted by the present author for several years and the results obtained to date have also been reported elsewhere (7, 8, 9). In the sand bank area, water depth at bank top is approximately 2.5 m; water depth in tidal channels is approximately 10 m. Tidal currents at the surface at strength reach speeds of 132 cm/sec on ebb and 105 cm/sec on flood; at the bottom, speeds reach 79 cm/sec on ebb and 66 cm/sec on flood. At all locations peak speeds are competent to move bottom sands which range from fine-grained (MD = 0.16 mm) to coarse-grained (Md = 0.66 mm).

FIELD METHODS, DATA REDUCTION, AND TRANSPORT CALCULATION

Current speed and direction in the entrance were measured at 24 stations with a Kelvin Hughes Direct Reading Current Meter suspended from an anchored ship. Observation periods at each station averaged 27 hours and thus included two ebb cycles and two flood cycles. The meter was calibrated in a flume and found to be accurate within 3 cm/sec. At each station the meter was raised successively through 11 different levels with a 4-minute observation period at each. Individual readings at each level were taken every 15 seconds.

In the office, averages of current speed for each 4-minute period for each depth were obtained. These speeds were then plotted versus the logarithm of the corresponding distance above the bed. Values for 11 standard depths were obtained by interpolation. These depth-corrected values of speed were then plotted for each standard level versus time. At standard times corresponding to every hour on the hour, readings of speed were obtained by linear interpolation. By this means a pseudo-synoptic data set for current speed was produced for each station.

The 11 time- and depth-corrected data points for speed versus depth were fitted by the method of least squares with a parabolic function. This function is an empirical velocity-defect law developed by Hama (2) and pertains to outer flow at distances greater than 0.15 d, where d is the thickness of a turbulent boundary layer, or water depth in the case of fully-developed flow in a uniform channel. The equation is

$$\frac{U - u}{u_*} = 9.6\left(1 - \frac{y}{d}\right)^2 \qquad (1)$$

whre U is the surface speed, u is speed at a distance y above the bed, and u_* is the friction velocity.

From the least squares curve fitting to the 11 data points, an estimate is obtained of τ_0, the boundary shear stress, according to the definition, $u_* =$

$(\tau_0 / \rho)^{1/2}$, where ρ is fluid density.

An algorithm proportional to bed sediment transport rate is taken from the work of Bagnold (1). His analysis of the energetics of sediment transport leads in the present study to a computation of flow power per unit area, $\tau_0 u_{100}$, where τ_0 is derived from u_* as explained above and u_{100} is the observed current speed at a point 100 cm above the bed. The units of this quantity are dyne-centimeters per second per square centimeter. Another publication (10) shows in detail how this quantity is reduced to mean tidal range, further corrected by subtracting 150 dyne-cm/sec/cm^2, a threshold value for initiation of sediment motion, and finally integrated separately over each ebb half-cycle and each flood half-cycle. The results are then averaged for ebb and flood. After integration and averaging, the units of the measure are dyne-centimeters per square centimeter per average ebb (or flood) half-cycle. Table 1 gives the magnitude and direction of the sediment-transport index, $\tau_0 u_{100}$, for 24 stations in the entrance area of Chesapeake Bay.

THE CHARTS OF BED-SEDIMENT MOTION

From the sediment-transport index described above at each of the 24 stations, it is possible to develop charts showing streamlines of near-bottom sediment transport for ebb and for flood half-cycles, and for the vector sum of ebb and flood sediment transport. This development is best viewed as a problem in vector field interpolation. A simple chart plot of the vectors for $\tau_0 u_{100}$ at the 24 stations for, say, ebb shows the need to obtain interpolated values between stations, values consistent in both vector direction and magnitude.

The first step in this procedure is to prepare a chart plot of the north-south component of $\tau_0 u_{100}$ for, say, ebb, and a separate chart plot of the east-west component of $\tau_0 u_{100}$. Isopleths are drawn on each chart. The two charts are then superimposed and at each intersection of isopleths the resultant direction is obtained and plotted. At contour intersections the resultant direction is obtained by trigonometry; the resultant magnitude is obtained by the pythagorean relationship. Any desired number or density of resultant directional arrows can be obtained by decreasing the isopleth interval of the component charts. Each directional resultant is plotted as a vector of unit length; only the directional information is used initially.

The unit vectors of the final dense field agree exactly with observed data at the 24 stations and are interpolated values elsewhere. Finally a streamline chart is prepared by drawing lines that are everywhere tangent to the unit vectors. Streamline charts for ebb, flood, and the vector sum of ebb and flood are shown in Figures 2-A, 3-A, and 4-A.

It is evident that lines of divergence and convergence occur in the sediment-transport field, particularly during flood flow. In the case of planar

TABLE 1.

Integrated Effective Flow Power Adjusted To Mean Tidal Range

Station Number	Water Depth m	Ebb $\tau_0 u_{100}$[1]	Ebb Dir °T	Flood $\tau_0 u_{100}$[1]	Flood Dir °T	Vector Sum $\tau_0 u_{100}$[2]	Vector Sum Dir °T
1-4	7.7	560	083	365	272	207	079
5	7.8	558	113	462	292	97	118
6	8.5	543	108	15	318	530	107
7	10.4	1003	069	660	271	462	037
8	8.3	582	095	294	273	288	097
9	10.1	1658	109	650	310	1077	097
10	7.2	92	089	1360	266	1268	266
11	6.1	1879	112	1038	318	1050	086
12	8.2	6352	097	1296	274	5058	098
13	3.3	1655	090	446	290	1245	083
14	10.7	751	134	1208	315	457	317
15	7.3	4004	078	562	281	3494	074
16	6.7	1846	079	828	283	1140	062
17	6.4	2538	129	28	327	2511	129
18	13.7	1707	125	2246	308	549	317
19	6.8	1514	101	76	285	1438	101
20	5.6	3253	119	262	322	3014	117
21	5.7	700	134	893	312	195	305
22	6.4	1243	118	380	299	864	118
23	10.0	2094	063	93	274	2015	062
24	8.4	145	081	538	266	394	269

1 Units are dyne-cm/cm^2/half cycle x 10^{-4}

2 Units are dyne-cm/cm^2/tidal cycle x 10^{-4}

potential flow of an imcompressible fluid, it is not possible for streamlines to join. However, in the case of three-dimensional flow, merging or joining of streamlines indicates ascending fluid motion; divergence of streamlines indicates descending fluid motion. A pattern of alternating zones of convergence and divergence is characteristic of spiral flow in a set of cells with horizontal parallel axes.

Sediment, unlike liquid, can be lost or gained during transport; hence sediment input in a stream tube does not necessarily equal sediment output. Sediment can be lost from transport by deposition or gained by erosion. Spacing

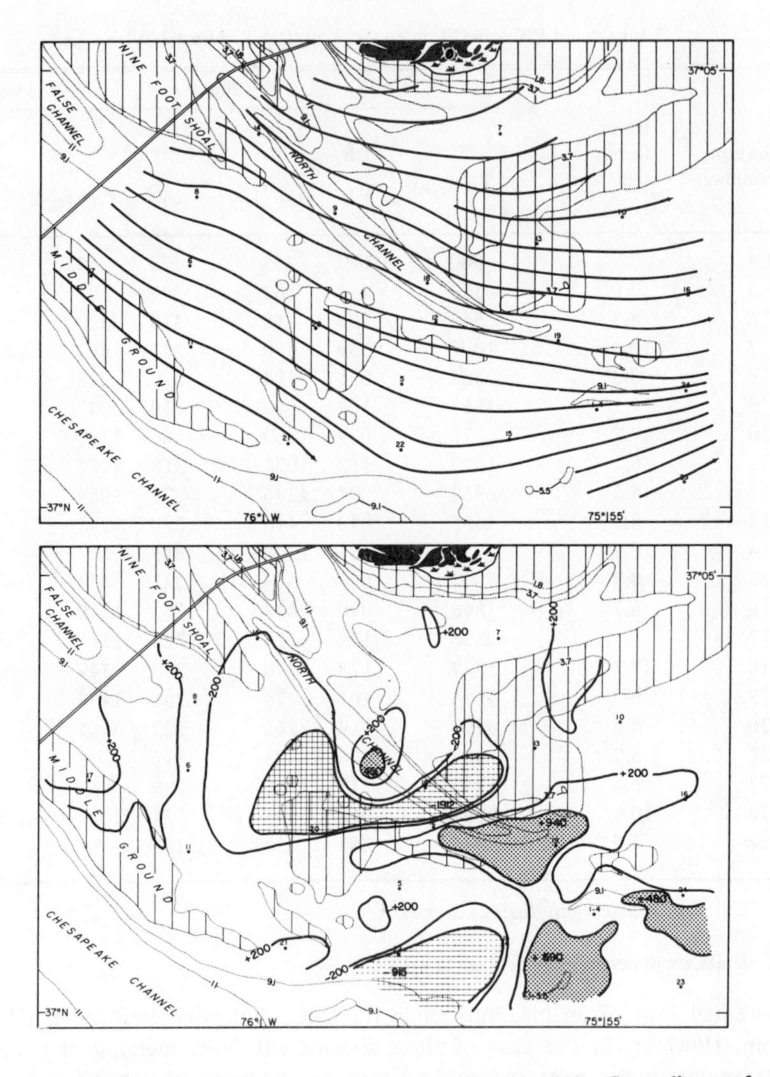

Figure 2. Ebb-directed sediment transport at the bed. (A). Streamlines of the sediment-transport vector $\tau_0 u_{100}$; depths are in meters; vertically ruled areas are shoaler than 5.5 m. (B). Erosion-deposition chart on which erosion is positive (+) and deposition is negative (-); units are dyne-cm per sq. cm per ebb half tidal cycle per 463 m of transport x 10^{-4}; cross-hatched areas indicate erosion intensity greater than -400 units; stippled areas indicate deposition intensity greater than +400 units.

between sediment-transport streamlines is not a measure of sediment-transport flux. There is no underlying stream function in the development presented here. What is assumed in the use of the streamline charts is that sediment transport is confined to a channel of unit width. The centerline of this channel is a streamline. The floor of the channel conforms to the bathymetry along the channel path. For the entire transport pattern, it is assumed that conditions are steady and non-uniform.

Now as a separate and ensuing procedure, it is possible to deduce the location of erosional areas and depositional areas. The usual sediment continuity equation is

$$\frac{\partial \eta}{\partial t} = - \varepsilon \frac{\partial Q}{\partial x}, \qquad (2)$$

where η is bed elevation relative to a datum plane, t is time, ε is a dimensional constant related to sediment porosity, Q is weight rate of bed-sediment transport per unit width of channel, and x is distance along a sediment-transport pathway or streamline in the direction of movement.

In what follows, Q is taken as proportional to $\tau_0 u_{100}$, the right hand partial derivative is approximated by a finite difference,

$$\frac{\partial Q}{\partial x} \approx \frac{\Delta Q}{\Delta x} = k \frac{(\tau_0 u_{100})_2 - (\tau_0 u_{100})_1}{x_2 - x_1}, \qquad (3)$$

and $x_2 - x_1$ is held constant arbitrarily at a value of 463 m, whence

$$\frac{\partial \eta}{\partial t} \alpha - \Delta \tau_0 u_{100}. \qquad (4)$$

Thus a decrease in transport rate along a transport path requires deposition; whereas, an increase requires erosion. Returning to the component charts and the streamline chart for, say, ebb, the three are finally superimposed and the magnitude of $\tau_0 u_{100}$ is determined at equispaced points along each streamline. Differences between adjacent pairs of values are taken and the distribution of erosion and deposition intensity is then mapped over the area of study. Positive values correspond to inferred deposition; negative values correspond to inferred erosion. The size of the positive or negative difference is proportional to the rate of deposition or erosion. Charts of this type have been prepared for ebb (Fig. 2-B), flood (Fig. 3-B), and the the vector sum of ebb and flood (Fig. 4-B).

RESULTS

With reference to the ebb streamline chart for bottom-sediment transport (Fig. 2-A), it is seen that the motion is not markedly deflected by local shoals probably because of low declivity of bottom slopes. Transport is, however, directed distinctly across North Channel. Transport is also up and across the end of Nine Foot Shoal, an ebb spit. In the southeast corner of the chart, the pronounced change in transport direction is a manifestation of the James River jet, the edge of which has spread more than half-way across the entrance width and into the study area. It is noteworthy that many of the elongated features of the area are more or less parallel to the ebb transport streamlines.

With reference to the erosion-deposition chart for ebb (Fig. 2-B), it is seen that there are two erosion areas characterized by intensity indices greater than -400. Maximum erosion on ebb is inferred to occur in a part of North Channel. The large erosion area may be an extension of False Channel. The other erosion area is positioned so as to suggest an escaping or easement flow from Chesapeake Channel, perhaps as a result of the blocking action of a large shoal at the mouth of that channel. Continued action in this pattern should cause an ebb bifurcation channel to form. The principal depositional areas on ebb are situated so as to suggest that eroded sediment is soon deposited to seaward in a patch. The area of maximum deposition at the seaward end of north Channel corresponds to an area where seaward-facing sand waves up to 4 m in height are prominently developed. In this area these waves migrate seaward at rates of 35 to 150 m/year (9).

With reference to the flood streamline chart for bottom-sediment transport (Fig. 3-A), it is seen that there is much less transport parallelism than for the ebb case. There are features of the pattern suggestive of lines of transport convergence and lines of transport divergence. Moreover, there is some indication that lines of convergence and divergence occur alternately as one moves across the chart from northeast to southwest. Flood transport in North Channel is parallel to the channel axis and into a prominent sinus in a flood delta. Ebb and flood transports are not opposed exactly; most of the intersection angles open to the north.

With reference to the erosion-deposition chart for flood (Fig. 3-B), it is seen that the intensity of both these processes is lower than for ebb conditions. Maximum erosion is -483 units; maximum deposition is +348 units. Some deposition occurs on Middle Ground and in North Channel. Erosion of note also occurs in the same channel.

With reference to the streamline chart for the vector sum of ebb and flood transport (Fig. 4-A), it is seen that most of the area is characterized by net ebb-directed transport. This is because much of the longer flood flow is

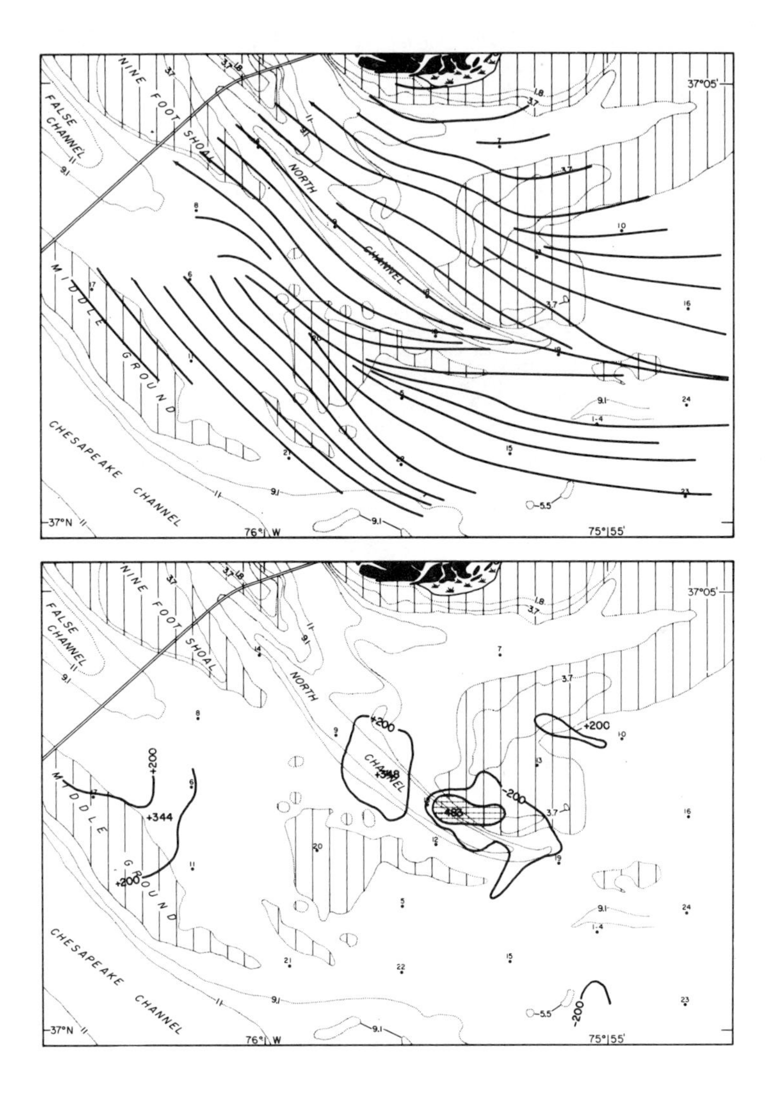

Figure 3. Flood-directed sediment transport at the bed. (A). Streamlines of the sediment-transport vector $\tau_0 u_{100}$; depths are in meters; vertically ruled areas are shoaler than 5.5 m. (B). Erosion-deposition chart on which erosion is positive (+) and deposition is negative (-); units are dyne-cm per sq. cm per flood half tidal cycle per 463 m of transport x 10^{-4}; cross hatched areas indicate erosion intensity greater than -400 units; stippled areas indicate deposition intensity greater than +400 units.

Figure 4. Vector sum of ebb and flood sediment transport at the bed. (A). Streamlines of the resultant sediment transport vector $\tau_0 \; u_{100}$; depths are in meters; vertically ruled areas are shoaler than 5.5 m. (B). Erosion-deposition chart on which erosion is positive (+) and deposition is negative (-); units are dyne-cm per sq. cm per tidal cycle per 463 m of resultant transport x 10^{-4}; cross hatched areas indicate erosion intensity greater than -400 units; stippled areas indicate deposition intensity greater than +400 units.

incompetent and because ebb shear stress is usually greater than flood shear stress (10). There are, however, 5 areas in which flood-sediment transport is dominant. The stations are 10, 14, 18, 21, and 24. Stations 10 and 24 are located in channels or sinuses on the seaward edge of the area and are open to the flood. Stations 14 and 18 are in North Channel in the landward part and are on the same streamline which leads into a major sinus in a flood delta. Station 21 is thought to occur on the edge of flood-dominated flow in the bottom of Chesapeake Channel. The fifth area is marginal to Nine Foot shoal and is based on station data not reported in this paper.

It must be borne in mind that the vector sum chart represents resultants of alternating processes and not of two processes operating simultaneously. Nevertheless, in this single chart representation, ebb-dominant areas can be distinguished and compared readily with flood-dominant areas. There is much yet to be learned about the nature of boundaries between the two flow areas. There is a tendency for shoals to be located in such a way that there is an ebb-dominated transport path on one side of a shoal and a flood-dominated transport path on the other. There are several examples of this relationship, but perhaps the clearest is seen in the case of stations 7 and 10. At station 7 transport is strongly ebb-dominated; at station 10 transport is strongly flood-dominated. There is a prominent sand ridge, or shoal, separating the two stations.

With reference finally to the erosion-deposition chart for the vector sum of ebb and flood transport (Fig. 4-B), one principal motif is evident. Moving in the direction of the streamlines, (Fig. 4-A), areas of erosion are followed by areas of deposition. The former often correspond to the up-current backs of existing shoals; the latter often correspond to the crests and down-current slopes of shoals. This relationship holds for ebb-dominated paths as well as for flood-dominated paths. It is also seen that erosion areas tend to be in existing deeper areas and to extend, in the form of branching arms, around existing higher shoal areas.

DISCUSSION

An essential, but not necessarily obvious, aspect of tidal flow in estuaries is that, at a fixed observation point, ebb and flood flows are not usually equal and not always oppositely directed. The concept of flow inequality has become embodied in terms such as net flow, time-velocity asymmetry, flow dominance, residual flow, and non-tidal flow. Various measures are in use to describe the inequality; direction and magnitude of the difference between peak ebb and peak flood velocity; difference in duration of flow; and difference in the areas under the ebb and flood curves on a time-velocity plot of tidal currents. This last measure has units of length and has been referred to as the difference in tidal excursion.

The causes of inequality in tidal flow are probably many, but at least three seem to be fundamental: (a) flow inequality caused in essence by mixing that occurs between river water and sea water, i.e., vertical entrainment due to breaking of internal waves at density interfaces, and vertical mixing due to turbulent exchange; (b) preferred flow location associated with coriolis deflection; and (c) flow inequality associated with the local mutual evasion of ebb and flood tidal current pathways.

The first-listed cause produces a characteristic gross circulation pattern in partially-mixed estuaries of moderate salinity in temperate climates. In this pattern net flow at depth is directed into the estuary; net flow in near-surface layers is directed seawards.

The second-listed cause is manifested in the northern hemisphere as flood-dominant flow on the left-hand bank and ebb-dominant flow on the right-hand bank looking down an estuary toward the sea.

The third-listed cause may show itself as ebb-dominated flow in a given channel and flood-dominanted flow in an adjacent channel or area. Several common channel patterns of mutually evasive tidal flows are known: the forked pattern; the flanking pattern; and the parallel offset pattern (8).

The three above-listed causes of flow inequality can co-exist; however, coriolis effects are usually overshadowed by other processes. The flow inequalities associated with gross estuarine circulation tend to be depth-related, at least in partially-mixed estuaries. In deep channels, headward net flow is expected; in shallow areas net flow to seaward is expected. In mutual evasion, it is the depositional history and present configuration of shoals and channels that are important factors along with tidal range and relation between tidal elevation and current speed, i.e., the presence of a progressive tidal wave in some estuaries and a standing tidal wave in others.

When bed-sediment transport dominance, as distinct from fluid-flow dominance, is considered, conclusions reached often depend on the assumptions made about threshold velocity for sediment transport and transport rate. Net transport of bed sediment is not simply correlated with the net transport of fluid near the bed. In the case of Chesapeake Bay entrance, although the net fluid flow at depth is directed headwards, taking the entrance section as a whole, the net sediment transport in some local areas is headward, and in other areas, at the same depth, is seaward. The pattern of bed-sediment-transport dominance near entrance shoals appears to be related more closely to local mutual evasion of tidal flows than to gross estuarine circulation or coriolis deflection.

Models of shoal configuration and flow dominance have been developed by others for inlets (3, 4). There are similarities and contrasts when these models are compared with shoaling patterns in a wide entrance. Near the seaward end of North Channel there is an arrangement of banks and channels similar in some respects to the ebb-tidal model of Hayes (3, 4). There are two prominent flood-dominated areas corresponding to channels heading landward from the sea. In between is a small shoal corresponding perhaps to the ebb bar; however, this present feature is bounded on one side by ebb-dominated flow and on the other side by flood-dominated flow, and not by flood-dominated flow on both sides, as in the Hayes model.

North Channel corresponds to the major channel in a tidal inlet and like such a channel is ebb-dominated in seaward parts and flood-dominated in landward parts. Nine Foot shoal and the immediately adjacent sand banks exhibit many if not all of the features of Hayes' flood-tidal delta including the flood ramp, flood channels, high areas, and ebb spits.

In the present instance, however, the sediment forming this major flood delta apparently first experiences a net motion across North Channel which interdicts some of the transported material and serves as a chute for sediment movement landward. The flood delta is formed where spreading of the current occurs and deposition takes place. The final completed flood delta is probably a quasi-stable form in balance amongst mutually evasive ebb-dominated and flood-dominated flows.

The finding of flood-dominated bed transport in North Channel and ebb-directed transport in the shallower marginal areas is similar to results reported by Hubbell, Glenn, and Stevens (5) for the Columbia River estuary, Washington, and by Wright, Sonu, and Kielhorn (14) for East Pass near Destin, Florida. Middle Ground of the latter report is a flood delta situated between two branches of the main channel. Bed transport is flood-dominated in both branches. Their flood delta lacks a central flood-dominated sinus when compared to Nine Foot shoal of the present study.

Among the charts in the present paper, including bathymetric, streamline, and erosion-deposition charts for ebb and flood, certain geomorphic forms, sediment-transport directions, and erosion-deposition patterns appear to be interrelated. A barchan-like, or parabolic-shaped, shoal is a recurring form usually seen compounded with other similar, but opposite-facing, parabolic shoals. The prominent flood delta of the area is an obvious example. In addition, the line of somwhat discontinuous shoals that begins at the north headland and extends southward three-fourths of the distance across the entrance is an example of linked alternate-facing parabolas. Flood-dominant transport occurs in those sinuses open to the sea; ebb-dominated transport occurs in the seaward end of channels leading out from the estuary. Characteristically, these latter channels leading out from the estuary. Characteristically, these latter channels display the ramp-to-the-sea shape described by Oertel (11) from the Georgia coast.

If a comparison is made of the principal features of a major interior flood delta, Nine Foot shoal, and features of the smaller flood deltas in the exterior line of shoals, it is evident that there are only minor differences, except for scale. Thus the exterior line of zig-zag, or alternate-facing, parabolic shoals is regarded as a single feature comprised of alternating ebb deltas and flood deltas linked together. Across the entire entrance there are four or five ebb-dominated and three or four flood-dominated pathways of bed-sediment transport. At the headlands, net flow is directed seaward.

It was seen that bed-transport streamlines are not deflected by shoals, and that the up-current backs of shoals are often the site of erosion. A likely conclusion is that it is the decrease in water depth with distance up the back of a shoal that causes a velocity and shear stress increase, an increase in flow power, a corresponding increase in sediment transport rate, and hence erosion. These relations are similar to those occurring on the backs of sand waves. Present findings thus indicate that the sand banks and shoals of the area may be formed by processes quite similar to those that form sand waves and cause them to migrate. It is noteworthy that the down-current sides and the crests of the entrance shoals are often sites of deposition. It is thus also inferred that the sand banks are migrating in the direction of bed-sediment-transport dominance.

With reference to the development of shoals, perhaps the essential difference between wide estuaries and inlets is the stability of flow through the opening. When the width of an opening exceeds some unknown limit, it may well be that the flow alters from frontal to fingered and a linked series of ebb and flood deltas are formed as a consequence, rather than single features as in an inlet. The development of spiral flow in long cells with parallel horizontal axes may be favored in flood flow because of weaker density stratification. The relationship of the proposed fingered flow and spiral flow is presently unknown.

SUMMARY AND CONCLUSIONS

The measurement of tidal currents, their velocity profiles, inferred shear stress at the bed, and flow direction, provide a basis for a deduction of relative bed-sediment-transport rate over an area of shoals in the wide entrance to Chesapeake Bay. Streamlines of ebb- and flood-directed bed-sediment-transport are little deflected by shoals. During flood there is some evidence of convergent and divergent transport suggestive of spiral flow. Ebb transport is across North Channel; flood transport is along the axis of the channel.

When the vector sum of ebb and flood transport is plotted, five areas of flood dominance appear on a chart that shows largely ebb dominance of bed-sediment transport. The five areas are mostly in or related to channels.

A decrease in transport rate with distance requires deposition; an increase requires erosion. Using this continuity concept, areas of deposition and erosion can also be deduced from data on tidal currents. The intensity of erosion and deposition is greater on ebb than on flood. It is seen that erosion on the upcurrent side of a shoal is soon followed, down the streamline, by deposition on the shoal crest or down-current side. The process is strikingly similar to that by which sand waves migrate. The parabolic, or barchan-shaped, shoal with a pronounced sinus between the arms appears to be a common quasi-stable form under reversing tidal flow. Best agreement with present shoal and channel configuration, and bedform-facing direction, is given by the streamline and erosion-deposition charts for the vector sum of ebb and flood transport.

Across the entrance to Chesapeake Bay there is an outer line of shoals and an inner major flood delta. The outer line is comprised of alternate-facing, ebb parabolas, or deltas, and flood parabolas, or deltas, linked together into a crude zig-zag pattern. The sinuses are alternately ebb-dominated and flood-dominated. The major inner flood delta is a simple barchan form.

ACKNOWLEDGMENT

The author appreciates the assistance of those who have helped with this work, particularly graduate students John T. Wells and Mitchell A. Granat. The effort was supported by the Office of Naval Research, Geography Programs Branch, under Contract Number N00014-70-C-0083.

REFERENCES

1. Bagnold, R. A.
1965 Beach and nearshore processes. Part I, Mechanics of marine sedimentation. In **The Sea,** Vol. 3, p. 507-528.

2. Hama, F. R.
1953 Discussion: The frictional resistance of flat plates in zero pressure gradient (by L. Landweber). **Trans. Soc. Naval Arch. and Marine Engineers,** 61: 27-28.

3. Hayes, M. O., Goldsmith, V., and Hobbs, C. H.
1970 Offset coastal inlets. **Proceedings of 12th Coastal Eng. Conf.,** p. 1187-1200.

4. Hayes, M. O., Owens, E. H., Hubbard, D. K., and Abele, R. W.
1973 The investigation of form and processes in the coastal zone. In **Coastal Geomorphology,** p. 11-41. (ed. Coates, D. R.) State Univ. New York, Binghamton, New York.

5. Hubbell, D. W., Glenn, J. L., and Stevens, H. H.
1971 Studies of sediment transport in the Columbia River estuary. In **Proceedings of 1971 Tech. Conf. on Estuaries of the Pacific Northwest,** Circular 42: p. 190-226. Eng. Expt. Station, Oregon State Univ., Corvallis, Washington.

6. Lauff, G. H.
1967 **Estuaries.** Am. Assoc. Adv. Sci., Publ. 83, Washington, D. C.

7. Ludwick, J. C.
1970a Sand waves in the tidal entrance to Chesapeake Bay: preliminary

observations **Chesapeake Sci.,** 11: 98-110.

8. Ludwick, J. C.
   1970b Sand waves and tidal channels in the entrance to Chesapeake Bay. **Virginia Jour. Sci.,** 21: 178-184.

9. Ludwick, J. C.
   1972 Migration of tidal sand waves in Chesapeake Bay entrance. In **Shelf Sediment Transport,** p. 377-410. (eds. Swift, D. J. P., Duane, D. B., and Pilkey, O. H.) Dowden, Hutchinson, and Ross, Stroudsburg, Pennsylvania.

10. Ludwick, J. C.
   1973 Tidal currents and zig-zag sand shoals in a wide estuary entrance. **Old Dominion Univ., Institute of Oceanography,** Tech. Rept. No. 7: 1-23.

11. Oertel, G. F.
   1972 Sediment transport of estuary entrance shoals and the formation of swash platforms. **Jour. Sed. Pet.,** 42: 857-863.

12. Robinson, A. H. W.
   1956 The submarine morphology of certain port approach channel systems. **Jour. Inst. Nav.,** 9: 20-40.

13. Shepard, F. P. and Wanless, H. R.
   1971 **Our Changing Coastlines.** McGraw-Hill, New York.

14. Wright, L. D., Sonu, C. J., and Kielhorn, W. V.
   1972 Water-mass stratification and bed form characteristics in East Pass, Destin, Florida. **Mar. Geol.,** 12: 43:58.

HIGH-ENERGY BEDFORMS IN THE NONTIDAL GREAT BELT LINKING NORTH SEA AND BALTIC SEA

by

Friedrich Werner and Robert S. Newton[1]

ABSTRACT

Water exchange between the brackish Baltic Sea and the marine North Sea is funneled through three channels that separate the Danish islands lying between the southern Danish peninsula and the southern tip of Sweden. The Great Belt, with a sill depth of 23 m, is the only one of these that is hydrographically important. Meteorologic anomalies over the two seas can lead to the development of strong, aperiodic pulses in the exchange between outflowing Baltic surface water and inflowing, heavier North Sea water. Currents related to these pulses are governed entirely by sea level and density differences; tidal forces play practically no role. The influence of these currents on the bottom is revealed through a wide range of bedforms developed in the thin sediment cover which overlies the glacial till basement. Sea-floor mapping with side-scan sonar showed the following bedforms to be present:

I. Longitudinal forms

(a) Comet marks – comet-shaped erosional scours down-current from obstacles; length from 1 to over 100 m.

(b) Sand ribbons – thin bands of residual or migrating sand; width 1 to over 100 m, length generally $>$ 100 m.

(c) Sand shadows – tails of sand deposited down-current from obstacles; length normally between 10 and 100 m.

II. Transverse forms

1. Geologisches Institut, Universitaet Kiel, Olshausenstrasse 40/60, 23 Kiel, West Germany.

(a) Megaripples – wavelengths from 60 cm to 90 m.
(b) Small-scale and large-scale ripples – wavelengths from 10 to 60 cm.

These current marks are all excellent indicators of flow direction. Their distribution pattern in the Great Belt shows a definite hierarchy related to current strength, starting with small comet marks and ending with wide sand ribbons and sediment-free hard ground. Based on this scheme, maximum inflow velocities lie slightly west of the channel axis; maximum outflow velocities are found in the shallow water along the eastern channel margin.

REGIONAL SETTING

The Great Belt (Store Baelt) lies between the Danish peninsula and the southern tip of Sweden and forms the critical connecting link between the full-marine waters of the North Sea and the brackish waters of the Baltic Sea. It is a crucial area because here the water column is constricted both vertically and horizontally, and here is found the chief mixing zone between brackish water and salt water. Lying to the east of the Great Belt is the tideless Baltic Sea, the world's largest body of brackish water, with surface salinities ranging from 0°/oo to 4°/oo in the Gulf of Bothnia to 20°/oo to 25°/oo in the western Baltic. It is the catchment basin for numerous Scandinavian, Russian, and German rivers; in fact, if only river inflow, precipitation, evaporation, and surface outflow were considered, the Baltic would rapidly become a fresh-water lake, as it was during the end of the Pleistocene era. However, to the northwest of the Great Belt is the North Sea whose full-marine water moves through the Skagerrak and Kattegat and finally through the Great Belt into the Baltic, compensating for the Baltic outflow (Fig. 1). This heavier water moves along the bottom and is funneled through three channels around the Danish islands that build the southern end of the Kattegat. Only one Great Belt – the central of the three channels – is of real importance for this water exchange; the others are too narrow and too shallow appreciably to alter the hydrographic picture.

Although the Great Belt fits only the loosest definitions of an estuary, it is an area where water from the open sea is measurably diluted and where inflow-outflow circulation builds bedforms that are at least in part typical of estuaries.

The northern end of the Great Belt has a sill depth of 26 m and the southern end a sill depth of 23 m. The block diagram (Fig. 2) shows the typical morphology of the southern portion of the Belt: on the western side, there is a narrow channel which in places is over 50 m deep, and east of this channel a fairly even slope from 23 m up to 10 m.

The inflow and outflow through the Great Belt is normally a rather low-keyed exchange. However, strong aperiodic current pulses caused, not by tides, but by meteorologic anomalies over the two seas can cause a strong surge of sea water

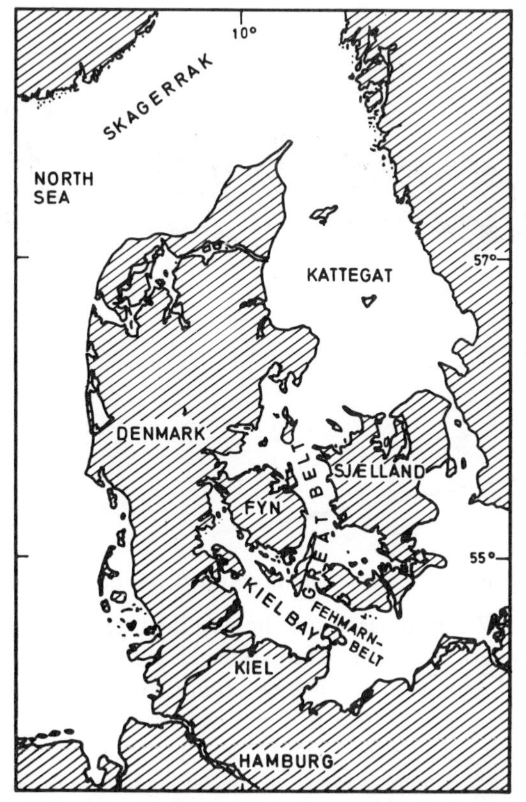

Figure 1. The position of the Great Belt in relation to the North Sea and Baltic Sea.

to enter the Great Belt. During such a surge, there develop bottom-current velocities which are considerably stronger than the average currents. Only one such surge has been measured by current meters (8). This pulse lasted several days and velocities of 120 cm/sec were recorded in 20 m depth in the Fehmarn Belt immediately south of the Great Belt. Although long-term current meter stations are planned, at present it is not known how often such pulses occur and their bottom velocities can only be estimated from the bedforms now present in the Great Belt.

BEDFORMS

Sea-floor mapping with side-scan sonar has revealed a much more dramatic picture than had been anticipated, particularly in relation to the longitudinal (current-parallel) features present. Older echo-sounder profiles had shown transverse forms, such as megaripples or sandwaves, but the side-scan sonar surveys revealed the domination of large-scale current lineations.

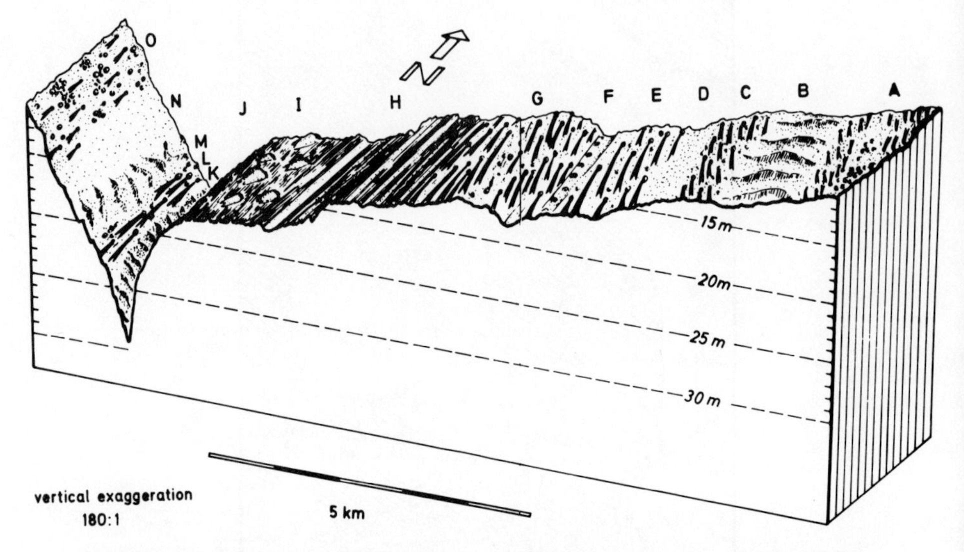

Figure 2. Typical morphology and bedform distribution in the southern Great Belt.

Figure 3 is a side-scan sonar record showing the typical current-swept bottom found in the southern Great Belt. The dark bands are strips of course sediment – in this case gravel with cobbles building a thin cover over the glacial till, which forms the basement in the Great Belt and western Baltic (4). This cover of lag sediment is usually only 10 to 30 cm thick. The lighter bands on the record are medium-grained sand which overlies the coarser lag sediment. The lineations here run north-south or parallel to the long axis of the Belt. From the analysis of all the side-scan sonar records, three categories of longitudinal current marks (comet marks, sand ribbons, sand shadows), as well as one category of transverse current marks (ripples), could be defined.

Comet Marks

To the left of "A" in Figure 3 it can be seen that the dark (acoustically rougher) areas all originate from individual points. These points are large cobbles or boulders up to 1 m in diameter which form an obstacle to the current flow. The dark streamers on the down-current side of the rocks are erosional scours where the finer sand has been removed, exposing the coarser sediment below . Because these scours are frequently comet-shaped, we have called them comet marks.

The size of comet marks ranges from very small and faint features with a length of 1 or 2 m up to forms over 100 m in length. The smaller comet marks are found along the eastern side of the Great Belt (Fig. 2, A); their size increases

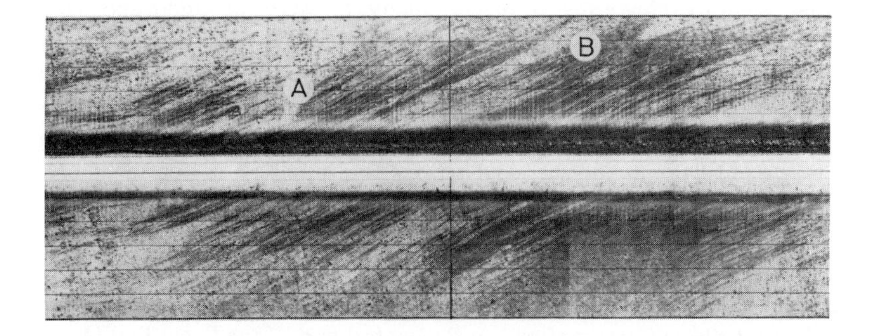

Figure 3. Side-scan sonar record from the southern Great Belt illustrating current-swept bottom. Record length 1150 m, width 160 m; water depth 17 m. Current flow from upper right (north) to lower left (south). Well-defined comet marks to the left of A; poorly defined, probably remnant sand ribbon at B. Further explanation in text.

toward the center of the Belt. Some are also found on the western flank of the deep channel (Fig. 2, L and O). With the exception of the easternmost margin of the Belt, all comet marks were formed by southward-moving inflow water. Although other types of erosional scours have been described from side-scan sonar surveying (3, 6), the comet mark seems to be uniquely good as an indicator of current direction.

As the comet tail is always coarser than the surrounding sediment, one can conclude that the obstacle is initiating a zone of instability and increased turbulence in the flow. Allen (1) has attributed the flow instability related to longitudinal current marks to the building of Taylor-Guertler vortices. However, regardless of the flow model, it seems safe to say that the increasing size of comet marks is directly related to increasing flow velocity. The fact that the comet marks build an orderly size sequence and are not randomly mixed is a further indication of their relation to a specific flow pattern.

Sand Ribbons

The side-scan sonar surveys also showed the presence of very sharply defined, alternating bands of coarse and fine sediment. We refer to the finer-grained strips which overlie coarser lag sediment as "sand ribbons," although it is not clear whether they are a stable remanant that has not been eroded or whether they are "streams" of migrating sand that would more closely fit Stride's (5) original definition of a sand ribbon.

The sand ribbons of the Great Belt normally show less than 30 cm of positive relief. They are in most cases over 100 m long and may reach several hundred meters in length. Their width varies considerably from 1 to over 100 m. In some areas they seem to show a degree of periodicity. Some of the sand ribbons have transverse forms, such as megaripples, superimposed upon them (Fig. 2, F). This would indicate that these particular ribbons are not quite as thin as normal and that the sand is moving. On the other hand, in some areas there appears to be a good transition from large comet marks to sand ribbons, thus pointing more toward a residual origin for the sand strips. This situation is seen in Figure 3 where "B" marks a poorly defined ribbon which is probably a remnant.

There is no clear relation between sand ribbons and minor changes in relief. The ribbons cross obstacles with up to 1 m relief without appreciable deviation – this has also been observed by Belderson (2) for sand ribbons in the Bristol Channel. Sand ribbons are confined to the central part of the Great Belt (Fig. 2, H).

Sand Shadows

Whereas with comet marks an erosional scour forms down-current from an obstacle, in the case of sand shadows just the opposite takes place; on the down-current side of an obstacle sand is deposited to form a dagger-shaped accumulation. Although both the depositional sand shadows and the erosional comet marks occur in roughly the same area (Fig. 2, I and J), there is a marked difference in the size of the generating obstacle. Sand shadows build out from behind barriers which are 1 to 3 m high and 10 t0 20 m long; comet marks form down-current from cobbles or boulders which seldom exceed 1 m in diameter. The barriers that precipitate sand-shadow deposition lie with their long axes normal to the current flow direction. The sand shadows themselves vary in length from 10 to about 100 m. In the Great Belt this type of current mark is confined to a narrow zone bordering and paralleling the deep channel.

Megaripples

In addition to the longitudinal current marks discussed above, there are other bedforms that are transverse to the flow direction. All the transverse current marks found in the Great Belt fall into the general category of ripples. In size they range from megaripples with wavelengths up to 80 m down to small-scale ripples with wavelengths of 10 to 30 cm. The megaripples (WL $>$ 60 cm) are the most useful indicators of current direction and strength.

Figure 4a is an echo-sounder profile over the megaripple field located at "B" in Figure 2. Megaripples located at "K" and "M" have similar crest orientation and asymmetry but maximum wavelengths of only 15 m. The smaller ripple size here is probably due to lower current velocity, however, at "M" the lack of loose sand may also play a role, because this field is interrupted by a patch of lag sediment or perhaps bare glacial till.

Where asymmetry is present, the steep, lee slopes of all the megaripples face south, indicating formation by inflowing water. At times the symmetry of only the crests of the higher megaripples has been reversed (Fig. 4a) giving a picture typical of tidal channels. This occurs only in the eastern part of the Great Belt and results from the Baltic outflow current. The crests of the megaripples, as a group, show more variation in direction than do the long axes of the longitudinal current marks. The megaripples may therefore be a more sensitive indicator of current-flow direction or, because they are higher than the longitudinal forms, they may be governed by shallower and more complex flow layers.

Megaripples occur sporadically on the western flank of the deep channel down to a depth of 35 m (Fig. 2, K) indicating that, although this channel is blocked to the south by a 23 m sill, it is at times subjected to strong currents.

To the southeast of the Great Belt, inflowing water is forced to turn eastward and flow through the Fehmarn Belt toward the Darss Sill and the main Baltic Sea (Fig. 1). Several fields of megaripples are found on the south flank of the Fehmarn Belt - again with all ripples showing and asymmetry resulting from inflow (eastward flowing) current (7). Although the authors have resurveyed these fields periodically over the past four years, no evidence has been found that the megaripples are migrating. Diving observations and vibrocores indicate that only the crests (down to a sediment depth of ca. 25 cm) are being restructured through small-scale ripple migration. An echo-sounder profile over one of these fields (Fig. 4b) shows the crests to be well rounded in comparison to the megaripples of the Great Belt 35 km to the north (Fig. 4a); in this case the rounding is an indication of old age.

CONCLUSIONS

(a) Side-scan sonar has provided a tool to map large-scale longitudinal and transverse bedforms. It brings to light a new dimension of bottom characteristics previously not seen because of the limited field of view provided by underwater TV, photography, or diving observations. The current pattern as well as the relative current velocities in the southern Great Belt can be deduced from an analysis of these bedforms.

(b) The Great Belt exhibits a rather clear axis of symmetry in relation to the distribution of current marks (Fig. 2). At the margins small comet marks and megaripples dominate; toward the center there is a zone of larger comet marks followed by sand ribbons, and finally by hard-ground bottom (glacial till) with few if any current marks.

(c) There is a hierarchy of bedforms which logically must be related to current strength. It seems reasonable to assume that longer comet marks are related to

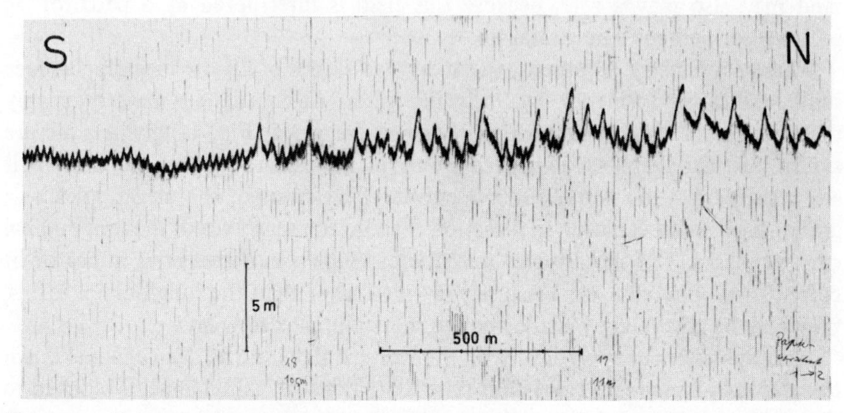

Figure 4a. Echo-sounder profile over megaripple field at "b" in Figure 2. Average water depth over ripple troughs (center) is 12 m.

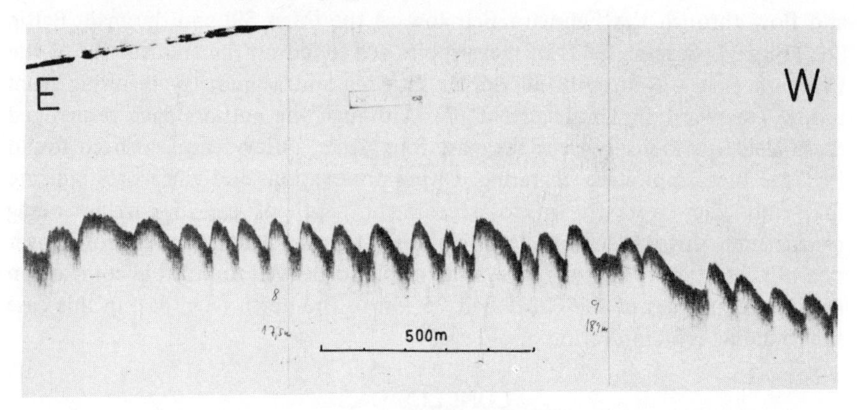

Figure 4b. Echo-sounder profile from megaripple field on the southern flank of the Fehmarn Belt. Average water depth over ripple creasts (center) is 16 m.

stronger currents. Using this line of reasoning, the above-mentioned sequence of bedforms shows that the inflow currents passing through the Great Belt increase in strength from the margins to the center and affect the bottom down to depths of over 35 m.

(d) The outflowing surface current from the Baltic is revealed only through small comet marks along the eastern margin of the Belt in a maximum water depth of 13 m. At this same depth the outflow current has sufficient velocity to reverse the crest symmetry of some megaripples. The outflow current apparently seeks the shortest route through the Great Belt, which means it is concentrated along the eastern margin where it is reinforced in its path by the coriolis effect.

(e) From the size and distribution of the bedforms found, we can infer that inflowing North Sea water recurrently sweeps the bottom of the Great Belt with velocities estimated to reach maximums of 100 to 150 cm/sec.

(f) The frequency with which the strong current pulses that are responsible for the large-scale bedforms take place is still unknown. The fresh appearance of some of the megaripples indicates that they may be rebuilt several times a year; however, sand-ribbon patterns in one survey line rerun after 14 months showed absolutely no change and the megaripple fields in the Fehmarn Belt have been essentially static over the past four years.

(g) The results presented in this paper should be considered preliminary. New side-scan sonar survey lines for the project were run in September 1973. A more comprehensive research paper on the area is in preparation and will be published soon.

ACKNOWLEDGMENTS

The authors are grateful to the captains and crews of the research ships *Alkor* and *Hermann Wattenberg* for their help at sea. Funding for the project was provided under a contract from the Fraunhofer Gesellschaft e. V., Munich.

REFERENCES

1. Allen, J. R. L.
   1971 Transverse erosional marks of mud and rock: their physical basis and geological significance. **Sediment. Geol.,** 5: 167-385.
2. Belderson, R. H., Kenyon, N. H., Stride, A. H., Stubbs, A. R.
   1972 **Sonographs of the sea floor.** Elsevier Pub. Co., Amsterdam, London, New York.
3. Newton, R. S., Seibold, E., Werner, F.
   1973 Facies distribution patterns on the Spanish Sahara continental shelf mapped with side-scan sonar. **"Meteor" Forschungsergebnisse,** C. (In press.)
4. Seibold, E., Exon, N., Hartmann, M., Koegler, F. -C., Lutze, G. F., Newton, R. S., Werner, F.
   1971 Marine geology of Kiel Bay. In **Sedimentology of parts of Central Europe,** p. 209-235. Guidebook VIII Int. Sediment. Congress, Heidelberg.
5. Stride, A. H.
   1963 Current-swept sea floors near the southern half of Great Britain. **Quart. Jour. Geol. Soc. London,** 119: 175-199.
6. Stride, A. H., Belderson, R. H., Kenyon, N. H.
   1972 Longitudinal furrows and depositional sand bodies of the English Channel. **Memoire du B.R.G.M.,** 79: 233-240.
7. Werner, F. and Newton, R. S.
   1970 Riesenrippeln im Fehmarnbelt (westliche Ostsee). **Meyniana,** 20: 83-90.
8. Wyrtki, K.
   1954 Die dynamik der wasserbewegungen im Fehmarnbelt. **Kieler Meeresforschung,** 10: 162-181.

Part II

ENGINEERING: 1.) USE OF VEGETATION IN COASTAL ENGINEERING
2.) ESTUARINE DREDGING PROBLEMS AND EFFECTS

Convened By:
Thorndike Saville, Jr.
Coastal Engineering Research Center
U. S. Army Corps of Engineers
Kingman Building
Fort Belvoir, Virginia 22060

THE INFLUENCE OF ENVIRONMENTAL CHANGES IN HEAVY METAL CONCENTRATIONS ON *SPARTINA ALTERNIFLORA*

by

William M. Dunstan and Herbert L. Windom[1]

ABSTRACT

No correlation was found between the heavy metal concentrations found in *Spartina* tissue and concentrations measured in the marsh sediment, river or estuarine waters of six major river systems of the southeastern coast of the United States. The ranges of metal concentrations in *Spartina* tissue are narrow in comparison to the ranges measured in the sediment and water and it appears that metal concentrations in the plants from the southeastern marshes are either "saturated" or the concentration of metals is controlled by the plant. The ratio of metal in the plant to metal in the marsh sediment ranges from less than one for iron to about one in most of the other metals. Mercury is an interesting exception up to four times as much being concentrated by *Spartina* as is found in the sediment.

Germination of *Spartina* was inhibited by high concentrations of methyl mercury while other metals had little effect. In experiments with seedlings toxicities resulted from high concentrations of copper, lead, and methyl mercury.

The southeastern marshes contain large amounts of trace metals necessary for *Spartina* growth and metal deficiency is probably never a limiting factor. On the other hand, *Spartina* is able to tolerate concentrations of heavy metals several times the average concentrations found in nature. Thus the changes in heavy

1. Skidaway Institute of Oceanography, P. O. Box 13687, Savannah, Georgia 31406.

metal concentrations caused by the deposition of dredge spoil and by the general increase in industrial effluents in rivers would not seem to be critical to *Spartina* at the present time.

INTRODUCTION

Man influences the salt marshes of the southeastern United States in two major ways. First, to maintain rivers and the inland waterway, he deposits large amounts of dredged materials onto them. Depending on the elevation and nature of the spoil material, a marsh may or may not be brought back into production but, whether it is or not, the chemical nature of its substrate is altered and this is of prime importance. Secondly, the increasing amount of industrial and municipal wastes that are brought to the marsh system by rivers and estuaries influences coastal marshes.

Both dredge spoil and industrial and municipal effluents contain heavy metals which, depending on the specific metal and the organism involved, can be either stimulatory or inhibitory and are, therefore, of importance to biological systems. The foundation of the salt-marsh biota in the southeast is the halophyte *Spartina alterniflora*. In this paper we discuss the relationship between this plant and the concentrations of seven heavy metals: iron, manganese, zinc, lead, copper, cadmium, and mercury. The area of primary consideration receives discharges from seven river systems on the coasts of South Carolina, Georgia, and Florida (Fig. 1) and contains over 400,000 hectares of salt marshes.

Five to ten stations in each river-estuary system were sampled bimonthly over a year. Salt marshes at the mouths of each river were sampled for sediment and plants during the early fall of 1972. Samples of plants and sediment were frozen immediately upon collection and analyzed by atomic absorption spectroscopy for total and leached metals (9). In addition to samples, laboratory plants were grown from seed collected in local marshes (Skidaway Island, Georgia), using standard techniques of sand and water culture (7) with modified Hoaglands medium (5).

PLANT-METAL RELATIONS IN NATURAL ENVIRONMENTS

When a volume of fine-grain sediment from a river or creek bottom which is generally reduced is deposited on top of an existing marsh, a series of abrupt alterations in pH and Eh take place and cause important changes in sediment-metal chemistry. Figure 2 shows the iron content of a marsh area before, right after, and two months after 40 cm of dredged material had been deposited on it. There is a threefold increase in the leachable (hydroxylamine hydrochloride-acetic acid) iron concentration caused by the addition and subsequent oxidation of the reduced dredged sediment dumped on the marsh.

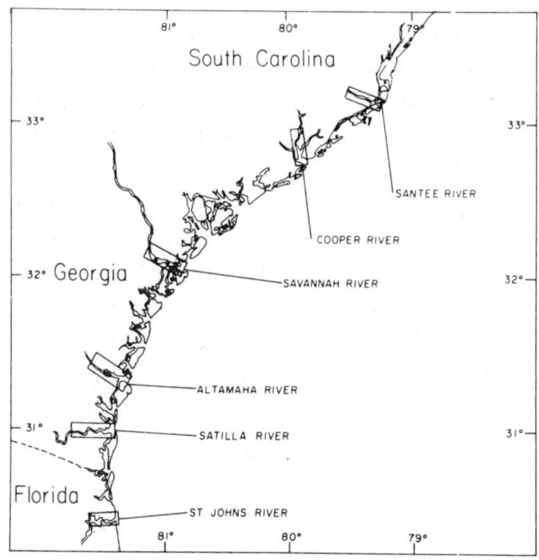

Figure 1. Seven river systems sampled for heavy metals in water, sediment, and **Spartina.**

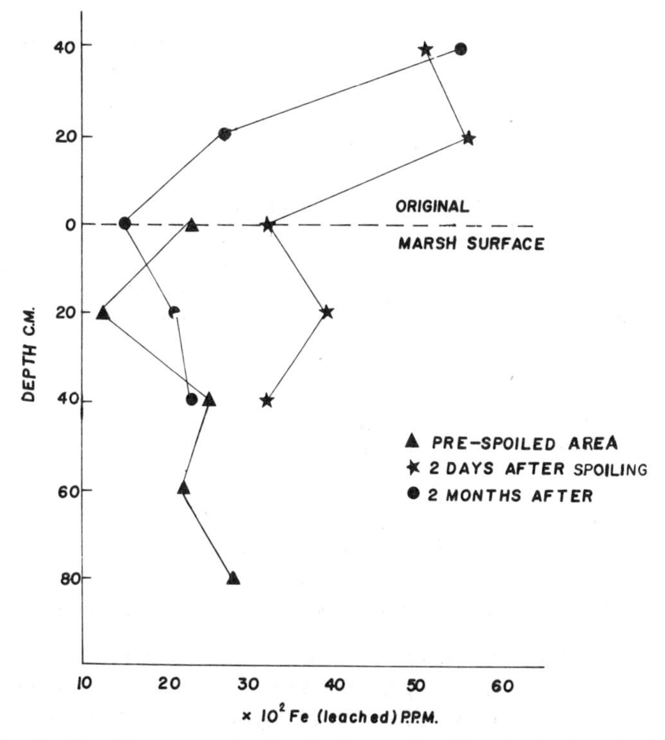

Figure 2. Profile of iron concentrations in the marsh sediment prior to and subsequent to the deposition of dredged material.

These changes are shown even as deep as 40 cm into the original marsh. The fractionation of iron by the leaching of various compounds, probably mobile iron monosulfide, from the dredged material leads to its concentration in the surface of dredged spoil areas due to subsequent oxidation. The schematic diagram in Figure 3, modified from Hodgson (6), indicates the forms in which a metal exists in marsh sediments and some of the competitive reactions which might limit or enhance a plant's uptake of a particular metal. With present techniques it is generally possible to measure only total concentrations in solution or the total soluble organic and inorganic fractions. In the case of methylated mercury, however, other metal species, e.g., methyl mercury (4), can be determined. It is clear that the total concentration of metals (at least iron) available to *S. alterniflora* is increased as a result of dredged-spoil deposition.

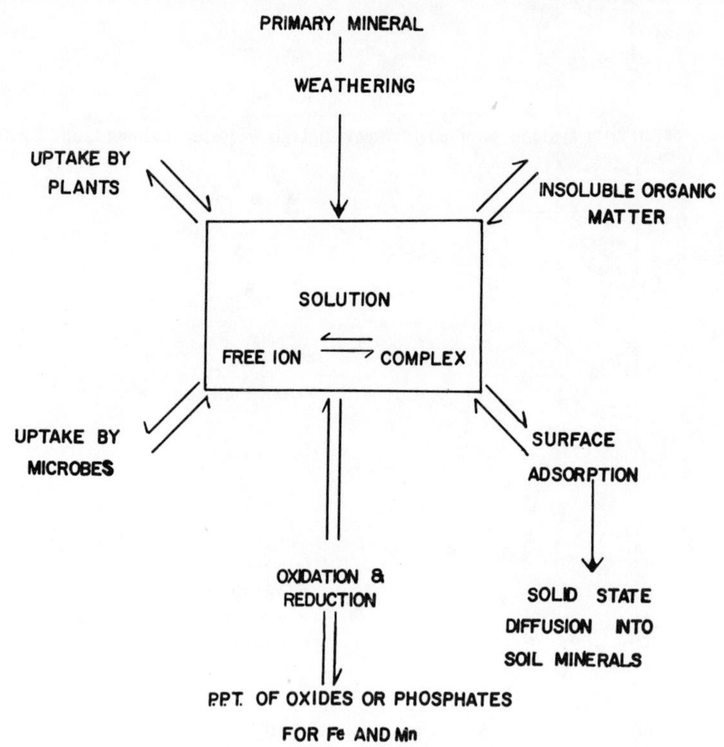

Figure 3. Possible sites for the competitive uptake of metals in the marsh system.

Table 1 summarizes the data for metals concentrations in salt marshes adjacent to the discharges of important rivers of the southeastern coast. It is obvious that, with the possible exception of Hg, there is generally an excess of metals in the sediments. The seven metals differ in their range of concentrations by some four or five orders of magnitude and these relative concentrations are

TABLE 1.

Average Metal Concentrations in **Spartina alterniflora**

and Sediments from Marshes Proximate to two Southern Rivers

		PPM						
		Fe	Mn	Zn	Pb	Cu	Cd	Hg
Santee	Sediment*	44,300	130	69.0	22.2	25.4	3.56	.09
	Spartina**	1,000	59.5	33.8	5.35	3.95	.85	.17
Cooper	Sediment	11,700	146	42.4	13.9	7.8	.79	.03
	Spartina	400	46.2	36.7	4.2	4.1	.60	.5
Savannah	Sediment	40,300	249	69.6	15.9	14.7	.69	.11
	Spartina	900	46.4	24.6	5.2	4.8	.49	.44
Altamaha	Sediment	89,800	253	44.2	19.2	8.9	1.2	.10
	Spartina	1,000	45.8	30.6	4.2	4.8	.73	.27
Satilla	Sediment	36,600	221	43.3	14.0	8.2	1.0	.07
	Spartina	350	35.3	39.8	2.6	3.8	.58	.27
St. Johns	Sediment	14,000	105	14.9	9.4	2.7	.56	.04
	Spartina	750	39.3	27.8	2.83	1.67	.38	.25

*Average of upper 40 cm μg/g dry weight.
**μg/g dry weight of above ground and rhizome portion of plant.

usually maintained in the water, sediment, and plant system. It is also noteworthy that the percent of metal in the more available leached form ranges from a low 7% through 17% for copper, 38% for manganese and cadmium, 42% for lead, to 54% for zinc. An important difference in the plant's response to heavy metals is the magnitude to which the plant concentrates the metal from the sediment (Fig. 4). There is the indication of an inverse relationship between the concentration of a particular metal in the sediment and the concentration factor of this metal in *S. alterniflora.* While Zn, Cu, and Cd are slightly concentrated in *S. alterniflora* compared to the leached sediment, mercury is highly concentrated by *S. alterniflora.* The relationship of mercury to *S. alterniflora* may be a significant step in the transport of mercury through the estuary to the coastal environment.

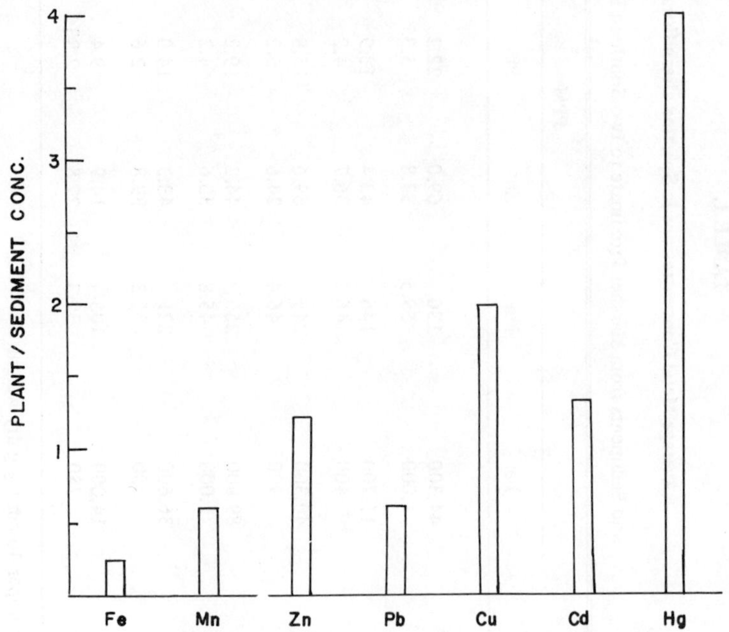

Figure 4. The ratio of plant to sediment (leached) concentrations of metals in order of decreasing concentration in the marsh sediment. Totals are averages of all rivers in the sample area.

In sampling the broad area of the southeastern coast with river inputs ranging from the relatively undistrubed Altamaha to the industrialized Savannah and Cooper rivers, one would expect to find differences in the concentrations of metals reflected in the water, sediment, and plants. When the concentrations in the river system that generally had the highest metal values were compared with those of the river that had the lowest, the plants and sediment followed in relative concentrations (Table 2). That is to say, all the values determined for plant and sediment content in the St. Johns River were lower than those in the Santee River. This is as far as any relationship goes, and we are not able satisfactorily to correlate either river or estuary sediment metal measurements with metal concentrations in the plant. In fact (Figs. 5 and 6) the range of concentrations of all metals found in *S. alterniflora* is quite narrow in comparison to the range for each metal in the water and marsh sediment. This may be interpreted as indicating that *S. alterniflora* is either "saturated" in tissue concentration of heavy metals or it possesses a mechanism that serves to control the levels of all heavy metals in its tissues.

RESPONSE OF *SPARTINA* TO METALS

IN LABORATORY SYSTEMS

In preliminary phases of laboratory studies our initial concern has been to evaluate the toxicities of heavy metals to *S. alterniflora*. This information is important in determining man's potential impact on salt marshes. The effects of various concentrations of metals on the germination of *S. alterniflora* seeds have been determined (Table 3). Methyl mercury has a significant effect on seed germination at higher-than-natural levels. Zn and Pb also show some effects at very high concentrations. Preliminary work with seed sprouts indicates that at this life stage methyl mercury, lead, and cadmium can influence a variety of developmental changes.

Data on *S. alterniflora* seedlings grown for eight weeks in 100 ppm of five inorganic metals and methyl mercury are given in Table 4. While both lead and methyl mercury cause high mortalities, only lead seriously affects the growth of the surviving plants. It has been shown, for example, in *Agrostis* that resistance to high metal concentrations is strain-specific and tolerant plants emerge (8). This response is also indicated for MeHgCl, where a percentage of the plants die but resistant individuals show no adverse effects. In the cases of lead and copper a very basic, general enzyme or membrane function may be affected, causing inhibition of growth or death in all experimental plants.

CONCLUSION

From the large number of field samples taken, it is obvious that there is abundant iron, manganese, and zinc available for *S. alterniflora* growth. Marsh sediments not only represent a large pool of these metals but also operate continually to trap high concentrations in suspended sediments brought down rivers into the estuary. *S. alterniflora* tissue contains a large amount of iron in comparison to the average for other plants (700 ppm versus 100 ppm), while its

TABLE 2.

Metal Concentrations in Solution, Suspension, Sediment and **Spartina alterniflora** for Southeastern Coastal Areas.

	Fe	Mn	Zn	Pb	Cu	Cd	Hg
				PPB in Solution (μg/L)			
River Water	144	18	—	2.3	4.5	.84	.07
Estuarine Water	17	22	—	2.2	3.3	1.13	.09
				PPM (μg/g)			
Suspended Sediment, River	48,000	1153	496	87	79	15.8	.81
Suspended Sediment, Estuarine	50,700	1139	471	82	58	8.7	.86
Marsh Sediment* (total)	38,800	194	49.8	16.6	11.4	1.19	.08
Marsh Sediment* (leached)	2,857	74	26.9	6.9	1.9	.45	—
S. alterniflora	700	45	32.2	4.1	3.9	.61	.32

*Upper 40 cm

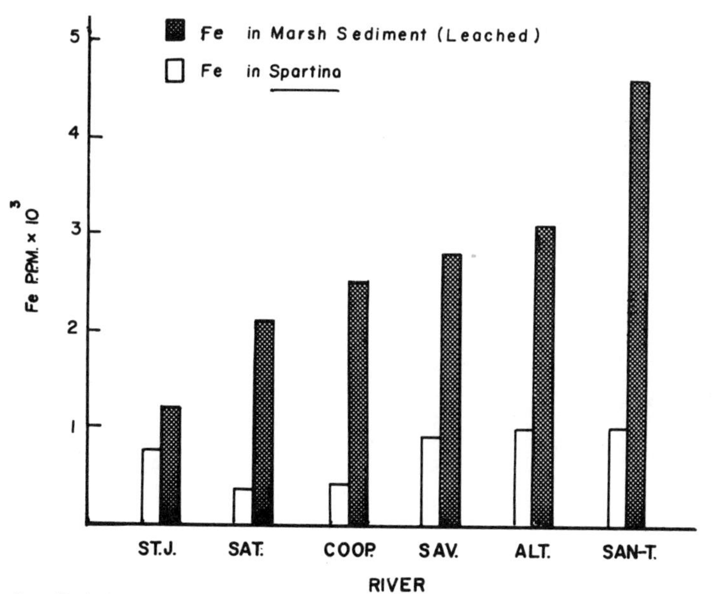

Figure 5. Relative concentrations of iron in the sediment and iron in **Spartina,** arranged in order of increasing concentration in the sediment of each river system.

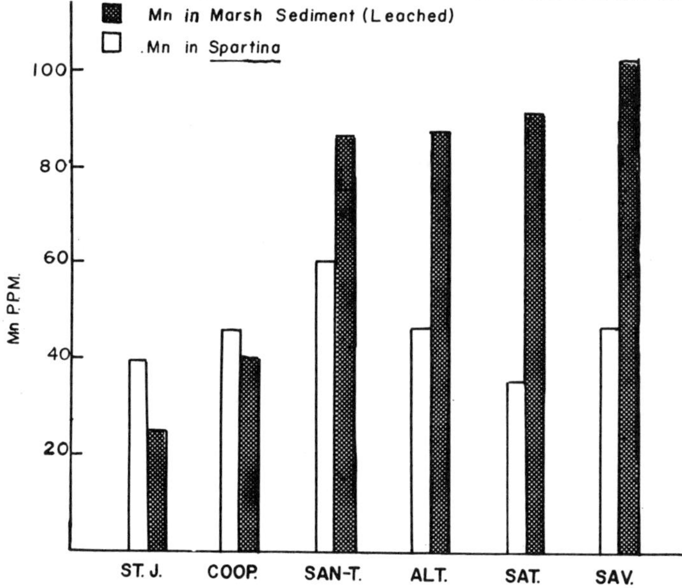

Figure 6. Relative concentrations of manganese in the sediment and manganese in **Spartina,** arranged in order of increasing concentration in the sediment of each river system.

TABLE 3.

Germination of **Spartina** seeds exposed to heavy metal concentrations

Metal	Percent Germination*			
	Concentration PPM			
	100	10	1	.1
Zn	76	96	—	96
Pb	74	88	—	88
Cu	--	92	—	--
Cd	92	100	—	92
$HgCl_2$	80	--	92	75
MeHgCl	24	—	56	92
Control		(100)		

*Seeds were selected as embryo containing

TABLE 4.

Growth of **Spartina** in high metal concentrations

	Mean Increase Per Live Plant		
	Dry Wt. Gms	Length Cm.	% Dead
Zn	.62	36.3	37.5
Pb	.18	14.3	62.5
Cu	—	—	100
Cd	.45	32.6	25
$HgCl_2$	.32	34.0	37.5
MeHgCl	.34	31.6	50.0
Control	.47	34.9	12.5

manganese and zinc concentrations are similar to other plants (45 Mn, 32.2 Zn versus 50 Mn and 20 Zn; Epstein, 2). In view of the vast metal pools in the sediment; only by being chemically unavailable to *S. alterniflora* uptake could these metals ever cause limitation of growth. Further research on the physiology of metal uptake by *S. alterniflora* is needed before we can predict the effects on metal availability caused by dredge-spoil deposition and increased industrial or municipal effluents in rivers.

Although preliminary in scope, the laboratory studies indicate that toxicities of lead, copper, cadmium, and inorganic mercury would require concentrations orders of magnitude higher than those found in the southeastern salt-marsh environment at present. Certainly further work is needed to define the toxic responses that result from changes in such other factors as organic pollution and inorganic sedimentation. The toxicity picture is also complicated by phenotypic plasticity (1) which is probably a characteristic of *S. alterniflora* and by synergistic effects which have been shown to be important in many other plant systems (3).

ACKNOWLEDGMENTS

This work was partially supported by a grant from the National Science Foundation, Office of the International Decade of Ocean Exploration (no. GX-33615).

REFERENCES

1. Bradshaw, A. D., McNeilly, T. S., and Gregory, R. P. G.
1965 Industrialization, evolution and the development of heavy metal tolerance in plants. In **Ecology and the Industrial Society.** (eds. Goodman, G. T., Edwards, R. W., Lambert, J. M.) p. 327-343. Oxford Univ. Press, London.

2. Epstein, E.
1972 **Mineral nutrition of plants: principles and perspectives.** John Wiley & Sons, New York. p. 392.

3. Foy, C. D., Fleming, A. L., and Armiger, W. H.
1969 Aluminum tolerance of soybean varieties in relation to calcium nutrition. **Agron. Jour.,** 61: 505-511.

4. Gradner, W.
1973 The transfer of heavy metals through the inner continental shelf to the open ocean. Annual report to the International Decade of Ocean Exploration. GX-33615. May 1973. Skidaway Institute, Savannah, Georgia.

5. Hoagland, D. R. and Arnon, D. I.
1950 The water-culture method for growing plants without soil. Calif. Expt. Sta. Circ. 347.

6. Hodgson, J. F.
1963 Chemistry of the micronutrient elements in soils. **Advances in Agronomy,** 15: 119-159.

7. Hewett, E. J.
1952 Sand and water culture methods used in the study of plant nutrition. Commonwealth Bureau of Hort. Plantation Crops. Tech. Commun. No. 22.

8. Jowett, D.
1958 Populations of **Agrostis** spp. tolerant of heavy metals. **Nature,** 182: 816-817.

9. Smith, R. and Windom, H.
1972 Analytical handbook for the determination of arsenic, cadmium, cobalt, copper, iron, lead, manganese, mercury, nickel silver and zinc in the marine and estuarine environments. **Ga. Mar. Sci. Cen., Tech. Rpt.** Ser. No. 72-6.

BIOTIC TECHNIQUES FOR SHORE STABILIZATION

by

Edgar W. Garbisch, Jr.,

Paul B. Woller, William J. Bostian, and Robert J. McCallum[1]

ABSTRACT

Biotic techniques for shore stabilization include establishing fresh, brackish, and salt-water marshes either on existing shores or on fill material deposited alongshore and graded to appropriate elevations. This paper presents a synopsis of such techniques currently under exploration.

One objective of the various projects under way is to identify methods for and limitations to establishing thirteen species of emergent marsh plants on various natural and artificial substrates within a tidal zone subject to different wave stresses, water salinities, and tidal amplitudes. Other objectives include determining the effects that vegetative establishment have on substrate stabilization, sediment accretion, estuarine invertebrate recovery, and successional characteristics in vegetated and unvegetated tidal flats.

Tentative conclusions derived from this research are: (a) substrate characteristics do not appear to limit vegetative establishment; (b) periodic fertilization may be essential for vegetative establishment on some substrates and in areas subject to high degrees of physical stress; (c) mammal and water fowl feeding on virgin artificial marsh areas may be extensive for specific plant species and may lead to the permanent removal of plants that do not flower the first season; (d) artificial intertidal areas appear to acquire within one year benthic invertebrate populations comparable to those in undisturbed control areas; (e)

1. Environmental Concern Inc., P. O. Box P, St. Michaels, Maryland 21663.

physical stress appears to limit the age of plant material that can be incorporated and the elevations at which it can become established within the tidal zone; (f) the most successfully established vegetation in salt-water and brackish-water areas have been *Spartina alterniflora, Spartina patens, Spartina cynosuroides, Distichlis spicata, Phragmites communis, Panicum virgatum,* and *Ammophila breviligulata* seedlings, and in fresh-water areas *Scirpus americanus* and *Spartina alterniflora* seedlings.

INTRODUCTION

Shoreline erosion and problems associated with the deposition of the sediment arising therefrom are confronted in all major lakes, rivers, and estuaries. Physical engineering techniques are known or can be designed to control nearly any shoreline-erosion condition. These techniques may be prohibitively costly and the controls may cause new erosion or maintenance problems elsewhere by restricting sediment transport or altering movements and volumes of littoral drift. Physical erosion controls alter to varying degrees the regional hydrological dynamics alongshore and in shallow nearshore subtidal areas. The marine organisms associated with these regions often are of high commercial value or significant trophic value to commercial species. Although the impacts that erosion-control structures have on nearby benthos have not been extensively examined, it is certain that such structures lead to habitat restructuring and concomitant adjustments to the abundance and diversity of benthic organisms.

Biotic alternatives to or combinations with physical approaches to shore stabilization often may exist. Marshlands are nature's counterpart to bulkheads, groins, and revetments for erosion abatement in areas not subject to direct ocean exposure. Their stabilities may be high if sedimentary conditions are optimal and stresses by waves, ice, and animals are moderate. Under such conditions, marshlands may provide long-term erosion control. Otherwise, periodic maintenance is necessary to sustain this natural protection.

The technology of establishing, restoring, and maintaining marshlands is beginning to evolve (1, 2, 4). As practical applications of marsh establishment to shore stabilization and other controls of marine environmental parameters are performed, a complete assessment of the environmental impact therefrom should be forthcoming.

Because of the rigors and variability of marine environments, marshland establishment is not likely to become standardized. Guidelines will have to be variable and influenced largely by the physical, chemical, and biological conditions prevailing at particular project sites. Consequently, broad practical experience is necessary for the design and execution of successful projects.

The establishment of aquatic vegetation on natural intertidal and supratidal shores constitutes the most straightforward and economical approach to shore stabilization. Bank sloping and vegetative stabilization thereof (3) is a refinement that should contribute more erosion control. Shores may be too low in elevation either to establish a significant band width of emergent marsh vegetation or reduce significantly the regression of the banked shoreline, should marsh vegetation be established. Dredge spoil or terrestrial quarry sediment may be deposited in these areas and graded alongshore to generate optimal elevations and slopes for rapid vegetation establishment and stabilization. This provides an attractive and economical option for spoil-deposition sites and appears to be a most constructive utilization of dredge spoil.

This paper provides a qualitative account of a variety of projects currently underway that are associated with the establishment of salt, brackish, and fresh-water tidal marshes on artificial tidal creek banks, natural shores, and on alongshore dredge-spoil and terrestrial-fill material. (Technical reports that provide quantitative biological, chemical, and physical data for each project will be available upon request.)

PROJECT SITE DESCRIPTION AND PROJECT OBJECTIVES

Information pertaining to some physical and chemical parameters associated with currently active research sites are collected in Table 1. Locations of the sites are indicated in Figure 1, except for the tidal stream restoration site (No. 7), which is at Ocean City, New Jersey.

Hambleton Island (Sites Nos. la through 1d)

Hambleton I. was selected as a logistically convenient site experiencing varying exposure. The island is eroding rapidly (55 acres in 1847 and <15 acres in 1970) and apparently is the principal source of mostly fine sediments for alongshore transport. The long-range objective of this project is to establish some two acres of marshland on natural shore sections along the rapidly eroding western shoreline and on artificial tidal areas within and on either side of a recently made breach through the island. As the banks of the island continue to regress, it is expected that the nucleus marsh will trap some of the sediment passing through the breach and thus increase in elevation and in area. This process should convert some fraction of the island to productive tidal wetland.

The more immediate objectives of this project are to identify methods of synthesis and limitations to brackish-water tidal-marsh synthesis applicable to the Chesapeake Bay and similar areas. Determinations under investigation with one or more of the following species of tidal-marsh plant species (*Spartina alterniflora, S. patens, S. cynosuroides, Distichlis spicata, Panicum virgatum, Ammophila breviligulata, Phragmites communis, Typha latifolia, Typha*

TABLE 1.

Project Site Descriptions

Project Site (location)[a]	Tidal Amplitude[b]	Salinity[c]	Substrate gravel	Composition[d] sand	mud	Major Fetches (km)		Elevation Range vegetated[b,e] min.	max.	Acres
Quarry Fill (la)[f]	43	8.7	4	79	16	SE (3.5)	NW (3.2)	5.5	69	1.5
Dredge spoil (1b)[f]	43	8.7	0	57	42	SE (3.5)	E (0.6)	12	46	0.2
Broad Creek Natural Shore (1c)[g]	43	8.7	0	11[h]	89[h]	NW (3.2)	SW (2.8)	21	37	0.01
San Domingo Creek Natural Shore (1d)[g]	43	8.7	0	18[i]	82[i]	SE (3.5)	E (0.4)	0.0	43	0.01
San Domingo Creek Natural Shore (1e)[f]	43	8.7	0	87	13	NE (1.1)	E (.6)	9.1	58	0.08
Rich Neck (2)[f]	37	8.2	0	93	7	S, SW (27) SE	N, NW (8.7) NE	24	82	0.16

TABLE 1 (continued)

Project Site (location)[a]	Tidal Amplitude[b]	Salinity[c]	Substrate Composition[d] gravel	sand	mud	Major Fetches (km)		Elevation Range vegetated[b, e] min.	max.	Acres
Tred Avon R. (3)[f]	43	8.7	68	30	2	(12) NW	(3.5) S, SW	6.1	79	0.05
Long Point I. (4)[f, g]	37	8.8	10	72[j]	18[j]	(4.4) N, NE	(2.4) E	-9.1	70	0.22
Sand Spit (5)[f]	37	8.9	0	96	4	(9.5)	(5.2)	-9.1	67	0.13
Susquehanna Delta (6)[f, k]	52	0.3	0	82	18	S, SE (9.6)	W, SW (5.9)	12	34	2.0
Tidal Stream (7) (Ocean City, N. J.)[f]	110	23	1	1	1	—	—	55	110	0.11

(a) See Figure 1. (b) In cm. (c) I^{n} ppt. Average of determinations on 6/2, 7/10, and 9/2/73. (d) In percent (e) Relative to MLW. (f) Planting accomplished using spade. (g) Planting accomplished using mechanical auger (h) Mostly clay. Soil compactness varied from 0.6 to 1.0 kg/cm^2. (i) Mostly silt. (j) Soil compactness varied from 0.1 to 1.6 kg/cm^2. (k) Planting accomplished using tractor and mechanical planter. (l) Substrate was either sand (quarry fill) or natural peat.

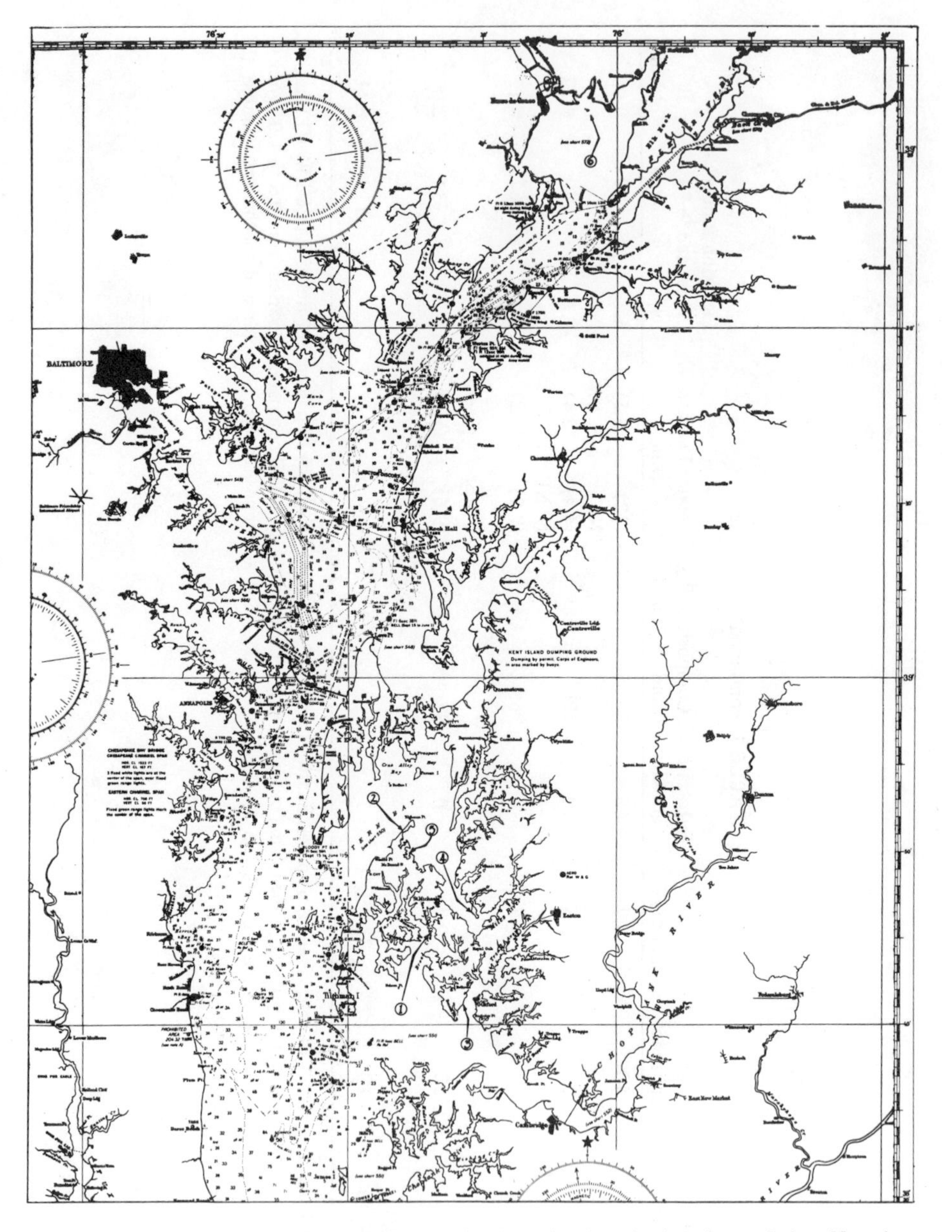

Figure 1. Northern part of the Chesapeake Bay showing the locations of sites Nos. 1 through 5 surrounding 38°50' N and 76°15' W and site No. 6 at 39°33' N and 76°2' W.

angustifolia, Scirpus robustus, Scirpus olneyii, and *Juncus roemerianus*) are: effects of (a) elevation, (b) substrate compaction, particle size distribution, and nutrient content, (c) artificial fertilization, and (d) wave stess on plant establishment, morphology, and production; relative potential for achieving plant establishment by seeding and by planting bare-root seedlings, peat-potted seedlings, and dormant rhizome sections; seasonal limitations to the various types of plantings; changes in elevation attributed to vegetative establishment; estuarine invertebrate recovery and successional characteristics in the vegetated and unvegetated artificial tidal flats as a function of elevation, substrate characteristics, and wave stress; and effects, specificities, and patterns of mammal and waterfowl utilization of established vegetation.

Natural deep sand shores (Sites Nos. le, 2, and 3)

These sites were selected (a) to identify limitations and potentials for vegetative stabilization of both open and groined natural sandy shores, and (b) to determine the capability of established vegetation to increase elevations through sediment entrapment and peat accumulation as a mechanism for abating shoreline erosion.

The three sites experience varying degrees of exposure and wave stress. Site No. 2 experiences year-round wave stress from the prevailing summer southwest winds and the winter dominant west to northwest winds. Waves achieve two to three-foot (one half to one meter) amplitudes during storm tides and winds. Site No. 3 experiences wave stress occasionally, at times preceding usual weather systems where winds develop from the east to southeast, and during times of tropical storms. Site No. 1e is not subject to significant wave stress at any time and is the only one of these sites that does not contain a riprap groin system.

Plant species used at these sites are *S. alterniflora* within intertidal regions, and *S. patens, S. alterniflora, D. spicata,* and *A. breviligulata* within supratidal regions. Sand-filled, peat-pot-cultivated seedlings three to four months old were used for both April and June plantings.

Long Point Island (Site No. 4)

The western shore of Long Point Island is subject to wave year-round stress, having directional exposure similar to site No. 2. In contrast to sites Nos. 1e, 2, and 3, site No. 4 has negligible supratidal shore area and 700 feet (210 meters) of wooden bulkhead. The unbulkheaded intertidal shore abuts steep banks of up to four feet (1¼ meters) in height. The shore consists mostly of a thin layer of sand followed within the planting zone by poorly sorted compacted sediment composed of sand, silt, and clay.

Objectives of this project are to determine shoreline erosion rates, the sources and sinks of sand-sized sediment, volumes of sand transported alongshore, optimal conditions for rapid establishment of *S. alterniflora* throughout intertidal shore areas, volumes of sand trapped by the established vegetation and concurrent elevational increases, and the feasibility of vegetatively stabilizing sediment along the foot of a bulkhead.

The basic idea being explored is that a twenty to thirty-foot (6-9 meter) wide band of intertidal *S. alterniflora* marsh seaward of an eroding banked shoreline is not likely to greatly abate shoreline erosion. However, if vegetative spread landward keeps pace with the bank regression, and if most of the new sediment supplied through erosion and other sources is retained within vegetation boundaries, then an erosionally stable condition should be achieved gradually at the expense of some mainland area.

Sand Spit (Site No. 5)

The dynamic nature of spits generally prohibits natural vegetative establishment and stabilization. Spits initiate at the terminus of a point of land and develop seaward in the direction of the predominant longshore sediment drift often towards another point of land, tending to enclose harbors, lagoons, and bays. Artificial or natural channels which intercept the migrating sediment may require periodic maintenance dredging. In some instances, spits may be vegetatively stabilized through the incorporation at optimal density of suitably developed live plant material. If this can be accomplished, fresh supplies of sand which normally would traverse the sand spit to the channels or deep-water depositories may be trapped and stabilized by the established vegetation and promote the expansion and development of the marsh.

The feasibility of vegetative stabilization of sand spits is being pursued at site No. 5. There, *S. patens, S. cynosuroides, A. breviligulata, P. communis,* and *P. virgatum* seedlings in various stages of development were planted throughout supratidal areas during late winter and mid-summer. Intertidal areas were planted with variously aged *S. alterniflora.* Shallow subtidal areas (2.5 to 5 cm below MLW) were planted with three-month-old *S. alterniflora* seedlings several feet in height. Portable plastic break-waters are being used during the first year of the project to reduce wave energies along the exposed eastern shore of the spit. Long-term elevational changes attributed to vegetative establishment are being determined and will indicate the degree of success of the project.

Susquehanna Delta (Site No. 6)

This project is designed to extend our experiences to a tidal fresh-water environment subject to year-round wave stress. In June, 1972, about 500 acres of intertidal sand delta were formed from fluvial sediments transported down the Susquehanna River by the flood waters resulting from tropical storm Agnes. A small tidal flat of approximately 10 acres was selected as a site to explore the feasibility of vegetative establishment and substrate stabilization through (a) winter planting of dormant rhizome sections, (b) spring seeding, and (c) spring and summer planting of both fresh- and salt-water emergent marsh-plant seedlings of various ages. Plant material used include *S. alterniflora, T. angustifolia, T. latifolia, S. robustus, S. americanus,* and *S. olneyii.*

Tidal Stream (Site No. 7)

This site represents an effort to restore a tidal stream that was filled in connection with a motel development near Ocean City, New Jersey. The stream was relocated through the fill area by ditching to specified depths and bank slopes and finally reconnecting the ditch to the natural undisturbed parts of the stream. The primary objective of the project is to incorporate *S. alterniflora* throughout the intertidal bank area to stabilize it against rain and current erosion.

QUALITATIVE RESULTS AND GENERAL CONCLUSIONS

The purpose of this report is to provide a synopsis of research underway together with general conclusions therefrom that are justified from data on record. (Open-ended draft technical reports for each project site, with the exception of site No. 7 are available upon request.)

Substrate Considerations and Limitations

Nonpolluted substrate characteristics do not appear to limit the establishment of *S. alterniflora* incorporated as live seedlings older than two months of age. *S. alterniflora* seedlings have been successfully established on substrates dominated by gravel, sand, clay, silt, and various combinations thereof. With or without artificial fertilization (see below), *S. alterniflora* production is lowest in gravel and well-sorted, uncompacted sand substrates with high percolation and low nutrient adsorption capacities, and is highest in poorly sorted and permeable substrates.

High soil compaction does not appear to limit initial *S. alterniflora* establishment, provided such substrates are auger-drilled to disrupt the soil continuity at the individual plant sites. Figures 2 and 3 show an area at site No. 4 in which *S. alterniflora* was planted with the aid of a mechanical auger.

Plant species other than *S. alterniflora,* as indicated earlier in this paper, have all been established in well-sorted uncompacted substrates dominated by sand-sized particles. Comparisons with other substrate compositions and

Figure 2. Section of site No. 4 after being planted in May 1973 with three-month-old **S. alterniflora** peat-potted sedlings.

compactions have yet to be explored.

Plant establishment of *S. alterniflora* by seeding in loose sand and mud substrates appears to be limited primarily by factors such as elevation and physical stress. Seeding of other plant species has yet to be pursued.

Need for and Effects of Fertilization

Requirements for fertilization depend upon specific plant nutrients in the substrate and tidal waters. Implementation of fertilization programs in newly planted areas may, on occasion, be vital to plant establishment and contribute to maximizing plant production.

The need for fertilization appears greatest in sand substrate areas subject to frequent wave stress, and least in mud substrate areas. In sand substrate areas, *S. alterniflora* nutritional requirements do not appear to be increasingly satisfied with increasing periods of tidal inundation.

Effects of fertilization with both fast-dissolving (surface broadcast) and slow-releasing (subsurface placed) fertilizers are least pronounced for

Figure 3. Same view as in Figure 2 but in September 1973. The well-developed plants define the area subject to surface fertilization. The foreground area containing poorly developed plants was not fertilized.

uncompacted, well-sorted sand substrates (sites Nos. 2, 3, 5, and 6) as opposed to compacted, poorly sorted substrates containing some percentage of fine-sized particles (site No. 4 and sections of site No. 3 are not indicated in Table 1). The best time to initiate a fertilization program appears to be in early summer when the concentrations of plant macronutrients in estuarine waters begin to decline. Also, low nutrient concentrations appear to lead initially to a more balanced development of roots and shoots of young seedlings than do excessive nutrient concentrations.

Figure 3 illustrates the difference between a fertilized area and a neighboring unfertilized area at a section of site No. 4 after one growing season (September, 1973). In the beginning the two areas had identical substrates and elevations. By the end of the first growing season, elevations near the center of the fertilized area peaked, through sand entrapment, to 30 cm higher than they were at the time of planting. No significant elevation changes were encountered in the neighboring unfertilized area. Three-month-old *S. alterniflora* seedlings in peat pots were planted in early May, 1973. Fertilizations were accomplished in June, July and August by depositing a portion of fast-dissolving fertilizer on the surface at each plant site, giving an application rate of 336 kg/ha of N (ammonium nitrate) and 74 kg/ha of P (super-phosphate). Figure 3 illustrates that fertilizer neither washed nor diffused to the neighboring unfertilized area, indicating rapid plant and/or substrate uptake of nutrients.

S. alterniflora and *S. americanus* at the fresh-water site No. 6 were relatively unresponsive to repeated surface fertilization with the above-mentioned fertilizer at the same application rate, or to subsurface-placed, slow-releasing fertilizer. Figure 4 shows a midsummer view of a section of site No. 6.

Figure 4. A midsummer view of a section of site No. 6. **S. americanus** (foreground) was incorporated as two-to-three-month-old, peat-potted seedlings. The background plants are mechanically planted **S. alterniflora** seedlings.

The chemical composition of the water associated with site No. 6 is greatly different from that of the waters associated with the other project sites listed in Table 1. It is not known at this time whether the unresponsiveness to fertilization of the plants at site No. 6 can be attributed to some basic chemical deficiency in the water of the area or to the high permeability and low compaction of the sand substrate. *S. alterniflora* is not an indigenous plant and natural *S. americanus* in the area is developing mostly in peat substrates. As specific physiological and nutritional requirements of the emergent marsh plants become identified, it is conceivable that special fertilizer applications may be required in certain areas in order to maximize initial plant establishment and development.

Surface fertilization of the dredge spoil used at site No. 1b did not significantly increase plant production and subsurface fertilization with both fast- and slow-releasing fertilizers led to plant fatalities. This was attributed to the already high nutrient content of the spoil and to its poor permeability.

The first-year production of *S. alterniflora* at site No. 1b is superior to that at any other project site. Figures 5 and 6 illustrate the *S. alterniflora* development at the end of the first growing season, which started in May, 1973, with three-month-old, peat-potted seedlings.

Figure 5. Section of site No. 1b being planted with three-month-old **S. alterniflora** peat-potted seedlings in May, 1973.

Effects of Stress

Stress factors are varied in the dynamic intertidal zone and may dictate the overall design of marsh-establishment projects. Effects of waves and currents,

Figure 6. Same view as in Figure 4 but in September 1973. Foreground plants were incorporated in July.

sediment turbulence, salinity, shading, deep freeze, ice scouring, tidal amplitudes, and animal excavations are but some of the stresses that influence marsh establishment and development. The natural environment is the perfect laboratory to explore these matters.

Site No. 2 provides an ultimate test of the tolerance of *S. alterniflora* to the stresses of waves and beach elevational changes (elevations changed 75 cm during one tidal exchange during 1973). Figure 7 shows one of six planted sections at this site in late July, 1973, where in June three-month-old *S. alterniflora* seedlings were planted 30 cm above and 18 cm below MHW and similarly aged *S. patens* and *A. breviligulata* seedlings were planted at higher elevations. At this site, *S. alterniflora* plant fatalities reached over 80% below MHW, and less than 20% at 0 - 30 cm above MHW. The fatalities of the other species planted at higher elevations were less than 2%.

For sites subjected to continued wave stress, initial marsh establishment at and above MHW appears feasible. However, the effects of winter stress on the continued development of such vegetated areas have yet to be determined.

Figure 7. A July 1973 view of one of the planted areas at site No. 2. Three-month-old **S. alterniflora** (center and right) and **S. patens** (left) peat-potted seedlings were planted in May. The center plants are situated approximately at MHW elevation. Lower elevation plantings suffered high fatality percentages.

Surveys of control and planted areas at site No. 2 indicate no effects of vegetative establishment on elevation during the first growing season.

Vegetative establishment within stone groin systems may be significantly more difficult than that along a neighboring open beach. At site No. 2, for example, *S. alterniflora* fatalities reached 100% below MHW and greater than 40% above MHW within three groin planting areas. However, other species planted above the *S. alterniflora* in elevation experienced less than 2% fatalities. The additional fatalities within that groin system as compared to the open beach can be attributed largely to the fact that greater and more frequent elevational changes occur within a groined area than occur on a neighboring open beach.

The turbulence of the gravel associated with the substrate at site No. 3 was largely responsible for the 38% fatality of *S. alterniflora* planted within the intertidal zone. In early spring, when growth rates are slow, the incorporated plants were torn by the large-sized sediment brought into suspension by the waves. No further fatalities, however, were encountered after June, indicating that reduced fatality percentages in such areas might be expected if the planting were done later in the season.

Another example where high plant fatality is attributed to physical stress occurs at site No. 4 where three-month-old *S. alterniflora* seedlings were planted along the foot of a wooden bulkhead. Plants were incorporated at elevations from MLW to 21 cm above MLW at the foot of the bulkhead. Within 2.5 months after planting in May, 63% plant fatalities were encountered, the majority being at the highest elevations adjacent to the bulkhead. It appears that the backwash occurring near the bulkhead created stress beyond the plants' tolerance level. Non-bulkheaded shore plantings at site No. 4 averaged 12% fatalities.

The dynamics associated with the intertidal area of site No. 5 (Fig. 8) were too severe for two-month-old, bare-root or peat-potted *S. alterniflora* seedlings. Essentially all the plant material incorporated was washed out within one month. However, at supratidal elevations, early-spring plantings of bare-root seedlings of *S. patens, S. cynosuroides,* and *P. communis* and late-spring plantings of three-month-old, peat-potted seedlings of these plant species and of *A. breviligulata* and *P. virgatum* were successful.

Figure 8. Aerial view of site No. 5 in February 1973.

Also successful, was the planting of three-month-old *S. alterniflora* seedlings throughout intertidal and shallow (to 5 cm below MLW) subtidal areas of the spit. Part of this area is shown in Figure 9. One advantage of planting more mature plants is that they can be planted deeper and the chance of plants washing out due to substrate elevational shifts is minimized. Also, because of the increased heights of the older plants, fewer plants are lost by being buried under the shifting sediment, and low-elevation plantings do not suffer from extended periods of complete inundation.

Figure 9. Ground view of site No. 5 in September 1973, looking landward from near the terminus of the spit. The **S. alterniflora** pictured was planted as three-month-old, peat-potted seedlings in June and July.

Surveys at site No. 5 indicate increases in both subtidal and intertidal elevations throughout the spit after planting. Vegetative establishment particularly of *S. patens* has increased the stability of certain supratidal areas of the spit.

The effects of elevation on net production of three-month-old seedlings planted in the early spring were determined in 1972 for six plant species at site No. 1a. Least tolerable of periodic inundation are *A. breviligulata* and *P. virgatum*. The latter was found to die at approximately MHW. With all elevations given relative to MHW, the following decreases in total plant net productions were observed: *A. breviligulata,* factor of 4.6 going from +26 to +2.4 cm; *D. spicata,* factor of 1.8 going from +22 to -17 cm; *S. patens,* factor of 2.8 going from +18 to -17 cm; *S. cynosuroides,* factor of 2.3 going from -0.6 to -22 cm; and *S. alterniflora,* no change upon going from -14 to -37 cm.

These data and those for 1973, which currently are being processed, show that *S. alterniflora* production is less sensitive to elevation changes throughout the entire tidal regime than the other four species investigated. *S. patens* and *S. cynosuroides* are best suited for establishment at and above MHW; whereas *D. spicata* is more tolerant of periodic inundation of short duration.

Animal utilization of new marsh areas may jeopardize a shore stabilization project. The only planted species that did not encounter notable animal grazing or excavation at all project sites during the first two years were high marsh plants *S. patens, D. spicata,* and *P. virgatum.* With the exception of *T. angustifolia* and *T. latifolia* seedlings most of which were destroyed at site No. 1a during the growing season by the blue crab, the most extensive animal damage occurred during the fall and winter months. Beginning in October, 1972, there were so many muskrat and raccoon excavations of the rhizomes of the *S. cynosuroides* planted at site No. 1a that all viable material was lost. Since this plant did not flower during the first growing season, it is not surprising that it did not reappear as seedlings the following year (*see* later discussion regarding *S. alterniflora*). Deer extensively harvested the aerial parts of *A. breviligulata* and to a lesser extent of *P. communis* at site No. 1a after the first growing season and at the beginning of the second. However, little permanent damage resulted.

Canada geese began excavations of *S. alterniflora* rhizomes in October, 1972, at site No. 1a and by early spring, 1973, only 20%-25% of the original planted material remained. The animal decimation of the vegetation at site No. 1a and the remarkable natural recovery are depicted in Figures 10-13. The planted *S. alterniflora* flowered profusely the first year and its seeds were sown by the waterfowl during their plant excavations. Seedlings emerged in early April from depths of down to 7.6 cm below the substrate surface. By September, the new seedlings had attained heights up to 90 cm and had flowered again (Fig. 14).

Animal devastions of virgin marsh, though possibly temporary, may hazard biotic approaches to shore-stabilization projects. Deterrents to and management of the major wildlife utilizers of marsh areas should be explored. One deterrent under investigation is the winter planting of well-developed units of plant roots and rhizomes, thereby eliminating visibility from above ground. Vegetative development during the following summer may be sufficient to deter excessive animal excavations of plant material. Another possible deterrent is to initiate vegetative stabilization at tidal elevations that will discourage maximum animal utilization. For example, it appears that Canada geese prefer to float while excavating *S. alterniflora.* At site No. 1a, the only *S. alterniflora* not extensively consumed was above MTL. Consequently, establishing this vegetation at and

Figure 10. View of site No. 1a in February 1972 before filling the area with 3000 cu yd of mostly sand.

Figure 11. View of site No. 1a in September 1972, after filling and planting the tidal area.

Figure 12. View of site No. 1a in March 1973 following animal consumption of approximately 70% of the established vegetation shown in Figure 11.

Figure 13. View of site No. 1a in September 1973 showing the natural recovery of the vegetation.

above MTL will decrease the time available to Canada geese for plant depredation.

Benthos Recoveries Within Artificial Tidal Flats

The fill material for site No. 1a was obtained from a terrestrial quarry and contained no marine organisms. The dredge spoil used at site No. 1b was stockpiled on the mainland for approximately six months prior to its use at the site, and did not contain any living marine organisms. By the end of the first summer at site No. 1a, the benthos found were comparable in species and in population to those inhabiting a suitable undisturbed control area.

At site No. 1b, from early June 1973 (three months following site preparation) to early September, benthic organisms per m^2 varied at elevations relative to MLW from no organisms at +21 cm and 150 individuals of one species at -7.3 cm to three species totalling 900 individuals at +21 cm and seven species totalling 3,450 individuals at -7.3 cm. No suitable control area for site No. 1b was identified; however, a natural tidal flat nearby yielded in early September four species totalling 1,250 individuals at +21 cm and eight species totalling 4,400 individuals at -7.3 cm. (Unpublished data from D. A. Fisch.)

Methods of Vegetative Establishment

In principle, establishment of marsh vegetation can be initiated by planting (a) seed, (b) bare-root plant stock from nurseries or natural marsh areas, (c) soil plugs from nurseries or natural marsh areas, and (d) peat-potted plants in varying stages of development. All planting methods can be mechanized.

Seeding is by far the most economical, but also the most risky method because (a) environmental conditions for optimum germination can never be guaranteed, (b) the optimum time for sowing seed is restricted to winter and spring, (c) retaining seed within the optimal (shallow) planting depths depends upon the absence of significant wave- or current-driven sediment transport, and (d) seedling development is highly restricted by wave and tidal (Inundation) stresses.

Planting peat-potted plants of various developments may be the most expensive method of vegetative establishment; however, it (a) disturbs the plant material minimally and thereby minimizes fatalities, (b) offers the greatest time flexibility for scheduling plant incorporation, and (c) provides the capability to fashion the age or development of the plant material utilized to the various stresses prevailing at the project site.

Planting bare-root plant stock is particularly advantageous for large vegetative establishments where logistics may be a major consideration. For us it has proven to be generally inferior, with regard to the benefits enumerated above, to planting peat-potted plants. While planting bare-root plant stock is better than planting soil plugs, principally because of labor and logistics advantages, these two methods appear to lead to comparable results.

ACKNOWLEDGMENTS

The work reported herein has been supported in part by grants from James W. Rouse (site No. 4) and Julian Miller (site No. 5) and by contract from the Coastal Engineering Research Center (site Nos. 1b, 1e, 2, and 3).

REFERENCES

1. Garbisch, E. W., Jr.
   1973 Tidal marsh synthesis in the Chesapeake Bay. **Potomac Appalachian Magazine,** 1: 17-27.
2. Jenkins, R. E.
   1972 Ecosystem restoration. In **Third Midwest Prairie Conference Proceedings,** Kansas State Univ., Manhattan, September 22-23.
3. Sharp, W. C. and Vaden, J.
   1970 10-Year report on sloping techniques used to stabilize eroding tidal river banks. **Shore and Beach,** April, 31-35.
4. Woodhouse, W. W., Jr., Seneca, E. D., and Broome, S. W.
   1972 Marsh building with dredge spoil in North Carolina. Agricultural Experiment Station, North Carolina State Univ. (Raleigh) Bull. No. 445.

SALT-WATER MARSH CREATION

E. D. Seneca[1], W. W. Woodhouse, Jr., and S. W. Broome[2]

ABSTRACT

Studies were designed and conducted to determine procedures and techniques for constructing salt marsh on sandy dredge spoil in North Carolina by seeding and transplanting *Spartina alterniflora* Loisel. After three years, other flowering plants have invaded the upper parts of the intertidal zone of an *S. alterniflora* planting on a low-salinity spoil island. Both length of tidal inundation and salinity must be considered when constructing the elevational gradient over which a *S. alterniflora*-dominated marsh is desired. Although there is a gradual decline in growth potential of heeled-in transplants with time, results indicate that transplants can be stored by heeling-in in the intertidal zone for ten weeks with little influence on survival and only about 30% loss in growth potential. Seeding at the rate of 100 viable seeds per square meter from April through May can result in complete cover by the end of the first growing season, with biomass accumulation approaching that produced by transplants by the end of the second growing season. By the end of two growing seasons, there was no difference in aerial biomass between plants resulting from April seedings and those resulting from June seedlings. Seed collected from populations of *S. alterniflora* from New England to Texas germinated in several alternating diurnal thermoperiods following storage for three months in estuarine water at 2-3 C. Plants grown from these seeds exhibited morphological and physiological differences, some of which probably have genetic bases. These findings along with other observations suggest that both seed for for direct seeding and plants for transplanting should be obtained from sources as close as possible to the prospective marsh construction sites.

1. Botany Department, North Carolina State University, Raleigh, North Carolina 27607.

2. Soil Science Department, North Carolina State University, Raleigh, North Carolina 27606.

INTRODUCTION

Earlier investigation in North Carolina (1, 2, 4, 5) have indicated that estuarine salt marsh can be constructed on sandy dredge spoil by either direct seeding or transplanting *Spartina alterniflora* Loisel. These studies emphasized that while seeding is possible on rather protected areas under favorable conditions, transplanting can be successful over a much wider range of environmental conditions. Initial establishment of plants by seeding is restricted to approximately the upper half of the intertidal zone. This paper presents results obtained by a continuation of our studies by: (a) following earlier plantings for 2-3 years, (b) comparing areas initially seeded with areas initially transplanted, (c) storing transplants in the intertidal zone for planting at a later time, and (d) comparing germination and seedling growth responses for selected population of *S. alterniflora* from New England to Texas.

TRANSPLANT STUDY AT SNOW'S CUT

In April, 1971, single culm (stem) transplants of *S. alterniflora* were hand-planted (4) about 1 m apart on a sandy dredge-spoil island near Snow's Cut, North Carolina (N lat 34° 07'; W long 77° 56'). Fresh spoil had been deposited on an old spoil island in the Cape Fear River estuary 60 days prior to planting. Spoil within the intertidal zone was 96% sand, 1% silt, and 3% clay and had a slope of 1.5-2.0%. The tide range was 1.2 m and salinity of the ground water and estuarine water was 8-10°/oo. The planting was sampled for aerial biomass in September for three consecutive years. Samples were harvested from four tidal zones based on the period of time they were inundated daily.

Results indicate that after increasing in the first and second years in all four zones, *S. alterniflora* production decreased in the upper two zones the third year (Table 1). This decrease in production was largely due to invasion of these zones by other flowering plants (Table 2, Fig. 1). Most of the invaders were plants characteristic of brackish and fresh-water marshes but some were common inland weeds. These plants began invading the planting during the second growing season, but did not become abundant until the third year, when they constituted 41% of the production in the uppermost zone (Table 1). Based on observations in other plantings over a wide range of conditions, we expected the planting to be slowly invaded by other plants, but we did not expect the invasion to be so large. This invastion by other plants was possible because of available seed sources already on the spoil island and the invading species' tolerance to relatively low substrate salinity and relatively short periods of inundation. *S. alterniflora* was able to maintain dominance in the lower two zones apparently because few invading plants could tolerate the longer period of tidal inundation (Table 1). The increase in *S. alterniflora* production from the

TABLE 1.

Mean above-ground biomass for four tidal zones for **S. alterniflora** and invading plants[1] over 3 years on dredge spoil at Snow's Cut, North Carolina

Period inundated daily (hr)	Aerial biomass (g dry wt/m^2)			
	1971[2]	1972[3]	1973	
	Spartina	**Spartina**	**Spartina**	Invaders[1]
2.5	–[4]	1180	408	282
5.5	–[4]	1298	726	104
8.5	203[5]	989	988	76
11.5	–[4]	790	1290	4

[1]**See** Table 2 for a list of the invading flowering plants.

[2]No invaders.

[3]Too few invaders to sample

[4]Not sampled.

[5]Most productive zone in 1971.

Figure 1. Three-year-old **S. alterniflora** planting at Snow's Cut, North Carolina, being invaded by other flowering plants.

TABLE 2.

Presence of flowering plants that invaded the **S. alterniflora** planting during the second and third years following initial transplanting in 1971 at Snow's Cut, North Carolina.

Scientific name	1972	1973
Aeschynomene indica L.	X	
Alternanthera philoxeroides (Mart.) Griseb.	X	X
Amaranthus cannabinus (L.) J. D. Sauer	X	X
Aster subulatus Michx.	X	X
Aster tenuifolius L.	X	X
Atriplex patula L.		X
Borrichia frutescens (L.) DC.	X	X
Cyperus polystachyos var. **texensis** (Torr.) Fern.	X	X
Cyperus strigosus L.	X	X
Daubentonia punicea (Cav.) DC.	X	X
Echinochloa walteri (Pursh) Heller	X	
Erianthus giganteus (Walt.) Muhl		X
Fimbristylis spadicea (L.) Vahl.	X	X
Iva frutescens L.	X	X
Juncus roemerianus Scheele	X	X
Panicum dichotomiflorum Michx.	X	X
Panicum virgatum L.		X
Phragmites communis Trin.		X
Pluchea purpurascens (Sw.) DC.		X
Polygonum pensylvanicum L.		X
Sabatia stellaris Pursh		X
Scirpus americanus Pers..		X
Scripus robustus Pursh	X	X
Scirpus validus Vahl.	X	X
Spartina patens (Ait.) Muhl.		X
Suaeda linearis (Ell.) Moq.	X	
Vigna luteola (Jacq.) Benth.		X

second to the third season in the lowest zone suggests that in areas of relatively low salinity the lower half of the tidal range provides the best habitat for this grass. Further, these results suggest that to maintain a *S. alterniflora* marsh, both salinity and length of tidal inundation must be considered when grading the substrate to the proper elevation prior to planting.

Observations over the past three years suggest that natural salt marsh would have been very slow to develop on this spoil island mainly because of its extreme exposure to wave action. Our planting efforts stabilized a large part of the intertidal zone, thus enabling native marsh plants to invade and become established.

TRANSPLANT STORAGE AT BEAUFORT

Public and private agencies have recently become interested in whether transplants of *S. alterniflora* can be stored prior to planting. Where marsh is to be altered or disturbed for several months, the question has been raised as to whether the existing plants can be removed, stored, and replanted at a later date. In April, 1973, we plowed plants out of a nursery area that we had established on sandy dredge spoil near Beaufort, North Carolina (N lat 34° 43'; W long 76° 40'). Several thousand transplants were heeled-in in the intertidal zone on this dredge spoil (Fig. 2). On five dates ranging from 11 to 74 days following heeling-in, some of these transplants were removed and planted alongside freshly dug transplants from the same location to compare their growth.

Figure 2. Transplants heeled-in in intertidal zone at Beaufort, North Carolina.

By the time of the September, 1973, harvest, freshly dug transplants had produced more above-ground biomass than the heeled-in transplants except for the first planting date (Table 3). We cannot explain the relatively poor growth of heeled-in transplants in the May 14 comparison, but it appears to be misleading as to the overall growth potential of heeled-in transplants. Heeled-in transplants stored for 4, 6, and 10 weeks produced 85%, 75%, and 69%, respectively, of the above-ground growth produced by freshly dug transplants (Table 3, Fig. 3). Although these results indicate a gradual decline in growth potential for the heeled-in transplants, it appears that transplants can be stored for up to ten weeks with only about a 30% loss in growth potential. Survival of these heeled-in transplants was only slightly lower or equal to that of freshly dug transplants except for the first planting date when their survival was higher (Table 3). These survival and above-ground biomass data indicate that, in the first growing season, about as good a stand can be obtained with heeled-in as with freshly dug transplants for the storage times tested. Further experiments are necessary to determine whether storage of transplants over the cooler months might produce different results from those observed from late April to mid-July in this study.

TABLE 3.

Mean above-ground biomass and survival by freshly dug transplants[1] compared to heeled-in transplant[2] on each of five planting dates by the time of the September 1973 harvest on dredge spoil at Beaufort, North Carolina.

Planting dates	Aerial Biomass (g dry wt/m^2)		Survival (%)	
	Freshly dug	Heeled-in	Freshly dug	Heeled-in
May 7	62	89	72	86
May 14	66	32	75	69
May 28	43	36	76	73
June 11	31	23	65	59
July 10	13	9	59	59

[1]Transplants dug and planted on the same day on each of 5 planting dates in 1973.

[2]Transplants dug on April 26, 1973, and heeled-in in the intertidal zone for 5 periods of time prior to planting.

SEEDING STUDY AT BEAUFORT

In April, 1972, following harvest, threshing, and storage of seed collected from Oregon Inlet, North Carolina (2, 4), 100 seeds per square meter were

Figure 3. Comparison between freshly dug (left) and heeled-in (right) treatments in September, 1973, following transplanting of both treatments on May 28, 1973.

sowed on a sandy, 1 ha dredge-spoil area within the Newport River estuary near Beaufort, North Carolina. Spoil material was deposited 6-8 months prior to seeding on a 1.5-2.0% gradient in the intertidal zone. Tide range was about 1.0 m and the salinity of estuarine water was 30 $^{o}/oo$-35 $^{o}/oo$. About the lower two-thirds of the April seeding was destroyed by erosion resulting from gale-force winds associated with an unusual storm in late May. Consequently, this area together with about 0.6 ha along the upper edge of the April seeding was seeded at the rate of 200 viable seeds per square meter on June 6, 1972.

The June seeding resulted in a dense stand of seedlings, but their yield was not equal to that of the April seedlings by the time of the October, 1972, harvest; however, there was no difference in production by the end of the second growing season (Table 4). By the end of the first growing season, production by seedlings from the April seeding compared favorably with that by transplants. By the end of the second growing season, there was essentially no difference between aerial biomass of the two seeding dates and these plants produced over 70% of the yield harvested from a transplanted area of equivalent age (Table 4). These results indicate that in certain areas salt marsh can be constructed about as quickly from seeding as from transplanting. Preliminary estimates indicate that seeding is much more economical.

ATLANTIC AND GULF COASTS POPULATIONS

OF *SPARTINA ALTERNIFLORA*

Interest in our initial findings by workers outside North Carolina motivated us

TABLE 4.

Mean above-ground biomass for seeded and transplanted[1] areas for two growing seasons on dredge spoil at Beaufort, North Carolina

Planting technique	Aerial biomass (g dry wt/m^2) 1972	1973
Seeded (April)[2]	354	713
Seeded (June)[3]	57	781
Transplanted (April)	313	1017

[1]Transplanted with single culms about 1 m apart in April, 1972.

[2]Seeded at rate of 100 viable seeds/m^2 in April, 1972.

[3]Seeded at rate of 200 viable seeds/m^2 in June, 1972.

to study selected populations of *S. alterniflora* from along the Atlantic and Gulf coasts (3, Fig. 4). The study was conducted to compare germination and seedling growth among the populations to determine whether our results based on genetic material from North Carolina might be applicable to populations throughout the range of *S. alterniflora.*

Following storage for three months in estuarine water (32 $^{o}/oo$) at 2-3 C, germination above 45% was obtained in all theromoperiods by all population groupings (Table 5). Although populations throughout the range of *S. alterniflora* may have some differences in their overall germination requirements, these results indicate that satisfactory germination can be obtained with storage and temperature conditions used. Germination generally occurred at a faster rate in the higher temperature treatments.

Results of the seedling-controlled environment and field studies indicated that: (a) flowering progresses from north to south, (b) vegetative growth is adapted to a progressively longer growing season from north to south, and (c) vegetative growth becomes progressively less influenced by photoperiod from north to south. Field observations of natural stands suggest that the range of flowering time is smaller from about Cape Lookout, North Carolina, northward to New England than it is southward of that point. When seedlings of these 12 populations were grown on dredge spoil at Snow's Cut, North Carolina, differences in growth form were observed among the populations. New England plants were shorter, dark green, with numerous small culms arising from rhizomes. Plants from Georgia and Florida populations were yellow-green, with large basal culms and wide leaf blades. Gulf coast plants were tall, dark green, with long, narrow, upright leaves that formed a more acute angle with the culm than did those of the other populations. Plants from the mid-Atlantic

Figure 4. Map of 12 populations of **S. alternifora.**

populations were similar to one another in growth, form, and color. In the first growing season plants from populations geographically closer to North Carolina grew better than those from populations farther away (Table 6). By the end of the second growing season, however, South Atlantic and Gulf coasts populations had produced from three to four times as much above-ground biomass as Mid-Atlantic populations which, in turn, had produced three times as much aerial biomass as New England populations.

Results from these controlled environment and field studies suggest that there is considerable morphological and physiological variation among populations of *S. alterniflora.* A considerable amount of this variation appears to be genetic. In any event, we should exercise a great deal of caution in obtaining either seed or transplants far from the intended site of marsh establishment. Local populations are probably better adapted to their environmental conditions than are populations farther to the north or south. It would be unwise to produce large

TABLE 5.

Mean germination percentage for four geographic population groupings in four alternating diurnal thermoperiods[1] following three month's storage in estuarine water at 2-3 C

Thermoperiod (deg C)	Germination (%) by population grouping New England[2]	Mid-Atlantic[3]	South Atlantic[4]	Gulf[5]
40-25	90	90	88	92
35-20	66	80	78	76
30-15	63	69	77	75
25-10	45	70	91	87

[1]Seeds remained in the higher temperatures of the thermoperiods for 17 $\pm$ 1 hr.and in the lower temperatures for 7 $\pm$ 1 hr.

[2]Massachusetts, Rhode Island, and Connecticut populations.

[3]New York, Virginia, and North Carolina populations.

[4]Georgia and Florida populations.

[5]Mississippi and Texas populations.

TABLE 6.

Mean above-ground biomass for two growing seasons for four geographic population groupings grown on dredge spoil at Snow's Cut, North Carolina

Population grouping	Aerial biomass (g dry wt/m^2) 1972	1973
New England[1]	15	114
Mid-Atlantic[2]	56	343
South Atlantic[3]	35	1419
Gulf[4]	16	1134

[1]Massachusetts, Rhode Island, and Connecticut populations.

[2]New York, Virginia, and North Carolina populations.

[3]Georgia and Florida populations.

[4]Mississippi and Texas populations.

quantities of seed and transplants from a single population to be used for marsh-building by persons along the entire Atlantic coast.

ACKNOWLEDGMENTS

This research was initiated with support of the Coastal Engineering Research Center, U. S. Army Corps of Engineers, and continued with support of that organization under contract no. DACW72-72-C-0012; the North Carolina Sea Grant Program, Office of Sea Grant, NOAA, Department of Commerce, grant no. GH-103, 2-35178 and 04-3-158-40; and the North Carolina Coastal Research Program. The authors wish to express appreciation to the North Carolina Department of Natural and Economic Resources and the U. S. Army Engineer District, Wilmington, North Carolina. Special thanks go to Messrs. D. M. Bryan, C. L. Campbell, Jr., and U. O. Highfill for their assistance in all phases of field work.

REFERENCES

1. Broome, S. W.
1972 Stabilizing dredge spoil by creating new salt marshes with **Spartina alterniflora. Soil Sci. Soc. North Carolina Proc.,** 15: 136-147.

2. Broome, S. W., Woodhouse, W. W., Jr., and Seneca, E. D.
1973 An investigation of propagation and the mineral nutrition of **Spartina alterniflora.** Sea Grant Publication UNC-SG-73-14. 121 p.

3. Seneca, E. D.
Germination and seedling response of Atlantic and Gulf coasts populations of **Spartina alterniflora. Am. Jour. Bot.,** (In press.)

4. Woodhouse, W. W., Jr., Seneca, E. D., and Broome, S. W.
1972 Marsh building with dredge spoil in North Carolina. North Carolina State Univ. at Raleigh Agric. Exp. Sta. Bull. 445. 28 p.

5. Woodhouse, W. W., Jr., Seneca, E. D., and Broome, S. W.
1973 Establishing salt marsh on dredge spoil. **Proc. World Dredging and Marine Construction,** 9(7): 16 (abstract).

SUBMERGENT VEGETATION FOR BOTTOM STABILIZATION

by

Lionel N. Eleuterius[1]

ABSTRACT

Three species of sea grass, *Thalassia testudinum, Cymodocea manatorum,* and *Diplanthera wrightii* were transplanted from natural habitats to barren submergent spoil areas and to control areas located adjacent to undisturbed sea-grass beds. Anchoring devices were developed to hold the transplants in place on the bottom regardless of water depth. *Diplanthera* had the greatest percentage of survival, its growth rate for exceeding that of *Thalassia. Cymodocea* did not succeed at all. Some characteristics of vegetative morphology and rate of growth were compared under varying conditions of sediment deposition and erosion. In view of its distribution, growth, tolerances, and low rates of deposition, *Diplanthera* was judged the best candidate for further transplant studies. Successful transplants formed established beds in control areas. Erosion and sedimentation rates exceeded the growth capacity of the plants although no successful transplants were obtained on spoil. Low temperatures and prolonged exposure to low-salinity waters apparently affect sea-grass beds and transplants adversely. Available plant nutrient levels of bottom (substrate) samples were not found to vary appreciably between vegetated and barren areas.

INTRODUCTION

Sea-grass beds in the Gulf of Mexico are composed predominantly of the

1. Gulf Coast Research Laboratory, Ocean Springs, Mississippi 39564.

marine angiosperms *Diplanthera wrightii* Ascherson (shoal grass), *Thalassia testudinum* Konig (turtle grass), and *Cymodocea manatorum* Ascherson (manatee grass). Their distribution and ecology in various regions of the northern Gulf of Mexico have been studied by Humm (8), Phillips (14, 15, 17), Strawn (24), Moore (12) and Eleuterius (4). Animal communities associated with sea-grass beds in the waters of Florida have been reported upon by Voss and Voss (28), Tabb and Manning (25), Stephens (22), and O'Gower and Wacasey (13). Gunter (6) reported that most of the pink shrimp production in Texas comes from the Gulf opposite Redfish Bay, which has one of the largest *Thalassia* communities on the Texas coast. Hoese and Jones (7) studied vegetated sea bottoms in Texas and the associated larger animals. Christmas et al. (3) made special reference to the abundance of postlarval penaeid shrimp in the grass beds near Horn Island, Mississippi. They pointed out that more shrimp were taken at sampling stations that included grass beds than at those without them. These studies have shown that sea-grass habitats support a large and varied community of marine animals in the Gulf of Mexico. Similar reports by Wood (27) and Fuse et al. (5) from other parts of the world indicate that vegetated sea bottom provides one of the richest and most productive of marine communities.

The ability of vegetation to prevent erosion of the soils of the earth has been realized for a long time, and its use in soil-conservation practices related to agriculture, forestry, and highway construction is well documented. For a number of years Europeans have used emergent vegetation along stream and river banks to stabilize them against erosive currents (21). More recently submerged vegetation has been used for bottom stabilization of American waterways (19, 20). These reports, of course, relate to fresh-water systems. The use of submergent marine flowering plants for bottom stabilization has had little study. Kelly et al. (9) transplanted *Thalassia* in Boca Ciega Bay, Florida, and Phillips (17) transplanted *Zostera marina* in Puget Sound, Washington. However, natural establishment of *Diplanthera* on submerged spoil areas from Port Isavel to Port Mansfield, Texas, has been reported by Breuer (1, 2). Stevens (23) reported sea grasses growing on spoil in Corpus Christi Bay, Texas. *Diplanthera* has also been reported to occur on spoil areas in Sarasota Bay, Florida (Gordon Gunter, personal communication). Strawn (24) reported that at Morehead City, North Carolina, the bay scallop fishery was partially restored by the increase of *Diplanthera.* Thus, successful establishment of sea grasses on barren spoil would have two beneficial effects: the stabilization of spoil and the provision of additional productive habitat.

There were three primary objectives in the present study: a. development of effective methods for transplanting sea grasses; b. determination of the best species for transplanting; and c. identification of some of the limiting ecological factors by studying existing grass beds and experimental areas. This project began in July 1971 and the first transplants were made in the fall of the same year.

Mississippi Sound is bordered on the south by four barrier islands, and four river systems contribute directly to the fluctuating salinity gradient across it. Most of the sea-grass communities are located north of the barrier islands and in the island passes. The grass beds in the shallow waters along the islands occupy areas of various size, thus their distribution here is discontinuous. In the deeper waters grass beds are more extensive in area. The spoil islands used as experimental areas in this study were created primarily as the result of maintenance dredging of the Pascagoula ship channel. General location of sea-grass beds and transplanting sites are shown in Figure 1.

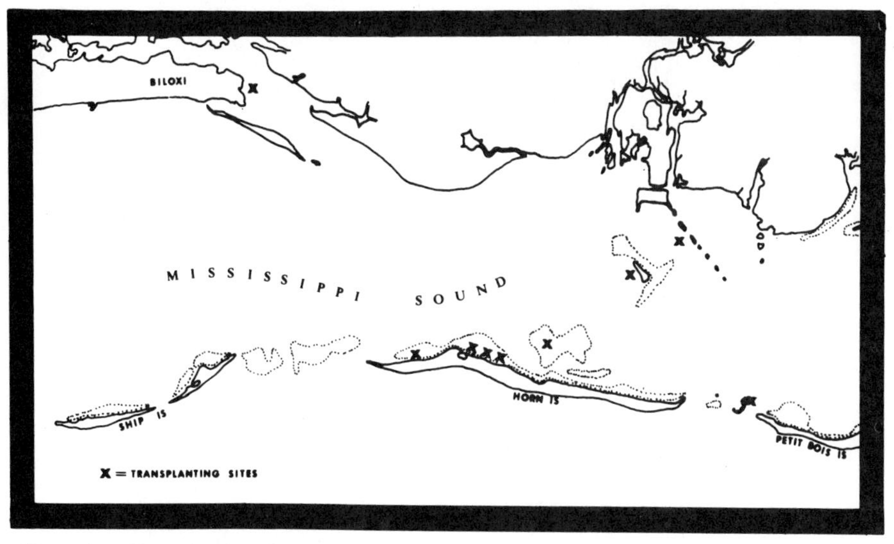

Figure 1. Transplanting sites (x) of sea grasses. The stipled area indicates general location of naturally established grass beds.

METHODS AND MATERIALS

The three common species of sea grass shown in Figure 2 were transplanted from natural sea-bed habitats near Horn Island, Mississippi, to spoil islands adjacent to the Pascagoula ship channel and to barren areas which served as controls adjacent to areas vegetated by sea grasses.

Two types of anchoring devices developed and used in this study are shown in Figures 2-D, 3-A, and 6-C. The specific details of development of these devices will be reported elsewhere (Eleuterius, in preparation). In order to determine the best time of year to conduct transplanting operations, transplants were made seasonally. Vegetative morphology and rate of growth were noted and compared under varying conditions of sedimentation and erosion rates. General

Figure 2. Three marine angiosperms found in Mississippi Sound used in transplant study. (A) **Diplanthera wrightii** (shoal grass), (B) **Cymodocea manatorum** (manatee grass), (C) **Thalassia testudinum** (turtle grass), (D) **T. testudinum** attached to wire-mesh anchor.

characteristics such as nutrient levels and developmental patterns of sea-grass beds established naturally and by transplantation were compared to barren areas.

Transplant material was dug from natural habitats on low tides and most of it was transported by boat to the laboratory some twenty-five miles away. During

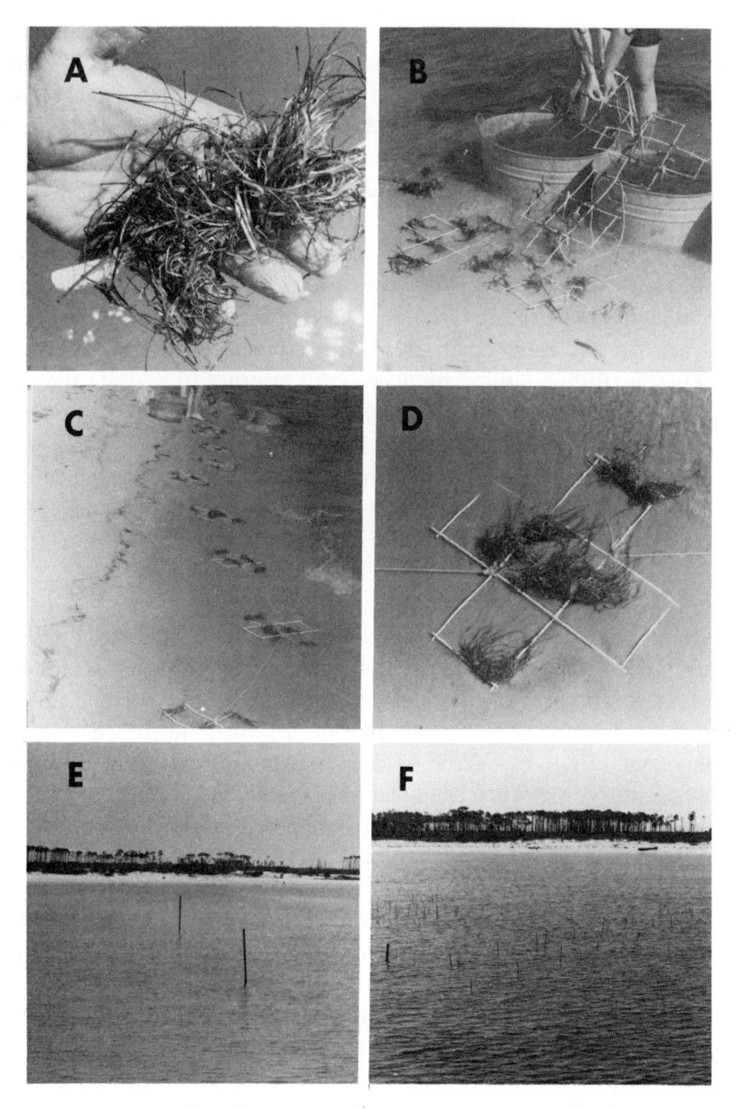

Figure 3. Transplants, anchors, planting techniques, and study areas along Horn Island. (A) **Diplanthera wrightii** tied to coated iron rod. Plastic-coated wire was used in attaching transplants to anchor. (B) Tying anchors together in trotline fashion. (C) Wire-mesh anchors arranged along shore to demonstrate transplanting technique. (D) Wire mesh with **Diplanthera wrightii** attached. (E) Study area along Horn Island. Pipes were used as permanent markers. (F) Hand emplaced transplants (attached to construction rods) on barren areas adjacent to grass beds at Horn Island.

the first year, transplants were moved to adjacent barren areas and replanted the same day they were dug from the habitats (Fig. 3-F). The sea grasses that were returned to the laboratory were stored at room temperature in aerated sea water. Transplants were then selected and attached to anchoring devices cut from consturction rod and heavy-gauge wire mesh.

Each wire-mesh anchor, approximately 45 cm x 45 cm (18" x 18") of 15 cm (6") mesh, weighed approximately 190 grams, and each iron rod, approximately 20 cm (8") in length, weighed 110 grams. The anchors were coated with white vinyl paint and transplants were attached to them with plastic-coated copper wire. The first year all transplants were emplaced by hand or broadcast over the study areas. During the second year transplant units were tied together in trotline fashion by nylon cord in order to relocate the anchors and record the condition of the transplants attached to each (Fig. 3-B, C, D, 6-D).

The transplants were laid out in rectangular belts along the experimental spoil islands and on barren areas adjacent to or within the sea-grass habitats (Fig. 4-A).

RESULTS

Transplants were made simply by sprigging initially, but all were pulled from the substratum by wave action and currents in control areas and spoil areas. A variety of anchoring devices, whose emplacement required a person to be in the water, were used during this project. The ones described herein were developed for employment in either shallow or deep waters and proved to be especially suitable for depths too great to hand plant by wading.

The inshore submerged spoil area composed primarily of mud was continously covered by highly turbid waters ranging in salinity from 0°/oo-15°/oo. The offshore spoil area consisted primarily of sand and was covered by waters ranging in salinity from 0°/oo-35°/oo. Both areas had clam shell intermixed, the proportion, however, was relatively low. No submerged vegetation existed on either area.

No appreciable difference in available nutrients was noted between barren spoil, sea-grass beds, and adjacent barren areas. Slightly higher averages in phosphorus and potassium were observed in the older portion of sea-grass beds, probably due to the accumulation of organic matter. Samples from the periphery of colonies had lower levels and approximated adjacent barren areas. Phosphorus ranged from 23 kg (50 lbs) to 80 kg (175 lbs) per acre. Potassium ranged from 46 kg (100 lbs) to 136 kg (300 lbs) per acre. Nitrogen was less than 4.5 kg (10 lbs) per acre for all areas sampled. Organic matter was recorded at approximately 2% in *Cymodocea* and *Thalassia* beds and slightly over 1% in *Diplanthera* beds. Adjacent barren areas varied from slightly less than 1% to slightly less than 2% . Hydrogen ion concentration (pH) ranged from 6.2 to 8.2

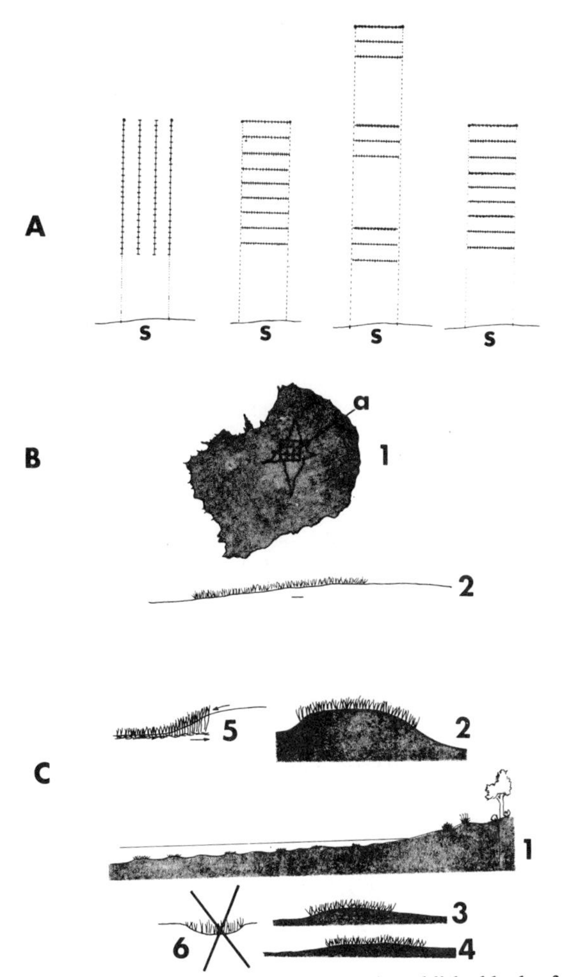

Figure 4. Aspects of transplanting, growth pattern, and established beds of sea grasses. (A) Planting schemes arranged in rectangular belts. Anchors of transplants were tied together with nylon cord and placed parallel and perpendicular to shore line (s). Wire mesh (x) relation to iron rod (-) anchors are indicated. (B) 1. Diagrammatic pattern of growth and centrifugal spread from wire-mesh anchor. (a) The inner outline indicates the general pattern and spread after six months' growth, and the outer indicates spread at eighteen months. 2. Profile diagram of established bed (B1). (C) Profile diagrams of naturally established grass beds. 1. Submerged grass beds. 1. Submerged grass beds along north shore of Horn Island (lagoon area). Note that most grass beds are on elevated areas (sand bars). These vegetated areas may range from almost flat to abrupt changes in elevation (2, 3, 4). **Thalassia** beds may grow into the side of sand bars with some sand washing onto the beds (5). No grass beds were found to occupy small shallow depressions (6).

with averages of 7.2 for grass beds; 7.6 for adjacent barren areas, and 7.3 for the barren spoil area at Horn Island.

None of the transplants made on spoil areas were successful. Mortality was due to the extremely unstable bottom. The spoil at Horn Island Pass was reequently exposed to heavy seas which reworked drastically the sandy substratum. From November 1971 through March 1972, approximately 60 cm (24") of sand were deposited over the transplants on the south, southeast, and eastern section of the spoil. No velocity measurements were made, but tidal currents through the passes are always considerable. These currents at times were so great that a person could not remain standing in the shallow waters of the spoil areas at Horn Island Pass. Transplants made on the west side of the spoil were quickly eroded away and no further transplants were made there.

Mortality of transplants on the spoil near the mouth of the Pascagoula River may have resulted from a number of causes; inadequate light due to highly turbid water, exposure to the great quantity of low-salinity fresh-water discharged from the Pascagoula River, or siltation. Survival and establishment were attained in transplants of *Diplanthera* and *Thalassia* in the control area near Horn Island. The number of successful transplants of *Diplanthera* was almost twice that for *Thalassia.* No success was achieved in transplants of *Cymodocea.* Survival data showing dates of transplanting and subsequent observation are shown in Table 1. Since most anchors of transplants made by the broadcast method could not be found, evaluation of each transplant was not possible. (At the time of writing this paper a U. S. Army surplus metal mine detector (model DT-44D/PRS-3) has been successfully employed to locate anchoring devices placed by broadcasting. This sensing device will be used to ascertain rapidly and more accurately those grass beds established from transplants attached to anchors. This location process is on-going and the results were not available for incorporation herein.) An effective method of tying the anchors together by nylon cord was developed to aid in relocating the units. Most transplants were killed by sediment deposition. The predominant sediment along Horn Island and spoil at Horn Island Pass was sand. The inshore area sediment was primarily mud. Some transplants were lost to erosion and many may have been killed by man-made distrubances. Approximately 100 transplants were found entangled in the nylon cord and anchors, obviously the work of vandals. Others were found strewn on high dunes near the study areas. Whether or not the transplants were killed before or at the time of this disturbance is not known.

Elevational changes due to sedimentation are illustrated in Figure 5. Deposition rates greater than 2.5 cm (1") per month appeared to exceed the growth rate of *Thalassia testudinum* and *Cymodocea manatorum* and deposition greater than 5 cm (2") per month apparently exceeded the growth rate of *Diplanthera wrightii.* The plants were obviously not able to adjust to rapid increases in substrate elevation caused by sand deposition and death resulted.

TABLE 1.

Surviving, successfully established and mortality of transplants of **Diplanthera wrightii, Thalassia testudinum** and **Cymodocea manatorum,** for the period April, 1972 through September, 1973.

Transplant Date	Location of Transplant Area	Species	Number of Anchoring Units used		Observation Date	Number of Transplants alive		Number of Transplants Dead		Number of units not located		Observation Date	Number of Trnasplants established	
			Rod	Mesh		Rod	Mesh	Rod	Mesh	Rod	Mesh		Rod	Mesh
3 Apr 72	Spoil-Horn Island Pass	**Diplanthera** (BC)	60*	30*	6 Jul 72					60*	30*			
		Malassia (BC)	60*	30*						60*	30*			
		Cymodocea (BC)	60*	30*						60*	30*			
1 May 72	Spoil-Horn Island Pass	**Diplanthera** (BC)	30	15	10 Jul 72					30	15			
		Thalassia (BC)	30	15						30	15			
		Cymodocea (BC)	30	15						30	15			
1 May 72	Spoil-Mouth Pas. River	**Diplanthera** (BC)	30	15	10 Jul 72					30	15			
		Thalassia (BC)	30	15						30	15			
		Cymodocea (BC)	30	15						30	15			
25 Jul 72	Spoil-Horn Island Pass	**Diplanthera** (BC)	60	30	20 Aug 72					60	30			
		Thalassia (BC)	60	30						60	30			
		Cymodocea (BC)	60	30						60	30			
3 Apr 72	Horn Island (lagoon area)	**Diplanthera** (BC)	30	15	6 Jul 72	20	10			10	20	6 Aug 75	14	8
		Thalassia (BC)	30	15	15 Aug 72	10	6			20	9		7	4
		Cymodocea (BC)	30	15		0	0			30	15			
25 Jul 72	Horn Island (Lagoon area)	**Diplanthera** (BC)	60*	15*	15 Aug 72									
		Thalassia (BC)	70*	15*										
27 Jul 72	Horn Island	**Diplanthera** (BC)	40*		15 Aug 72									
		Thalassia (BC)	38*											
27 Jul 72	Horn Island	**Diplanthera** (T)	25		15 Aug 72	0		25						
		Thalassia (T)	25			0		25						

* Two groups placed in separate area.
(BC) = Broadcast method
(T) = Anchors tied together with nylon cord.

TABLE 1. (continued)

Transplant Date	Location of Transplant Area	Species	Number of Anchoring Units used		Observation Date	Number of Transplants alive		Number of Transplants Dead		Number of units not located		Observation Date	Number of Transplants established	
			Rods	Mesh		Rod	Mesh	Rod	Mesh	Rod	Mesh		Rod	Mesh
10 Aug 72	Horn Island	**Diplanthera** (T)	25		6 Aug 73	0		25						
		Thalassia (T)	25			0		25						
1 Sep 72	Horn Island	**Diplanthera** (T)	25		6 Aug 73	0		25						
		Thalassia (T)	25			0		25						
		Cymodocea (T)	25			0		25						
19 Sep 72	Horn Island (West end)	**Diplanthera** (T)	75		6 Aug 73	0		75						
		Thalassia (T)	75			0		75						
		Cymodocea (T)	75			0		75						
6 Feb. 73	Horn Island	**Diplanthera** (T)	10	10	6 Aug 73	0	0	10	10					
		Thalassia (T)	10	10		0	0	10	10					
		Cymodocea (T)	10	5		0	0	10	5					
12 Feb 73	Biloxi Bay	**Diplanthera** (T)	10	10	10 Apr 73	0	0	10	10					
		Thalassia (T)	10	10		0	0	10	10					
		Cymodocea (T)	10	5		0	0	10	5					
19 Jun 73	Horn Island	**Diplanthera** (T)	100	10	6 Aug 73	61	4							
		Thalassia (T)	100	15		69	7							
3 Jul 73	Horn Island	**Diplanthera** (T)	75		6 Aug 73	63								
		Thalassia	100			41								
3 Jul 73	Horn Island	**Diplanthera** (T)	100	15	6 Aug 73	94	12							
16 Aug 73	Horn Island	**Diplanthera** (T)	100											

* Two groups placed in separate area.
(BC) = Broadcast method
(T) = Anchors tied together with nylon cord.

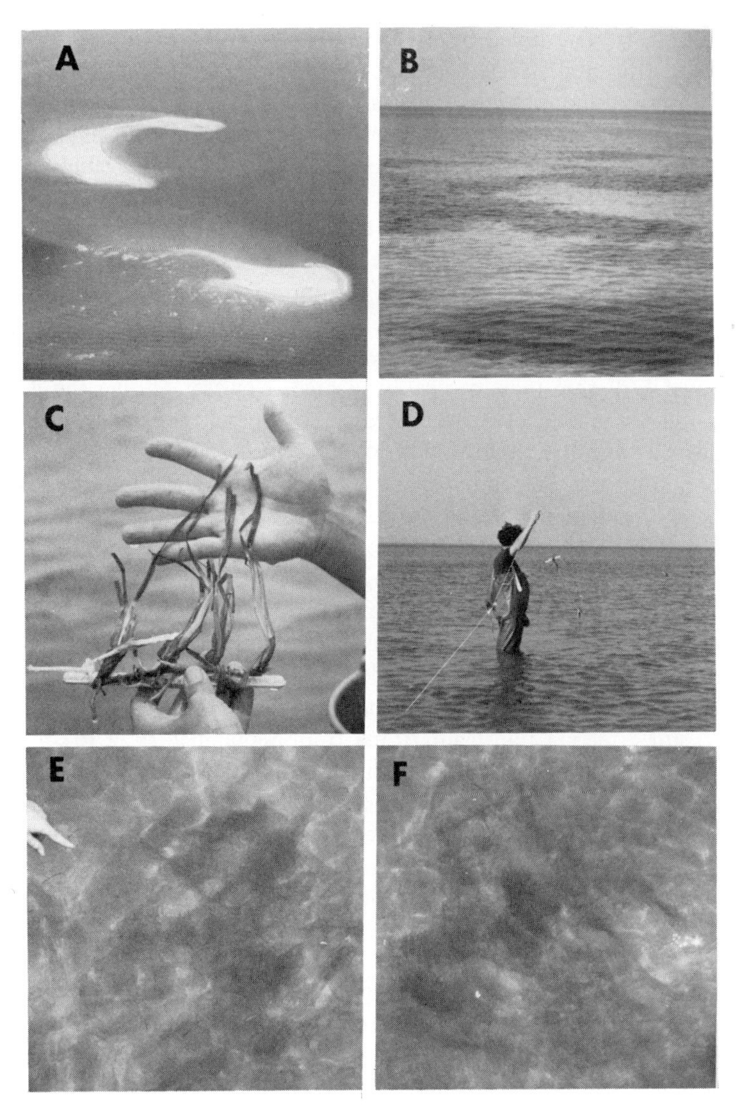

Figure 5. Sequential stages in the establishment of a bed of **Diplanthera wrightii** from an anchored transplant. (A) Transplant attached to iron-rod anchor lies on surface of substratum initially. (B) Anchor becomes covered by sand soon after transplanting. (C) Anchor covered with 5 cm (2 inches) of sediment deposition. (D) Established bed of **Diplanthera,** approximately six months from time of transplanting. (E) Well-established bed of **Diplanthera;** note increase in elevation of bed due to sediment deposition in this example. Successful transplants and established beds, however, were achieved with relatively less sediment than illustrated here.

Few transplants were killed by erosion in the control areas along Horn Island. However, detrimental effects related to transplanting during unfavorable seasons were obviously responsible for transplant mortality. Some trnasplants died from unknown causes. Since sediment deposition and erosion were not excessive it appears that detrimental physiological effects are caused by digging, preparing, attaching, or holding the transplants. However, during the first year of this study, from four to six transplants from grass beds were immediately placed in adjacent barren areas. All of them died. Transplants have been held affixed to anchoring devices in the laboratory for up to six weeks and growth was apparent, especially on *Diplanthera.*

During the winter of 1971-72 the sea grasses in Mississippi Sound did not lose their leaves and were therefore easily located. However, during the winter of 1972-73 all sea grasses observed lost their leaves and the beds were not easily distinguished from barren areas. The loss of leaves was apparently a result of a combination of low temperatures, and unusual fresh-water discharge from local rivers, and the opening of the Bonnet Carré Spillway from mid-February to mid-June 1973. Only short stubble remained on the surface of the beds. No transplants were made during the spring of 1973 although considerable time was spent in search of suitable transplant material. Most beds observed appeared to be dead or dying. However, complete death and erosion of these beds did not occur and new leaves began to appear in late June 1973. Obviously the environmental conditions mentioned above as affecting sea-grass beds were also a major factor affecting transplants made from September 1972 to June 1973. Similar leaf loss from sea grasses occurred furing the winter of 1969 and was attributed at that time to low temperatures since rainfall was not excessive.

During periods when the open (covering) waters were "fresh" ($0^{o}/oo$-$10^{o}/oo$) the soil waters had consistently higher salinities. Perhaps higher soil-water salinities protected the roots and rhizomes from rapid changes in salinity. Exposure of roots and rhizomes of transplants to this low-salinity water may have killed them.

The most remarkable and apparent difference between the two successful species, (*Diplanthera* and *Thalassia,*) was the prolific growth of the former. The best example of growth was achieved by a transplant made in April 1972. Rhizomes had spread centrifugally approximately 124 cm (48") by July 1972 and approximately 92 cm (60") by late August 1972. This entailed an average growth rate of 27 cm (10.5") per month. Thus one rhizome grew approximately 14 cm (5") per month. Kelly et al. (9) reported that *Thalassia* grew at an annual rate of 20 cm (8") or less in Boca Ciega Bay, Florida. By the end of November 1972 the bed was estimated to be 2.6 meters (8.5') in diameter. Observations made on this same established bed on 6 August 1973 showed that its centrifugal spread had increased to 3.7 meters (12') across the greatest width and 2.4 meters (8') across the narrowest. This is illustrated diagrammatically in Figure

4-B. The anchoring device was located approximately 28 cm (11") beneath the surface of the substratum.

Naturally established grass beds of *Thalassia* and *Cymodocea* are generally found in deeper and far less turbulent waters than those of *Diplanthera*. In deep, calm water continuous stands are formed, but the species are practically never intermixed. In shallow water, especially along sandy beaches and bars (shoals), *Diplanthera* generally occupies the shallowest depths and always extends nearest the shoreline, being therefore exposed almost continuously to turbulent conditions. *Thalassia* and *Cymodocea* beds were found only where sediment deposition rates were low and generally in areas where erosion was apparent. Shoal grass was found in areas of heavy sand deposition and also in areas of heavy erosion. Practically all vegetated areas are elevated as shown in Figure 4-C, especially near the lagoon area of Horn Island. However, extensive beds at the middle ground and in deeper waters along Horn Island are also found on long gentle slopes.

The ability of *Diplanthera* to occupy bottom exposed to relatively high energy is due in part to the extensive rhizome and root mass produced. The proliferation of these fine rhizomes and roots binds the substratum and contibutes directly to stabilization. Water movement across stabilized beds carries away little sediment and in near-shore areas over which waves break there is generally a deposition of sediments when leaves are present. Shoals (vegetated and non-vegetated) are built up to a certain elevation, to which they are restricted. Thus, they do not get above a certain elevation in relation to mean low water (MLW). These shoals are generally exposed on tides lower than MLW and of course especially at extremely low tides. Some of the transplants extended over and between sand bars, and those on the highest elevations did not become covered with much sediment (Fig. 6-E, 6-F). In many such instances 5 cm (2") of sediment was deposited within two months after transplanting and no further deposition appears to have occurred. On barren areas between or on the side of sand bars various amounts of deposition were recorded and were always greater than that found on top of the bars. Some bars have less elevation than others, and appear to be building. Establishment of vegetation on these lesser bars increased the building rate.

The peculiar morphology of *Diplanthera* also favors the growth of the plant in these high-energy areas. Waves and currents apparently exert less frictional force on its narrow leaf blades than they do on the wide blades of *Thalassia* and the relatively large terete blades of *Cymodocea*. These larger-leaved species were observed to be quickly pulled from the substratum when transplanted to areas constantly or periodically exposed to high energy. Another characteristic contributing to the success of *Diplanthera* was the flexibility of the blades. They bend easily, whereas those of *Thalassia* and *Cymodocea* are less flexible and more easily broken.

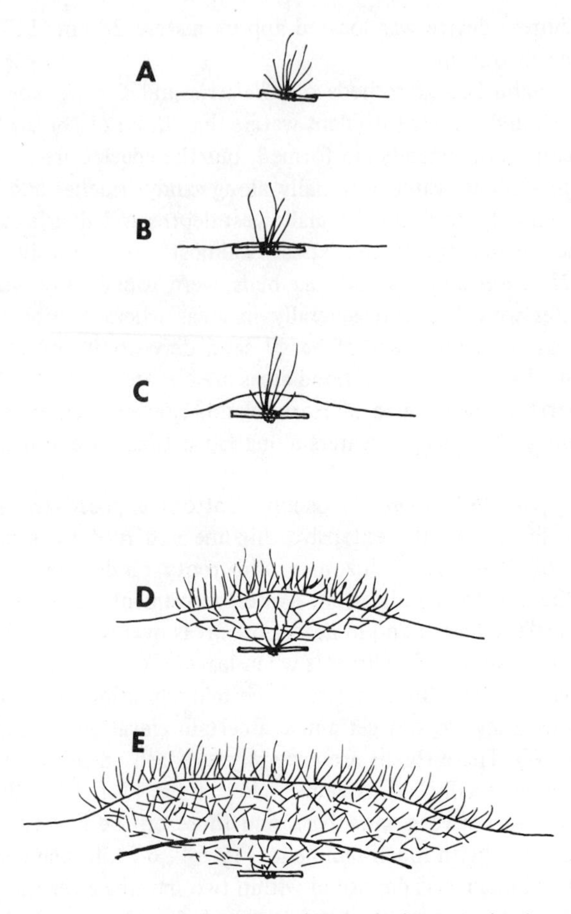

Figure 6. Spoil islands, naturally established grass beds, transplanting techniques and established beds of **Diplanthera wrightii** from transplants with wire mesh anchors. (A) Barren spoil islands at Horn Island Pass. The submerged spoil shown around the island in the lower portion of the photograph was used as transplanting site. (B) Naturally established sea grass beds north of Horn Island. The beds appear as dark patches on the bottom. (C) **Thalassia testudinum** tied to coated iron rod anchor. Plastic coated copper wire was used to tie transplants to anchors. (D) Transplants and anchors tied together in trotline fashion to aid in rapid relocation of units for evaluation. (E) F (). **Diplanthera wrightii** bed established from wire meshes. The beds are shown approximately six months after transplanting. These beds have subsequently tripled in area. (Note outline of wire mesh anchors.)

The rhizomes and roots of *Thalassia* and *Cymodocea* are also larger than those of *Diplanthera,* but their proliferation is not as great and their network of interweaving does not bind the substratum as well.

DISCUSSION

The environmental conditions that existed during the winter of 1971-72 were quite different from those found during the winter of 1972-73. The leafless, non-flowering condition found in the spring of 1973 was in sharp contrast with the luxurious leaf growth and flower production in the spring of 1972. The great outflow of fresh water and resulting low salinities probably affected natural grass beds and transplants more than any other factor through July of 1973 and contributed to the inability to duplicate in 1973 the successful results obtained in the spring and summer of 1972.

The low survival rate of transplants shows that the successful transplanting and establishment of sea-grass beds for bottom stabilization or as a habitat is not a simple matter. However, its survival and rapid growth qualify *Diplanthera* as a better candidate for more detailed study than *Thalassia.*

Diplanthera is found along the Atlantic coast from North Carolina to Florida and from Florida to Texas, but *Thalassia* and *Cymodocea* do not occur on the Atlantic coast, except for a short extension northward around the tip of extreme south Florida. *Diplanthera* apparently has wide environmental tolerances. McMillan and Moseley (11) showed in a comparative study in Texas that *Diplanthera* had a greater tolerance to salt than any other species tested. In a previous study (4) *Diplanthera* was found to occur over a larger areaa of bottom than *Thalassia* or *Cymodocea* in Mississippi Sound. Of the four marine flowering plants studied only *Diplanthera* extended into low-salinity waters along the mainland shore. It also occurred in areas exposed to highest salinities, for example, in the shallow waters near the barrier islands and in the open passes. *Diplanthera* alone is found in the island passes exposed to powerful waves and strong tidal currents. These characteristic tolerances allow wide distribution and indicate greater application of research results throughout the south Atlantic and northern Gulf of Mexico than would be possible with *Thalassia* or *Cymodocea.*

The particular set of environmental conditions that must exist during the time of natural establishment may be present only briefly each year or not at all for several years. Exactly what those conditions are is not known, but it has herein been ascertained that they are of paramount importance if grass beds on natural or man-made sea bottoms are to be successfully established.

There is no doubt that once established, vegetation suppresses erosion of sea bottom within limits, but heavy deposition of sand and other sediments eroded from adjacent areas may suppress plant growth or kill the beds quickly. Slight accumulation of sediments in *Thalassia* and *Cymodocea* beds appears to be characteristic, but heavier rates were found in *Diplanthera* beds. The rapid growth of *Diplanthera* apparently enables it to grow under greater rates of sedimentation. The spoil upon which transplants were made was not a suitable habitat for sea grasses, even though a large *Diplanthera* bed is found nearby. Perhaps more time is needed for the particular spoil to stabilize to physical forces before conditions within the growth capacity of *Diplanthera* are reached. Sea bottom areas with slight disturbance, if only seasonal, appear to be more favorable to the establishment of vegetation.

ACKNOWLEDGMENTS

This project was funded by the Corps of Engineers, U. S. Army, under contract no. DACW01-71-C-0105. I thank Dr. Gordon Gunter and Mr. Charles Eleuterius for their review and criticism of the manuscript.

REFERENCES

1. Breuer, J. P.
1960 Life history studies of the important sports and commercial fish of the lower Laguna Madre. Texas Game and Fish Commission Marine Laboratory. Coastal Fisheries Project Reports. 5 p.

2. Breuer, J. P.
1961 Life History studies of the marine flora of the Lower Laguna Madre area. Texas Game and Fish Commission Marine Laboratory. Coastal Fisheries Project Reports. 2 p.

3. Christmas, J. Y., Gunter, G., and Musgrave, P.
1966 Studies of annual abundance of postlarval penaeid shrimp in the estuarine waters of Mississippi, as related to subsequent commercial catches. **Gulf Research Reports,** 2(2): 177-212.

4. Eleuterius, L. N.
1971 Submerged plant distribution in Mississippi Sound and adjacent waters. **Jour. Miss. Acad. Sci.,** 17: 9-14.

5. Fuse, S-I., Habe, T., Harada, E., Okuno, R., and Miura, T.
1959 The animal communities in the submerged marine plant vegetations. **Bull. Mar. Biol. Str.,** Asamushi Tohoku Univ., 9(4): 173-175.

6. Gunter, G.
1962 Shrimp landing and production of the State of Texas for the period 1956-1959, with a comparison with other Gulf states. **Publ. Inst. Mar. Sci. Univ. Tex.,** 8: 216-226.

7. Hoese, H. D. and Jones, R. S.
1963 Seasonality of larger animals in a Texas turtle grass community. **Pub. Ins. Mar. Sci.,** 9: 37-47.

8. Humm, H. J.
1956 Seagrasses of the northern Gulf Coast. **Bull. Mar. Sci. Gulf Carib.,** 6(4): 305-308.

9. Kelly, J. A., Jr., Fuss, C. M., and Hall, J. R.
1971 The transplanting and survival of turtle grass, **Thalassia testudinum,** in Boca Ciega Bay, Florida. **Fishery Bulletin,** 69(2): 273-280.

10. Marmelstein, A. D.
1966 Photoperiodism, and related ecology, in **Thalassia testudinum.** M.S. thesis, Texas A & M Univ. 54 p.

11. McMillan, C. and Moseley, F. N.
1967 Salinity tolerances of five marine spermatophytes of Redfish Bay, Texas. **Ecology,** 48: 503-506.

12. Moore, D. R.
1963 Distribution of the sea grass, **Thalassia,** in the United States. **Bull. Mar. Sci. Gulf Carib.,** 13(2): 329-342.

13. O'Gower, A. K. and Wacasey, J. W.
1967 Animal communities associated with **Thalassia, Diplanthera,** and sand beds in Biscayne Bay 1. Analysis of communities in relation to water movements. **Bull. Mar. Sci. Gulf Carib.,** 17: 175-210.

14. Phillips, R. C.
1960 Environmental effect on leaves of **Diplanthera** Du Petit-Thouars. **Bull. Mar. Sci. Gulf Carib.,** 10(3): 346-353.

15. Phillips, R. C.
1960 Observations on the ecology and distribution of the Florida seagrasses. Fla. St. Brd. of Consv. Mar. Lab. Prof. Papers, Ser. no. 2: 1-72.

16. Phillips, R. C.
1962 Distribution of seagrasses in Tampa Bay, Florida. Fla. St. Brd. of Consv. Div. of Salt Water Fisheries, Special Scientific Report No. 6:12 p.

17. Phillips, R. C.
1967 On species of the sea grass, **Halodule,** in Florida. **Bull. Mar. Sci. Gulf Carib.,** 17(3): 673-676.

18. Randall, J. E.
1965 Grazing effect on sea grasses by herbivorous reef fishes in the West Indies. **Ecology,** 46(3): 255-260.

19. Ree, W. O.
1949 Hydraulic characteristics of vegetation for vegetated water ways. **Agricultural Engineering,** 30(4): 184-189.

20. Ree, W. O. and Palmer, V. J.
1949 Flow of water in channels protected by vegetative linings. Soil Conservation Technical Bulletin no. 967, U.S.D.A. Soil Conservation Service. 115 p.

21. Seibert, P.
1968 Importance of natural vegetation for the protection of the banks of streams, rivers and canals. In **Freshwater Pub. by Council of Europe,** p. 35-67.

22. Stephens, W. M.
1966 Life in the turtle grass. **Sea Frontiers,** 12(5): 264-275.

23. Stevens, H. R., Jr.
1960 A qualitative survey of floral types in Corpus Christi Bay. Texas Game and Fish commission Marine Laboratory Coastal Fisheries Project Reports. 2 p.

24. Strawn, K.
1961 Factors influencing the zonation of submerged monocotyledons at Cedar Key, Florida. **Jour. Wildl. Management,** 25(2): 178-189.

25. Tabb, D. and Manning, R.
1962 A checklist of the flora and fauna of northern Florida Bay and adjacent brackish waters of the Florida mainland collected during the period of July 1957 through September 1960. **Bull. Mar. Sci. Gulf Carib.,** 11(4): 552-649.

26. Williams, A. B.
1958 Substrates as a factor in shrimp distribution. **Limnol. Oceanogr.,** 3(3): 283-290.

27. Wood, E. J. F.
1959 Some east Australian sea grass communities. **Proc. Linn. Aoc. N. S. W.,** 84(2): 218-226.

28. Voss, G. L. and N. A.
1955 An ecological survey of Soldier Key, Biscayne Bay, Florida. **Bull. of Mar. Sci. Gulf Carib.,** 146: 204-229.

VEGETATION FOR CREATION AND STABILIZATION OF FOREDUNES, TEXAS COAST

B. E. Dahl, Bruce A. Fall, and Lee C. Otteni[1]

ABSTRACT

Padre Island, Texas, a subtropical, semi-arid barrier island, was chosen as the study site for development of technical specifications and methodologies to rebuild and/or stabilize deteriorated foredunes as natural barriers against storm surges. The necessary attributes of a plant useful for beach or foredune plantings are a fibrous root system and the abilities to grow in the salt-spray zone, to trap sand, to continue growth during rapid sand accumulation, to spread laterally, and to survive prolonged drought. Of the species tested, bitter panicum, proved best adapted for transplanting. Sea oats were less desirable only because transplanting survival was slightly poorer and they were harder to handle while planting. Other native species tested were unsatisfactory for one reason or another, as were thirteen species exotic to the area. Both of the preferred species were successfully planted from December through May. January plantings were most successful for sea oats, but bitter panicum transplants survived well during any of these months. Successful beach plantings only 15 m wide trapped all the available sand, therefore wider plantings would be unnecessary. Extended drought and sea water impounded on new plantings from seasonal storm surges were the two obvious reasons for planting failures. Transplants were unable to survive soil salinity levels of 4300 μ mhos/cm for more than a few days. Therefore prolonged inundation by sea water (conductance in excess of 13,340 μ mhos/cm) if not accompanied by rain, proved fatal to transplants. Experimental plantings with no more than 8% survival have accumulated as much as 42.4 m^3/m of beach in 50 months. Other plantings with 20% transplant

1. Range and Wildlife Management Department, Texas Tech University, Lubbock, Texas 79409.

survival have accumulated 48.9 m^3/m of beach in 50 months.

The Gulf of Mexico coastline bordering the United States in 2610 km in length, extending from the mouth of the Rio Grande to Key West, Florida. This coastline varies from mainland sand beaches and marine marshes of offshore barrier islands. The barrier islands are discontinuous from Florida to the mid-Texas coast, but they are continuous along the south Texas coast.

These sandy barrier islands, built by natural processes, afford protection to the mainland from seasonal high tides, storm surges, and hurricane-generated waves. Barrier islands in the humid regions are covered with a variety of vegetation including trees, shrubs, and grasses, whereas those in the semi-arid regions have little woody vegetation. Although considerable erosion may occur naturally along the shore face and in low areas overtopped by rising water, erosion is accelerated in areas that lack a vegetative cover. The combined effects of drought, storm surges, and man's destructive influence - such as overgrazing, improper burning, and dune-buggying, have denuded large expanses (1). At such areas storm surges breach the foredunes allowing excessive amounts of sand to be transported into the interior of the islands onto the existing lowland vegetation (2).

There is no evidence to indicate that primordial Padre Island was other than fully vegetated. A vegetated dune barrier will prevent or reduce flooding from hurricane storm surges, as well as from high tides from high pressure systems. Such a surge on February 13, 1969, washed out the road on south Padre Island and the approaches to bridges over hurricane passes on Mustang Island (5). Observations by the U. S. Army Corps of Engineers, Galveston District, (10, 11, 12) for hurricanes Carla, 1961; Beulah, 1967; and Celia, 1970, revealed that barrier islands of the Texas coast were breached in approximately 100 places or, on the average, every 5 to 6 km along the barrier island chain. In each case, sizable volumes of sand were transported from the beach and foredune across the Padre Island. Vegetated dunes 3-5 m above sea level suffer minimal erosion and loss of sand during these hurricanes, as attested by an unbroken 16 km segment of dune along central Padre Island. The barrier dune's potential to absorb surge energy and prevent its penetration into the bays, estuaries, and rivers is immeasurable.

For a number of reasons, the U. S. Army Coastal Engineering Research Center was obliged to study ways and means whereby well-vegetated primary dunes could be restored on the barrier islands along the Texas coast. Accordingly they contracted with the Gulf Universities Research Consortium of Galveston, Texas, to establish technical specifications and methodologies for cost-effective use of natural grasses to build and/or stabilize foredune ridges as natural barriers against storm surges. Ultimately, Texas Tech University, a member of the consortium, was charged with formulating plans for, and carrying out, on-site research into ways to accomplish this objective. The authors have for five years pursued these goals on north and south Padre Island (5, 7, 8, 9).

DESCRIPTION OF THE STUDY AREA

Padre Island has a subtropical, semi-arid climate with hot, long summers, mild winters, and short fall and spring seasons. Freezing temperatures are infrequent. Corpus Christi experiences 0°C or below about ten times each year. Temperatures of areas closer to the water are seldom below freezing. Freezes in Port Isabel average one every three years (4). Annual rainfall on north Padre Island during this study ranged from 643 to 1138 mm. Records are incomplete on south Padre Island, but those we have indicate amounts only slightly less than those to the north. Precipitation is highly irregular, but greatest amounts usually occur in summer and fall months.

The soils of the island developed on recent marine and eolian sands. Sand particle size is dominantly in the fine ($\leq$ .10mm) category. Silt and clay fractions never exceeded a trace. The highest organic matter obtained was 0.1%. Shell fragments ranged from a trace to 15% (8).

Because this research concerns only the back-beach area and the primary dunes, native vegetation adapted to these beach areas is of most concern. Saltmeadow cordgrass (*Spartina patens*), seashore dropseed (*Sporobolus virginicus*), and sea oats (*Uniola paniculata*) are the dominant back-beach grasses, with sea purslane (*Sesuvium portulacastrum*), fiddleleaf morning glory (*Ipomoea stolonifera*), and railroad vine (*I. pes-caprae*) the most prevalent forbs.

On the fore slope of the primary dune sea oats are also dominant, and fiddleleaf morning glory and gulf croton (*Croton punctatus*) are conspicuous inhabitantts. Saltmeadow cordgrass, railroad vine, and seashore dropseed occur as frequently here as on the back-beach area. The grasses, gulfdune paspalum (*Paspalum monostachyum*) and thin paspalum (*Paspalum setaceum*) and the forbs, beach ground cherry (*Physalis viscosa*), western ragweed (*Ambrosia psilostachya*), broom groundsel (*Senecio riddellii*), and beach evening primrose (*Oenothera drummondii*) are among the plants found on the dunes but not on the back-beach. When no cattle are grazing, bitter panicum (*Panicum amarum*) is co-dominant with sea oats in both zones, but it is not abundant on either Padre Island or Mustang Island because it is a preferred cattle forage.

RESULTS AND DISCUSSION

What to Plant

Planting experience soon dictated that necessary attributes of a plant useful for beach or foredune plantings were a fibrous root system and the abilities to grow in the salt-spray zone, to trap sand, to continue growth during rapid sand accumulation, to spread laterally, and to survive prolonged drought.

Although they did accumulate sand, non-grass species adapted to these zones made unsuitable planting material. Generally, they lack the fibrous root system found necessary to bind accumulating sand in place for a stable dune. They die back to ground level following freezes, leaving the sand surface bare and subject to wind excavation in the winter. They are mechanically hard to handle during a transplanting operation.

The four native species believed most promising for dune construction are sea oats, bitter panicum, saltmeadow cordgrass, and seashore dropseed. Survival comparisons were made among these species from monthly plantings from October 1969 through May 1970. Plantings were at study sites on both the north and south of Padre Island (Table 1).

Planting dates, species, and interactions varied from site to site and produced significantly different results. Bitter panicum had the highest survival among all species tested. Also, it handled easily while being transplanted and made excellent growth without fertilization or irrigation. Although sea oats were harder to handle while planting and survival rates were lower than for bitter panicum, the species is difinitely desirable. Lack of vigor, lack of sand-trapping potential, and inconsistent survival made saltmeadow cordgrass undesirable. Although more seashore dropseed plants survived than sea oats, it was considered an undesirable species for dune-building because its small size prevented it from trapping sand as efficiently as the larger species.

Shoredune panicum (*P. amarulum*), a species so closely related to bitter panicum that recent investigation by Dr. P. Palmer indicates it probably should be considered a subspecies thereof, was not included in these comparisons. However, later plantings indicate comparable results to those of bitter panicum. Its major deficiency is that it is a bunch grass and its ability to spread laterally is poor. All studies currently underway include bitter panicum, sea oats, and in special situations, shoredune panicum.

A variety of species, all but one exotic to Padre Island, were tested during the study with mostly negative results. Both seeds and plants were obtained for trial plantings. Species planted were: *Agropyron distichum, Elymus giganteus, Sporobolus fimbriatus, Calamovilfa gigantea, Panicum havardii, Digitaria macroglossa, Dactyloctenium australe, Myrica cordifolia, Desmostachys bipinnata, Indigofera miniata, Indigofera pseudotinctoria, Scaevola thunbergii,* and *Hemarthria altissima.* The only plants that emerged from the seeds were *I. miniata,* the one species native to the area. Of the transplants, only *I. pseudotinctoria* showed promise for beach plantings. None of the others tried persisted.

Most dune work along the Atlantic seaboard was with American beachgrass (*Ammophila breviligulata*). W. W. Woodhouse of North Carolina State University sent us plant material of this species and, although it survived for a year or so, it

TABLE 1.

Survival of grasses from monthly plantings for North (N) and South (S) Padre Island. Sample size is 100 plants.

	Sea Oats										Ave.	Saltmeadow cordgrass				Seashore dropseed	
	68-69		69-70		70-71		71-72		72-73			68-69		69-70		69-70	
	N	S	N	S	N	S	N	S	N	S		N	S	N	S	N	S
Oct.	1/	1/	15	0	1/	1/	7	29	0	1/	10	1/	1/	23	0	7	4
Nov.	1/	1/	4	2	12	0	46	3/	0	5	9	1/	1/	12	0	26	1
Dec.	1/	1/	38	6	28	0	2/	76	2	26	25	1/	1/	9	0	7	25
Jan.	1/	1/	69	62	8	9	64	73	26	43	44	1/	1/	32	10	30	13
Feb.	3/	3/	27	29	2	2	3/	61	60	81	26	3/	3/	13	49	27	44
Mar.	46	1	2	5	0	0	76	59	0	0	19	14	2	2	0	10	32
Apr.	46	1	8	1	4	2/	27	0	34	4	14	54	1	1	3	65	9
May	78	1	30	1/	1/	1/	75	22	8	15	33	93	1	85	1/	41	1/
Jun.	33	1/	1/	1/	1/	1/	1/	1/	49	41		77	70	1/	1/	1/	1/
			Bitter panicum														
Oct.			39	11	1/	1/	34	28	0	1/	22						
Nov.			66	26	8	32	22	3/	30	0	23						
Dec.			54	61	37	7	2/	39	24	12	33						
Jan.			60	72	10	32	37	21	0	40	34						
Feb.			17	69	0	22	84	26	34	33	36						
Mar.			20	10	2	34	88	35	62	38	36						
Apr.			15	10	10	2/	58	0	68	46	30						
May			67	1/	1/	1/	80	51	70	40	62						
Jun.			1/	1/	1/	1/	1/	1/	86	97							

1/ Not planted or data unavailable.

2/ Planting destroyed by vehicular traffic.

3/ Planting destroyed by salt-water inundation.

has not persisted.

When to Plant

Survival of sea oats transplants from 1969 to 1973 indicates that the best planting season is from December through February. Bitter panicum plantings during those months also were successful, but many excellent stands were obtained from plantings from October through May. In fact, May appears to be a good planting month for both species (Table 1). Summer (June-August) plantings of both species were highly successful in 1973; comparative data from previous years are not available.

Reasons for transplant survival and death are as yet largely a mystery. Salt-water inundation, drought, soil temperatures at planting time, and health of the transplants are all suspect. For example, bitter panicum planted during the fall and winter of 1972-73 had high initial survival followed by high late-spring mortality. Consequently, by July, survival was only mediocre. Four days of salt-water inundation in April probably contributed, but this is not a complete explanation because March and April plantings, also inundated, had high survival rates. Conversely, March sea oats transplants died a few days after planting, well before the April inundation. Apparently environmental and physiological factors affecting survival are important long after the original planting, even as long as six months.

Salt-water inundation when not accompanied by significant rainfall has been the most obvious single reason for poor transplant survival. Extended drought prior to and during plantings has been the next most obvious reason for poor survival. Transplants put in during 1970 and 1971 had overall the lowest survival rates (Table 1). While 1970-71 was not an exceptionally dry year, the period November 1 through July 31 had only one month, April, with more than 225 mm of rain on north Padre Island. Rainfall amounts were only slightly more on south Padre Island for this period.

How to Plant

Either grass transplants or grass seeds can be used to vegetate bare areas. However, bitter panicum seldom, if ever, produces viable seeds, and insects and ground squirrels feed on the seeds of sea oats to such an extent that few remain for harvest. In addition, since two growing seasons are usually required for full establishment of seedlings, direct seeding is probably not useful on areas such as fence-built dunes that require rapid establishment. In most situations, vegetative transplants harvested from existing plants are required.

Either hand or mechanical planting can be used. Where the dunes have been totally eroded to beach elevations, mechanical transplanting equipment may be used. We successfully used a tobacco transplanter. This machine opens a 15 cm furrow, two men put the transplants into place, and packer wheels close the furrow and pack the sand against the sprigs. When planting on steep slopes, such as on dunes built by wooden-picket fences (sand fence or snow fence), planting has to be done by hand with a shovel.

Beach-level plantings require no special techniques. However, stabilizing fence-built dunes with vegetation requires extra effort. On elevated dunes, irrigation greatly facilitated the hand-planting as digging in dry surface sand was very laborious. Other than for preparing the surface, irrigation benefited survival very little. Experience revealed that on this kind of dune, plantings not protected from wind were eroded and the transplants excavated. Two materials were used as sand binders, 210 g burlap and an open weave (5 x 6.4 cm) netting (mulchnet). The open-weave netting proved most satisfactory, as it was cheap and it allowed leaves of transplants to grow through without raising the netting off the ground. Burlap cost $0.24/m and netting $0.07/m. The mulchnet has proven 100% effective in preventing wind excavation of the plants. However, where sand tended to accumulate, the netting was detrimental because it flattened the transplants, which if erect, would have better withstood the accumulating sand without suffocating.

How Wide to Plant

Because the low survival of transplants can make stabilization of areas devoid of foredunes expensive, it is desirable to delimit the maximum amount of transplants needed to trap the available wind-transported sand. In February 1972, bitter panicum was planted in two sections, 15m by 92m and 30m by 92m, on 60cm centers in an area that had no foredunes.

Fifteen months after planting, the 15m-wide planting had accumulated 8.8 m^3/m of beach, whereas the 30m-wide planting had accumulated 12.0 m^3/m of beach. However, the 15m portion of the latter planting nearest the Gulf had accumulated 10.5 m^3/m *vs.* only 1.5 m^3/m for the 15m portion on the bay side (Fig. 1). We concluded that a 15m-wide planting is sufficient to trap the sand available.

What To Do With Hurricane Overwash Channels

The salinity levels left in the soil by periodic storm surges are more likely toxic to transplants on low-elevation hurricane overwash channels than to those on higher areas of the beach. Therefore, if these passes are to be closed by vegetated dunes, soil salinity must be lowered sufficiently to allow transplant survival. This can be accomplished by elevating the planting surface to a height not susceptible to frequent salt-water inundation. Parallel 60 cm sand fences

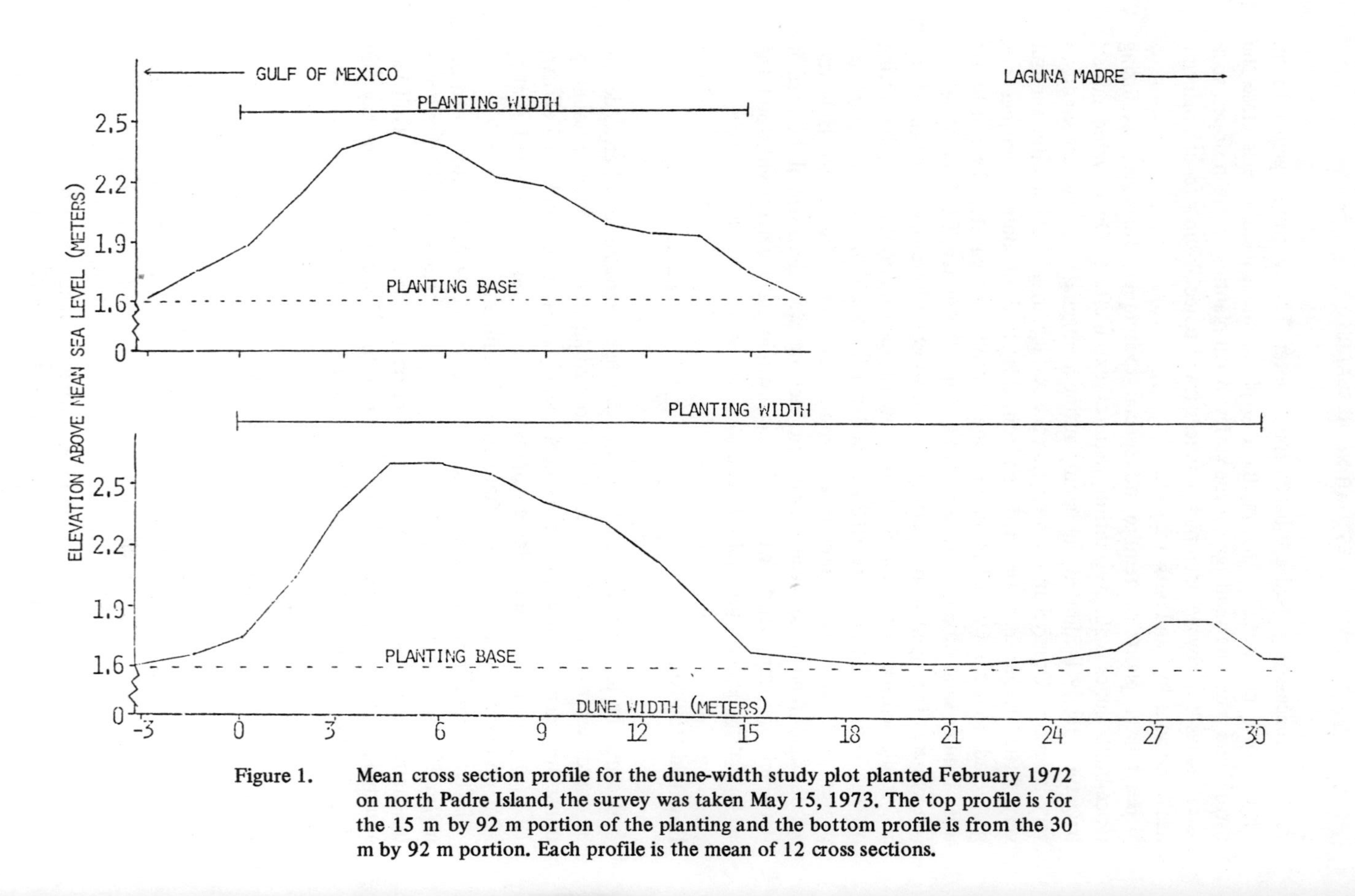

Figure 1. Mean cross section profile for the dune-width study plot planted February 1972 on north Padre Island, the survey was taken May 15, 1973. The top profile is for the 15 m by 92 m portion of the planting and the bottom profile is from the 30 m by 92 m portion. Each profile is the mean of 12 cross sections.

placed 2.4 m to 3.7 m apart to a total width of 15 m catch sufficient sand within a year to yield a flat planting surface relatively free of salt. Bitter panicum transplants, planted on a 60 cm elevated surface in February 1973 had 48% survival. A year earlier, the same site was planted to bitter panicum on the original (unelevated) surface; after 9 months and two salt-water inundations, survival was less than 1%. Mean salinity levels for this period were 500 μ mhos/cm for the elevated surface compared to 2800 μ mhos/cm for the original elevation. Over 10% survival has proven sufficient to allow for successful dune growth and stabilization, whereas 1% survival has not.

Salinity Tolerances

Hurricanes and seasonal storms increase soil salinity through increased spray and from evaporation of the sea water that remains behind the storm berm, but only in such extreme situations as when sea water is impounded on transplants does salinity reach a level detrimental to transplants in beach sand. In one greenhouse study, bitter panicum tolerated soil salinity up to 4300 μ mhos/cm (a 1:2 solution, grams dry sand: distilled H_2O) before wilting and necrosis of the leaves became apparent. A current greenhouse study shows all bitter panicum plants surviving at 2700 μ mhos/cm but little, if any, growth. Only half of them have survived 3.5 months of 3400 μ mhos/cm and these are necrotic. None survived even two weeks at 4300 μ mhos/cm. Bitter panicum growing in soils with salinity levels of 1500 μ mhos/cm or less appear healthy and are growing.

Sea-water conductance exceeds 13,340 μ mhos/cm, but inundation of a beach surface usually results in soil salinity levels only slightly over 2,000 μ mhos/cm (1:2 dilution). Although salt-water inundation was responsible for several planting failures, the failures occurred when storm surges caused sea water to pond on transplants and no rainfall accompanied the surge, so that soil salinity reached toxic levels. However, if rain accompanied the surge or followed immediately, then sea water caused little plant mortality. Otherwise, salinity of beach sand seldom reached sufficiently high levels to cause death of transplants. Other investigators also concluded that beach soils ordinarily contain no higher concentrations of salt than do the average cultivated soils of the area, and that beach plants need not be obligate holophytes (3, 6).

Dune Construction

To test the capabilities of grass to reconstruct the primary dunes of Padre Island, a number of experimental plantings were made throughout the study. Three of these will be described by way of illustration. On south Padre Island, three sections, each 122 m x 15 m, were planted with sea oats in February 1969. On February 13, 1969, they were inundated by salt water from a storm surge that left only 8% of the plants alive in the south section. Less than 1% of the sea oats plants survived in the other two sections.

On north Padre Island, a 366 m planting of sea oats and saltmeadow cordgrass was also made in March, 1969. All of the saltmeadow cordgrass died and some of the sea oats, resulting in less than 20% survival in this planting.

A third major planting, this time with bitter panicum, was made in February 1970, and 17% of the transplants survived. Sand accumulation as of May, 1973 for all three major plantings is tabulated below.

	Sand Volume (m^3/meter)
South Padre Island sea oats plantings (after 50 months)	
South section-122 m long (8% survival)	42.4
Mid section-122 m long (1% survival)	12.0
North section-122 m long (1% survival)	19.8
North Padre Island plantings	
Sea oats-saltmeadow cordgrass-366 m long (after 50 months)	48.9
Bitter panicum-335 m long (after 39 months	
North 168 m	31.3
South 168 m	20.8

Composite cross-section profiles for the north Padre Island plantings are illustrated in Figures 2 and 3.

Time-Cost Analysis

Experimental plantings provided information about the time required to plant a given number of hills of sea oats, bitter panicum, and saltmeadow cordgrass. Transplants were dug near the areas to be planted, so that a minimum of time, usually 30 to 60 minutes, was involved in transportation. Times given include: training the crew for selecting material to be planted; separating plant material and trimming; and actual machine planting. A crew of five men could plant about 10,000 hills of saltmeadow cordgrass, or 9,000 hills of bitter panicum, or 5,300 hills of sea oats in an eight-hour period. The man-hours required to make a beach planting 15 m wide and 1.6 km long, with plants spaced 60 cm apart, were 264, 287, and 500, respectively, for the three species.

For a variety of reasons, sand fences are used to accumulate sand on islands such as Padre Island. Commonly, they are used to prevent sand accumulation on access roads, highways, etc. These fence-built dunes can be vegetated, but extra

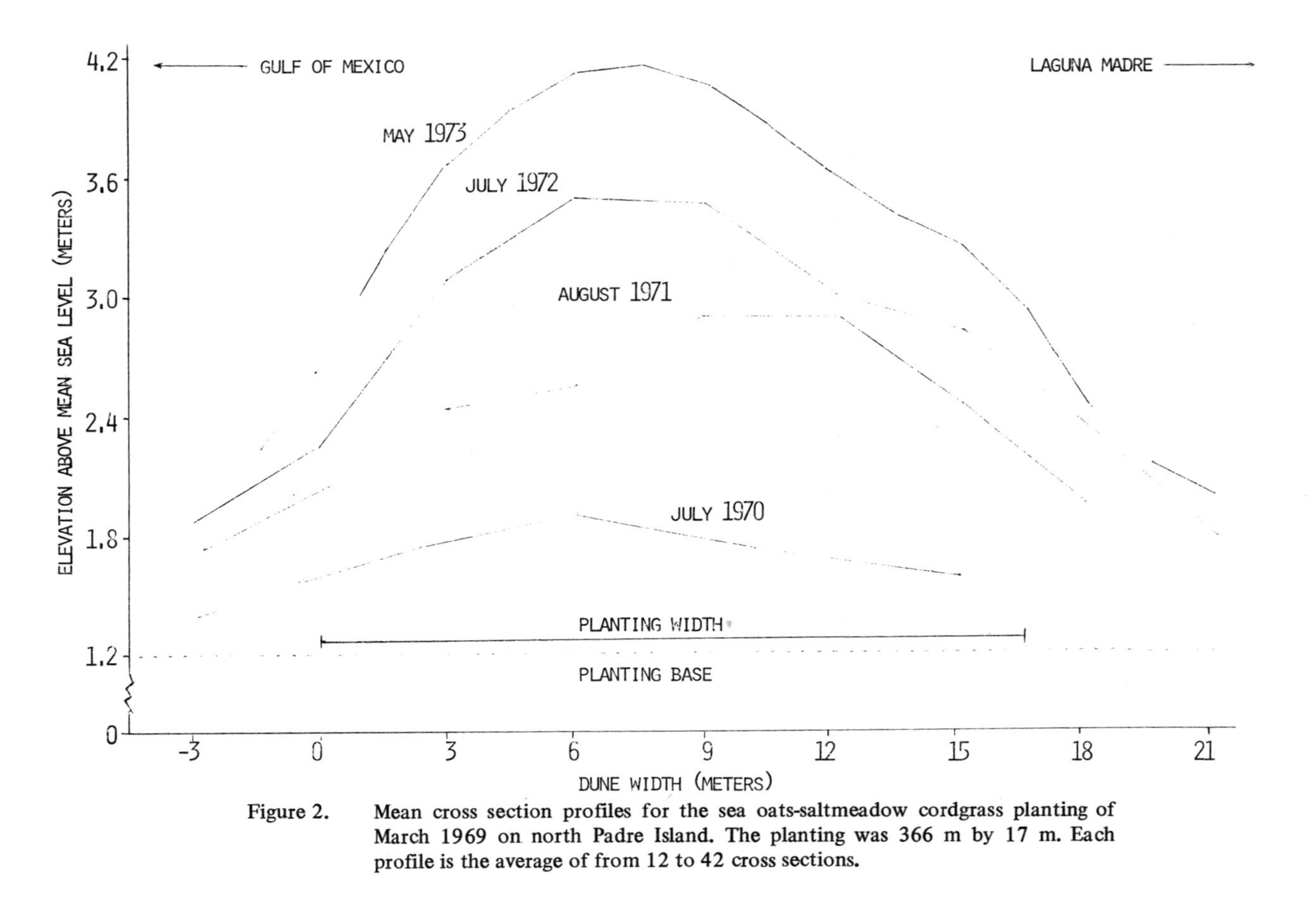

Figure 2. Mean cross section profiles for the sea oats-saltmeadow cordgrass planting of March 1969 on north Padre Island. The planting was 366 m by 17 m. Each profile is the average of from 12 to 42 cross sections.

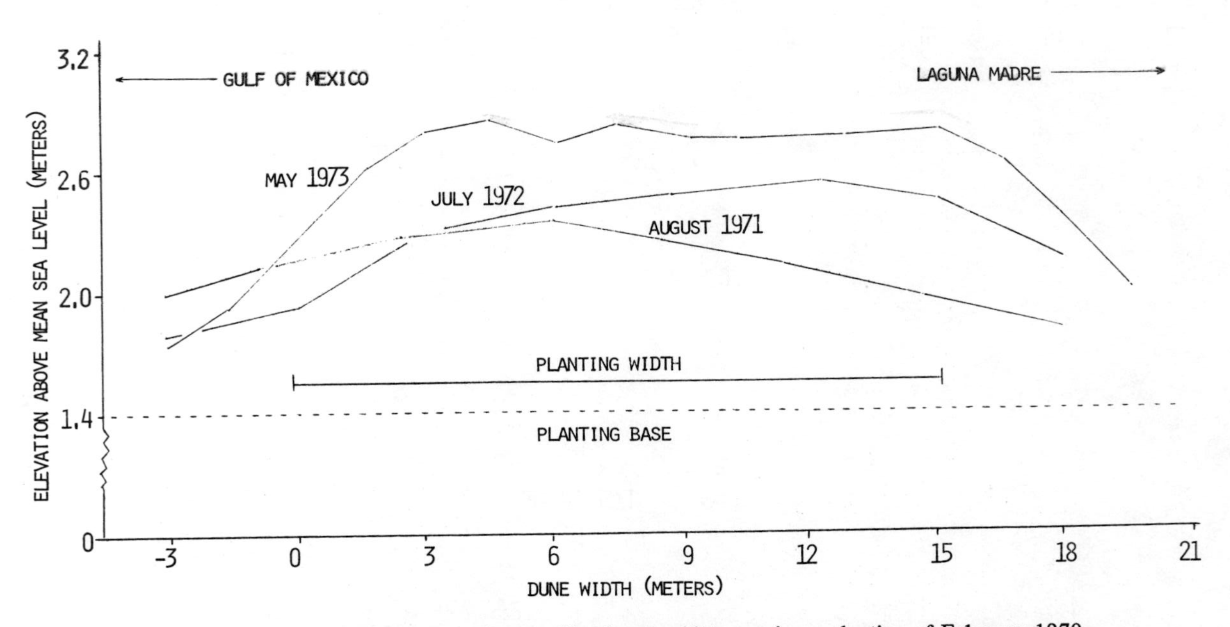

Figure 3. Mean cross section profiles for the bitter panicum planting of February 1970 on north Padre Island. The planting was 335 m by 15 m. Each profile is the average of from 9 to 11 cross sections.

labor, mulching, and sometimes irrigation are required. Although we are keeping cost information, man-hour requirements similar to the above are not available at this time.

Miscellaneous

The effort expended in establishing a planting can be obliterated by several phenomena besides those already covered. Insect damage has occurred in some few areas, but, to date, damage has been minimal. The offending species were collected but they have not yet been identified. Beetle damage to sea oats seed is ubiquitous, but this is considered no problem unless seed production is desired.

Rabbit damage was no problem on established plantings. However, on numerious occasions, jack rabbits (*Lepus californicus*) destroyed plantings of sea oats by jerking the transplants out of the ground within a few days after planting. Surprisingly, they have ignored the bitter panicum.

Range cattle were definitely a problem to plantings, before they were removed. They usually avoided the beach, but when wet weather and warm temperatures brought multitudes of mosquitoes, the cattle came to the beach to escape them. The cattle liked bitter panicum, shoredune panicum, and sea oats in that order. They ate sea oats if neither of the more palatable species was available. Therefore, all plantings had to be fenced. The fencing served a double purpose in that it not only eliminated cattle damage but it lessened human vandalism to the plantings.

ACKNOWLEDGMENTS

The authors are indebted to a number of federal agencies, academic institutions, and private corporations for their contributions to this study. All cooperating organizations have a common interest in either creating or stabilizing dunes on barrier islands, but their interests vary from beach erosion to control of blowing sand. We are particularly grateful to the U. S. Army Coastal Engineering Research Center for funding the study, and to Mr. R. P. Savage and Mr. D. W. Woodard of the agency for their guidance throughout the study. Excellent cooperation was also received from the Galveston District, Corps of Engineers, which is conducting an associated study using sand fence to create dunes. Special appreciation is due Mr. Roger Baker for his part in collecting the data and for the invaluable assistance of personnel of Padre Island National Seashore. Mr. Alan Lohse, Mr. Dan Feray, and Mr. T. Bilhorn of the Gulf Universities Research Consortium were especially helpful in coordinating the project.

REFERENCES

1. Harris, V. M.
   1965 History of cattle raising on Padre Island. **South Texas Agr., 1**: 1-16.
2. Hayes, M. O.
   1967 Hurricanes as geological agents: Case studies of hurricanes Carla, 1961 and Cindy, 1963. Bur. Econ. Geol. Report Invest. No. 61.
3. Kearney, T. H.
   1904 Are plants of sea beaches and dunes true halophytes? **Bot. Gaz., 37**: 424-436.
4. Orton, R., Haddock, D. J., Bice, E. G., and Webb, A. C.
   1967 Climatic guide: the lower Rio Grande Valley of Texas. Texas Agr. Expt. Sta. MP-841.
5. Otteni, L. C., Dahl, B. E., Baker, R. L., and Lohse, A.
   1972 The use of grasses for dune stabilization along the Gulf Coast with initial emphasis on the Texas Coast. Gulf Universities Research Consortium, Report No. 120.
6. Randall, R. E.
   1970 Salt measurement on the coast of Barbados, West Indies. **OIKOS, 21**: 65-70.
7. Woodard, D. W., Dahl, B. E., Baker, R. L., and Feray, D. E.
   1969 The use of grasses for dune stabilization along the Gulf Coast with initial emphasis on the Texas Coast. Year-end report (1968-69) to Dept. of the Army, Corps of Engineers, Coastal Engineering Research Center.
8. Woodard, D. W., Dahl, B. E., Baker, R. L., and Feray, D. E.
   1970 The use of grasses for dune stabilization along the Gulf Coast with initial emphasis on the Texas Coast. Year-end report (1968-69) to Dept. of the Army, Corps of Engineers, Coastal Engineering Research Center.
9. Woodard, D. W., Otteni, L. C., Dahl, B. E., Baker, R. L., and Bilhorn, T. W.
   1971 The use of grasses for dune stabilization along the Gulf Coast with initial emphasis on the Texas Coast. Gulf Universities Research Consortium, Report 114.
10. U. S. Army Corps of Engineers
    1962 Report on Hurricane Carla, September, 1961. Galveston District.
11. U. S. Army Corps of Engineers
    1968 Report on Hurricane Beulah, 1967. Galveston District.
12. U. S. Army Corps of Engineers
    1971 Report on Hurricane Celia, July 30-May 5, 1970. Galveston District.

MANAGEMENT OF SALT-MARSH AND COASTAL-DUNE VEGETATION

by

D. S. Ranwell[1]

ABSTRACT

The broad aims of management of coastal vegetation are preservation, information, education, and exploitation.

Site survey is an essential precursor to effective choice of aims and subsequent surveys to monitor change act as continuous feedback to management.

There are six main options concerned with implementing management: restoration, protection, modification, augmentation, extermination, and diversification.

Close attention must be given to the aims, feasibility, and objectives of management if we are to safeguard coastal wildlife resources effectively.

Techniques for planting *Zostera,* for field establishment of genetic stocks of *Spartina* variants, and for the establishment of *Phragmites* by aerial seeding are described. Experimental studies on salt-marsh grazing and turf cutting, and the general management of salting pasture for stock and wildfowl are discussed. Control of *Spartina, Agropyron pungens,* and *Juncus maritimus* with herbicides or mowing techniques are outlined. Potentialities for the design of new salt marshes are considered in relation to reclamation proposals.

Techniques for the restoration of damaged dune systems are illustrated.

1. Coastal Ecology Research Station, Insitute of Terrestrial Ecology, Colney Lane, Norwich, NOR 70F, England.

Preliminary results on treatments for producing dune grass seed stocks of high germination capacity are given. Experiments in progress to record effects of trampling, mowing, fertilizing, and burning on dune swards are described. Management of the invasive dune scrub species *Hippophaë rhamnoides* is considered in the context of a national management plan for this species based on a detailed knowledge of its biology. The use of dunes for afforestation and recreational purposes (e.g. as golf links) is considered in relation to preservation of dune wildlife resources.

INTRODUCTION

Management is a generalized term concerned with control and organization of groups for specific purposes. In any particular case, such as the management of coastal vegetation, it is important to be clear at the outset what its aims and options are. We will start, then, with an attempt to define the essential features of management in relation to salt-marsh and coastal-dune vegetation.

Next come a series of illustrated examples of management activities on salt marsh related especially to its functions as an aid to coast defense and a food resource for stock and wildfowl.

Finally, examples of management activities on sand dunes are outlined. These are concerned not only with the maintenance of dunes as sea defences, but with the recreational impacts to which they are subject, their use as golf links and for afforestation.

Salt marshes and sand dunes are of value for the study of ecological processes and of interest for the wealth of plants and animals they contain. Any management that is undertaken should be applied, so far as possible, in such a way as to maintain and enhance these values. This last consideration, rather than management aimed at exploitation (whether passively in the form of sea defense, or actively in the form of a cropping policy), is the main theme of this contribution.

MANAGEMENT AIMS AND OPTIONS

The urge to manage, to act, frequently takes precedence over clear thought about the aims, options, and feasibility of management.

Broad aims

The broad aims of management of any habitat seem to be preservation, information, education, and exploitation.

A site survey of wildlife resources is an essential precursor to the effective choice of one or more of these aims. In the absence of a survey, the only rational policy is one of laissez faire that is to say continuation of the existing management (or lack of it) that, together with past management, presumably created the site.

Equally essential is an effective means of monitoring change, for example, by rapid air surveys at suitable intervals. Monitoring surveys act as continuous feedback to management.

The *preservation* of coastal vegetation barriers as an aid to sea defense is clearly of over-riding importance where life and property are at stake. However, the principle of developing coastal protection *behind* highly unstable sections of the coast, rather than on the actual coastline, deserves more consideration than it has received. Preservation of the flora and fauna of coastal marshes and dunes also depends on allowing space and freedom for the natural physiographic processes to operate to create their habitats.

Where the primary aim is to obtain *information* by research there must be effective control over human use and disturbance. The particular communities involved should be large enough to prevent edge effects from influencing experimental results and sampling from destroying something worth preserving.

Educational use requires carefully controlled penetration into the resource via nature trails. Assessments must be made of human carrying capacity in relation to the stability and viability of the wildlife resources.

Successful *exploitation* depends on the maintenance of sustained yields, and correct management is of critical imporatnace in such unstable habitats as salt marshes and sand dunes.

Within the borad aims of management outlined there are much more specific aims to which management activity is actually applied, for example, preservation of a particular community; information of a particular type, exploitation for a special crop. A limited number of options in biological terms are available to achieve these specific aims.

Options

The six main options that are concerned with implementing management may be summarized by the activities: restoration, protection, modification, augmentation, extermination, and diversification (Fig. 1).

The reconstitution of a coastal dune destroyed by military activity is *restoration.*

Protection implies action against a threat, such as the ploughing of a fire-break around a dune woodland.

Modification means changing the balance of populations by, for example, altering the grazing regime on a salt marsh.

Augmentation relates to management concerned with introducing desirable species, *extermination* with getting rid of undesirable species.

Diversification is the creation of new habitats to attract a greater variety of species.

Management aims for wildlife resources are not as clear-cut as those concerned with direct cropping. They depend on such illogical factors as the appeal of larger animals and birds and rare species to human beings, and such intangible concepts as the attraction that variety in wildlife has for us. It is recognized that the management options given are not always mutually exclusive. Nevertheless, most management can be fitted into the system suggested, and it serves a useful purpose if it helps us to think more clearly about objectives and the means of achieving them.

Some practical examples in the management of salt marshes and sand dunes will help to demonstrate this.

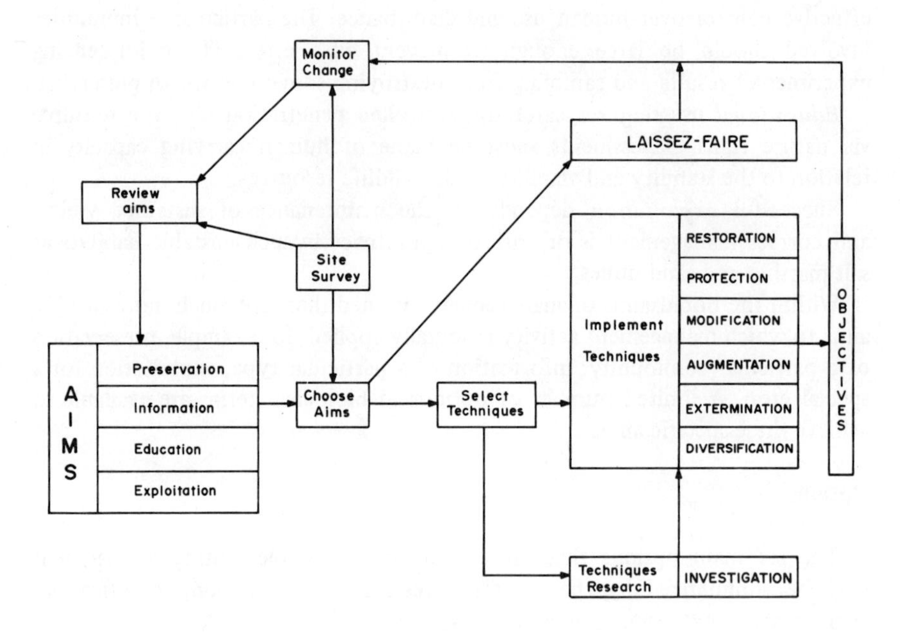

Figure 1. Management options.

SALT-MARSH MANAGEMENT

Techniques for planting *Zostera*

While not strictly salt marsh plants, intertidal species of *Zostera* do, in some cases, overlap with the lower limits of growth of such salt marsh plants as *Spartina anglica* and act like algae as precursors of salt marsh growth. It is relevant therefore to start with an account of trial plantings with *Zostera*.

The aims of these trials is to provide information on the feasibility of promoting *Zostera* growth in order to provide more food resources in protected areas to offset the losses, through reclamation, of feeding areas of brent geese.

Trials with *Zostera* planting in the United States had met with limited success, so an area was selected at Breydon Water, in Norfolk, in which *Zostera marina* var. *angustifolia* already existed and *Zostera noltii* was beginning to establish – in other words optimum conditions for *Zostera*. In March, 1972 initial plantings were made with twenty 22 x 15 x 10 cm depth clumps of *Zostera noltii* dug with an iron frame and transported wet in plastic boxes lined with muslin. These were lifted out by the muslin and set into holes in the mud, flush with the surface. In September 1972 all plantings were living, several had increased in area by a factor of 10, and 95 percent were flowering. The winter was mild, and although the plantings were unprotected, they did not suffer from grazing. All plantings survived through to 1973. Mud-flat oscillations of $\pm$ 3 cm occurred in a single day's gale, but over the year not more than $\pm$ 5 cm in the generally low-wave-energy area characteristic of *Zostera* sites.

A field-scale trial involving 1 hectare and planting at 3 m intercepts was set up in March 1973 at the same site, using student and prison labour. Wooden duck boards were used to get people out to the site. Establishment has so far been generally good, but it is evident that careful planting of clumps flush with the surface is critical for establishment.

Establishment of genetic stocks of *Spartina*

Spartina anglica has been widely and successfully planted in at least ten countries since its first appearance as a natural hybrid in Southampton Water about 100 years ago (8). Techniques of planting are well known. However, in some cases sterile forms of *Spartina* have been introduced (for example in New Zealand and along parts of the Australian coast), through ignorance of the nature of planting stock. We have been concerned with the preservation of genetically distinct stocks of *Spartina* for future use as planting or breeding material. It was found that to build up stocks of pure strains it was necessary to transplant large clumps in high density. Initially an area of high-level *Spartina* marsh was cleared using 2, 4 - dichlorophenoxy acetic acid (Dalapon) at 50 lbs/acre (57 kg/ha) to enable stocks to be established near to trial mud-flat planting sites. From this establishment site, blocks of pure strains have been planted on mud-flats at two different sites in high density to ensure their purity for many years. These are being studied biometrically to determine their performance and are available as sources of planting material.

Establishment of *Phragmites* by aerial seeding

Similar studies on genetic stocks of *Phragmites communis* have been carried out in Holland by Professor D. Bakker and his colleagues (2). Large quantities of seed were harvested from selected stocks and sown by air on newly reclaimed mud-flats with the object of preventing weed establishment in the southeast Flevoland polder. Subsequently, when drainage had been improved and salt had leached from the soil, the reeds were ploughed in and weed-free land was produced for agricultural crop production. This work is mentioned because it is little known and yet is one of the largest and most successful examples of biological engineering management yet accomplished. A total of 35,000 hectares of *Phragmites* were established during this operation. An interesting by-product was a great increase in rare marsh harrier birds which bred in the reed beds. So while the ultimate aim was towards exploitation, the management adopted also aided in the preservation of a rare species.

Management of salting pasture

With the advent of *Spartina anglica* to European shores, large areas of a new type of marshland appeared round our coasts and partially replaced some of our existing salt marshes and salting pasture. Unlike *Spartina alterniflora* on American coasts, which is a valued food plant of snow geese, no herbivorous wildfowl have become adapted to feeding on *Spartina anglica* in Europe (though black swans graze on it in Australia). Stock however do graze on it.

In order to get information on the effects of sheep-grazing on *Spartina anglica,* a controlled grazing experiment was set up in the Bridgwater Bay National Nature Reserve on the Bristol Channel coast (3).

It was found that at the normal open-range grazing level of two to three sheep to the acre (0.4 hectares), high-level *Spartina* marsh was rapidly converted to a low sward in which *Puccinellia maritima,* palatable to both sheep and wildfowl, could establish. In the absence of grazing, *Spartina* marsh gradually became converted to tall *Phragmites* and *Scirpus maritimus* marsh suitable as breeding territory for a number of bird species. This experiment was of considerable educational value to the many school parties and other visitors to the Reserve. Current management of *Spartina* in the Reserve is to leave a centre block of marsh ungrazed and to graze the two end blocks thus enhancing diversity. It should be emphasized that these marshes are on the mesohaline part of the estuarine gradient. *Phragmites* will not survive the higher salinities in polyhaline marshes nearer to the seaward end of an estuary.

It is, of course, on the mesohaline marshes in the middle zone of estuaries that really large areas of salt marsh suitable for exploitation as grazing develop. Oligohaline marshes towards the landward end of the estuary are generally of smaller extent and rapidly absorbed by reclamation into agricultural land. They do, however, carry a considerable variety of species compared with the more saline marshes to seaward and the few that survive undisturbed round our coast deserve preservation. One example, on the Fal estuary in Cornwall, shows a complete transition to deciduous tidal woodland that contains more than 120 species of flowering plants and a rich insect fauna.

Close sheep-grazed salting pasture also provides valuable winter grazing for widgeon duck and white-fronted geese. It requires intensive grazing to maintain its quality as pasture. When grazing is reduced during changes in ownership, for example, less palatable tussocky growths of *Festuca rubra* occur. Palatable species, such as *Puccinellia maritima,* near their upper limit of survival are quickly excluded by the sharp increase in silt accretion and shading caused by taller growth.

An experiment has been set up to see if it is possible to reverse this succession by turf-cutting. Cut salting turf is much in demand for lawns and bowling greens, so this type of management could have the dual aim of preserving good quality salting pasture and exploiting the turf as a crop. Normally turf-cut areas re-colonize in three to five years, so the salting pasture and turf cutting can be worked on a rotational basis.

Control of unpalatable species

It will be recalled that *Spartina anglica* can invade the upper limits of high-level *Zostera* beds. At Breydon Water, Norfolk, where we are carrying out the *Zostera* transplant experiments, there is only a small amount of *Spartina,* less than 1 acre (0.4 hectares). Since this threatens to invade the *Zostera* beds, we have decided to try to exterminate it. Following detailed survey of the *Spartina* locations, spot treatment with a pelleted substituted urea herbicide, 3 - phenyl - 1, 1 - dimethyl urea (Fenuron) at 50 lbs/acres (57 kg/ha) was used with 100 percent success. Sprays are less effective in controlling low-level *Spartina* and pelleted herbicides are less affected by tidal disturbance. Thus in this site positive management (*Zostera* planting) is operating simultaneously with negative management (*Spartina* extermination).

Unpalatable species, the grass *Agropyron pungens* and the sea rush *Juncus maritimus*, may be controlled in a different way. *Agropyron* can be controlled in salting pasture by regular mowing. Both species were effectively controlled by ploughing and subsequently cutting the natural re-growth, which included a number of palatable species.

Design of new salt marshes

We are currently interested in the design of new salt marshes on the silt and spoil which will accumulate near large reclamation areas proposed for The Wash and the Thames estuary. Left to themselves, these areas would probably be colonized by *Spartina anglica.* It could be ten or twenty years before such *Spartina* marshes were ripe for conversion to salting pasture, during which time herbivorous wildfowl would suffer from loss of feeding habitat. Our aim is to promote rapid growth of more generally palatable species like *Puccinellia maritima* before the *Spartina* can invade. This is a very difficult task, but we are encouraged by the fact that since *Spartina anglica* is very late in ripening seed, only in seasons with late summers, which occur once in five or ten years, is it likely to be a problem. We cannot predict these, of course, but timing will obviously be a critical factor in successful management.

SAND DUNE MANAGEMENT

Dune vegetation restoration

Management of sand-dune vegetation is much more concerned with direct human impacts than is that of salt marshes. In contrast to salt marshes, easily accessible sandy shores attract people in large numbers and they can create extensive secondary mobility by trampling out dune vegetation. For example, at Camber dunes in Sussex, in the height of the tourist season, as many as 17,000 people a day may descend on only 140 acres (57 hectares) of dune. This system had already been much disturbed by military activity and, as tourist trampling increased during the past twenty years, sand movement got out of hand and began to overwhelm buildings and roads. Three principles were applied in the successful restoration of this system: (a) it was accepted that restoration of the system was to be undertaken as a whole (and not piecemeal); (b) public access to the shore was strictly controlled by fencing; (c) to begin with, remedial measures (sand-trap fencing and planting) were concentrated at the seaward edge of the system, the shoreline.

These three principles are essential in my opinion to the success of any scheme for restoring dune vegetation and may be summarized as overall planning; control of public access; and shoreline restoration.

We have used successive air photo cover brought to scale and a grid system to monitor the success of dune restoration at Camber. Conventional planting techniques with unselected stocks of *Ammophila arenaria* and hydraulic seeding with turf grasses were used, and it is planned eventually to replace the fences controlling access with planted shrub barriers. In comparison with the critical experimental work on planting-densities and fertilization carried out on dune grasses in the United States by Brown and Hafenrichter (3), Jagschitz and Bell (5) and Woodhouse and Hanes (10) for example, little attention has been given to selection and controlled experimental work with European dune grasses such as *Agrophron junceiforme, Ammophila arenaria* and *Elymus arenarius.* However, a useful summary of recent work on dune restoration in the United Kingdom is given in Anon (1).

Seed Germination of dune grasses

Under natural conditions, it is advantageous if germination of dune grasses occurs sporadically at different times of the year, because simultaneous germination can result in total failure of the year's crop if conditions for establishment happen to become unfavourable after germination. It is found that germination of *Ammophila arenaria* and *Elymus arenarius* does occur very sporadically with freshly harvested seed. However, since we are interested in producing maximal germination at times when we recognize conditions are likely to be good for establishment, studies are directed towards producing seed stocks of high simultaneous germination capacity or even pregerminated seed.

Elymus arenarius seed is ripe for collection in late August on the English coast and can be stored dry at room temperature without loosing viability for a year or more. Maximum germination was obtained by soaking seeds in distilled water at 5°C and changing the water daily. This treatment destroys a water-soluble germination inhibitor in the seed coat and reduces infection of newly germinated seed. Following this treatment germination occurs most freely in fully saturated conditions at temperatures around 20°C to 25°C.

Preliminary studies on the germination of *Ammophila arenaria* show that germination capacity improves markedly from 7% to 28% in newly harvested seed to 40% to 60% after one year's air-dry storage. Pre-treatment of three weeks in distilled water changed every three days and stored at 4°C improves germination capacity from 25% (no pre-treatment) to 50%. Very poor results were obtained with fixed temperatures for germination, and an alternating warm/cool temperature regime seems to be necessary.

Dune sward management

When the rabbit populations was lowered by myxomatosis disease in the 1950s, the increased growth of dune grasses reduced the floristic variety and eliminated some terrestrial lichen communities on many of our dune systems. In addition, increasing trampling pressure from people, car-tyre wear, and greater incidence of fire associated with more people are all making inroads into dune wildlife resources.

Experiments are now in progress to record the effects of trampling, mowing, fertilizing, and burning on dune vegetation. For example, at Winterton, Norfolk, a controlled trampling experiment is combined with a fertilizer experiment to see if fertilizing can to some extent compensate for the effects of trampling. At Morfa Dyffryn, Merionethshire, changes in the strand-line flora are being recorded on transects, and the numbers of people crossing the area are being measured on an electronic counter. Work carried out by Mr. M. Liddle from the University of North Wales at Bangor, Caernarvonshire, shows that the flowering of attractive herbs like *Thymus drucei* is rapidly inhibited by quite moderate amounts of car-tyre wear. Foot trampling at 40 times per month is not quite sufficient to break through dune sward at Winterton.

Mowing experiments are in progress at Newborough Warren, Anglesey, and Holkham, Norfolk. It does not seem likely that mowing would be a practical technique for maintaining dune swards, but since it has been observed that most people avoid walking in dune grass 30 cm or more high, it might be used to encourage use of specified pathways through such vegetation. Cattle-grazing does not produce the desired herb-rich short turf, and sheep-grazing breaks the sward open and induces erosion. However, there are possibilities in pony-grazing for the maintenance of short dune swards, and we are setting up a controlled pony-grazing experiment at Whiteford Burrows, Glamorganshire, to explore this. A most interesting preliminary study of pony range behaviour and grazing on these dunes has been carried out by Glyn (8). Among much useful information, it shows that there is very noticeable clumping in the distribution and grazing of ponies associated with movements of family parties. A notable concentration of animals occurs along the junction between sand dunes and salt marsh where drainage seepages emerge. Heavier nocturnal, as opposed to diurnal, use of the dunes, which provide better shelter than salt marsh, was observed.

Management of dune scrub

We have made a special study of the biology and management of the invasive dune shrub species *Hippophaë rhamnoides* (9). The object was to bring together available knowledge of the ecology of this species, to give a balanced assessment of the advantages and disadvantages the species confers on the dune environment, and to recommend specific management policies that might be adopted for selected sand dune areas round the British coast. This work is an approach to unified dune management at the national level.

Hippophaë grows to about 8 metres in the British Isles and forms dense, inpenetrable thickets particularly on lime-rich dune systems. It has fleshy fruits and its seeds are spread by birds, especially in hard winters. It is a native species and has been introduced to many dune systems because it helps to stabilize sand and its bright organge fruits make it attractive as an amenity planting. However, since the reduction in rabbit-grazing it has begun to spread invasively and is shading out virtually all the great variety of dune sward species which dune nature reserves were created to protect.

Management policy proposes that establishment of this species should be prevented by uprooting seedlings when they first appear on new sites, or by exterminating by cutting, or by herbicide treatment where this is still practical. Where the shrub is strongly established it is proposed that there should be enough control to maintain habitat diversity in the dune system but that some stands should be left to develop naturally to maturity. Ultimately these old stands may act as nurse crops to a more varied dune shrub flora.

Dune Afforestation

The largest dune system in the United Kingdom, at Culbin, Morayshire, in Scotland is 7650 acres (3096 ha). Only 26 dune systems are larger than 1000 acres (405 ha). There are, therefore, rather limited opportunities for afforestation on British dunes compared, for example, with Les Landes in France where 250,000 acres (101,075 ha) were afforested during the nineteenth century. In fact, up to 1954 in Great Britain, only some 10,000 acres (4,047 ha) of dunes had been afforested, chiefly with *Pinus nigra* var. *calbrica,* some *Pinus maritima,* and *Pinus sylvestris,* according to Macdonald (6), and there is no evidence to suggest there has been any really large increase in acreage since then.

At a number of sites, for example Tentsmuir, Fife, and Newborough, Anglesey, parts of the dune system are afforested with conifers, and other parts are national nature reserves. The spread of pine seedlings into the nature reserves presents a management problem; they have to be regularly cut to maintain the dune vegetation. Another and more serious problem is that the water table of the dune system is lowered by extensive afforestation and this has adverse effects on the rich flora of sand-dune slacks.

Plantations of young conifers on dunes soon shade out underlying vegetation and are prone to fire which could result in serious erosion where large areas of bare sand are exposed. This depends very much on the orientation of the system in relation to prevailing and dominant winds. For instance, transect studies of regeneration following fire in conifers on dunes at Holkham, Norfolk, showed hardly any significant sand movement and rapid regeneration of *Ammophila arenaria.* This east coast system has the prevailing and dominant winds in opposition and the burnt area was sheltered from prevailing wind by surviving tall conifer plantations. On a west coast system where prevailing winds are also the dominant winds the consequences of fire could be much more serious.

In the special case of the British Isles, where dune resources are limited and are mostly in the form of small dune systems of high recreational value, it seems logical to discourage afforestation as a management policy, except where it is a well-established practice.

Use of dunes as golf links

One-third of the sand dune systems in the United Kingdom are used as golf links. A recent study of the flora of an existing golf course on dunes compared with an adjoining undisturbed dune site in Jersey, Channel Islands, showed that golf courses seriously impoverish the flora and other wildlife interest in those parts where they are intensively managed. For example, plant communities with thirty to forty species per 25 m^2, rich in rare and floristically attractive plants, are reduced to communities of five to ten species per 25 m^2 consisting of grasses and common herbs in close-mown fairways. These are subject to some of the most intensive management in existence. On 25 acres (10 ha) of fairway and greens at this site some 14 tons (14.2 kilos) of fertilizer per year may be used, fairways are treated twice a year and greens three times year with 2, 4-D Mecoprop type herbicide. In addition, 150 tons of light soil were sifted and spread by hand on 15 fairways and the fairways mulched every four or five years with 400 to 500 bags of peat. Fairways are regularly mown, hard-raked in autumn and winter months, and slit-tined to improve aeration while greens are irrigated whenever there is drought. In spite of this, there are considerable areas of relatively undisturbed dune vegetation within the golf course bounds but they are always liable to future disturbance in the interest of golfers.

CONCLUSIONS

It is clear that the management of such valuable coastal resources as salt marshes and dunes needs to be examined on a national rather than a regional basis. People concerned for wildlife resources have a great deal to learn from traditional management patterns developed by those exploiting these resources, and much to offer in improving management for whatever purpose.

ACKNOWLEDGMENTS

I am most grateful to my colleagues Dr. L. A. Boorman and Miss J. M. Pizzey and to my former colleague, Mrs. J. Bramwell, for their kindness in allowing me to quote from unpublished work.

REFERENCES

1. Anon
1970 Dune Conservation 1970. North Berwick Study Group Report. East Lothian County Planning Department, Haddington, East Lothian, Scotland.

2. Bakker, D.
1960 A comparative life-history of **Cirsium arvense** (L.) Scop. and **Tussilago farfara** L., the most troublesome weeds in the newly reclaimed polders of the former Zuiderzee. In **Biology of Weeds,** p. 205-222. (ed. Harper, J. L.). Blackwell Scientific Publications, Oxford.

3. Brown, R. L. and Hafenrichter, A. L.
1948 Factors influencing the production and use of beachgrass and dune-grass clones for erosion control. **Jour. Amer. Soc. Agron.,** 40: 512-521; 603-609; 677-684.

4. Glyn, P. J.
1972 Some aspects of the ecology of ponies on Whiteford National Nature Reserve. M.Sc. (Conservation) thesis. University of London.

5. Jagschitz, J. A. and Bell, R. S.
1966 American Beachgrass (Establishment - Fertilization - Seeding). Bull. 383 Agric. Expt. Station, Univ. Rhode Island, Kingston, Rhode Island.

6. Macdonald, J.
1954 Tree planting on coastal sand dunes in Great Britain. **Adv. Sci.,** 11: 33-37.

7. Ranwell, D. S.
1961 **Spartina** salt marshes in Southern England. I. The effects of sheep grazing at the upper limits of **Spartina** marsh in Bridgwater Bay. **Jour. Ecol.,** 49: 325-340.

8. Ranwell, D. S.
1967 World resources of **Spartina townsendii (sensu lato)** and economic use of **Spartina** marshland. **Jour. appl. Ecol.,** 4: 239-256.

9. Ranwell, D. S. (ed.)
1972 The management of Sea Buckthorn **(Hippophaë rhamnoides** L.) on selected sites in Great Britain. Report of the **Hippophaë** Study Group. Nature Conservancy, London.

10. Woodhouse, W. W. and Hanes, R. E.
1967 Dune stabilization with vegetation on the Outer Banks of North Carolina. Tech. Mem. U. S. Army Coastal Engineering Research Center, Washington, D.C.

SOME ESTUARINE CONSEQUENCES OF
BARRIER ISLAND STABILIZATION

Paul J. Godfrey and Melinda M. Godfrey[1]

INTRODUCTION

The large, shallow lagoons behind the North Carolina Outer Banks constitute one of the most outstanding estuarine systems on the east coast of the United States. Intertidal salt marshes fringe the back sides of the barrier islands which make up the Outer Banks, and also form the shorelines of rivers emptying into the sounds. The various modes of formation of these marshes have a great deal to do with their contribution of organic matter to the estuary. These modes are determined by rising sea level and the geomorphic processes associated with retreat and submergence of the shore.

On the mainland side of Core, Pamlico, and Albemarle sounds, marshes develop as the sea slowly rises and drowns the land; tree stumps in these marshes have been cited as evidence for this mode of marsh formation (1). Marshes behind the barrier islands develop by two additional processes – inlet closure and overwash – which reflect both the rise of the sea level and the retreat of the islands. The significance of these two processes has been largely overlooked in the ecological literature, but geologists have recognized their role in moving sand back during barrier island transgression. Fisher (8) pointed out that most of the area known as the Outer Banks shows evidence of having been formed as inlets opened and closed, leaving telltale flood tidal deltas which subsequently became salt marshes; he lists at least thrity historical inlet sites between Bogue Inlet, North Carolina, and the Virginia line. Studies by Kraft (14) show inlet and overwash facies as significant portions of the stratigraphic profile of the Delaware barrier islands. Belt (2) demonstrated clearly that marsh islands near

1. National Park Service Cooperative Park Studies Unit, Department of Botany and Institute for Man and Environment, University of Massachusetts, Amherst, Massachusetts 01002

Fire Island, New York, were formed on flood tidal deltas of the last century. The formation of salt marshes behind elongated spits has been well documented for Barnstable Harbor, Massachusetts (16), and in Great Britain (3).

The ecological significance of overwash as a source of sediment for new marshes behind barrier islands was described by Godfrey (9), and the interaction between inlets and overwash as a major means of salt marsh formation on the Outer Banks was discussed by Godfrey and Godfrey (12). Subsequent field studies have underlined the importance of this mode of marsh formation behind rapidly retreating barriers on high energy shorelines. The flood tidal delta behind an inlet becomes the primary substrate when its elevation is raised to the low tide level. As the inlet closes, or a spit lengthens in front of the delta, the shoals are invaded by *Spartina alterniflora* to form extensive marsh islands behind the barrier. As the beach retreats, overwash drives sand over the island and connects these delta islands.

Inlets appear to be the most important source of new sediment for marsh development; this kind of substrate underlies all the marsh islands investigated. These marshes have relatively thin layers of peat compared to the deep peat deposits found where submergence has been the primary geological change. Overwash is of secondary importance in that it provides sediment which connects marsh islands and creates a marsh fringe along the back side of the barrier beach. Such marshes have thin peat layers, since they are only a few decades old. Submergence of the mainland is of the least importance in marsh formation in this area, although it is definitely significant where islands have been stable for several centuries and unbreached by either inlets or overwash. Both the depth of the peat and the character of the underlying sediments can be used as indicators of marsh age and time of formation.

Before the 1930's, the Outer Banks were under the influence of coastal geomorphic processes without significant interference by man. The Banks have been settled since the 1700s, the residents living in scattered villages on the wider sections of the islands, where their chief occupations were fishing, piloting, and raising livestock. Inlets opened and closed, each major storm produced overwash, and dunes migrated without hindrance. All evidence suggests that such conditions have long been part of the natural environment of the Outer Banks, and indeed, necessary for their survival (7, 10). In some places, overgrazing may have resulted in more rapid retreat of the barriers and increased overwash. There is no evidence to suggest that such pressure led to what some have called the degradation of the barrier island system; nevertheless, the low, frequently overwashed nature of most of the Outer Banks was blamed on overuse by livestock, and during the 1930s major stabilization projects were begun along what is now Cape Hatteras National Seashore (5).

The primary goal of these projects was to create a high, stable dune line that would stop sand movement by wind or water. By the 1950s nearly all of Cape

Hatteras National Seashore had a continuous stable dune line in what was once the active beach zone. During the 1950s and early 1960s dunes were built on Ocracoke Island, the last island to be stabilized (17). Once the dune was created, permanent roads were built the length of the barrier chain from Kitty Hawk to Ocracoke Island, and villages began expanding toward the sea. The dune line created major ecological changes, in that the area that was once part of the natural wide berm became covered with vegetation, new plant communities appeared, and the processes of overwash, inlet formation, and dune migration ceased (11).

In contrast to these developments on Cape Hatteras National Seashore, the Outer Banks from Ocracoke Inlet to Beaufort Inlet were left wild, except for fishing camps and small-scale dune building projects in the early 1960s. All livestock was removed from these islands in the late 1950s, and in the following years, natural grassland vegetation increased dramatically. Whether this was a function of livestock having been removed or of a relatively long period of storm quiet is not known; probably it was a combination of both. In any case, there are no artificial continuous dune lines of any consequence on what is to become the Cape Lookout National Seashore. Here, overwash occurs during major storms, and inlets are free to form as they have in the past. Based on our studies of this southern section of the Outer Banks, we became convinced of the importance of natural geophysical processes in maintaining the barrier island ecosystems under conditions of rising sea level and high storm frequency.

Since it became clear that inlets and overwash provided the primary substrate for new salt marshes, and since intertidal marshes build vertically by peat accumulation, we became concerned about the condition of salt marshes behind Cape Hatteras National Seashore, where these processes have been to a great extent interrupted by man-made dunes. Previous observations suggested that these marshes were in a senescent condition and eroding along their sound-side margins, except near active inlets or where recent overwash had breached the man-made dune and filled sound waters. The purpose of this present preliminary survey was to compare the general condition of intertidal salt marshes behind stabilized barriers in Cape Hatteras National Seashore with that of marshes behind unstabilized sections of the same seashore and behind the islands of Cape Lookout National Seashore. In both sections of the Outer Banks, there are areas of extensive natural dune zones which prevent overtopping or inlet formation; marshes behind these naturally stabilized regions were also sampled.

STUDY AREA

The Outer Banks are a relatively thin strip of sandy barrier islands dividing extensive sounds from the Atlantic Ocean between latitude 36°30'N and 34°30'N, longitude 75°30'W and 76°40'W (Fig. 1). The barrier chain runs

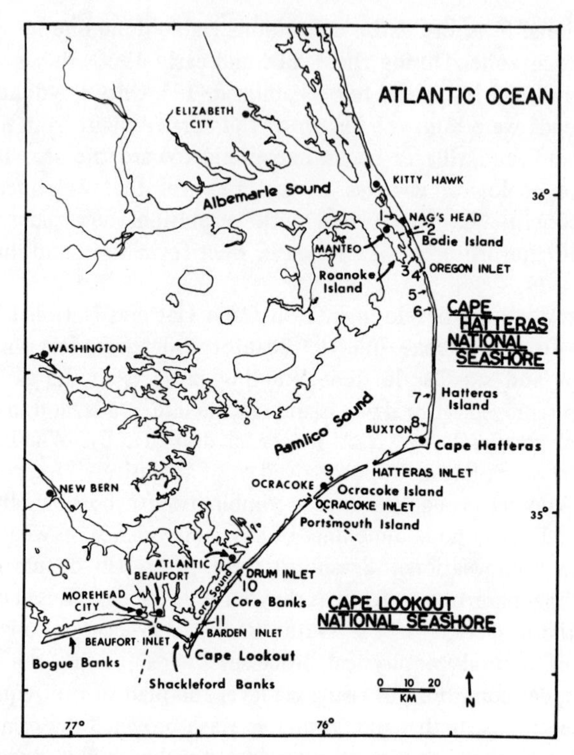

Figure 1. The Outer Banks showing locations of marsh study sites: 1 - Mann's Point; 2 - Whalebone Junction; 3 - Bodie Island; 4 - Oregon Inlet; 5 - Goose Island; 6 - New Inlet; 7 - Avon, Mile 33.5; 8 - Buxton Fill; 9 - Ocracoke; 10 - Core Banks near New Drum Inlet; 11 - Codds Creek.

approximately 280 km south from the North Carolina-Virginia border to Beaufort, North Carolina. The barrier islands south and west of Beaufort are close to the mainland and do not experience the same environmental conditions as the Outer Banks; they do not front extensive lagoons, nor are they as prone to storm attack. Winds on the Outer Banks are predominantly from the southwest in summer and northeast in winter, so that the orientation of the islands and sounds provides long fetches with subsequent wave attack, on both seaward and lagoonal shores. Tidal ranges on the ocean side of the Banks averages about 1 m, with lesser ranges in the sounds, depending on distance to the inlets. Wind tides are frequently more important than lunar ones on the sound shores. The Outer Banks have a history of more than 150 recorded hurricanes since 1585. Winter cyclones (northeasters) move north along the Banks, causing as much, and frequently more, damage than hurricanes. The passage of storms is closely linked to overwash frequency and the formation of inlets. In recent years man-made dunes and facilities along the developed coast have been badly damaged, but little harm is done by these storms on undeveloped coasts (4). Long-term figures suggest an average rate of sea level rise of about 30 cm/century (15), while more recent data have shown a rise of about 7.5 cm in the past decade (13).

METHODS

The majority of our sample sites were behind Cape Hatteras National Seashore, with one on Core Banks near Drum Inlet. The general nature of marshes on Core Banks was described in an earlier paper (12). Marshes to be sampled had to have exposure to lagoonal wave attack, be as close as possible to the main barrier island, have a west-facing shoreline, and have *Spartina alterniflora* as the primary indicator of intertidal salt marsh. Each sample marsh was chosen so as to include sites whose general history was known and which suggested age sequences from very old to very recent. We also chose sites with varying degrees of natural as well as artificial stabilization. Certain marsh sites could be reached conveniently only by boat from the lagoon side, so the exact site sometimes depended on access. When a relatively uniform marsh edge with *S. alterniflora* had been selected, a 30 m tape was stretched along the shore, and lines at right angles to the tapes were selected at random. On extensive marshes, time permitted only two or three random lines to be run. Each transect was profiled with a surveyor's level, using the bottom of the lagoon adjacent to the marsh as datum. Most vegetation sampling was done by means of 0.0625 m^2 quadrats, located at each meter from the edge of the marsh up to 15 meters back. On wide marshes, the intervals between samples were greater. Density was determined for each quadrat and species were recorded. The heights of the five tallest *S. alterniflora* culms were averaged for the purpose of profiling. Frequencies of species present were determined from the total quadrats in each marsh sample. Five or more samples of standing crop were taken at all but two sites. The general characteristics of each marsh site were recorded. The length of the transect was determined by the zones in which the *S. alterniflora* was present; thus, the lines terminated in what is generally referred to as the high marsh. Peat depth was determined by pushing a steel rod down through the marsh until contact was made with the sand base. The reliability of this method of peat depth sampling was verified by excavating where the peat layer was thin.

It was necessary to make two rapid surveys at Buxton and Ocracoke in October because there had not been time to do them at the end of August. At the Buxton site, a single profile was made in a representative section of marsh, and five samples of standing crop were taken at random in the marsh. At Ocracoke, ten samples of standing crop were taken over a wide area, with density, height of tallest grasses, and weights being determined. All standing crop data are expressed as oven-dry weight.

Low altitude infrared photographs, aerial reconnaissance, and ground checks were used to survey the general characteristics of as many other marsh areas as possible.

RESULTS

Data from the Outer Banks salt marsh transects are presented in three ways. Profile diagrams drawn to scale, both vertically and horizontally, contain data on density, standing crop, and species distribution. The height of the marsh grass is also drawn to scale on the topographic profile. Five profiles were made at all but two of the primary sample sites, but only three are drawn. All density data from transect lines are averaged for each station and presented as a series of bar graphs (Fig. 2). Productivity and density on a square meter basis are averaged for each station and presented in tabular form (Table 1). Lack of space prevents a complete discussion and evaluation of all the data gathered on these marshes. However, such material will be included in a more complete treatment of this survey at a later date.

Oregon Inlet

This was the youngest marsh sampled, having developed on shoals and overwash deposits behind a spit that has grown out into Oregon Inlet since 1962 (6). The marshes behind Oregon Inlet have some of the widest zones of intertidal *S. alterniflora* we found, in spite of their exposure to oceanic overwash and to flood tides from the sound (Fig. 3 and 4). The edge of the marsh is expanding rather than eroding, and the entire marsh slopes gradually upward from the sound bottom to the high tide mark 82 m back on the sample profile (Fig. 5). The tall form of *S. alterniflora* dominates the first 10 m, and the short form the next 70 m; other salt marsh species begin to appear well back from the marsh edge but are of little importance until the higher levels are reached. Organic accumulation, here more a mat of roots than true peat, is about 20 cm deep.

New Inlet - Recent Surface

This inlet was open in the 1930s and later closed (Fig. 6 and 7). Its tidal shoals developed into marshes, and grass was planted along the beach to encourage dune growth. The 1962 Ash Wednesday storm put overwash sand into the old inlet channel, creating a gradually sloping sand surface later colonized by *S. alterniflora.* In spite of a relatively wide intertidal zone, the productivity of this marsh was considerably below that at Oregon Inlet. The New Inlet marsh was also distinctive in that more other species were included in the *S. alterniflora* zone than in other marshes sampled (Fig. 8). These characteristics may be due

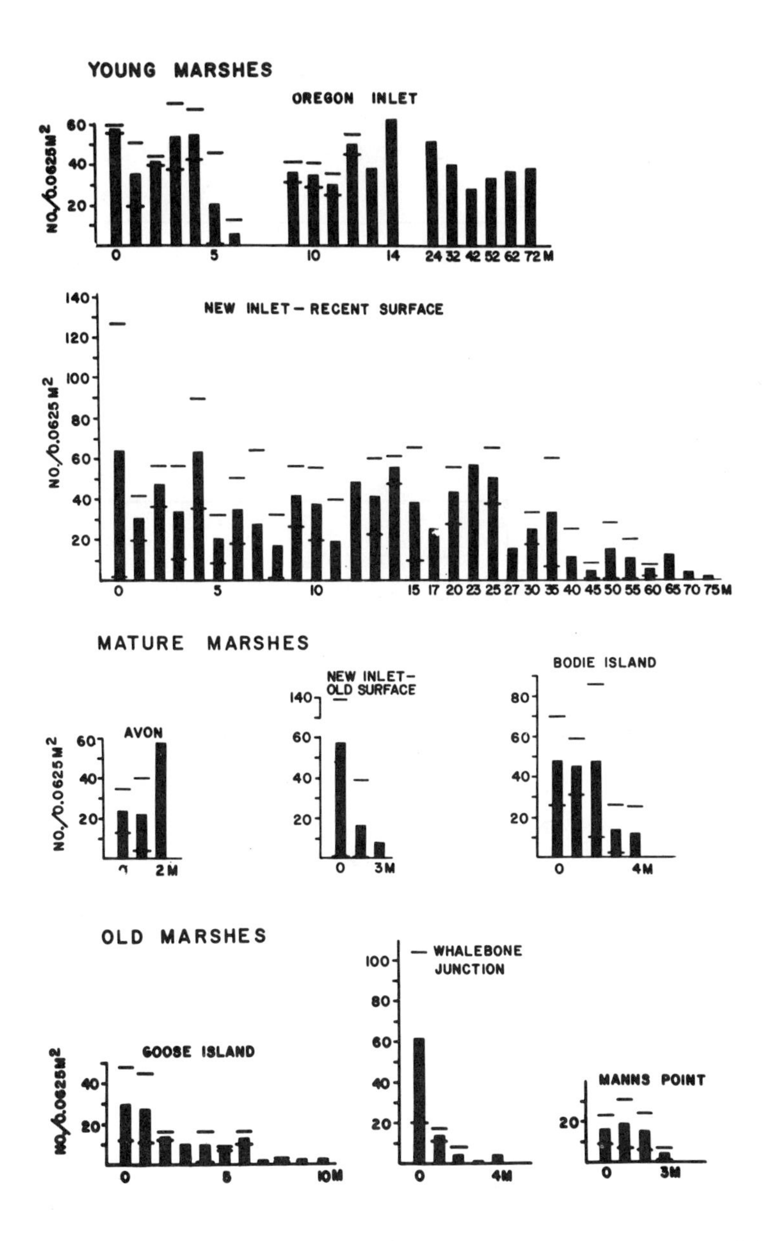

Figure 2. Average density data along each marsh profile, as taken in the field (on a 0.0625 m^2 basis). Bars indicate mean density values at each meter. Horizontal lines above and below the bars indicate one standard deviation; where lines are not present only one sample was available.

TABLE 1.

Standing crop and density of **Spartina alterniflora.** Summation of all transects at each station. Parentheses indicate number of samples used in computations.

	Standing crop, grams/m^2			Average Density, **Spartina alterniflora** zone Stems/m^2	Width of **Spartina alterniflora** zone, meters
	Low Marsh (Tall)	Low Marsh (Short)	Overall Mean		
Young Marshes					
Oregon Inlet	446.4 ± 193.6 (5)	457.6 ± 176.0) (7)	457.6 ± 171.1 (12)	625.8 ±272.0 (30)	72
New Inlet-Recent Surface	186.7 ± 110.4 (20)	59.2 ± 54.4 (8)	151.5 ± 112.0 (29)	590.4 ± 304 (50)	60
Core Banks	1172.8 ± 555.2 (5)				36
Codd's Creek	1900 (5)				45
Ocracoke	875 ± 475 (10)			1200 ± 475 (10)	270
Buxton Fill	1278 ± 677.5 (5)			917.5 ± 577.5 (5)	8

TABLE 1 (continued)

	Standing crop, grams/m^2			Average Density, **Spartina alterniflora** zone Stems/m^2	Width of **Spartina alterniflora** zone, meters
	Low Marsh (Tall)	Low Marsh (Short)	Overall Mean		
Mature Marshes					
Bodie Island	725 ± 528 (4)	292.8 ± 86.4 (2)	580.3 ± 467.2 (6)	752.0 ± 392 (15)	4
Avon	900 ± 760 (6)			416 ± 272 (11)	1
New Inlet-Old Surface	964.8 ± 1305 (8)			657.6 ± 1060.8 (8)	0-2
Old Marshes					
Whalebone Junction	760 ± 203 (5)			654.4 ± 625.6 (9)	2
Mann's Point				267.2 ± 145.6 (15)	2-3
Goose Island				283.2 ± 216 (19)	3-6

Figure 3. Oregon Inlet marshes behind the southward growing spit; view is to north. Dark patches are marshes dominated by **Spartina alterniflora.** Arrow indicates sample site.

Figure 4. Ground view of the Oregon Inlet study area looking from the marsh edge eastward toward the ocean. Low dunes appear in the distance beyond the bridge. The cordgrass is tall and easily flooded by the tides.

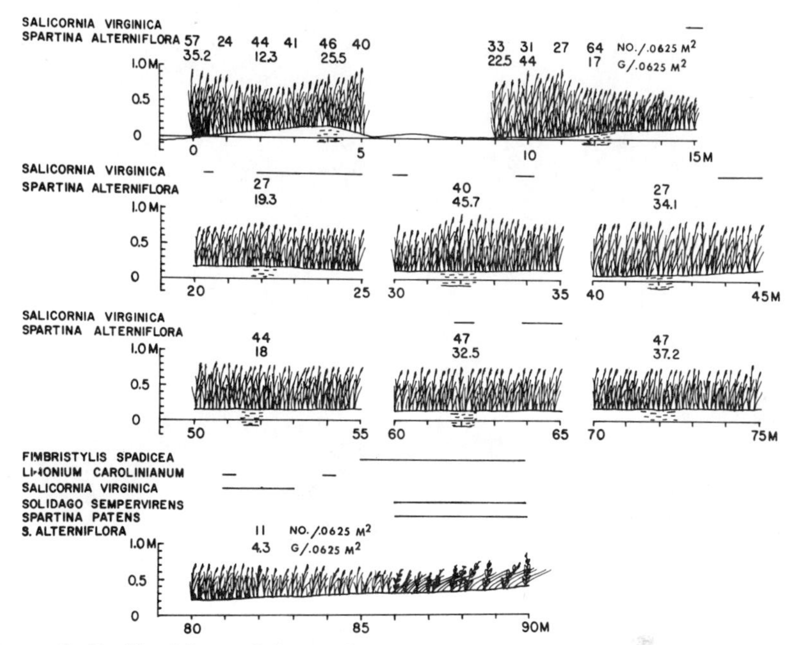

Figure 5. Profile of Oregon Inlet marsh.

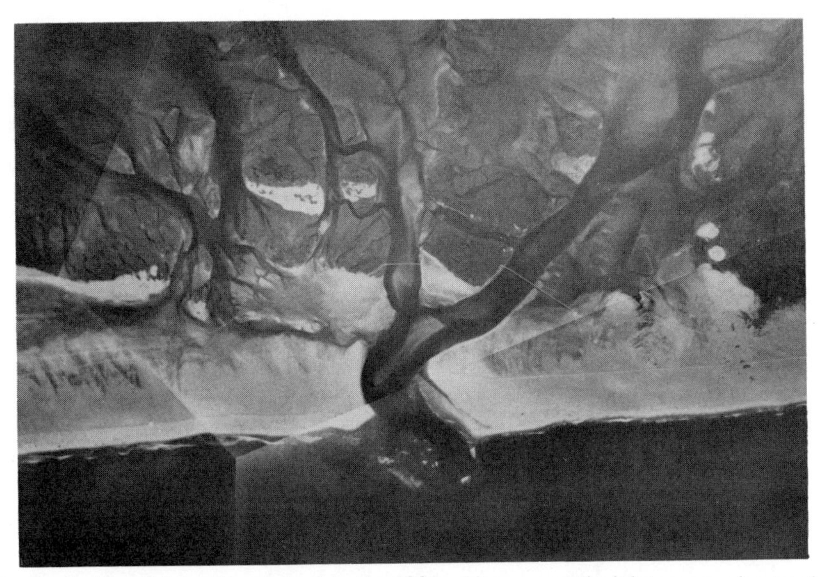

Figure 6. New Inlet as it appeared in the 1930s. Lines across the inlet are wooden trestle bridges. (N. P. S. photo)

Figure 7. New Inlet as of 1972. Note bridge remanants. Sample sites are indicated by arrows: 1 - recent overwash surface; 2 - old surface. Much of what was open water and shoals in the previous photograph is now salt marsh or grassland. The area is now part of the Pea Island National Wildlife Refuge. (N.A.S.A. photo)

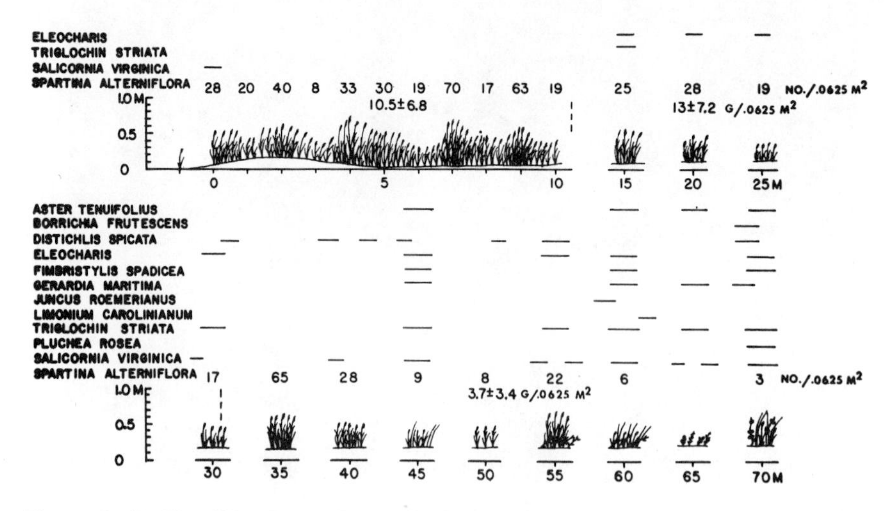

Figure 8. Profile of New Inlet – Recent Surface marsh.

to this marsh being 11.7 km from the nearest open inlet, so that it lacks the high tidal range and vigorous flushing that promote the best growth of *Spartina.* There was no accumulation of organic matter.

Ocracoke

In the 1960s, dunes were built on what were then broad, bare flats seaward of Ocracoke Village. After what was, in this case, excessive overwash stopped, grasses and shrubs colonized the barren sand. A salt marsh zone, now 270 m wide, appeared on the gradual intertidal slope on the sound side. Although this marsh was sampled only minimally and a month later than the other stations, the system is clearly a highly productive one (Fig. 9 and 10).

Buxton Fill

The Ash Wednesday storm opened an inlet just north of Buxton, near Cape Hatteras. This inlet was soon plugged with hydraulic fill and artificial stability was restored to that section of the island, but no planting was done on the intertidal zone (Fig. 11). Small patches of *S. alterniflora* have invaded spontaneously, and in some places are growing vigorously and preventing erosion (Fig. 12). In spite of the fact that, along with Ocracoke, this area was sampled late, it is apparent that, although the grass grows in only a narrow zone, its densities and standing crops are comparable to those of marsh grass growing on natural substrate (Fig. 13).

Core Banks - Drum Inlet

For comparison with the Cape Hatteras marshes, a marsh area on Core Banks was examined. The marsh is about 36 m wide and is expanding into tne sound at about 1 m/yr. (Fig. 14). It rises gradually from the bay, as do other recent marshes. Until the Corps of Engineers created New Drum Inlet close by in 1972, the nearest opening to the sea was Old Drum Inlet, 4.5 km to the north, so the only possible source of sediments for this marsh must have been overwashes of the early 1960s.

New marshes forming behind Cape Lookout, both on overwash sediments and along the backside of the elongating Cape Lookout spit, as well as behind lower Core Banks, are similar to the marshes described above, with a wide *S. alterniflora* zone, a gradual slope, growing margins, and high productivity (12).

The next marshes to be described were of the mature type, dating at least from before 1960 and sometimes back to the last century. These have peat layers up to 0.5 m thick, distinct zones of short and tall *S. alterniflora,* and frequently, eroding edges. There is rarely any evidence of recent overwash.

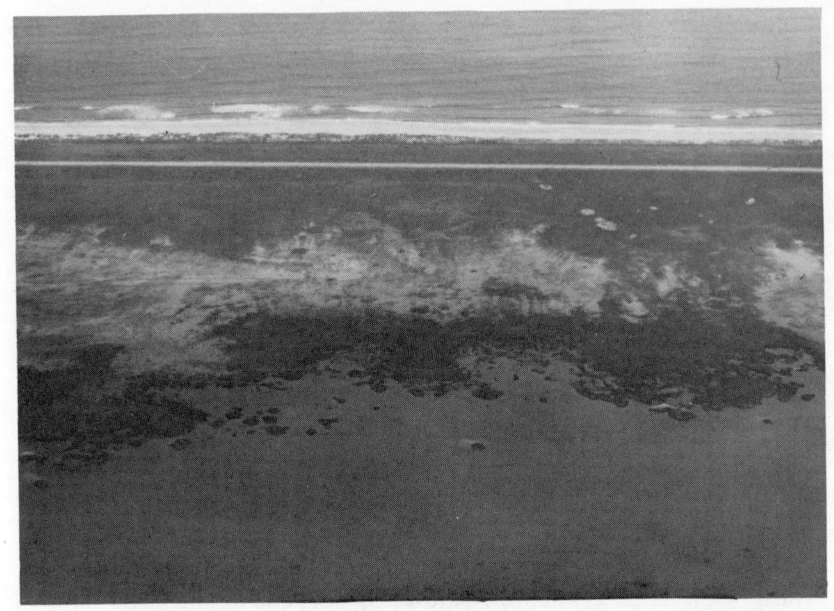

Figure 9. Ocracoke Island, west of Ocracoke village, showing new marshes developing on former barren flats. View is from sound side toward ocean. Highway and man-made dune appear at the top of the photograph. Dark zone in foreground is **S. alterniflora** marsh. Light patch between marsh and grassland is open sand flat with dense stands of **Salicornia.** It is basically a large panne.

Figure 10. Ground view of the extensive Ocracoke marsh looking from the sound side toward the ocean. The cordgrass is tall throughout and the marsh easily flooded. There is no evidence of erosion as the marsh slopes gradually soundward.

Figure 11. Site of the 1962 inlet at Buxton on Hatteras Island. The inlet was filled by dredging sand from shoals behind the island leaving the straight, dark patches visible in the photograph; these indicate deep water.

Figure 12. New marsh growing on part of the Buxton Fill. Other areas along the soundside are barren sand and eroding.

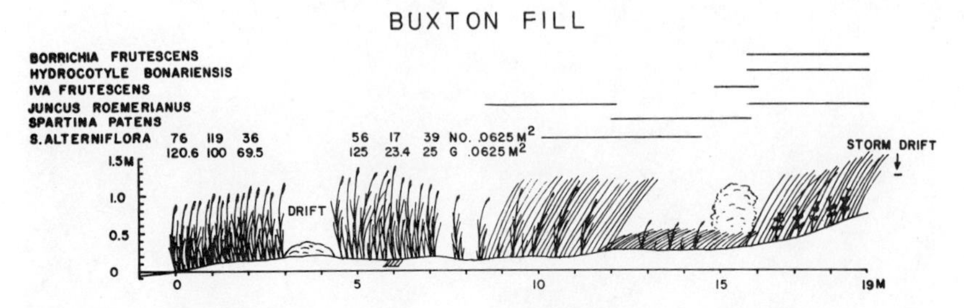

Figure 13. Profile of Buxton Fill marsh.

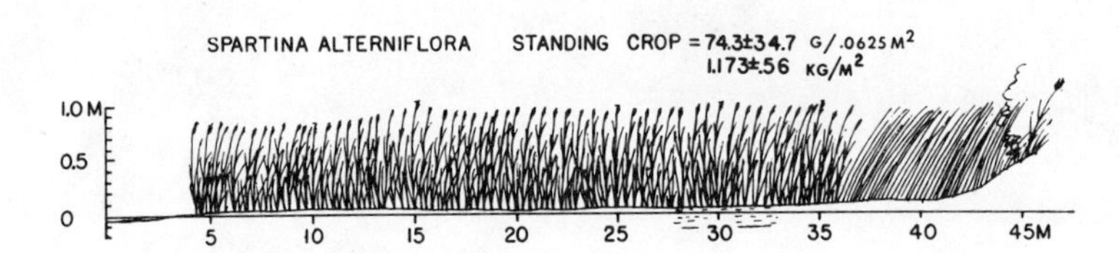

Figure 14. Profile of Core Banks marsh.

Avon, Milepost 33.5

The marshes here are typical of the sound side of Cape Hatteras National Seashore, in that the intertidal *Spartina* zone is only a meter or two wide, scarped, and eroding (Fig. 15 and 16). Drifted sea grass frequently buries the *Spartina*, causing bare spots. Cover ranges from sparse to relatively luxurious in scattered locations close to the water. There is evidence that these rather tattered marshes are remnants of more extensive stands that have eroded away.

At this site there were apparently no overwashes in the years immediately before stabilization, and overwash has since been cut off by the barrier dune. The 1962 storm, however, nearly did break across, and after it hydraulic fill was pumped from a borrow channel behind the island onto the dune and beach (Fig. 17). This channel is still steep-sided and about 3 m deeper than the surrounding bottom. Creeks leading into the island were filled at this time and are now marked by piles of sand.

New Inlet - Old Surface

Aerial photos taken in the 1930s show that this marsh grew up on what was a bare sand flat on the north side of New Inlet when it was open (Fig. 6 and 7).

Figure 15. Ground view of marshes at the Avon site. The intertidal zone is narrow, eroding, and heavy layers of sea-grass drift mark the high tide line. A typical view along most of these Banks.

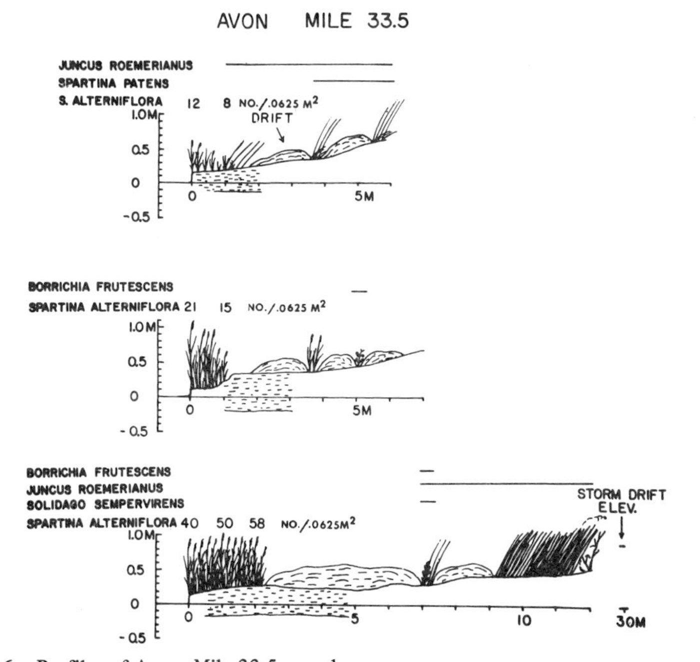

Figure 16. Profiles of Avon, Mile 33.5, marsh.

Figure 17. Avon – Mile 33.5 marsh system. The island is narrow here and fronted by a barrier dune with the road near the back side of the island. The marshes are much reduced. The dark line behind the island is a borrow channel from which sand was dredged to build up the dune and island surface. Piles of fill show in what were once marsh creeks. This operation was done after the 1962 storm. The arrow indicates the sample area.

Marsh growth and the accumulation of 0.5 m of peat began after the inlet closed. Today the marsh edge is eroding where it has not been overwashed, and the marsh surface rises rapidly from the scarped edge to a high marsh dominated by *Juncus roemerianus* and *Spartina patens* (Fig. 18). The *S. alterniflora* zone is only about 1 m wide, where it exists at all. Nevertheless, the standing crops in some quadrats were the highest of any area sampled.

Bodie Island

The marsh sampled here lies north of the lighthouse, along a bay extending into the central part of the island. Patterns of marshes and creeks suggest that this section of the island was formed by a series of overwashes (Fig. 19). The barrier dune has evidently stopped any recent oceanic overwashing. Peat depth is about 0.5 m. The *S. alterniflora* zone is 0 to 4 m wide, with height, density, and standing crop averaging out to values similar to those from other marshes (Figs. 20 and 21).

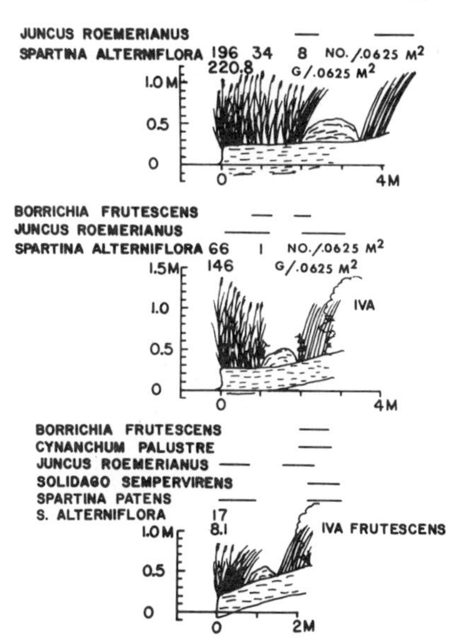

Figure 18. Profiles of New Inlet - Old Surface marsh.

Figure 19. Bodie Island marshes north of the lighthouse; view is to the northeast. The arrow indicates the sample area. Shrubs are common around and in the marshes; **Juncus** dominates most.

Figure 20. Bodie Island marsh survey area. The intertidal fringe is narrow and rises to a sand ridge with shrubs. Poorly drained marshes lie behind the ridge with **Juncus** gaining dominance. View is to the north with a water tower in the distance.

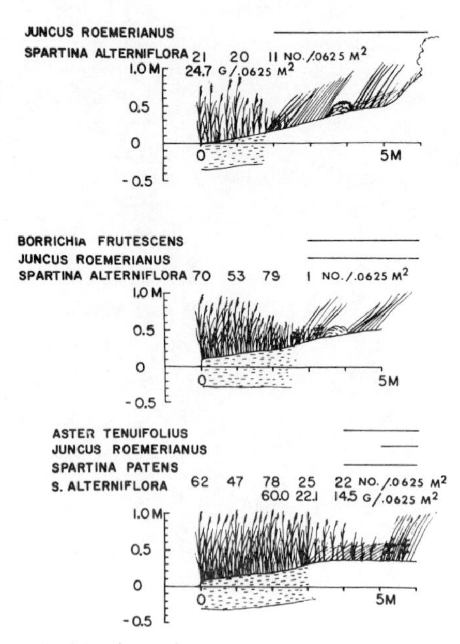

Figure 21. Profile of Bodie Island marsh.

The marsh surface rises rapidly to a low ridge built by sound overwashing and drift, dominated by *S. patens, Juncus,* and shrubs. Behind the ridge the marsh contains short *S. alterniflora* being invaded by *Juncus.* Tidal flushing is hampered by the ridge and the small tidal range. Local residents recalled that the marsh border was once much wider but has steadily eroded.

The rest of the marshes sampled probably date from before the 1800s. They could be identified by depth of peat (over 0.5 m), the degree of stability seaward, historical information, and their vegetation.

Goose Island

This marsh was somewhat farther out in Pamlico Sound than others sampled, since marshes attached to the main island were inaccessible in this area. As part of Pea Island National Wildlife Refuge, the marsh has been influenced by management projects to the seaward, being connected to the main island by dikes. Its edge is badly eroded. Peat was between 0.7 m and 1.0 m deep, and the *S. alterniflora* zone was 0 to 5 m wide. The marsh surface rises slightly from the edge of the scarp to become a high marsh of *Juncus, S. patens,* and *Borrichia;* short *S. alterniflora* extends back into the marsh along former creeks but is seldom reached by the tides (Fig. 22).

Before the dikes were built, there were open sand flats between Goose Island and the main island. These and the surface on which the marsh formed probably originated as shoals from a one-time inlet. Judging by peat depth in the marsh, this inlet must have closed several hundred years ago.

Whalebone Junction

This is the site of an inlet that closed in about 1800 (8). The marshes here are typical of those that develop on inlet shoals (Fig. 23). As usual on an old site, the marsh edge is scarped and badly eroded (Fig. 24). The *S. alterniflora* zone appears only here and there and is very narrow; the grass grows densely but its shortness results in low productivity. Most of the marsh is dominated by *S. cynosuroides, Juncus,* and *Borrichia,* with patches of *Iva* and *Baccharis* shrubs. Peat depth was between 0.6 m and 0.8 m (Fig. 25).

Mann's Point

The marshes here grow on the most stabilized section of the Outer Banks north of Hatteras Island. The land is 3.1 km wide, and extensive, stable, natural dunes afford protection to seaward (Fig. 26); the marshes have doubtless not been overwashed for hundreds of years. Peat is up to 2 m deep and is, of course, eroded by waves from Albemarle Sound. The vegetation is in the last stages of

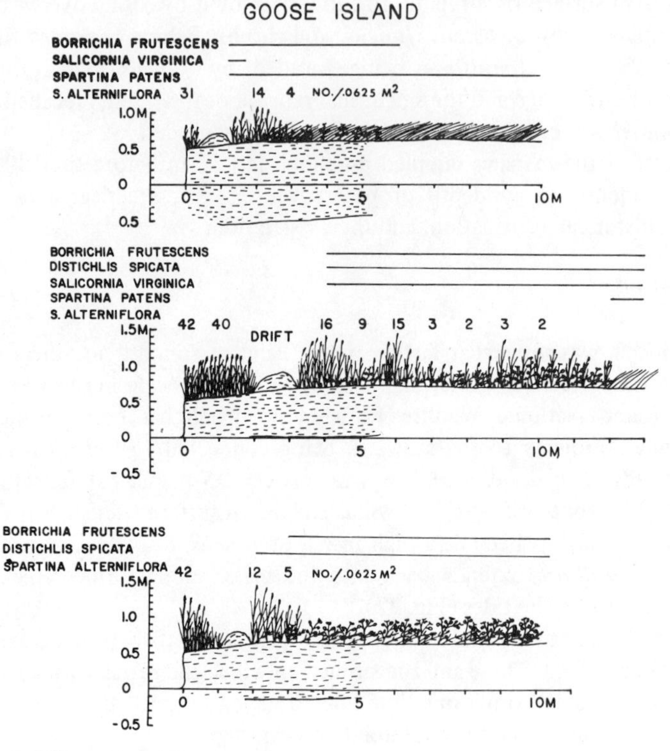

Figure 22. Profiles of Goose Island marsh.

marsh succession, with *S. patens, S. cynosuroides,* and *Juncus* predominating and thickets of *Iva* and *Baccharis* coming down to the water's edge (Fig. 27). The *S. alterniflora* zone is only about 3 m wide, the grass sparse and short and, therefore, of low productivity (Fig. 28).

Figure 2 presents mean density data for each set of transects. In all cases but Oregon Inlet and New Inlet-Recent Surface, the values at the beginnings of the transect are means of five samples. Toward the ends of the transect lines, the number of samples may be less, because the width of the *S. alterniflora* zone on each line varied. Thus averages were made only within the boundaries of that zone. At Oregon Inlet two transects were made, while at New Inlet-Recent Surface three lines were laid out. It is clear from these data that the densities of grass within each transect were highly variable, but were all within the same range from site to site, except for Mann's Point and Goose Island. Where the densities drop to low values at the end of the transects, the *S. alterniflora* zone grades into another vegetation type.

Table 1 summarizes the standing crop data taken on the marsh transects, the average density of grass culms within the *S. alterniflora* zone, and the width of

Figure 23. Whalebone Junction, the reported site of Roanoke Inlet, which closed about 1800. View is northeast toward South Nags Head. Marshes are dominated by **Spartina cynosuroides, Juncus roemerianus, Iva frutescens** and **Baccharis halimifolia.** The arrow indicates one small stand of **S. alterniflora** selected as a study site.

Figure 24. Ground view of the Whalebone Junction marsh fringe; view is south. The shoreline is very irregular with eroding peat hummocks and a very narrow **S. alterniflora** zone.

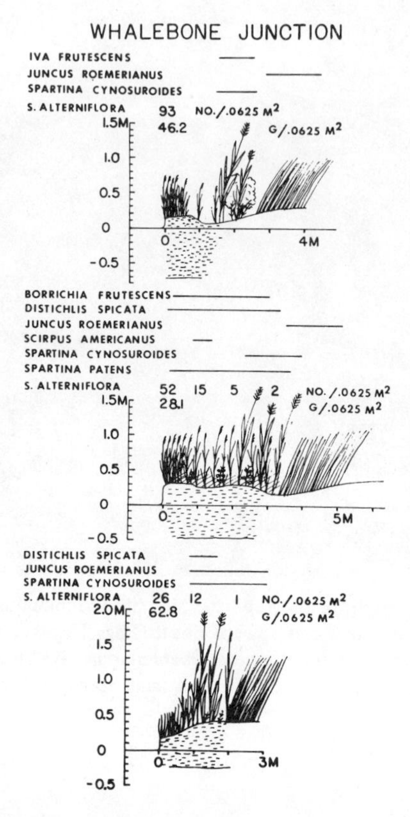

Figure 25. Profiles of Whalebone Junction marsh.

that zone. No standing crop data were taken at Mann's Point or Goose Island. Data for Codd's Creek on Core Banks are from an earlier paper (12). Where data for low marsh standing crop are lacking, the predominant grass form was of the tall type; data for the short form were not collected. The standing crop data show considerable variability, as indicated by the large standard deviations. Because of such ranges in values, no real correlations can be drawn between marsh age and standing crop of the intertidal edge, although in some cases the values for young marshes are well above those for the older (especially the Codd's Creek site), but at the same time some young marshes are well below the old (New Inlet-Recent Surface). Such descrepancies are not easily explained at this point. Likewise, average densities show great variation and have mean values that are of the same order of magnitude regardless of marsh age. The main difference was found in the oldest marshes, which had considerably lower densities, and thus, although not sampled, would have correspondingly lower productivity. The major difference between marshes is in the width of the *S.*

Figure 26. Mann's Point marshes looking east to Nags Head. Pine hummocks are scattered about in the largely **S. cynosuroides - Juncus** - shrub dominated marshes. Arrow indicates study site.

Figure 27. Mann's Point marsh edge and study area. The **S. alterniflora** is narrow and eroding in most places. Density and height are low for the cordgrass. **S. cynosuroides** and **Iva** shrubs dominate in the background.

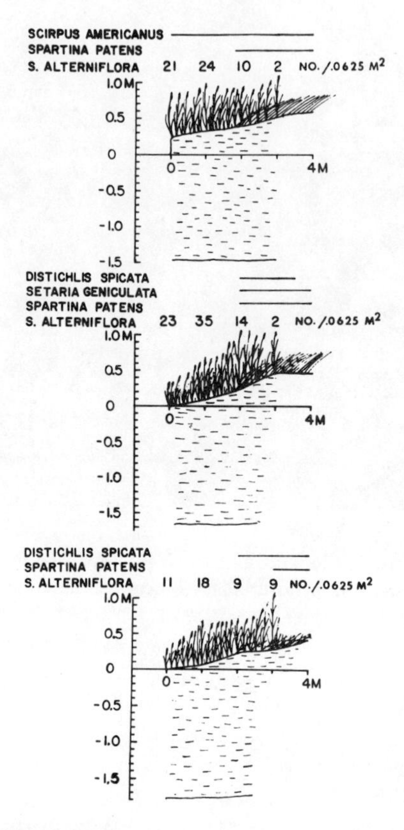

Figure 28. Profiles of Mann's Point marsh.

alterniflora dominated low marsh, and this becomes a significant factor in overall productivity.

DISCUSSION AND CONCLUSIONS

This preliminary survey documents Outer Banks salt marsh succession and its relationship to the stability of the islands. Using the presence of *S. alterniflora* and the ability of tides to flush its production from the marsh as indicators of relative importance to estuarine productivity, these marshes may be grouped into three types: young, with wide zones of highly productive *S. alterniflora*; mature, with narrower zones of *S. alterniflora* of lower productivity, and eroding edges; and old, with *S. alterniflora* lacking or greatly restricted, productivity very low, edges eroding badly, and a predominance of supratidal species.

The youngest marshes – Oregon Inlet, Ocracoke Island, New Inlet, Core Banks – show marked structural similarities as well as differences in productivity not easily explained. These marshes grow on recent substrate laid down by water – inlet shoals or overwash sediment – and perhaps on some wind deposits, sloping up from the lagoon to a high marsh or dune zone. The intertidal sand is colonized by *S. alterniflora,* which soon forms a low marsh, readily flooded by diurnal tides. Despite exposure to sound waves, most of these marshes are expanding rather than eroding. Some are bordered by a low ridge where waves have piled up sand, but such ridges do not prevent tidal flooding. Waves dissipate energy by rolling up into the marsh along the gentle slope and then into the resilient grass. Except for *Salicornia,* other salt marsh plants rarely appear in well-developed *S. alterniflora* stands; these plants become increasingly important at higher levels.

If the marsh is protected by seaward dunes as it develops, peat accumulates fast enough to keep ahead of sea level rise and the surface builds up to the point where less and less material can be carried out by the tides, as seen at the Avon, Bodie Island, and New Inlet-Old Surface marshes. The leading edge of the marsh begins to erode as waves break against the peat-sand interface. The short form of *S. alterniflora* dominates the inner marsh, with the taller form growing only along the edges where tidal exchange is better; *S. patens, S. cynosuroides,* and especially *Juncus roemerianus* begin to invade with vigor.

Marshes seem to reach this intermediate stage in twenty to thirty years, and may persist in it for a hundred years. Mature marshes dominate the estuarine system behind the Outer Banks, particularly Cape Hatteras National Seashore, but also behind some islands of Cape Lookout National Seashore which are relatively wide and stable.

If no drastic geomorphic changes occur, the marsh eventually enters the "old" or senescent stage, as far as nutrient contribution to the estuary is concerned. The marshes at Goose Island, Whalebone Junction, and Mann's Point are in this category. Senescent marshes are usually found behind long-stable barrier dunes. Peat accumulation has built the surface so high that it is rarely flooded even by storm tides; the barrier dune, by discouraging opening of inlets, has reduced tidal amplitude and thus further narrowed the *S. alterniflora* zone. Where the *S. alterniflora* fringe exists at all it is very narrow and much less productive than in the younger marshes. Most of the marsh is taken over by *Juncus, S. patens, S. cynosuroides, Borrichia frutescens,* and thickets of *Iva frutescens* and *Baccharis halimifolia;* the shrub stage is persistent and could be called a "climax" of sorts. Erosion continues, as sound waves expend their energy directly on the scarped and undercut peat.

In these latter stages, the contribution of nutrient containing organic matter from the marsh's intertidal zone is minimal. Density and standing crop of the *S. alterniflora* fringe of a mature or old marsh may equal or even exceed those of a younger marsh, as shown in Table 1, but the point is that a zone of grass a meter or two wide must have far lower total productivity than one many tens of meters in width. Thus older marshes are of reduced nutritional value to the estuarine system, although of course they have intrinsic value from other viewpoints.

Figure 29 summarizes marsh succession. The cycle can only begin again when an inlet opens and creates new shoals, or when the barrier island retreats over the old marshes to the extent that sand is carried over the island and into the lagoon behing. In these ways the salt marsh system is rejuvenated and contribution to the estuary remains high.

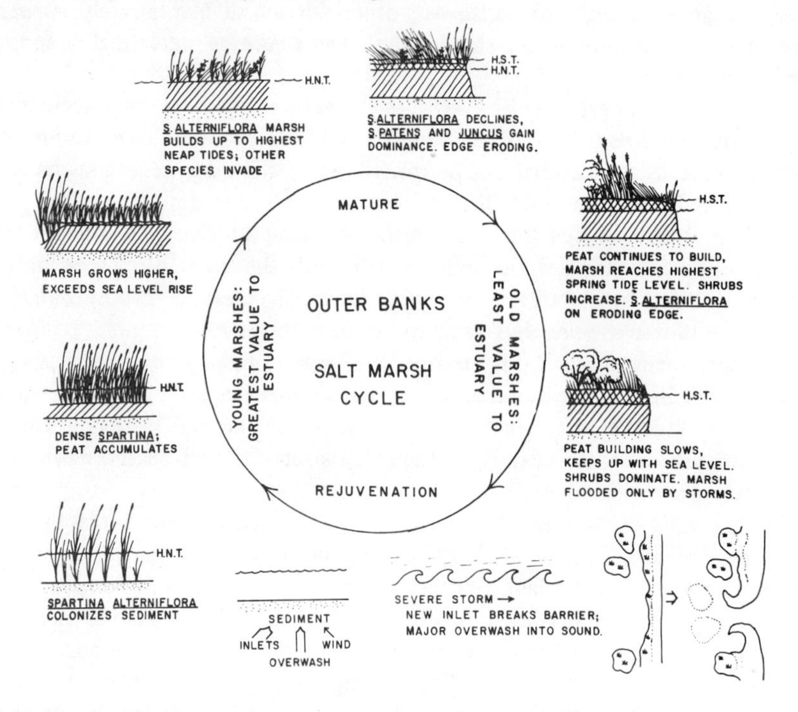

Figure 29. Hypothetical marsh succession cycle for the Outer Banks and similar barrier islands based on surveys made in this paper. Illustrations are not drawn to scale and indicate only general patterns.

Patterns of marsh development and succession vary. The model presented here applies to the Middle Atlantic coast north of Beaufort, North Carolina, and to some extent to the Northeast. J. Dowhan has found similar patterns of marsh succession behind Fire Island, New York, and ties them to the same processes at work on the Outer Banks (personal communication, 1973). From central North Carolina southward, wave energy is lower, and tidal range and inlet frequency are higher; successional patterns are correspondingly different, although some features of the present model may appear in certain areas.

Impact of stability

Marsh succession as described above obviously requires a certain degree of stability to the seaward. In areas near active inlets or experiencing frequent overwashing, succession will stay in the early, productive stages; organic production and the rate of its export into the estuary are high, which is consistent with succession theory.

Long-term stability, either naturally or artificially created, will lead to senescence. The intertidal marsh becomes an upland ecosystem, flooded only by storm tides, and undesirable from the standpoint of estuarine productivity. Again consistent with theory, the pioneer species, *S. alterniflora,* alters its environment in ways that are detrimental to its own survival but creates conditions favorable to the species that follow in the successional sequence. The pioneer stage persists only where environmental conditions supersede the self-destructive tendencies of the pioneer species.

Some stability is desirable for the early stages of marsh development. What is wanted is some overwash but not too much. Natural or man-made dunes in the proper locations can help the marsh by keeeping overwash from being too heavy and frequent. This seems to have been the case on parts of Ocracoke Island and on small sections of northern Core Banks, where there are man-made dunes, as well as on other parts of the Outer Banks, where natural dune fields have slowed down overwash that might have overwhelmed the establishment or recovery of marsh vegetation. These dunes have served their purpose and will continue to do so even as they retreat.

Unfortunately, stable barrier dunes which last too long, be they natural or man-made, become a liability for the marsh. They prevent influx of sand over the marsh and into the sound, so there is no new surface for new growth, and we are left with a senescent, eroding, under-productive marsh. This has become the case along most of Cape Hatteras National Seashore, and the situation is made worse by a lack of inlets and consequent smaller tidal range.

Where artificial stability or controlled retreat is deisred, fill might be used to create new marsh substrate behind a narrow barrier and thus mimic natural processes. The practicality of creating salt marsh has been shown in this symposium and elsewhere (18); even where no planting is done, marsh grass will eventually come in on its own.

The problems facing artificially stabilized systems such as Cape Hatteras National Seashore can be attacked with at least the three following sets of options, depending on local conditions.

Letting the islands remain in or return to their natural state is desirable where possible. Such a course could be tolerated on much of Cape Hatteras National Seashore and on all of presently undeveloped Cape Lookout National Seashore.

Where development has halted natural processes, problems arise. One option here is to hold the line against beach erosion and strive for long-term stability. This course will lead to increasing problems on both sides of the islands and ever-mounting expense, and could be taken only for small areas. It will never be a final answer.

A compromise approach that may have some application is controlled retreat, in which overwashes are allowed to pass into dune fields and back across the island. Where practical, overwash into the sound may be permitted, or artificial marshes could be built to simulate the natural process.

Survival of barrier islands and their salt marshes depends on interplay of inlet dynamics, overwash, dunes, vegetation, and a certain degree of short-term stability. Long-term stability leads to undesirable changes and is not good for estuarine productivity. Thus, management of barrier islands must take all these natural interactions into account, as well as social and economic considerations, and draw information from all sources, to provide for the continued health of the outstanding estuaries behind the Outer Banks.

ACKNOWLEDGMENT

This research was supported by the Office of the Chief Scientist, National Park Service and by National Park Service Southeast Region contract no. 00031710. We wish to thank the staffs of Cape Hatteras National Seashore and Cape Lookout National Seashore for supplying certain data and materials and for the use of aircraft, facilities, and equipment.

We wish to acknowledge the assistance in the field of Ms. Cheryl McCaffrey, Mr. Richard Travis, and Dr. Alan Niedoroda. Mr. Richard Nathhorst prepared all the photographic material in the paper. Infrared aerial photographs were supplied by the National Aeronautics and Space Administration, Wallops Flight Center.

REFERENCES

1. Adams, D. A.
1963 Factors influencing vascular plant zonation in North Carolina salt marshes. **Ecology**, 44(3): 445-456.

2. Belt, E.
1971 Geological history of Sedge Island, Tiana Shore, Suffolk County, New York. Unpublished report to the Trustees of the Township of Southampton, New York.

3. Chamman, V. J.
1960 **Salt Marshes and Salt Deserts of the World.** Leonard Hill (Books), Ltd., London.

4. Dolan, R. and Godfrey, P.
1973 Effects of Hurricane Ginger on the barrier islands of North Carolina. **Geol. Soc. Amer. Bull.,** 84(4): 1329-1334.

5. Dolan, R., Godfrey, P. J., and Odum, W. E.
1973 Man's impact on the barrier islands of North Carolina. **Amer. Scientist,** 61(2): 152-162.

6. Dolan, R. and Glassen, R.
1973 Oregon Inlet, North Carolina: a history of coastal change. **Southeastern Geograph,** 13(1): 41-53.

7. Dunbar, G. S.
1958 **Historical Geography of the North Carolina Outer Banks.** Louisiana State Univ. Studies, Coastal Series No. 3. Louisiana State Univ. Press, Baton Rouge.

8. Fisher, J. J.
1962 **Geomorphic Expression of Former Inlets along the Outer Banks of North Carolina.** M. A. thesis. Univ. North Carolina.

9. Godfrey, P. J.
1970 **Oceanic Overwash and its Ecological Implications on the Outer Banks of North Carolina.** Office of Natural Science, National Park Service, Washington, D. C..

10. Godfrey, P. J.
1973 Ecology of barrier islands influenced by man. In **Symposium on Human Impact on the Atlantic Coastal Zone,** A.A.A.S. Annual Meeting, Washington, D. C. 19 pp. (Unpublished manuscript.)

11. Godfrey, P. J. and Godfrey, M. M.
1973 Comparison of geological and geomorphic interactions between altered and unaltered barrier island systems in North Carolina. In **Coastal Geomorphology,** p. 239-258. (ed. Coates, D. S.) Publications in Geomorphology, State Univ. New York, Binghamton.

12. Godfrey, P. J. and Godfrey, M. M.
1973 The role of overwash and inlet dynamics in the formation of salt marshes on North Carolina barrier islands. In **Ecology of Halophytes.** (ed. Reimold, R.). Academic Press, New York. pp. 407-427.

13. Hicks, S. D.
1971 As the ocean rises. **NOAA Bull.,** 2(2): 22-24.

14. Kraft, J. C.
1971 Sedimentary facies patterns and geologic history of a holocene marine transgression. **Geol. Soc. Amer. Bull.,** 82(8): 2131-2158.

15. Milliman, J. D. and Emery, K. O.
1968 Sea levels during the past 35,000 years. **Science,** 162(3858): 1121-1123.

16. Redfield, A. C.
1972 Development of a New England salt marsh. **Ecol. Monogr.,** 42(2): 201-237.

17. Woodhouse, W. W., Jr. and Hanes, R. E.
1967 **Dune Stabilization with Vegetation on the Outer Banks of North**

Carolina. U. S. Army Coastal Engr. Res. Center Tech. Memo. No. 22.

18. Woodhouse, W. W., Jr., Seneca, E. D., and Broome, S. W.
1972 **Marsh Building with Dredge Spoil in North Carolina.** Bull. 445. Agr. Sta., North Carolina State Univ., Raleigh.

WHERE DO WE GO FROM HERE?

Donald W. Woodard[1]

INTRODUCTION

"Where Do We Go From Here?" I accepted this title with some apprehension. After consulting my crystal ball, I concluded that sufficient information was available, as today's papers indicate, to allow me to discuss some thoughts about where future research would be fruitful. While I was preparing this paper, the prolonged heat and pollution alert in the Washington, D.C., metropolitan area may have obscured my view of the crystal ball. Nonetheless, I do have some thoughts about where we should go from here.

The information presented by the previous speakers summarizes much of what is presently known about the use of vegetation for engineering purposes in estuarine and coastal areas of the United States. Vegetation will trap eolian-transported sand to build dunes; soil surfaces may be stabilized by the establishment of vegetation; vegetation can reduce water velocities, thereby filtering fine sediments; vegetation scavenges nutrients from the water to make them available eventually to other organisms; vegetation attenuates short-period wave attack to reduce shoreline erosion; and vegetation in low-lying areas provides short-term water storage that may diminish some storm surges. Concomitant benefits of habitat restoration or creation, detritus production, and other environmental benefits are well known to this group so there is little need to develop these further.

I have divided the remaining material into two sections: vegetation and application. The vegetation section has been divided into three subsections based upon the zone of growth, terrestrial, emergent, and submergent vegetation.

1. U. S. Army Coastal Engineering Research Center, Kingman Building, Fort Belvoir, Virginia 22060.

VEGETATION

Terrestrial Vegetation

Terrestrial vegetation has been used primarily for two purposes; to stabilize unconsolidated soil surfaces and to trap sand to form dunes. Grasses are the best plants for stabilizing soil surfaces and trapping sand on coastal and estuarine sites. Commercially available grasses suitable for stabilization require sustained maintenance. I believe we have overlooked several promising low-growing species that tend to form a mat-like cover, increase through rhizomes or stolens, are native to the near-shore environment, and are adapted to such stresses of the estuarine and coastal environment as salt spray and infrequent inundation by sea water, sand, or silts. These species include *Distichlis spicata* (L.) Greene, *Panicum amarum* Ell., *P. repens* L., *Paspalum monostachyum,* Vasey, *P. vaginatum,* Sw., and *Sporobolus virginicuus,* (L.) Kunth. The material presented by today's speakers indicates the progress that has been made in the use of some alternative species for stabilization. However, information is still needed on transplanting, sprigging, and seeding techniques; maintenance requirements; palatability and suitability for use in grazing systems; and susceptibility to and rate of recovery from trampling. I might add at this point that in discussions with property owners about the use of certain native plants for stabilization, their greatest concern seems to be the necessity of planting a low-growing lawn grass "so snakes will stay out".

For the other purpose of terrestrial vegetation, i.e. trapping sand to form dunes, beach grasses are generally used. Dunes increase the elevation of coastal areas to provide safety from frequent flooding and short-term protection from wave-caused damage to property landward of them. Protective stabilized sand dikes have been in use for centuries in low areas bordering the English Channel and the North Sea.

The most successfully and widely used beach grasses are two species of *Ammophila*; *A. arenaria* (L.) Link, and *A. breviligulata,* Ferald. European beach grass or marram grass (*A. arenaria* (L.) Link) is widely used along the coasts of Europe, Africa, Japan, Australia, New Zealand, and the western United States. American beach grass has been used successfully almost entirely within the limits of its geographic range, the mid and upper Atlantic coast and the Great Lakes region. The only notable extension of its use from its natural distribution has been southward into the Carolinas. Recent work along the mid and south Atlantic coasts and the Gulf coasts within the United States has expanded the list of useful beach grasses to include panic beach grass (*Panicum amarum,* Ell.) and sea oats (*Uniola paniculata* L.). In general, the agronomic techniques for propagating, establishing, and maintaining these four species of beach grasses are known.

However, not enough is known about design factors and environmental interactions. Needed design factors include the dune width and elevation necessary to maintain the integrity of a foredune system during a ten-year storm, a fifty-year storm; long term trapping rates and accompanying growth of foredunes; variation in trapping rates with geographical area, changes in the directional trend of beaches, or beach grass species; and the dimensions of foredune necessary along accreting or eroding shorelines. Simultaneously, cost information must be developed for the establishment and maintenance of foredune systems so that prudent, economic decisions can be made. Environmental considerations concerning the role of foredunes and overwash have already been well presented in this symposium.

Let me summarize by asking; What impact does sand storage in foredunes and the cessation of overwash have on mid- and back-island sites, sound and estuarine shorelines, human activities and use? Is grazing compatible with the protective function of foredunes? How much trampling or recreational activity will foredunes tolerate? Is it possible to have a deferred rotation system, that is, is an alternating grazing or recreation use and rest cycle, for barrier island areas?

Continued monitoring of existing and developing foredune systems will provide design information. An opportunity now exists along mid and south Padre Island, Texas, to investigate the development of a natural foredune system and its long-term influence on mid- and back-island areas and the adjacent Laguna Madre. The area is north of Mansfield Channel, which traverses Padre Island and provides access between the Gulf of Mexico and Port Mansfield or the Intercoastal Waterway. The first time I saw this area, it was nearly devoid of vegetation, a result, I believe, of the combination of drought, grazing, and storms. Grazing has been discontinued on adjacent grasslands to the north, so cattle will no longer wander periodically into this area to crop pioneer vegetation. Recent inspection of the area revealed that numerous dunelets are forming and growing. Study of this incipient foredune field would provide information useful to understanding present dunefield configurations and growth rate, and would allow an evaluation to be made of foredune-beach interactions. The continuous, well-developed foredunes and grasslands to the north would allow comparable studies. I have selected Padre Island, Texas, as my example because of my familiarity with it. Similar opportunities probably exist on other barrier islands and spits.

Emergent Vegetation

Intensive field studies to develop agronomic techniques for emergent vegetation have only recently been started. Emergent vegetation is here defined as those species that are flooded infrequently. To date, the studies have been mostly with *Spartina alterniflora,* Lois. and *S. patens* (Ait) Muhl, but they should be expanded to include two other species of *Spartina, S. foliosa* Trin. and *S. cynosuroides* (L.) Roth, and other genera; *Borrichia, Juncus, Phragmites, Typha,* and *Scirpus* are the most obvious. Investigation of transplanting and seeding techniques for establishment of mangroves is underway by personnel of the State of Florida's Department of Natural Resources.

The next important step will be to find out what duration and intensity of wave attack or current a given species on a particular substrate can tolerate without being irrecoverably damaged. The answer to this question must be known before reliable recommendations can be made about when and where marsh and shoreline vegetation can be used. The development of such criteria will require numerous field trials with adequate environmental monitoring. Controlled experiments in wave tanks, tidal basins, and flumes will be essential to the development of the necessary design criteria.

Questions concerning environment and economics must also be pursued. How much bottom area or water column can be utilized for emergent habitats? What trade-offs are acceptable? And again, what is the price of doing nothing, of doing something? What are the costs for establishment and maintenance? What is a reasonable life expectancy of a created marsh or shoreline under prevailing environmental conditions? I believe by now you will agree we have only just begun to grapple with many of these problems.

Submergent Vegetation

One of this morning's speakers has indicated how difficult it is to establish sea grasses. To my knowledge no one has induced any sea-gras colonization of sufficient size to develop operational guidelines. Even though investigations have been conducted during the last two decades, survival of transplants is variable and the factors responsible for success or failure are still conjectural.

This would seem to be an area for rewarding research. Numerous bottom areas appear to be susceptible to colonization but remain bare. Questions to be answered include propagative techniques, transplanting techniques including anchoring schemes, and determination of environments susceptible to colonization. Once we have answers to these questions, we will be able to investigate the stabilization and filtering capabilities of submergents on estuarine bottoms modified by engineering activities. As in the previous section, experiments in wave tanks and tidal basins may be necessary to obtain adequate design criteria once the practicality of using submergents for engineering purposes has been demonstrated.

APPLICATIONS

The next step in this presentation seemed to me to be: what opportunities will there be to utilize the information gained by the research recommended above? Let me start by reminding you that sand dikes and foredunes have been created to halt the encroachment of sand onto adjacent lands and to provide protection from flooding and wave attack.

As has been pointed out by another speaker, overwash may have some beneficial effects. The thought arises – if beneficial, could overwash be simulated by the deposition and distribution of dredged materials onto barrier islands?

A project currently underway in San Francisco Bay will, if successful, result in the creation of a marsh. Dredged material is being pumped into a former salt-evaporation pond to form the substrate for the establishment of an estuarine marsh. The material is being placed at an elevation necessary for transplanting native species. This study could point the way to one acceptable method for the disposal of dredged material from navigational maintenance projects.

A similar situation exists along the western shoreline of Lake Erie. It is proposed to construct a dike to replace a barrier beach that formerly protected a large marsh. When the dike has been built, the marsh is to be rejuvenated.

Estuarine shoreline erosion concerns many. Several of today's speakers have indicated how vegetation can be used to abate shoreline erosion caused by wind-wave attack.

Looking further into the future, if mainland superports become a reality, sizeable volumes of dredged material will be excavated. This non-contaminated material will be available for productive use; beach and shoreline nourishment, simulation of overwash, fill for creation of offshore protective structures, or even the formation of an estuarine-like body of water by the construction and placement of a series of sea islands. Whatever the fate of this material, all subaerial exposures will afford opportunities to establish vegetative cover as an alternative to all concrete surfaces. Small offshore islands constructed for industrial and transportation functions will probably afford few opportunities for vegetative establishment. Stress and wave energy will exceed the tolerances of most plants.

I am now shifting to a somewhat different application to complete this presentation: the determination of the relative value of one estuarine area to another. If it has been decided that the environment should be altered in some way – that a channel should be dredged, a spoil-disposal site should be created, or some facilities should be installed – what ecological guidelines should be followed in determining which part of the estuary is to be so modified? Which habitats are needed to maintain the diversity and viability of the estuary? Indeed, can the productivity of one habitat be increased to offset the loss of another? These are important questions to which answers are needed before rational decisions can be made.

SUMMARY

Today's presentations indicate progress in the manipulation of vegetation to augment, rehabilitate, and function in concert with engineering activities in order to maintain environmental conditions capatible with the resources, needs, and desires of the citizenry. Most progress has been with terrestiral species, but work on emergent and submergent species is off to a good start. In particular, the design and cost information needed before reliable, practical, and prudent use of vegetation can unquestionally be incorporated into engineering development and maintenance projects is lacking.

ACKNOWLEDGMENT

This report was prepared as a part of the research program of the U. S. Army Coastal Engineering Research Center of the U. S. Army Corps of Engineers. Permission was granted by the Chief of Engineers to publish this information.

AN OVERVIEW OF THE TECHNICAL ASPECTS OF THE CORPS OF ENGINEERS NATIONAL DREDGED MATERIAL RESEARCH PROGRAM

Conrad J. Kirby, John W. Keeley, and John Harrison[1]

ABSTRACT

The Chief of Engineers was authorized by section 123(i) of Public Law 91-611 to conduct a comprehensive research program related to dredging and the disposal of dredged materials. Phases I and II of the four-phase study identified the various problems associated with dredging and disposal, and developed a comprehensive plan of research to address them. The research effort (Phase III) was approved by the Office of Management and Budget in February 1973 and is being inplemented by the Office of Dredged Material Research (ODMR) of the Waterways Experiment Station. The ODMR has the responsibility of program planning, management, and research supervision of a comprehensive study designed to assess the significance and magnitude of the impact dredging and disposal operations have on the environment. Additionally, alternatives that have potential to enhance the environment will be developed, tested, and implemented.

INTRODUCTION

The vital role that the navigable waterways of the United States continue to play in the nation's economic growth is reflected in the fact that in the 20-year period from 1950 to 1970 total waterborne commerce increased by some 85 percent, and now exceeds 1.4 billion metric tons per year. In fulfilling its mission in the development and maintenance of these waterways, the Corps of

1. Waterways Experiment Station, Corps of Engineers, P. O. Box 631, Vicksburg, Mississippi 39180.

Engineers is responsible for dredging a current average of about 230,000,000 cu m in maintenance dredging operations and approximately 61,000,000 cu m in new-work dredging annually. The total annual cost exceeds $150,000,000. Dredging entails removing sediments from the bottoms of streams, lakes, and coastal waters, transporting them via ship, barge, or pipeline, and discharging them into water or onto land. It is usually done for the purpose of maintaining, improving, or extending navigable waterways or of providing construction materials such as sand, gravel, and shell. In recent years, as sediments in waterways and harbors have become polluted, a wide variety of questions has arisen over the nature and significance of the environmental impact that results from dredging and disposal operations.

Much of the concern over the actual dredging process is related to the possibility that benthic communities which are known to play an important role in the aquatic ecosystem, might be destroyed, as might such commercially valuable species such as oysters and clams. Although the direct effects of dredging on benthic organisms may appear to be obvious, there is little information available that permits prediction or assessment of their overall extent, significance, and duration. In addition to the concern with regard to the direct effects of dredging operations, there is concern over the possible indirect effects on aquatic communities. The potential for indirect effects is usually attributed to physical alterations of the environment, such as changes in bottom geometry and bottom substrate which trigger subsequent alterations in current patterns, salinity gradients, and the exchange of nutrients between bottom sediments and the overlying water. Any one of these physical changes can, either singly or in combination, initiate varying resonses within the biological communities. As an example, a change in the salt-water gradient may be beneficial for young fish and crab transport but detrimental to oysters because it allows predators to penetrate deeper. The current state of knowledge does not always make it possible to assess such effects definitely or to judge whether they are adverse, neutral, or beneficial.

Most of the concern over the dredging-disposal process is directed toward the effects of open-water disposal on water quality and aquatic organisms. It has long been known that, depending on individual circumstances, bottom sediments are continuously being resuspended by natural processes. Thus, under cursory examination, the open-water disposal of bottom sediments may be viewed as an extension of natural processes. However, in contrast to the natural phenomenon of sediment resuspension, open-water disposal often results in the resuspension of large volumes of sediments over a relatively short period of time in a relatively small area. Further, the dredging and redeposition of certain types of polluted sediment may convert a localized problem in a noncritical location to a serious regional problem, as pollutants are dispersed by currents and/or carried to such critically important areas as oyster grounds and coral reefs. The effects of open-water disposal are, therefore, often similar to those resulting from normal resuspension, but their intensity and range may be greater.

Another possible effect of open-water disposal is sediment buildup, which could result in the smothering of benthic organisms, changes in spawning areas, less diversity of habitat, and less or changed vegetative cover. Furthermore, increased levels of suspended solids reduce light penetration, which in turn interferes with biological productivity, decreases the availability of food, and alters the chemistry and temperature of the water. Finally, because some of the sediments of the nation's waters have become contaminated, there is concern that man-induced resuspension of such sediments may increase the possibility of these contaminants adversely influencing biological communities. Because of the poor understanding of the possible consequences of these changes, definitive research is needed to assess all aspects of open-water disposal of dredged sediments.

Primarily because of the concern over open-water disposal of polluted sediments, a trend toward land disposal has developed. Yet without definitive research it is not possible to determine, from an overall environmental viewpoint, in which cases land disposal is in fact a wise alternative to open-water disposal. Land disposal often involves marshlands or other wetlands, which are among the most biologically productive areas on earth. The effects of disposal on the role of marshlands as breeding areas, nurseries, and zones of high biological production are only marginally understood. Besides the rather special case of marsh disposal, there are other environmental concerns that are common to all types of land disposal and these must be considered. One of the more intensive concerns involves the possible pollution of ground water and its subsequent effects on man. Land disposal could alter vegetation assemblages and local relief, thereby triggering changes in drainage patterns and wildlife migration. The relocation of sediments from one biotype to another could be an alien intrusion of significant ecological concern. Finally, as is always the case, each of these alterations could initiate further sequences of events in both the terrestrial and aquatic regimes.

A consideration of the dredging-disposal process in even such broad philosophical terms readily points out a need for a more comprehensive understanding of the precise nature of the problems and a need to fill numerous gaps in knowledge regarding the significance of known or suspected environmental effects of dredging and disposal.

The Corps of Engineers was authorized by Section 123(i) of the Rivers and Harbors Act of 1970 (Public Law 91-611) to initiate a comprehensive program of research, study, and experimentation related to dredged material. The problem identification and the research effort developed and currently being implemented by the Office of Dredged Material Research of the Waterways Experiment Station are dedicated to these goals (1). An outline of the Dredged

Material Research Program is presented in Table 1 and a detailed discussion ot the initiated research follows in the text.

Aquatic Disposal and Its Impact on the Environment

In this age of "environmental awareness," which is evidenced by the public's concern over the wise utilization of the nation's resources, it is of little wonder that many questions have arisen concerning the dredging and aquatic disposal of some 191,000,000 cu m of bottom sediment per year. Indeed, the sheer volume appears sufficient to warrant research designed to determine what impact open-water dredging and disposal have on the environment. Unfortunately, many of the bottom sediments throughout the United States reflect the cultural pollution of the nation's waters to the extent that, by dredging, the Corps of Engineers is involved in waste management, treatment, and disposal. However, the concerns voiced to date go far beyond consideration of just the quantities or the pollution aspects involved in open-water dredging and disposal. Many of the concerns are now directed toward the effects dredging operations may have, either directly or indirectly, on the structure and function of biological communities. Both dredging and disposal are involved; as might be suspected, disposal is the subject of the greatest amount of concern.

For the purpose of this paper, open-water disposal operations are defined as those operations that result in the disposition of dredged material in the open ocean, bays, estuaries, inlet rivers, and lakes. From the standpoint of water-quality effects, this definition includes materials placed on beaches, marshes, along river edges, or any other type of unconfined land disposal in which the placed materials are subject to the influence of tides, river-stage fluctuations, or are readily washed back into the water by rainfall. Increased levels of suspended solids, sediment buildup, and oxygen depletion are the environmental effects of aquatic disposal that have most frequently been documented. However, many other potential problem areas have not been well documented. Concern over the existence of or potential for long-term or indirect adverse effects on aquatic communities often stems from observed phenomena that indicate the possibility of such adverse environmental effects. In other cases, such concern is often largely the result of conceived hypotheses. Regardless of the basis for concern, the potential for long-term adverse effects resulting from open-water disposal of dredged materal does exist. It is therefore of paramount importance that studies be conducted to provide a better understanding of the causative mechanisms involved and of the nature and significance of any adverse environmental effects. One of the greatest concerns is the possibility of long-term effects on water quality and associated flora and fauna resulting from the disposal of materials classified as polluted. As mentioned above, such short-term effects as the decrease in dissolved oxygen

TABLE 1.

Outline of Dredged Material Research Program

	Research Area		Research Task
1.	Environmental Impact of Open Water Disposal	A.	Evaluation of Disposal Sites (1)*
		B.	Fate of Dredged Materials (1)
		C.	Effects of Dredging and Disposal on Water Quality (1)
		D.	Effects of Dredging and Disposal on Aquatic Organisms (1)
		E.	Pollution Evaluation
2.	Environmental Impact of Land Disposal	A.	Environmental Impact Studies (1)
		B.	Marsh Disposal Research (1)
		C.	Containment Area Operation Research (1)
3.	New Disposal Concepts	A.	Open Water Disposal Research (2)
		B.	Inland Disposal Research (3)
		C.	Coastal Erosion Control Studies (3)
4.	Productive Uses of Dredged Material	A.	Artificial Habitat Creation Research (1)
		B.	Habitat Enhancement Research (2)
		C.	Land Improvement Research (3)
		D.	Products Research (2)
5.	Multiple Utilization Concepts	A.	Dredged Material Drainage/Quality Improvement Research (2)
		B.	Wildlife Habitat Program Studies (1)
		C.	Disposal Area Reuse Research (1)
		D.	Disposal Area Subsequent Use Research (2)
		E.	Disposal Area Enhancement (2)
6.	Treatment Techniques and Equipment	A.	Dredged Material Dewatering and Related Research (2)
		B.	Pollutant Constituent Removal Research (1)
		C.	Turbidity Control Research (1)
7.	Dredging/Disposal Equipment and Techniques	A.	Dredge Plant Related Studies (3)
		B.	Accessory Equipment Research (2)
		C.	Dredged Material Transport Concept Research (4)

*Numbers in parentheses indicate the beginning year of the research task.

have been documented. However, these effects usually result from organic materials (such as sewage sludge) mixed or incorporated into the bottom sediment. The possibility of long-term effects is usually attributed to the presence of other constituents in the sediments, such as biostimulants and toxins. Unlike sewage sludge, which is incorporated into the sediment and on mixing can exert a demand for dissolved oxygen, biostimulants and toxins are often chemically or physically sorbed within the sediment matrix. It is generally agreed that constituents sorbed on the sediment particles are not as readily available to the food chain as are dissolved materials. The question then becomes, "Under what circumstances and to what extent are the constituent-to-sediment attachment mechanisms altered in ways that could cause the release of constituents to the water?"

It has been estimated that dredged material accounts for 80 percent by weight of all materials disposed of in coastal waters and that the vast majority of this amount is placed in water less than 30 m deep. Of particular concern are the questions regarding the effects of coastal disposal on water quality and the subsequent effects of spoil sites on the structure and function of benthic communities. Because of these concerns, various alternatives have been proposed, one of which is that dredged materials should be dispersed instead of confined to a delineated site. Other suggestions include transporting the materials to deeper waters or confining them on land. However, there is little information against which alternative procedures can be compared and evaluated in all the nation's coastal waters. Therefore, research is being initiated to determine the magnitude and extent of effects of spoil sites on organisms, on the quality of surrounding water, and on the rate, diversity, and extent of colonization of such sites by benthic flora and fauna. It is generally agreed that, on account of seasonal factors and the subtle nature of the changes in benthic communities, on-site, long-term field studies are needed if cause and effect relations are to be adequately documented and assessed. However, because of the high cost of long-term field studies and the great number and diversity of coastal sites, a preliminary survey of all known sites is being conducted. The results of this survey will be used to consolidate all currently available information pertient to the coastal dumping of dredged material and to design monitoring studies for a few (perhaps six) representative disposal sites. Emphasis will be placed on assessing the physical, chemical, and biological factors that are currently thought to serve as indices of or control factors for benthic colonization.

In order to assess the environmental impact of open-water disposal on a case-by-case basis, methods are needed to predict the physical fate (location as a function of time) of dredged material. To aid in predicting the long-term effects of open-water disposal, methods are needed to predict subsequent resuspension of dredged materials and their distribution. Several mathematically based simulation procedures designed to predict physical fate of various materials disposed of in open waters are in existence. However, few of them have been developed, modified, or used in connection with the disposal of dredged materials. Similarly, very little has been done to develop techniques for predicting resuspension and subsequent transport of disposed dredged materials. To address this problem, research has been initiated to determine the fate of dredged materials by developing techniques for measuring their spatial and temporal distribution as a function of various hydrological regimes. This research will first be directed toward assessing presently available dispersion models and delineating areas for further developmental research. Refinements will be made to present models, and, where needed, additional techniques will be developed. Long-term research efforts will be directed toward the development of techniques for predicting sediment resuspension and transport with emphasis placed on determining what factors control or are related to resuspension.

As mentioned earlier, one of the primary concerns regarding open-water disposal is the possibility of water pollution. Information is needed on the quality and quantity of dredged materials, the method of their disposal, and the nature of the aquatic media in which they are disposed. In addition, fundamental information is needed on the constituent-to-sediment attachment mechanisms. Finally, methods need to be developed to predict, prior to disposal, the nature and significance of the effects on water quality.

In order to determine what effects the disposal of dredged materials have on water quality, research has been initiated to determine, on a regional basis, the short- and long-term effects the disposal of dredged material containing various contaminants has on water quality. Quantitative chemical analyses need to be modified or developed to investigate varying types of sediments. Such analyses include: determination of sediment sorption capacities, determination of ionic and cationic sediment exchange capacities, sediment elemental partitioning studies, and development of sediment leaching procedures. Particular emphasis will be placed on the development of laboratory leaching procedures, verified by field pilot studies, that will enable the effects of water quality to be predicted prior to dredging.

In addition to concern over the adverse effects that changes in water quality caused by the discharge of contaminated dredged material may have on aquatic organisms, there is concern over the physical effects that such operations may have on the structure and function of aquatic communities. Such effects may be caused by physical alterations in bottom geometry, bottom substrate, and current patterns. The current state of knowledge does not allow one to make a definite assessment of these effects or to judge whether they are adverse, neutral, or beneficial. Research has been initiated to determine what effects the physical impact of dredged-material disposal has on the structure and function of biological communities. Again, initial emphasis in the research program is being placed on laboratory studies designed to define and measure the influence of open-water disposal of dredged materials on biological communities. These laboratory studies will be used to provide experimental design information for subsequent highly controlled field studies. In all laboratory and field studies, particular emphasis will be placed on gathering information that will define, insofar as possible, cause and effect relationships.

Because of the concern over the detrimental effects that open-water disposal of contaminated dredged material might have on water quality and aquatic organisms, the Congress has recently passed legislation requiring the Environmental Protection Agency and the Corps of Engineers to establish criteria that regulate the disposal of such materials. While there is complete agreement over the need for criteria, there is currently little definitive information available on the effects of disposal of dredged material. Therefore, regulatory agencies faced with the legislative requirement of establishing dredged-material criteria must strive to establish meaningful ones based on the best possible knowledge, and avoid the tendency to set forth criteria that precede the current state of the art. Ideally, the Corps of Engineers envisions criteria that will provide the best possible guidance as well as serve as a vehicle to generate information that will (a) provide a basis for quantitative evaluation of water quality and aquatic organisms in terms of various use requirements; (b) provide information that would aid resource managers in viewing dredged material as a potential resource; (c) provide base-line conditions of value to the scientific community, thereby, it is to be hoped, reducing costs of future research investigations; and (d) provide a basis for making policy decisions. Recent research directed toward determining the location of pollutants within sediments (elemental partitioning) has indicated that the pollutants are closely associated with fine-grained sediments and organic matter. However, there is only a poor understanding of the specific location and form of pollutants in varying sediment types. Research has been initiated to provide information on the changes that occur in pollutant partitioning during the dredging process as well as the transport and deposition following open-water disposal. Although initially this research will be of a fundamental nature, its results will be used to update criteria for dredged-material pollution and to aid in determining the environmental impact associated with dredging operations. Research has also been initiated to develop a leaching test designed to measure that amount of any chemical constituent that, because of dredging, migrates from the solid phase (such as sorbed or crystalline matrix) to the dissolved phase. Such a test is being designed to measure the change in availability of sediment contaminants that might result from dredging and disposal operations.

Land Disposal and Its Impact on the Environment

Decisions concerning land or open-water disposal have, in the past, been based primarily on economic considerations; however, land disposal has more recently been recommended as the preferred method for disposing of polluted dredged material because of the potentially adverse environmental effects of open-water disposal. The land disposal alternative has been selected in most instances without any knowledge of what its environmental effects might be. In order to ensure that land disposal is an acceptable alternative to open-water disposal and to make it more readily acceptable to the public, research to identify and evaluate the broad, basic relationships between the disposal site and all aspects of the surrounding environment has been started.

The disposal of dredged material on coastal marshes is decreasing because of adverse environmental consequences. Since it is generally recognized that marshes and wetland systems serve as nurseries, breeding areas, and zones of high biological productivity for coastal zones and near-shore areas, the disposal of dredged material in these environments is discouraged. There are situations, however, where wetland disposal of dredged material may be the only practical method. In order to provide the district and division offices of the Corps of Engineers with guidance on disposing of material so that it will have minimal adverse effects, research into the biological, social, and economic implications of disposal on marshes and wetlands is underway.

Examination of existing confined disposal areas and operations has shown that the intent of confinement has rarely been met, regardless of the desired goals of efficiency, economy, safety, and environmental control. More effective long-term solutions are sought through research aimed at optimizing facility size and shape, weir design, location, and operation, and filling rates and patterns. Both interim and permanent solutions to short- and long-term needs in containment area management and environmental control will be explored, tested, and implemented.

Productive Uses of Dredged Material

Mounting evidence indicates that dredged material can and should be considred a manageable resource of environmentally compatible disposal operations with documented evidence (4, 5). Environmentally compatible disposal of dredged material in open water could include marsh creation, spoil-island development, beach nourishment, and substrate enhancement.

The loss of coastal salt marshes and wetlands through the combined efforts of both nature and man is fully expected to be a continuing process. The value of these coastal marshes as high-productivity food sources, wildlife habitats, areas for fishing and hunting, and as protection against storms and coastal erosion is only now becoming fully understood. A research program that has been designed will provide more definitive information on using dredged material to create marshes and spoil islands. The deliberate creation of marshes and spoil islands becomes an attractive alternative to either open-water disposal or disposal on marsh surfaces, because a definite need has been established for these types of habitats. These appear to be the most promising of the concepts that have evolved since they would involve a volumetrically significant amount of dredged material. Most of the research completed was done on relatively clean sand in protected areas of estuaries (5); however, this research effort will attempt to expand the basic concepts to areas in which the consistency of the dredged material or the conditions of the surrounding environment are less than desirable.

The disposal of dredged material in upland areas, on wetlands, and in navigable waterways in many cases results in altered physical and chemical characteristics of the substrate. While there are known and suspected cases of consequential damage or undesirable changes in indigenous flora and fauna, there are also known cases of beneficial changes and increased biological productivity (2, 3). Many of the unique situations that exist in terrestrial, wetland, and water-bottom habitats lend themselves to enhancement. For example, flood-control structures in many riverine and coastal wetland areas have restricted or prevented overbank flooding, with the result that there has been deprivation of natural nutrient and sediment replenishment; the controlled disposal of dredged materials might well be a method of alleviating this situation to some extent. The possibility that water bottom could be enhanced by planned alteration of the substrate is also particularly attractive. Besides improving sports and commercial fishing through a process analogous to artificial-reef construction and the enhancement of benthic communities, it might be possible to improve bottom topography and facilitate crop harvesting by coating polluted or undesirable substrates.

Multiple-Utilization Concepts Involving Confined Disposal Areas

It is estimated that approximately 200 active Corps of Engineers dredging projects rely in whole or in part on the confined disposal of dredged materials. Additionally, 2,875 hectares of new land are acquired each year to contain the volume of material that is generated solely during maintenance-dredging operations. Mounting evidence indicates that the percentage of dredged material that will have to be confined on land will increase rather substantially in the foreseeable future. A basic assumption and theme of this research program is that, in those cases where confined disposal is essential, adverse public reaction as well as undesirable environmental qualities can be considerably mitigated by carefully conceived, planned, and executed multiple-utilization schemes. In keeping with the philosophy of the research program, a series of practical alternatives should be available for consideration in the decision-making process regarding the disposal of dredged material.

Multiple-utilization involving wildlife and fisheries conservation appears to be a feasible and most desirable concept at this time, particularly in regard to the nation's coastal zone. The nature and location of many confined disposal areas make them quite amenable for use as wildlife habitats. The state of the art of game and wildlife management is sufficient to demonstrate adequately the benefits that can be derived through such techniques as water-level control and the creation of food, nesting or breeding grounds, and shelter or refuge. Confined disposal areas can and already have, through largely unplanned efforts, provided these desirable conditions (J. F. Parnell and R. F. Soots, personal communication, 1972). These disposal areas often are large tracts of undeveloped and unpopulated land in the midst of spreading urbanization and industrialization. A research effort currently in progress will attempt to optimize the planned utilization of these areas for a wide spectrum of wildlife enhancement, while maintaining campatibility with requirements for dredged-material disposal. Particular attention will be devoted to determining the physical environmental needs of various species' life functions such as breeding, nesting, feeding, resting, and predator protection, the ultimate goal being to design and test several concepts for producing habitats in various environmental settings.

The development of public recreational areas emerges as an attractive utilization concept that would be applicable to a number of places and would permit cognizance of regional needs, distribution of dredging projects, and legal and policy constraints. These areas could include waterfront or island parks, amusement parks, camping and picnic grounds, playgrounds, parking areas, and a variety of other uses or activities that are well sutied for staged construction on relatively poor foundations. Research efforts are designed to provide a choice of economical, practical, and aesthetically acceptable retention structure designs.

Treatment Techniques and Equipment

Although little is known of the specific nature of polluted dredged materials and, consequently, of the effects of land or aquatic disposal, there doubtless will remain many cases where treatment of dredged materials is necessary. Such treatment is at once both similar to and vastly different from the treatment of domestic or industrial sewage. The pollutants found in dredged materials are, for the most part, the same types found in most domestic and industrial waste. Consequently, the fundamental basis of the processes needed to treat dredged materials is essentially the same as that for the treatment of conventional wastes. The major differences lie in the operational procedures. This is primarily because conventional treatment facilities can be designed for usually uniform loading rates and waste characteristics. In the case of most dredging operations, however, tremendous volumes with highly variable characteristics are produced in a short period of time. Full-scale field pilot tests are being designed to determine the technical, economical, and ecological feasibility of treatment in connection with various types of aquatic and land disposal operations. Presently available chemical, physical, and biological treatment processes are being assessed, and those that indicate promise for treating dredged materials will be selected for subsequent modification and development. Also, laboratory bench studies are early efforts designed to provide information on the basic effectiveness of various known treatment processes on arying types of dredged materials. Design information necessary to scale up to field pilot studies will be provided.

SUMMARY

Realization that dredging and disposal processes are highly complex, the urgency of providing definitive information on the environmental impact of current and potential disposal methods, and the need to develop techniques for mitigating potential adverse effects have led to a comprehensive five-year research program.

Because dredging and the disposal of dredged materials occur in such highly variable environmental situations, it was generally accepted that a universally applicable methodology cannot be satisfactorily developed. Consequently, it was concluded that a broad-based program of research was required to provide definitive information on the environmental impact of dredging and of dredged-material disposal and to develop technically satisfactory, environmentally compatible, and economically feasible dredging and disposal alternatives, including consideration of dredged material as a manageable resource.

This research effort will provide solutions to many problems within two years and to some of the more difficult and complex ones within five years. The plan is by necessity a dynamic one and is being continuously adjusted as more information is developed from early research under this study or from other sources.

REFERENCES

1. Boyd, M. B., Saucier, R. T., Keeley, J. W., Montgomery, R. L., Brown, R. D., Mathis, D. B., and Gruice, C. J.
   1972 Disposal of dredged spoil, problem identification and assessment and research program development. Technical Report H-72-8, U. S. Army Engineer Waterways Experiment Station, Vicksburg, Mississippi.
2. Cronin, L. E., Gunter, G., and Hopkins, S. H.
   1971 Effects of engineering activities on coastal ecology. Report to

Office, Chief of Engineers, U. S. Army.

3. Howell, B. R. and Skelton, R. G. S.
   1970 The effect of china clay on the bottom fauna of St. Amstell and Mevagissy bays. **Jour. Mar. Biol. Ass.,** 50: 593-607.

4. Windom H. L.
   1972 Environmental response of salt marshes to deposition of dredged materials. **Amer. Soc. Civil Eng. Nat. Water Resour. Eng. Conf.**

5. Woodhouse, W. W., Seneca, E. D., and Broome, S. W.
   1972 Marsh building with dredge spoil in North Carolina. Bulletin R-2-72, U. S. Army Coastal Engineering Research Center, Washington, D. C.

ASPECTS OF DREDGED MATERIAL RESEARCH IN NEW ENGLAND

by

Carl G. Hard[1]

ABSTRACT

The New England Division of the Corps of Engineers has had a dredging research program going for several years. Early emphasis was on gross bathymetry, turbidity, and the repopulation of benthic organisms. Present orientation is towards offshore physical and soil mechanics and mathematical modeling pointing towards improved disposal equipment and methods. The program is understaffed but receives good cooperation from other agencies and from the National Science Foundation.

RESEARCH

I was invited to present any results or preliminary conclusions of general interest stemming from our research on dredge materials in New England. Upon considering this, I was inclined to agree with Leonardo da Vinci, who said, "Nothing can be written as the result of new researches" (1). The reports that we compiled in our initial efforts resemble pages of scattered notes rather than coherent literature. The semanticists Arthur Korzybski and Rudolf Carnap expressed the idea that a person who is bound up within a system of ideas cannot explain things with the degree of understanding that he would have if he could stand aside and view things from without. I must confess to having been bound up largely within the engineers' spectrum for the better part of the past

1. Project Manager, New England Division, Corps of Engineers, 424 Trapelo Road, Waltham, Massachusetts 02154.

two decades and now I am in the process of trying to transcend it. Queries such as, "what is the environmental impact?" and "what are the alternatives to the proposed action?" encourage such a basic change in orientation. I think that it is even harder for an engineer than it is for an industrialist to have to make such a change. Both are pragmatists while scientists are of the inquiring mind, and we have been forced to try to answer questions which never before have been addressed soberly.

The New England Division of the Corps of Engineers did not elect to embark on a research program if, indeed, our work in this area can be called a program at all. It is a fire-fighting operation and we have had to seek out specialists, most of them in the field of marine biology, for our answers. Some of these scientists are on this conference program and can speak very well for themselves regarding their findings.

Our initial efforts have been studies in the areas of mortality and repopulation of benthic organisms, topography and surficial character of spoil mounds, turbidity, rate and degree of recovery, potential toxicity of harbor sediments, physical and chemical oceanography, and fisheries and nutrient studies. Random findings, although not all are conclusive, are that benthic organisms repopulate dumping grounds at fairly predicable rates, density, and diversity; the best commercial lobstering grounds tend to be relict or recent dumping grounds; spoil piles in open ocean may be fairly stable below wave base; quickly established bio-fabric of spoil surfaces provides erosion resistance; so called "sludges" (in the New England area) are often black muds 4,000 or so years old, rather than sewage; clamshell-dredged mud is apt to drop in clumps rather than disperse all over the place; the worms in harbor mud tend to improve sport fishery by chumming fish when put in the dumping ground; and localized nutrients from scow discharges are not apt to trigger massive red tides.

In the past three years, the New England Division has sponsored three conferences on ocean disposal. The first one, at Woods Hole, seems to have been used to establish basic consideration for ocean dumping criteria in the eyes of the regulatory agencies, hence has some value as a first cut. People tend however, to take it too literally, as though it were an end point rather than just a start. Our most recent conference at the Maine Department of Sea and Shore Fisheries, uncloaked a desire for a whole-systems, quantitative approach to ocean-dumping research, and to regional oceanographic rather than site-oriented background studies. Such an integration certainly poses a problem for the imagination because most marine studies follow a routine station-data approach, and the principal investigator may be suspected of hoping that data prolixity will mask any lack of dynamic, creative interpretation. This is a criticism of the field in general. It in no way applies to such creative work as that of Saul Saila, Don Rhoads, or any of the other researchers in our program.

I feel that our initiation stage is at an end. We are trying to put together a study program that will involve bioassay of selected organisms in a sedimentation flume whose turbidity is regulated automatically and is spiked with selected types of clay or slit. Correlated studies on similar organisms are to be run in a controlled-nutrient environment. Then, when the results have been standardized, turbidity concentrations are to be varied with combinations of nutrients to produce synergistic effects that will give a basis for predicting the worst effects that dumping known harbor sediments will have on the organisms of the area. Other plans call for studies of the continental shelf, roughly fifty kilometers off the mainland, for the purpose of modeling dredge-disposal sites, based on modification of the basic equations appropriate to the diffusion-advection model discussed by Pritchard (2). The purpose of this is to determine how large an area is influenced and to what degree by a given type of dumping. If this materializes, it will be a large endeavor and we hope to get support from the national Dredged Material Research Program for support. Field work far offshore, as you know, is quite expensive and it will have to be extensive. It is hoped that the study will compare the effects of scow discharge from the water surface with those of discharge near the ocean bottom, in order to determine how future dredging equipment should be modified. The program stresses soil mechanics as they relate to transfer from dredge site to scow, to the dumping event, and to the physics of bottom displacement, dispersal phenomena, and the state of the dumping ground after the placement of large quantities of spoil. Tentative sites are at thirty and sixty meters water depth. The thirty-meter site is an active trawling ground and will require a related fisheries study. Not the least problem is how, other than hydraulically, to dump mud near the bottom without getting into astronomical equipment costs. This problem has been referred to the Marine Design Division of the Corps of Engineers in Philadelphia. A reasonably inexpensive solution might be to have net-encased polyethylene containers in conventional bottom-dump scows made to open near the bottom by means of a ripcord attached to the scow. Needless to say, such a program will have to be tied to a major dredging job that involves the appropriate sediment type. We have selected organic silt for two reasons. The first is that it is the material that we dredge most often in New England (which explains why we can't use it on land); the second is that it is apt to contain pollutants. Furthermore, it is more dispersive than coarse, granular materials. I think that covers the items that concern us most at this time, but we are making a start on a small scale in some other areas, such as sulphide fixation of metals in marine sediments, and flume tests of erodibility to see if Hjulstrum's curve is really true, since many use it.

MANAGEMENT

As talk spread that we had a research program, critics complained that we had no coherent overview. This was true. It also was true that we had no specific funding allocations, no internal organization to manage a research program, and no opportunity to hire help. The same situation more or less prevails today, except that the research activity has been dignified by being taken out of the soils lab and placed in the Project Management Branch of the Engineering Division. In the absence of a staff we have managed to adopt our various contractors as sounding boards, both informally on a daily basis, and in our annual ocean-disposal conferences. For scientific guidance we rely also on the advice of the Interagency Sub-Committee on Dredging and Ocean Dumping. It consists of four men, each appointed by his agency's regional Director. The agencies are EPA, NOAA, the Interior Department, and the Corps of Engineers. This has greatly improved coordination. State agency members sit in as sub-committee participants when appropriate. The New England Division program is being coordinated with relevant activity of the National Science Foundation.

REFERENCES

1. da Vinci, Leonardo
Circa 1500 A.D. The Notebooks of Leonardo da Vinci, Vol. I, p. 63. (ed. MacCurdy, E.) George Braziller, New York.

2. Pritchard, D. W.
1965 Dispersion and flushing of pollutants. In **Evaluation of Present State of Knowledge of Factors Affecting Tidal Hydraulics and Related Phenomena.** VIII. p. 1-39. (ed. Wicker, C. F.) Committee of Tidal Hydraulics, Corps of Engineers, U. S. Army, Vicksburg, Mississippi.

EFFECTS OF SUSPENDED AND DEPOSITED SEDIMENTS

ON ESTUARINE ENVIRONMENTS[1]

J. A. Sherk[2]

J. M. O'Connor[3]

and

D. A. Neumann[3]

ABSTRACT

Static bioassays conducted with fuller's earth suspensions on white perch, spot, silversides, bay anchovies, mummichogs, striped killifish, and menhaden showed that significant mortality in five of the seven species could be caused by concentrations of natural suspended solids typically found in estuarine systems during flooding, dredging, and spoil disposal. Lethal concentrations ranged from a low of 0.58 g/l fuller's earth (24 hr LC_{10}) for silversides to 24.5 g/l fuller's earth (24 hr LC_{10}) for mummichogs. Fishes were classified as either tolerant (24 hr $LC_{10} > 10$ g/l), sensitive (24 hr $LC_{10} < 10 > 1.0$ g/l), or highly sensitive (24 hr $LC_{10} < 1.0$ g/l) to fuller's earth. Generally, bottom-dwelling species were most tolerant to suspended solids; filter feeders were most sensitive. Early-life stages were more sensitive to suspended solids than adults.

Exposure to sublethal fuller's earth concentrations significantly increased

1. Contribution no. 575 of the Natural Resources Institute, University of Maryland.

2. Natural Resources Institute, University of Maryland, Chesapeake Biological Laboratory, Solomons, Maryland 20688.

3. QLM Laboratories, Inc., Hudson River at Burd Street, Nyack, New York 10960.

hematocrit value, hemoglobin concentration, and erythrocyte numbers in the blood of white perch, hogchokers, and striped killifish, but not of spot and striped bass. Evidence of O_2-CO_2 transfer interference during exposure to sublethal concentrations of fuller's earth was exhibited by the gills of white perch, which showed tissue disruption and increased mucus production.

Suspensions of fuller's earth, fine sand, and Patuxent River silt (> 250 mg/l) caused significant reductions in ingestion of radio-labelled ($NaHC^{14}O_3$) *Monochrysis lutheri* by the copepods *Eurytemora affinis* and *Acartia tonsa.* Differences in uptake between the two species may have been related to their different life habitats, although both are non-selective suspension feeders.

INTRODUCTION

Particulate organic and inorganic material can be introduced to or resuspended in the estuarine environment by nature (floods, storms, winds, tidal scour) or by man (dredging, dumping, filling, sewage and industrial discharges). The complex physical and chemical properties of suspended and resuspended sediments and substratum changes associated with deposition can have both direct and indirect effects on estuarine organisms. Even though the effects of these particles, or substances associated with them, on estuarine organisms are poorly understood, mortality, decreased yield, and interference with energy flow have been observed at estuarine areas selected for sediment-producing activities. Generally, these geographic sites have inherent physical, chemical, and biological limits beyond which significant effects will occur (26). The biological limits may be related to growth, survival, or reproductive aspects of various life-cycle stages in response to the quantity and quality of factors under investigation, length of exposure, and their interactions.

We have been investigating the effects of particle size and concentration of suspended solids on estuarine organisms independent of, and in addition to, complicating factors associated with natural sediments (12, 16, 30). These factors might include sorbed toxic metals, high biochemical oxygen demand, and nutrient enrichment. Our approach has been first, to identify lethal and sublethal biological effects of mineral solids similar in particle size to sediments likely to be found in, or added to, estuarine systems; and second, to study the effects of natural sediments in identical experiments.

LETHAL EFFECTS OF SUSPENDED SOLIDS ON ESTUARINE FISHES

Concentration-dependent mortality rates were determined for seven estuarine fish species by static bioassays with fuller's earth (Table 1). The norm for response, the LC_{50} (5) varied by a factor of 20 among them. Of these species, the mummichog, striped killifish, and Atlantic silverside commonly inhabit the

TABLE 1.

Lethal (10, 50, 90% mortality) concentration (g/l) for fishes exposed to fuller's earth (<0.5 μmed. size, 82%<2 μ), Patuxent River silt (0.78 μmed. size, 72%<2 μ), and natural sediment from the Chesapeake and Delaware Canal. Tests were run at 25 C, 5.5°/oo salinity, except for Acnhovy (18 C, 22°/oo), Silversides (18-22 C, 22°/oo), Bass Larvae (20 C, 1.0°/oo), and Perch Larvae (15 C, 1.0°/oo).

Species	Duration	LC_{10}	LC_{50}	LC_{90}	Sediment
Spot	24 hr	13.09	20.34	—	fuller's earth
Spot	24 hr	68.75	88.00	112.63	Patuxent silt
Spot	48 hr	1.14	1.89	3.17	fuller's earth
White Perch (adults)	24 hr	3.05	9.85	31.81	fuller's earth
White Perch (larvae)*	24 hr	—	3.73	—	natural sediment
White Perch (adults)	48 hr	0.67	2.96	13.06	fuller's earth
White Perch (larvae)*	48 hr	—	1.55	—	natural sediment
Striped bass (larvae)*	24 hr	—	4.85	—	natural sediment
Striped Bass (larvae)*	48 hr	—	2.80	—	natural sediment
Striped Killifish	24 hr	23.77	38.19	61.36	fuller's earth
Striped Killifish	24 hr	97.20	128.20	169.30	Patuxent silt
Common Mummichog	24 hr	24.47	39.00	62.17	fuller's earth
Atlantic Silversides	24 hr	0.58	2.50	10.00	fuller's earth
Menhaden (juvenile)	24 hr	1.54	2.47	3.96	fuller's earth
Bay Anchovy	24 hr	2.31	4.71	9.60	fuller's earth

*Morgan, Rasin, and Noe (17).

shore zone of the Patuxent estuary. The two killifish were the most tolerant, 50 percent surviving almost 40 g/l fuller's earth and 90 percent surviving more than 23 g/l. The silverside was the most sensitive species with 10 percent mortality occurring in 24 hours at 0.58 g/l fuller's earth, and 50 percent mortality at 2.50 g/l.

A wide range of sensitivity to suspended solids was evident among the white perch, spot, and bay anchovy, which are generally widely distributed and often found in deep water. The 24-hour LC_{50} for spot was above 20 g/l, white perch 24-hour LC_{50} values were below 10 g/l, and anchovies had an LC_{50} value of 4.7 g/l. LC_{10} of the spot was 13.1 g/l. Forty-eight hour LC_{50} values for white perch and spot were similar, 2.96 g/l and 1.9 g/l, respectively. These values differed by 50 percent, whereas the 24-hour LC_{50} values differed by more than 100 percent. While the 24-hour LC_{50} of white perch was less than for spot, the sensitivity was reversed for the 48-hour value. The relationship of the LC_{10}

values from perch and spot was more consistent for the 24-hour vs 48-hour exposure in that the 10 percent mortality value for perch was less than that for spot. However, as in the 24-hour LC_{10} determination, the differences in tolerance of the two species were not as great at the LC_{10} level as at the LC_{50} level.

Lethal effects of natural muds were assayed in perch, killifish, and spot. LC_{10}, LC_{50}, and LC_{90} values for exposure to natural sediments were generally greater than their respective values in fuller's earth. Also, larval white perch were more sensitive to natural sediment concentrations than the adults.

While LC_{50} values may serve as estimates of toxicity or lethality (5), the 50 percent survival norm is, in reality, insufficient as an estimator of the real effect on fishes of suspended solids or any other potentially lethal substance. The major influence of mature fishes on standing stock is in the number of juveniles produced (24). Mortalities far less than 50 percent among stocks of adult fishes, when considered in addition to natural mortality and mortality due to fishing, may have adverse effects on the potential of fish stocks (2, 24).

Concentrations of suspended solids capable of causing 10 percent and 50 percent mortality among estuarine fishes can be maintained by natural estuarine systems, near dredging operations, or during times of excessively high run-off (14, 15, 28). Assuming that the species used in our experiments were representative of adult estuarine fishes, we have established tentative classes of fish according to their capacity to tolerate suspended solids concentrations based upon our work with fuller's earth. The classification is subjective and is based upon LC_{10} values, as we consider 10 percent motality in addition to natural mortality rates a more realistic maximum than the 50 percent mortality limit.

In our studies, tolerant species were the mummichog, perch, and striped killifish (Table 2). The cunner, four-spined stickleback, and sheepshead minnow may also be considered tolerant (25). Other tolerant species included several that we have tested for suspension tolerance, but for which concentration-mortality curves have not been determined (oyster toadfish, hogchoker, and cusk eel). A common feature of these tolerant species is their habitat preference for the mud-water interface, where suspended-solids concentrations tend to be higher than elsewhere in the water column (14, 15). The killifish, hogchoker, and cusk eel frequently burrow into the substrate and remain covered for extended periods of time (10). The oyster toadfish is a bottom dweller and a relatively inactive organism.

Sensitive species were menhaden (juvenile), white perch, and bay anchovy. Common biological characteristics were difficult to ascertain for them, although their habitat preferences were quite similar to the suspension-tolerant spot. Three commercially important species, striped bass, croaker, and weakfish, may be in this class, but concentration-dependent mortality studies on them have not been completed. Mortality values of white perch and striped bass larvae exposed

TABLE 2.

Tentative suspended solids classification for estuarine fish based on 24-hour LC_{10} values from exposures in fuller's earth suspensions (28) (test temperature in parentheses).

CLASS I. Tolerant Species (> 10 g/l fuller's earth)	CLASS II. Sensitive Species (> 1.0 < 9.9 g/l fuller's earth)	CLASS III. Highly Sensitive Species (≤0.9 g/l fuller's earth)
Common Mummichog (25°C)	Adult White Perch (25°C)	Atlantic Silverside (18-22°C)
Striped Killifish (25°C)	Larval White Perch (15°C)**	Juvenile Bluefish (100% mortality 18 hr, 0.8 g/l, 25°C)
Spot (25°C)	Adult Striped Bass (25°C)	Young-of-the-Year White Perch (100% mortality, 20 hr, 0.75 g/l, 18°C)
Oyster Toadfish (25°C)	Larval Striped Bass (20°C)**	
Hogchoker (25°C)	Croaker (25°C)	
Cusk Eel (25°C)	Bay Anchovy (18°C)	
Cunner (15°C)	Juvenile Menhaden (25°C)	
Four-spined Stickleback (12°C)*		
Sheepshead Minnow (19°C)*		

*Tested in Kingston Harbor (Rhode Island) silt (25).

**Tested in natural sediment from the Chesapeake and Delaware Canal, Maryland (17).

to fine sediments from the Chesapeake and Delaware Canal tentatively place them in this sensitivity class (17).

The Atlantic silverside was classified as a highly sensitive species; juvenile and young-of-the-year life-history stages are particularly sensitive (27). Juvenile bluefish and young-of-the-year white perch have been classified as highly sensitive forms.

These data strongly support our previous observations (27) that the lethal effects of suspended solids on fishes differ according to the life stages of a given species. Juvenile and larval white perch are more likely to be killed by lower concentrations of suspended solids than are adult perch. The basis for age-specific differences in tolerance to suspended solids is not known.

We observed that when fishes were exposed to lethal concentrations of fuller's earth, their gill filaments and secondary lamellae acted as a sieve to entrap particles. Since the physical dimensions of a fish's gill increase as the fish grows (18), the size of the openings in the "gill filter" also increases. The lethal effect of a given concentration of suspended solids would decrease for larger fish because fewer particles would become entrapped in the gill. Also, small fishes, which demand much more oxygen per unit body weight than do large fishes (28), may not be able to tolerate the same relative intensity of gill clogging as the large fish. The combined effect of higher metabolic rate and a finer, more efficient, filter would render larval and juvenile life stages of a species to be highly sensitive to suspended solids, regardless of the tolerance shown by the adult.

SUBLETHAL EFFECTS OF SUSPENDED SOLIDS ON THE HEMATOLOGY OF ESTUARINE FISHES

Exposure of white perch to 0.65 g/l fuller's earth for five days resulted in significant increases in hematocrit (17%), hemoglobin concentration (15%), and the number of red blood cells (29%) (Table 3). The ionic concentration of the blood (plasma osmolality) did not change in response to the suspended solids.

In the hogchoker, a five-day exposure to 1.24 g/l fuller's earth increased red blood cell counts by 30.4%; hematocrit increased by 27.6%. Unlike the white perch, where the proportional increase in red cells was much greater than the increase in hemoglobin and hematocrit, the hematocrit increase (27.6%) observed in the hogchoker was proportionately the same as the increase in red cells (30.4%).

Striped killifish exhibited a significant relative increase in hematocrit (29.7%) when exposed to 0.96 g/l fuller's earth for five days.

Although these species had similar responses to sublethal concentrations of fuller's earth, they differed markedly in the concentration of fuller's earth that

TABLE 3.

Hematocrit, hemoglobin, erythrocyte count, and plasma osmolality changes of estuarine fishes exposed to suspensions of fuller's earth. Values are percent increase over control values.

Species	Time (Days)	Hematocrit Value	Hemoglobin Concentration	Erythrocyte Count	Plasma Osmolality	Fuller's Earth
White Perch	5	17.7**	15.1*	29.0**	2.8	0.65 g/l
Hogchoker	5	27.6*	—	30.4**	—	1.24 g/l
Striped Killifish	5	29.7**	—	—	—	0.96 g/l
Spot	5	-3.9	6.7	5.4	—	1.27 g/l
Striped Bass	11	0.5	-6.8	5.5	-4.4	0.60 g/l
Striped Bass	14	25.2**	—	—	5.7*	1.50 g/l

*$p < 0.05$

**$p < 0.01$

was necessary to kill them. We have been unable to generate an LC_{50} response curve for the hogchoker, perhaps because this species has a very high tolerance for suspended solids. The killifish showed the highest 24-hour LC_{50} (38.19 g/l) of the seven species tested in fuller's earth. White perch, on the other hand, have been classified as a sensitive species (24 hr LC_{10} value was below 10 g/l) (*see* Table 1). Such sublethal effects as hematological alteration may be induced by low concentrations of suspended solids, even though the species exposed to the suspension is relatively tolerant to this material. Also, the hogchoker, a species with high tolerance to suspended solids, showed a significant increase in energy utilization during a five-day exposure to 1.24 g/l fuller's earth (28).

Experiments with striped bass and spot suggested that low concentrations of suspended solids may have little effect on their basic hematological parameters. There were no significant differences between values derived from experimental and control groups of spot. After 11 days' exposure to 0.60 g/l fuller's earth, no significant changes occurred in hematological parameters of striped bass; however, after 14 days in 1.5 g/l fuller's earth, there were significant increases in hematocrit and blood osmolality. The increased hematocrit may reflect simple concentration of blood components caused by body water loss.

The hematological responses to sublethal concentrations of suspended solids in white perch, hogchokers, and striped killifish was clinically identical to the responses in goldfish and trout exposed to extremely low concentrations of dissolved oxygen for periods of from 4 to 25 days (19, 20, 21).

EFFECTS OF SUBLETHAL CONCENTRATIONS OF FULLER'S EARTH ON THE GILL TISSUE OF WHITE PERCH

In order to justify the contention that the exposure of fishes to sublethal concentrations of suspended solids reduced the available oxygen at the gill, histological evidence must demonstrate that suspended solids can affect gas transport across the secondary lamellae, which is the prime site of respiratory gas exchange. Since function is largely dependent on structure in living systems, suspended solids can reduce the availability of oxygen to the fish at the gill surface by disrupting that surface and making the tissue partially dysfunctional.

Control fish showed typical teleost gill structure and moderate concentrations of mucus goblet cells on the anterior margin of each gill filament (Fig. 1). More mucus goblet cells appeared on the gills of perch that had been exposed to 0.65 g/l fuller's earth for five days (Fig. 2). In some cases mucus cells were the only visible cellular component at the anterior margin of the filaments. The proliferation of mucus goblet cells was confined to the anterior margin, which is the first to come in contact with the stream of water that irrigates the gills. Little increase in concentration of mucus cells was observed elsewhere in the gill, and no increase in the size of these cells was apparent in fishes exposed to fuller's earth.

Secondary lamellae from perch exposed to 0.65 g/l fuller's earth showed pronounced abnormalities in structure (Fig. 3) and a swollen condition when compared to the lamellae of control fish (Figs. 3 and 4). The epithelium had become separated from the pilar cell tube. This tube was intact in most instances, but in some cases the pilar cell structure had been disrupted.

Epithelial cells in fishes exposed to fuller's earth were enlarged, and formed a thicker covering than in control fish (Figs. 3 and 4).

The effects of fuller's earth on the gill tissue of white perch were similar to the effects of diatomaceous earth on the gills of rainbow trout (29) and the effects of high (6) and low (8) concentrations of china clay mining waste on the gills of brown trout. The gills of those fishes showed separation of the epithelium from the lamellar structure, thickening of the epithelium, and occasional disruption of the pilar cell structure of the lamella. These effects were induced using concentrations of suspended solids (0.4 to 0.81 g/l) roughly similar to ours. The effects of different particle sizes on gill tissue have not been evaluated. However, there is a definite concentration effect associated with the broad size range of silt-clay particles. In the white perch, concentrations of fuller's earth well below the 24-hour LC_{10} value (*see* Table 1) may adversely affect the structure of gill tissue in a period of five days.

Gill damage caused by suspended solids has not been positively proved harmful to fishes in terms of their overall survival rates. Many species of

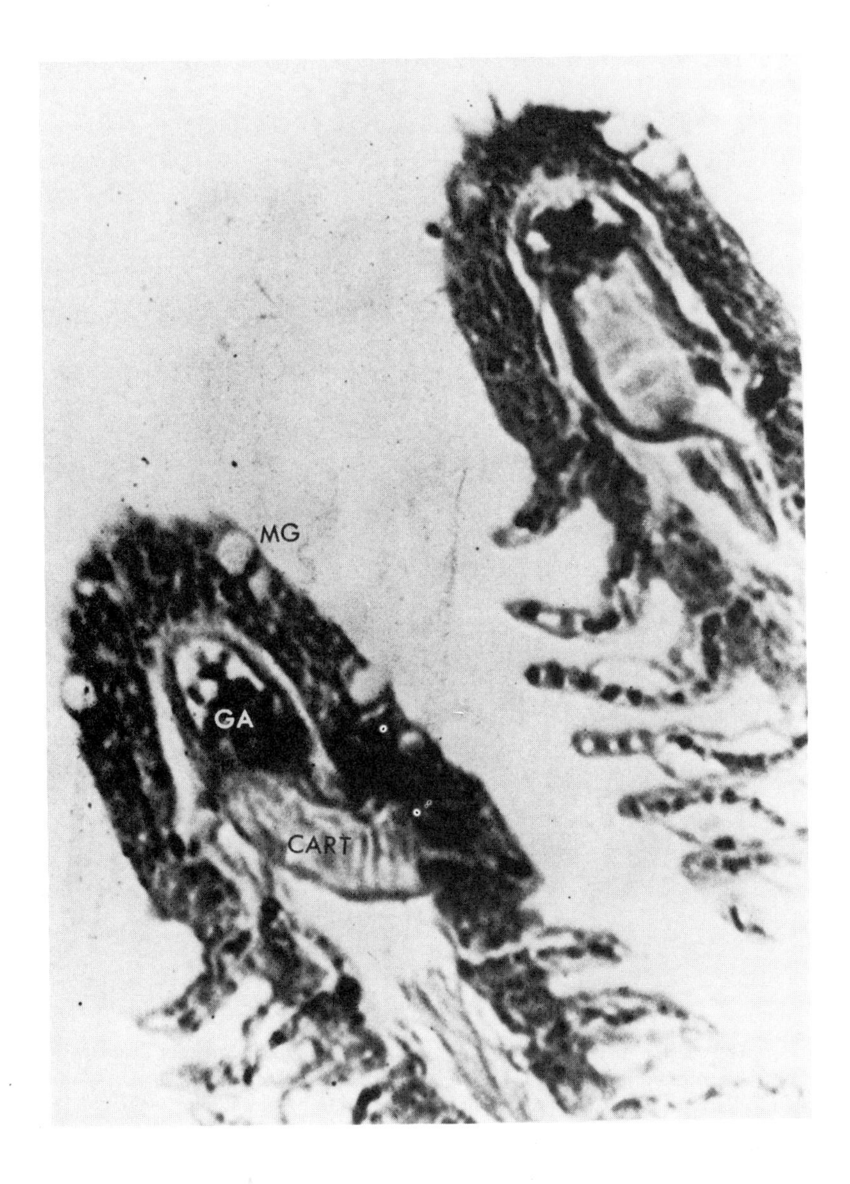

Figure 1. Section of gill of white perch held five days in clean water. Secondary lamellar structure is undisturbed, and epithelium is tightly applied to the pilar cell structure. Note mucus goblet cells in the region of the anterior margin. Photo taken at 300X, Bouin's, Gomori's trichrome (CART = cartilage of gill ray, MG = mucus goblet cells, GA = gill artery).

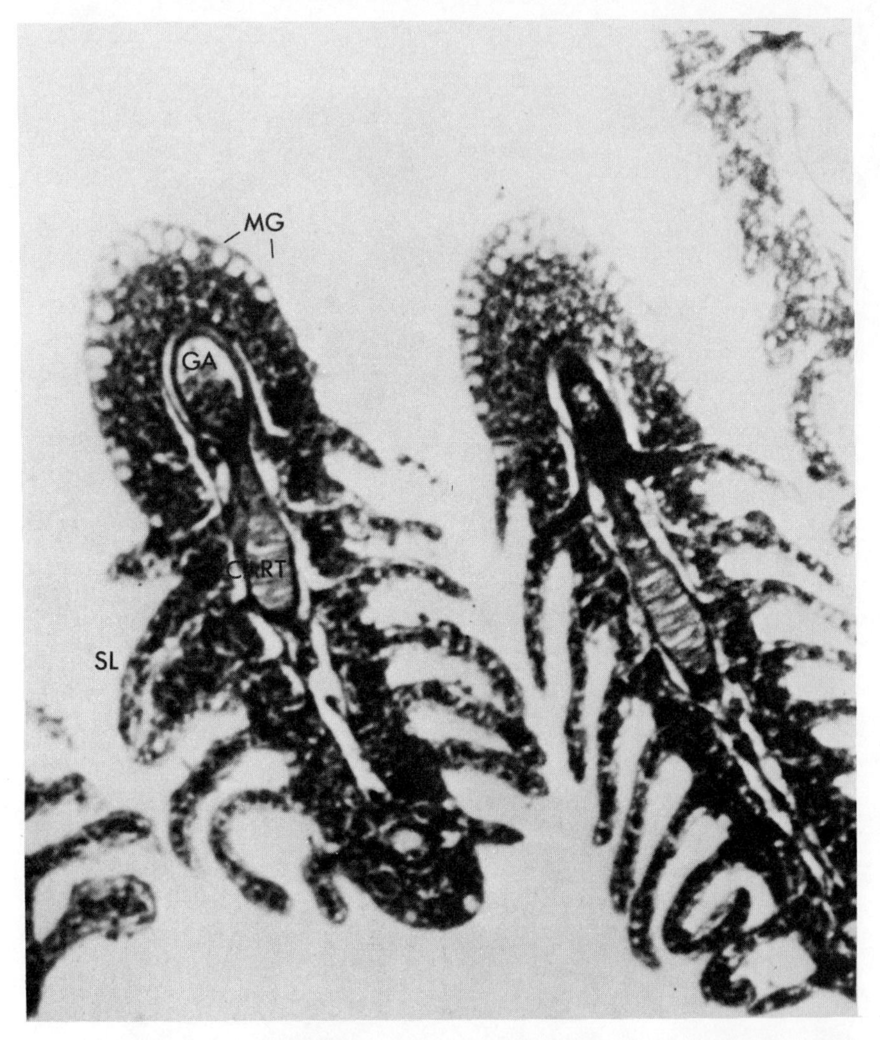

Figure 2. Gill section of white perch exposed to 0.65 g/l fuller's earth for five days. Note the marked proliferation of mucus goblet cells on the anterior margin of the gill filament. Areas at base of lamellae appear to be fused. Photo taken at 150X, Bouin's, iron hematoxylin-triosin (MG = mucus goblet cells, SL = secondary lamella, GA = gill artery, CART = cartilage of gill ray).

fresh-water fishes have survived for several weeks in highly turbid conditions (6). In these cases, there may be compensatory reactions that enable fishes to survive despite the damage to the gill. Shunt mechanisms are commonly used by fishes under normal conditions so that not all of the gill surface is used for respiration (22), and is probably being held in reserve. Use of the reserve surface area during

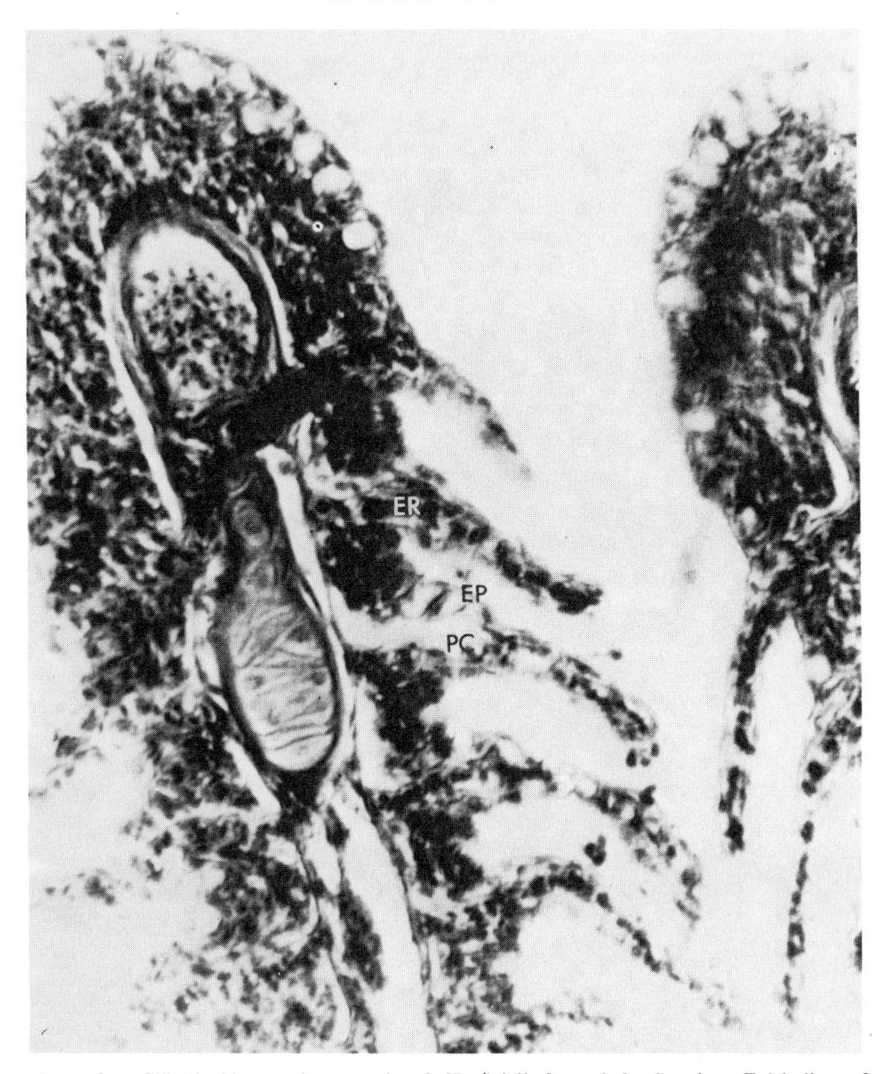

Figure 3. Gill of white perch exposed to 0.65 g/l fuller's earth for five days. Epithelium of secondary lamellae has separated, leaving space between pilar cells and epithelium. Epithelial cells are swollen. Photo taken at 300X, iron hematoxylin-triosin, Bouin's (EP = epithelium, PC = pilar cell tube, ER = red blood cells).

prolonged exposure to suspended solids, may give the fishes sufficient functional, though damaged, gas-exchange surface to survive. Also, the functional decrease in gill surface area caused by suspended solids may be offset by compensatory increases in the gas-exchange capacity of the blood (Table 3).

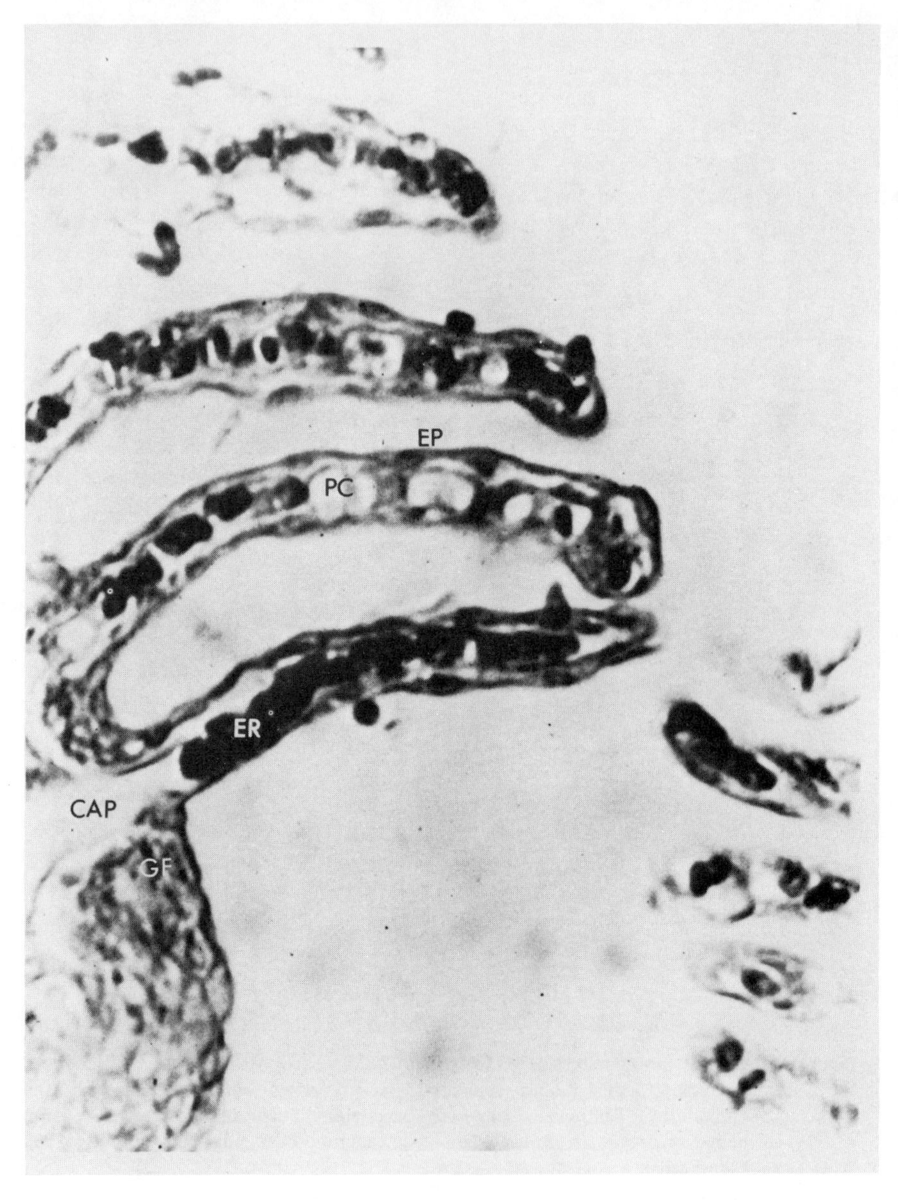

Figure 4. Typical lamellar structure in gill section of white perch after five days in clean water. Photo taken at 600X, Bouin's, Gomori's trichrome (PC = pilar cells, ER = red blood cells, GF = body of gill filament, CAP = capillary connection to artery, EP = epithelium).

SEDIMENT EFFECTS ON *EURYTEMORA AFFINIS* AND *ACARTIA TONSA*

The mechanical or abrasive action of suspended silt and clay is important to suspension-feeding organisms with respect to gill clogging, impairment of proper respiratory and excretory functioning, and feeding activity. Suspension-feeders living in estuaries are sometimes exposed to suspended sediment concentrations which tend to be quite high at times. In water that is strongly agitated by, for example, storms or dredging activities, a substantial amount of the suspended material in the water column is apt to consist of fine sand and silt-sized particles that have been resuspended from the bottom. Necessarily, high concentrations of suspended solids result for suspension-feeders in reduced rates of water transport because of their filters being clogged (11). For example, drastic reductions in pumping rates of adult American oysters have been observed at 100 mg/l (13). Apparently, oysters living in areas where the background values are persistently high may pump at reduced rates throughout most of their lives; examples of such areas are the upper Chesapeake Bay, Galveston Bay, and the Louisiana marshes, where background values have been recorded as 20 to 100 mg/l, 200 to 400 mg/l, and 20 to 200 mg/l, respectively (1, 14, 15).

The growth and survival of clam and oyster eggs and larvae were significantly affected at suspended sediment concentrations as low as 125 mg/l (3, 4, 13). While estuarine concentrations of suspended material should never exceed 100 mg/l if severe larval mortality is to be prevented, survival and growth even at concentrations as high as 4000 mg/l demonstrated a remarkable tolerance to the turbid nature of estuaries (4).

Considerable quantities of inorganic material along with particulate food can interfere with the suspension-feeding activity of the copepods *Eurytemora affinis* and *Acartia tonsa.* Additions of two different particle size distributions of fine sand, fuller's earth, and Patuxent River silt to a cell suspension of heavily labelled ($NaHC^{14}O_3$) *Monochrysis lutheri* (50,000 cells/ml) significantly reduced the maximum ingestion rate of these cells by *Eurytemora affinis* at solids concentrations in excess of 250 mg/l (Fig. 5). At 100 mg/l a significant enhancement of uptake was evident in fuller's earth. Non-significant reductions at this concentration occurred for all other particle types. Stimulation of pumping activity in the American oyster at low particle concentrations (13), increased activity of *Doliolida* and *Salpida* by the presence of suspended particles (11), and tripling of the ingestion rate of algal cells by *Artemia* sp. upon the addition of fine sand to the cell suspension (23) lend support to our observation with *E. affinis.*

The effects of Patuxent River silt caused a larger reduction in ingestion of radiocarbon-labelled cells than did other particle types over all concentrations tested, except at 50 mg/l (Fig. 5). The apparent enhancement effect of the silt at 100 mg/l was smaller than for all other particle types.

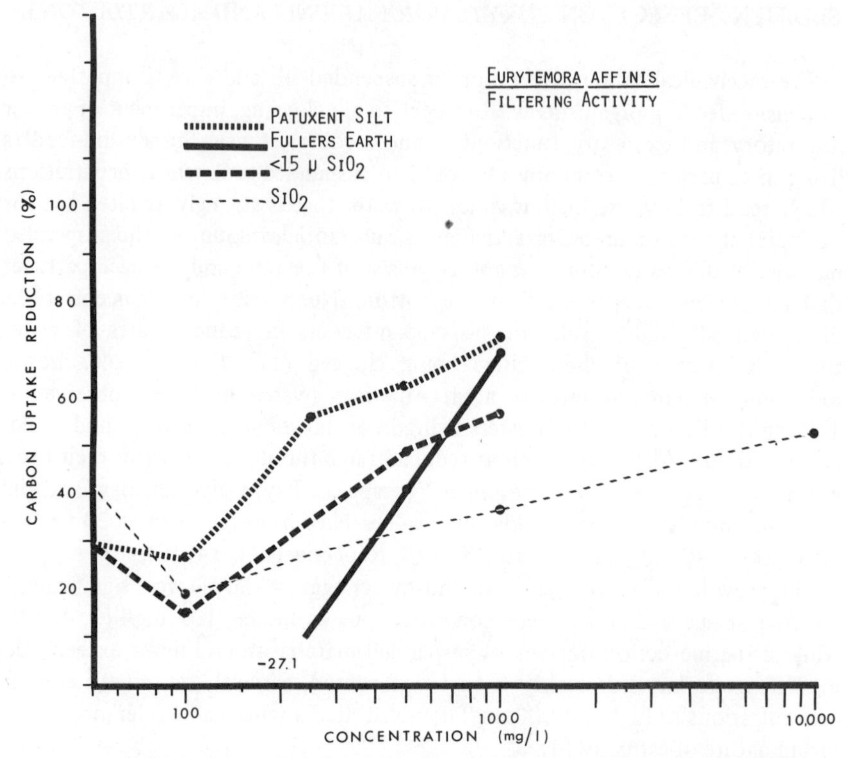

Figure 5. Reduction in maximum ingestion rate (20°C) of adult **Eurytemora affinis** (mixed sexes) feeding on labelled ($NaHC^{14}O_3$) **Monochrysis lutheri** (50,000 cells/ml) at increasing concentrations of Patuxent River silt (0.78 μ med. size, 72%<2 μ), fuller's earth (< 0.5 μ med. size, 82%<2 μ), SiO_2 (17 μmed. size, 6%<2 μ), and < 15 μ SiO_2 (6.2 μ med. size, 13%<2 μ) (Time to max. ingestion = 10 min., 40 organisms at each point). All reduction values above 30% are significant. Increase (-27.1%) at 100 mg/l fuller's earth is significantly higher than control.

Acartia tonsa exhibited drastic, significant reductions in maximum ingestion at all concentrations of all particle types tested (Fig. 6). The effects increased as the concentration increased. However, with this species the effect of Patuxent River silt was lower than fuller's earth or SiO_2 particles (< 15 μ), except at the highest concentration (1000 mg/l).

The differences in the shapes of the uptake curves may be related to the habitats in which these species are usually found. *E. affinis* is usually found in upper, more turbid estuarine areas (9), and evidently is stimulated to increase ingestion by the low concentrations of suspended solids normally to be found in these areas. Low concentrations of suspended solids may be indicative of the presence of food for this species, and in turn the organism is stimulated to begin

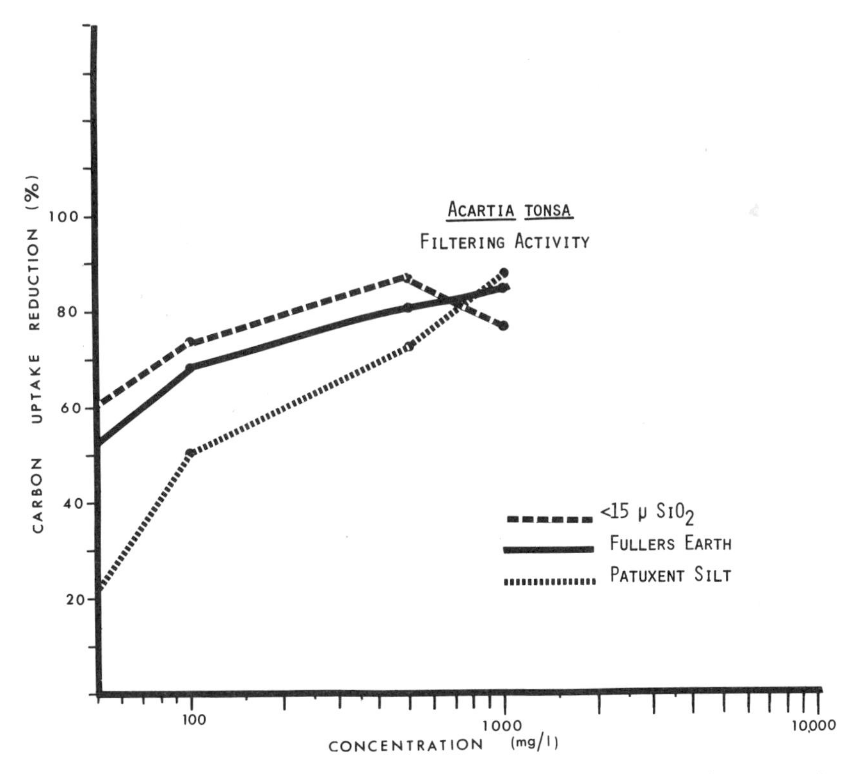

Figure 6. Reduction in maximum ingestion rate (20°C) of adult **Acartia tonsa** (mixed sexes) feeding on labelled ($NaHC^{14}O_3$) **Monochrysis lutheri** (50,000 cells/ml) at increasing concentrations of Patuxent River silt (0.78 μ med. size, 72% < 2 μ), < 15 μ SiO_2 (6.2 μ med. size, 13% < 2 μ), and fuller's earth (< 0.5 μ med. size, 82% < 2 μ) (Time to max. ingestion = 5 min., 40 organisms at each point). All reduction values are significant at all test concentrations.

feeding or to increase its rate of ingestion. The survival value in this case is evident. On the other hand, *A. tonsa* is usually found in open, less turbid waters of the Chesapeake Bay system. Apparently, this species is not stimulated to increase feeding activity by low-particle concentrations which may be uncharacteristic of its natural habitat (9). Because these species are non-selective suspension-feeders, the much reduced uptake of radioactive phytoplankton observed with increasing concentrations of suspended solids can be accounted for simply by the ingestion of increasing numbers of unlabelled particles because the gut of both species was full during all experimental treatments.

ECOLOGICAL IMPLICATIONS

Our results show that suspended solids have both lethal and sublethal effects on estuarine fishes, and significant sublethal effects at concentrations typically found during flooding and in the vicinity of dredging and disposal operations. Estuarine organisms with very different life habits have different sensitivities (tolerances) to particle size, distribution, and concentration. Differing sensitivities were evident even for the same species, but as different life stages.

The use of lethal concentration levels (LC_{10}, LC_{50}) to establish criteria for suspended solids ignores the biologically significant sublethal effects that suspended solids have on estuarine organisms. Therefore, in establishing criteria for the protection of estuarine organisms, the sublethal effects of suspended material on the most sensitive biological components (important species or life stages) must be taken into account at estuarine sites selected for environmental modification. Adequate knowledge of local conditions at these sites is absolutely essential. These would include at least life-history stages, sediment types, sediment concentrations, species (seasonal and resident), duration of exposure, and habitat preference.

ACKNOWLEDGMENT

The laboratory study was made possible by U. S. Army contract no. DACW72-71-C-0003. We wish to extend our appreciation to Dr. Kent S. Price, University of Delaware, for his kind permission to supplement our studies at the Bayside Laboratory, Lewes, Delaware.

Photomicrographic equipment was supplied by Dr. Ray Morgan, and photographs were prepared by Mr. Mike Reber of the Chesapeake Biological Laboratory. Slides for histological study were prepared by Ms. Elaine Drobeck.

REFERENCES

1. Biggs, R. B.
1970 Geology and hydrology. In **Gross physical and biological effects of overboard spoil disposal in upper Chesapeake Bay,** p. 7-15. Univ. Maryland, Natural Resources Inst. Spec. Rep. No. 3.
2. Breverton, R. and Holt, S.
1956 A review of methods for estimating mortality rates in fish populations with special reference to sources of bias in catch sampling. **Rapp. Conseil Expl. Mer.,** 147: 67-83.
3. Davis, H. C.
1960 Effects of turbidity producing materials in sea water on eggs and larvae of the clam **(Mercenaria mercenaria). Biol. Bull.,** 118(1): 48-54.
4. Davis, H. C. and Hidu, H.
1969 Effects of turbidity producing substances in sea water on eggs and larvae of three genera of bivalve mollusks. **The Veliger,** 11(4): 316-323.

5. Doudoroff, P., Anderson, B., Burdick, G., Galtsoff, P., Hart, W., Patrick, R., Strong, E., Surber, E., and Van Horn, W.
1951 Bioassay methods for the evaluation of acute toxicity of industrial wastes to fish. **Sewage and Industrial Wastes,** 23: 1380-1397.

6. EIFAC.
1964 Water quality criteria for European freshwater fish. Report on finely divided solids and inland fisheries. EIFAC working party on water quality criteria for European freshwater fish. **EIFAC Tech. Paper,** (1): 1-21.

7. Ellis, M.
1937 Detection and measurement of stream pollution. **Bull. U. S. Bur. Fish.,** 48: 365-473.

8. Herbert, D. W. M., Alabaster, J. S., Dart, N., and Lloyd, R.
1961 The effect of China clay waste on trout streams. **Int. Jour. Air Water Poll.,** 5: 56-74.

9. Herman, S. S., Mihursky, J. A., and McErlean, A. J.
1968 Zooplankton and environmental characteristics of the Patuxent River estuary 1963-1965. **Chesapeake Sci.,** 9(2): 67-82.

10. Hildebrand, S. F. and Schroeder, W. C.
1928 Fishes of Chesapeake Bay. **Bull. U. S. Bur. Fish.,** 43: 1-366.

11. Jøgensen, C. B.
1966 **Biology of suspension feeding.** Pergamon Press, New York.

12. Lee, G. F.
1970 Factors affecting the transfer of materials between water and sediments. Eutrophication Info. Prog. Water Resour. Center, Univ. Wisconsin. **Lit. Rev.,** 1: 1-35.

13. Loosanoff, V. L.
1961 Effects of suspended silt and other substances on rate of feeding of oysters. **Science,** 107(2768): 69-70.

14. Mackin, J. G.
1961 Canal dredging and silting in Louisiana bays. **Publ. Inst. Mar. Sci. (Texas),** 7: 262-314.

15. Masch, F. D. and Espey, W. H.
1967 Shell dredging – a factor in sedimentation in Galveston Bay. Center for Research in Water Resources, Univ. Texas. Tech. Rept. No. 7.

16. May, E. B.
1973 Environmental effects of hydraulic dredging in estuaries. **Alabama Marine Resources Bulletin,** No. 9: 1-85.

17. Morgan, R. P., Rasin, J. V., and Noe, L. A.
1973 Effects of suspended sediments on the development of eggs and larvae of striped bass and white perch. Final Report. Army Corps of Engineers contract no. DACW61-71-C-0062.

18. Muir, B. S.
1969 Gill dimensions as a function of fish size. **Jour. Fish. Res. Bd. Canada,** 26: 165-170.

19. Ostroumova, I. N.
1964 Properties of the blood of trout during adaptation to different oxygen and salt contents of the water. In **Fish physiology in**

Acclimatization and Breeding (ed. Privol'nev, T. I.) **Bull. St. Sci. Res. Inst. Lake and River Fish.,** 58: 24-34.

20. Phyllips, A. M., Jr.
1947 The effects of asphyxia upon the red blood cell content of trout blood. **Copeia,** 47: 183-186.

21. Prosser, C. L., Barr, L. M., Ping, R. D., and Lauer, C. V.
1957 Acclimation to reduced oxygen in goldfish. **Physiol. Zool.,** 30: 137-141.

22. Randall, D. J.
1970 Gas exchange in fish. In **Fish Physiology,** Vol. IV, p. 253-292. (eds. Hoar, W. S. and Randall, D. J.) Academic Press, New York and London.

23. Reeve, M. R.
1963 The filter-feeding of **Artemia.** II. In suspensions of various particles. **Jour. Exp. Biol.,** 40: 207-214.

24. Ricker, W. E.
1958 Handbook of computations for biological statistics of fish populations. **Bull. Fish. Res. Bd. Canada,** 119: 1-300.

25. Rogers, B. A.
1969 The tolerances of fishes to suspended solids. M.S. thesis. Univ. Rhode Island.

26. Sherk, J. A.
1972 Current status of knowledge of the biological effects of suspended and deposited sediments in Chesapeake Bay. **Chesapeake Sci.,** 13(Suppl.): S137-S144.

27. Sherk, J. A. and O'Connor, J. M.
1971 Effects of suspended and deposited sediments on estuarine organisms, PHASE II. Ann. Rept. to the U. S. Army Corps of Engineers, 1971. NRI Ref. No. 71-4D.

28. Sherk, J. A., O'Connor, J. M., and Neumann, D. A.
1972 Effects of suspended and deposited sediments on estuarine organisms, PHASE II. Ann. Rept. Project Year II. NRI Ref. No. 72-9E.

29. Southgate, B. A.
1962 Water pollution research, 1959. In **River Pollution II. Causes and Effects,** p. 272. (ed. Klein, L.) Butterworth, London.

30. Windom, H. L.
1972 Environmental response of salt marshes to deposition of dredged materials. Preprint in **1972 Amer. Soc. Civil Eng. Nat. Water Resour. Eng. Conf.,** 1612: 1-26.

WATER-QUALITY ASPECTS OF DREDGING AND DREDGE-SPOIL DISPOSAL IN ESTUARINE ENVIRONMENTS

by

Herbert L. Windom[1]

ABSTRACT

The dominant effect that dredging has on estuarine water quality results from chemical exchanges between the water and dispersed sediment. When first dispersed, estuarine sediments release high concentrations of ammonia. This increases microscopic plant production which, in turn, is followed by increases in pH, dissolved oxygen, and BOD.

In general, the first effect of dredged sediments being dispersed is that the heavy metals in the water are depleted. In time, however, some metals may be released from the sediment, and their concentration in the water becomes greater than it was to begin with.

INTRODUCTION

Many port facilities are situated in estuaries and a great deal of the commercial shipping along the U. S. coast travels by way of the extensive intracoastal waterways, all of which traverse estuaries. In order to provide access by water to those facilities and to keep the coastwise traffic moving, there have to be channels through the estuaries, and channels have to be dredged. Every year more than a half million cubic meters of sediment are dredged from such

1. Skidaway Institute of Oceanography, P. O. Box 13687, Savannah, Georgia 31406.

channels along the southeastern Atlantic coast and more than one million cubic meters from the Gulf of Mexico.

Dredging can lead to the degradation of valuable nursery grounds and food sources for commercial fisheries. This is particularly true along the southeastern U. S. coastline, where harbor channels and waterways traverse extensive salt-marsh estuaries. Probably the most destructive activity associated with dredging is the disposal of dredge spoil. Overboard disposal may destroy important bottom habitats. In some cases, marshes are used as disposal sites and these valuable habitats are thus degraded. Such problems can often be minimized by the use of upland spoil-disposal sites.

Regardless of the site or technique used for dredge-disposal, another effect dredging may have effect on estuarine environments in water-quality impairment. Most dredging in estuaries is done with hydraulic dredges, which transfer the dispersed sediment to the disposal site by pipeline. The mere act of dispersing the sediment can result in degrading the water quality because it increases turbidity. Also, dispersion can bring about a chemical exchange between the sediment and the water column. Many maintained channels contain polluted sediments which may release substances that are toxic or that lead to decreased water quality when dredged. In the case of overboard disposal the potential hazards are obvious. Disposal of spoil from hydraulic dredges in upland, confined or unconfined areas, however, may also impair water quality because the water used to carry the sediment to the spoil site eventually returns to the estuary. In such cases, water quality is impaired because dilution is not as great and run-off from the spoil area acts as a point source for given pollutants.

Since it is likely that society will continue to need ports and shipping channels, dredging in estuaries will necessarily continue. With this realization, much research has recently been conducted to determine what effects different dredging practices have on water quality. The goal of this work has been not only to predict the impact of dredging but to identify ways of minimizing its effects on water quality. This paper considers various water-quality parameters and reviews results of studies designed to understand processes responsible for their changes during dredging.

TURBIDITY

The most obvious change in water quality during dredging is increased turbidity. This is particularly true in estuarine waters that characteristically have low suspended-sediment levels, such as along the southern coasts of Florida. In estuaries that receive large sediment loads from rivers, such as those of the southeastern Atlantic and Gulf Coasts, increased turbidity during dredging is less obvious. For example, a study of maintenance dredging of the intracoastal waterway through salt-marsh estuaries (17) showed that, in the vicinity of dredging activity, suspended sediment did not increase significantly above the rather high ambient level. For that study the sediment was placed on salt marshes and no difference was noted whether the spoil was or was not confined. In Mobile Bay (10), elevated levels of suspended sediment were found to persist for only a few hundred feet from the point of overboard discharge.

In some estuaries, especially those along the southeastern Atlantic and Gulf coasts, where turbidity is naturally high, increased turbidity does not pose the problem that it does for low-turbidity waters. High natural turbidity tends to limit phytoplankton productivity because light-extinction coefficients are high. Even when increased turbidity caused by dredging is persistent, such as was observed in the upper Chesapeake Bay (2), no significant change in productivity was indicated (11). Many estuarine organisms also appear to be adapted to high suspended solids. Oysters were apparently not harmed by high turbidity during dredging operations in the intracoastal waterways of the southeastern Atlantic Coast (8, 9). Only when high levels of suspended sediments are maintained for long periods are oysters harmed (16). Spat settlement also appears to be unaffected by moderately increased turbidity.

In most cases, increased turbidity caused by dredging is a transient condition in estuaries. Because the electrolyte content of estuarine waters is high, fine-grained materials such as clays do not exist as discrete particles but aggregate into large particles with faster settling rates. Consequently, when dredging ceases the increased turbidity produced in the area diminishes to background levels within a few hours (2). This is true even when the dredging has been in progress for an extended period of time. The increased settling velocity of suspended particles also accounts for the limited area of increased turbidity surrounding a dredging operation.

NUTRIENTS

Both polluted and unpolluted sediments contain high concentrations of soluble phosphorous and nitrogen (5, 12). When sediments are suspended during dredging these substances may be released, increasing the nutrient level in the surrounding waters. Phosphorous and nitrogen have been found to increase from 50 to 100 times above ambient levels in the immediate vicinity of the site of overboard spoil-disposal during dredging operations in the Chesapeake Bay (3). However, since phosphate can be adsorbed by particulate matter (7), the concentration of this nutrient in dredging sites might be decreased.

Nitrogen is apparently the limiting nutrient in coastal waters (13) and is therefore of particular interest in relation to the effects dredging have on water quality. Increases in nitrogen released from discharged sediment during dredging are followed by increases in the microscopic plant population (18).

To evaluate the magnitude and duration of the release of nutrients from dispersed sediments during dredging operations, four areas were studied. Two of these were in Charleston Harbor, South Carolina, and the third was the Cape Fear River, North Carolina, and the fourth Terry Creek, near Brunswick, Georgia. In Terry Creek and Charleston Harbor, samples of the discharge from hydraulic pipeline dredges were collected in large vats and the nitrogen levels were subsequently monitored. In these areas the dredge spoil was placed in confined areas so that the water run-off from the spoil had a measurable residence time prior to discharge to the receiving waters. Samples collected at the discharge point of the spoil area showed increased nitrogen levels, primarily as ammonia with nitrate and phosphate remaining at levels similar to those in the receiving waters (18). In the Cape Fear River sediment samples were collected in the channel to be dredged and dispersed in samples of the overlying water, simulating the dredging operations. Here the nitrogen concentrations were also followed periodically as in the other three cases.

After sediments have been dispersed by dredges, at first ammonia is greatly increased in the water while nitrate concentrations change very little (Fig. 1). Both the areas studied in Charleston Harbor show a decrease in ammonia with time as a result of uptake by microscopic plants. This decrease is followed by an increase in dissolved oxygen and chlorophyll *a* (18). Although no decrease in ammonia can be seen in the results from Terry Creek, there were increases in the microscopic plant population (18), suggesting that the ammonia was the stimulant. Samples from Charleston Harbor showed that nitrate increased with time, possibly because of heterotrophic nitrification (15).

Phosphate variations in the water after the dispersion of sediments are not easily explained (Fig. 2). In some cases, phosphate in the water is initially increased, while in others it is decreased, possibly by adsorption on sediment (7). However, variations in the phosphate which result from the release of this nutrient from sediments are apparently less significant than those of nitrogen compounds. This is clearly due to the latter being limiting in estuarine environments.

The large amount of ammonia released from dispersed sediments demands important consideration when the effects of dredging are evaluated. In the case of overboard disposal of dredge spoil the large concentrations of reduced nitrogen released from the sediment are delivered to the natural phytoplankton community in a more dilute form than those delivered from confined dredge-spoil disposal sites. In the latter case, the microscopic plant community may be dominated by benthic algae, which can grow in abundance and be discharged to the estuary in the run-off from the spoil area. The consequences of these two ways in which excess nutrients might be assimilated in estuarine ecosystems deserves further consideration.

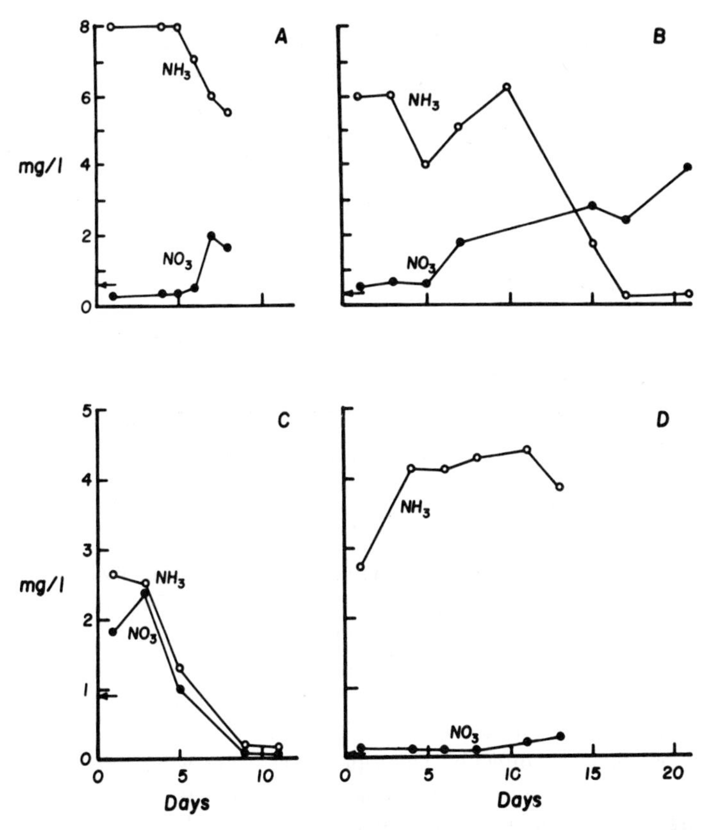

Figure 1. Changes in ammonia and nitrate in the water of the water-sediment system of discharged spoil with time after discharge. (A) Upper Charleston Harbor, (B) Lower Charleston Harbor, (C) Cape Fear River and (D) Terry Creek. Arrows indicate ambient level of NO_3. NH_3 was always less than this value.

DISSOLVED OXYGEN

Most estuarine sediments contain high concentrations of oxygen demanding materials, particularly organic detritus. Although very high oxygen demands of estuarine sediments are common, it is clear that during dredging the sediment is not exposed to the water long enough to allow the total oxygen demand to be exerted. In several cases, however, decreases in oxygen concentrations during dredging have been observed (1).

To assess the oxygen demand of dispersed sediment during dredging an approach similar to that described above for nutrients can be followed. Results of these studies indicate that at first, oxygen is depressed somewhat below ambient levels (Fig. 3). With time the oxygen concentrations increase as a

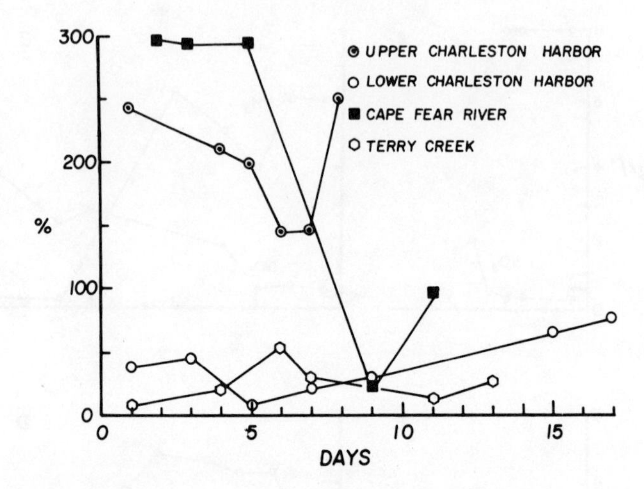

Figure 2. Changes (% above ambient) in phosphate in the water of the water-sediment system of dishcarged spoil with time after discharge.

consequence of the increase in photosynthesis brought about by the release of ammonia described above. The uptake of CO_2 is also manifested in the increase in pH with time. Although the dissolved oxygen increases, increases in the biomass of the microscopic plant population is reflected as an increase in the biochemical oxygen demand. The changes in these three parameters indicate the processes responsible for changes in dissolved oxygen resulting from the dispersion of dredged sediments. These processes occur whether the material is deposited overboard or in confined areas, except that in the former case the changes are not as obvious because of dilution by the large body of receiving water. The processes responsible for the changes shown in Figure 3 can be described by the following equations:

$$H_2O + CO_2 \rightleftharpoons CH_2O + O_2 \quad (1)$$

$$H^+ + HCO_3^- \rightleftharpoons H_2O + CO_2 \quad (2)$$

Equation 1 represents photosynthetic production of oxygen, with the microscopic plants being the catalyst. The reverse would represent the biochemical oxygen demand resulting from the bacteriological breakdown of carbohydrates. As shown in equation 2 the decrease in CO_2 also leads to decrease in hydrogen ions, thus increasing pH.

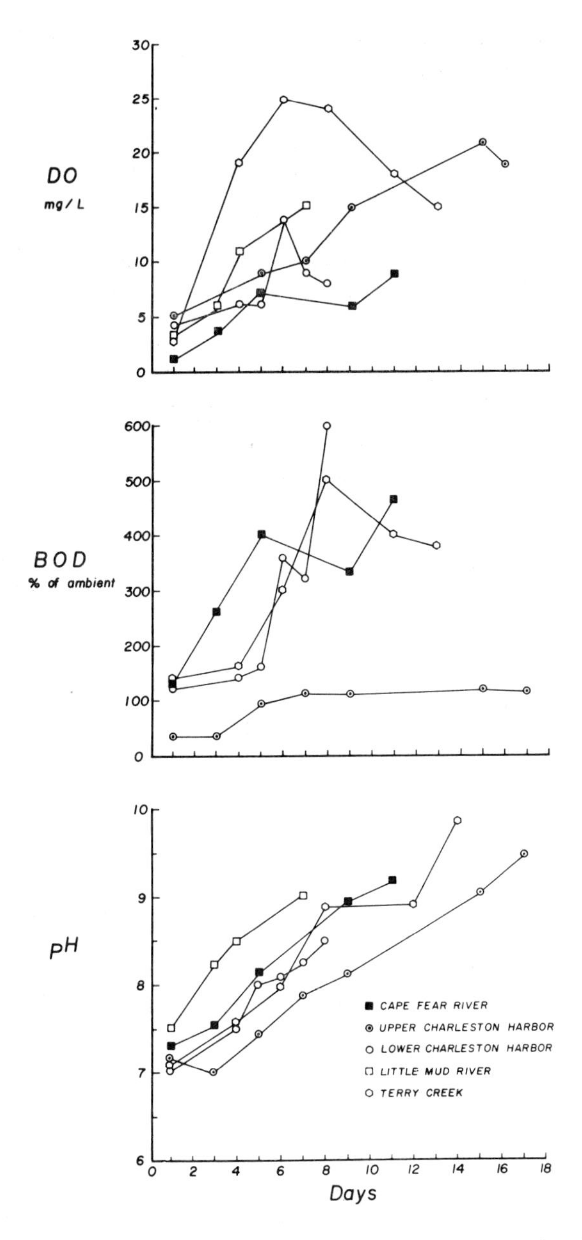

Figure 3. Changes in dissolved oxygen, biochemical oxygen demand, and pH in the water of the water-sediment system of discharged spoil with time after discharge. All samples except those from the Cape Fear River were collected directly from the dredge pipeline.

HEAVY METALS

Sediments of polluted estuaries often contain high concentrations of heavy metals resulting from industrial discharges. These areas are usually the sites of much dredging activity. Even with the great concern over heavy-metal pollution, relatively little research has been done to identify the effects of dredging on the levels of these substances in estuaries. This concern has, however, led to the establishment of criteria for the concentrations of heavy metals in sediments acceptable for overboard disposal (4). The criteria for several metals are given in Table 1.

TABLE 1.

Metal Content of Estuarine Sediments[1]

Location	Fe %	Cd ppm	Cu ppm	Pb ppm	Hg ppm	Zn ppm
Cape Fear River	2.1	1.5	21	82	0.2	115
Charleston Harbor						
Upper	2.5	4.0	35	30	0.3	105
Lower	7.5	1.4	45	41	0.3	150
Savannah River[1]	3.5	0.5	20	15	0.2	58
Little Mud River[1]	2.8	2.5	25	12	1.0	270
Brunswick Harbor	3.4	0.1	10	16	0.4	6
Mobile Bay						
Beacon 24	3.6	0.8	19	39	0.3	120
Beacon 33	3.7	5.7	19	28	0.8	220
Beacon 39A	1.5	0.4	5	12	0.2	62

[1] All values are on a dry-weight basis and averaged from a analysis of several sediment samples.

Obviously, the implication of a direct relationship between metal concentrations in sediments and water quality in the vicinity of dredging in geochemically oversimplified. A high concentration of a metal in a given sediment does not necessarily mean that large concentrations of that metal will be released upon dispersion of the sediment. In fact, in some cases the opposite is true (6).

To evaluate the relationship between metal concentrations in sediments and their release upon dredging, several dredging operations in estuaries throughout the southeastern United States were studied (18). These include Charleston

Harbor, Savannah River, Little Mud River (an unpolluted Georgia estuary), and Terry Creek and Brunswick Harbor, in Georgia. In all of these areas samples of the effluents from the dredge pipeline were collected and the water was periodically analyzed for metals. In other areas where dredging was to occur (Mobile Bay and the Cape Fear River) sediment samples were collected and dispersed in overlying water samples in the approximate proportions encountered in the effluent from dredge-discharge pipelines. This allowed the exchange of metals between the water and the dispersed sediment to be followed in a closed system similar to a spoil-confinement area. These data, of course, are also applicable in evaluating the exchange of metals from sediments where overboard disposal is employed.

The concentrations of some metals in the sediments from most of the areas studied exceed the EPA criteria (Table 2). This is true even of polluted estuaries (Little Mud River). Even so, changes in metal concentrations in the water subsequent to dispersion of sediment follow similar trends (Fig. 4). Subsequent to dispersion of the sediment, iron concentrations in the water sometimes decrease after an initial slight increase above ambient. After several days iron concentrations increase to ambient levels or higher. Copper generally follows a similar trend, as does lead. Mercury apparently does not increase significantly above ambient levels. After an initial increase, cadmium does not usually appear to decrease with time in the water associated with the dispersed sediment (Fig. 4). In some areas zinc also appears to decrease with time, while results from the Savannah River show an initial decrease followed by continual increase back to ambient levels. Samples from Mobile Bay show an initial increase three times above the ambient level followed by a decrease, with subsequent increases again to much higher values. The high concentrations of zinc in Mobile Bay sediments is apparently reflected in this large increase.

TABLE 2.

EPA Region IX Metal Criteria for the

Suitability of Sediments for Overboard Disposal

	ppm (dry weight)
Cadmium	2.0
Copper	50.0
Lead	50.0
Mercury	0.5
Zinc	75.0

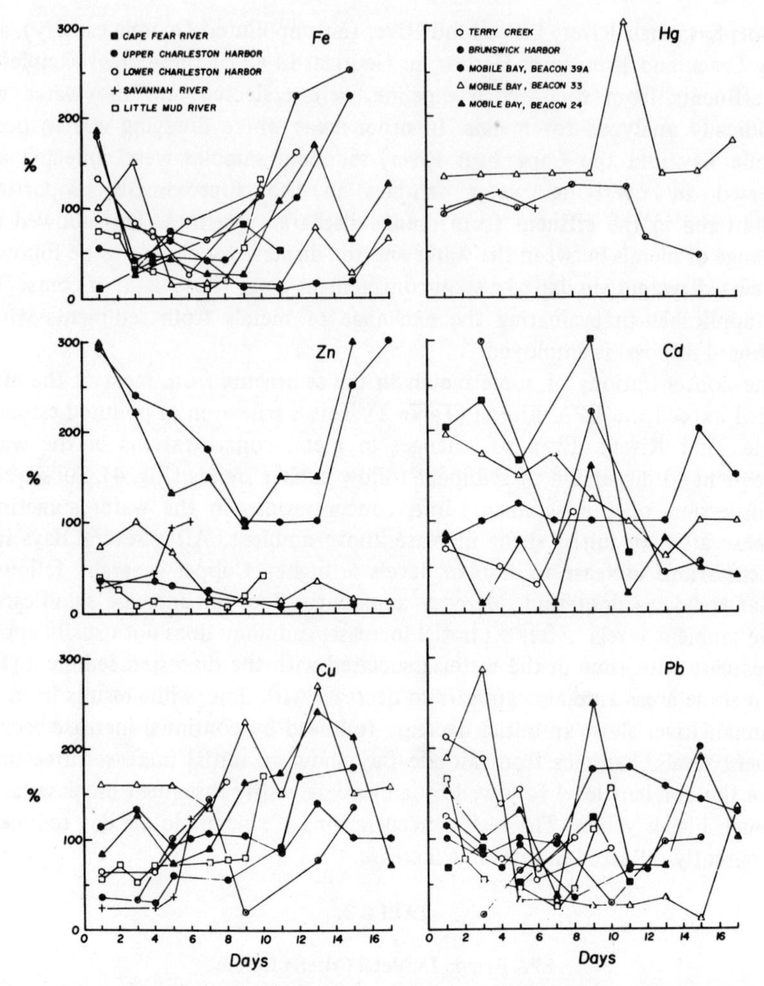

Figure 4. Changes (% above ambient) in metal concentrations in the water of the water-sediment system of discharged spoil with time after discharge.

It has been previously proposed (17, 18) that the oxidation of reduced iron in sediments during dredging controls the metal variations observed. For example, most non-detrital iron in estuarine sediments is in the reduced state. Upon dispersion of these sediments in the overlying water the iron is oxidized and forms insoluble hydrated iron oxides. This material has a great ability to scavenge other metals such as zinc, copper, cadmium, and lead, decreasing their concentrations in the water column. Clay minerals and other constituents of the sediment may also scavenge metals (6). Upon deposition of the iron hydroxide floc, iron is again reduced. In this deposit, high in sulfide, the metals may be expected to remain even though the iron hydroxide floc has broken down. It has been found however, that iron hydroxide flocculation efficiently scavenges organic complexes (14). If these are very stable, upon reduction of the iron hydroxide in the sediment, they may be released, leading to concentrations in excess of those originally present. The variations observed in the metal concentrations resulting from dredging activities may therefore depend primarily on the portion of the metals occurring as stable organic complexes. If a large part of the metals in the sediment is in the form of soluble organic complexes which have formed in the sediment, these may be irreversibly released upon dredging. Some of the metals released, however, may be in the form of metastable sulfides, which is probably the case for iron.

DISCUSSION

One of the most important parameters in predicting what changes a proposed dredging operation will have on water quality is the amount of soluble ammonia in the sediment. The release of this nutrient appears to be instantaneous upon dispersion of the sediment. The amount of ammonia released per volume of dredged sediment should provide sufficient information to judge the magnitude of the water quality effect of dredging on plant productivity and oxygen demand.

In evaluating the potential metal release of polluted sediments during dredging, a bulk chemical analysis of the sediment to be dredged is totally inadequate. The most realistic approach to evaluating the impact of a dredging operation on metal concentrations is to simply establish the amount that will be released both when the sediment is first dispersed and after it has been redisposed.

ACKNOWLEDGMENTS

The research described in this paper was partially supported by U. S. Army Corps of Engineers contracts DACW21-71-C-0020 (South Atlantic Division) and DACW01-73-C-0136 (Mobile District). The author wishes to thank Ralph Smith, Frank Taylor, and Elizabeth Waiters for their assistance in this work.

REFERENCES

1. Brown, C. L. and Clark, R.
1968 Observations on dredging and dissolved oxygen in a tidal waterway. **Water Resources Res.,** 4(6): 1381-1384.

2. Briggs, R. B.
1967 Overboard spoil disposal; I - Interim report on environmental

effects. In **National Symposium on Estuarine Pollution, Proc.,** p. 134-151 (eds. McCarty, P. L. and Kennedy, R.) Stanford Univ. Press, Stanford, California.

3. Briggs, R. B.
1968 Environmental effects of overboard spoil disposal. **Jour. Sanitary Eng. Div.,** 94(SA3): 477-487.

4. Environmental Protection Agency, Region IX
1972 Proposed guidelines for determining the acceptability of dredged spoils to marine waters. San Francisco, California, 9411.

5. Gahler, A. R.
1969 Sediment-water nutrient. **Proc. Eutrophication-Biostimulation Assessment Workshop.** (eds. Middlebrooks, E. J., Maloney, T. E., Powers, C. F. and Kaack, L. M.) Univ. California Press, Berkeley, California.

6. Gustafson, J. F.
1972 Beneficial effects of dredging turbidity. **World Dredging Mar. Const.** 9(13): 44-72.

7. Jitts, H. R.
1959 The adsorption of phosphate by estuarine bottom deposits. **Aust. Jour. Mar. Freshwater Res.,** 10: 7-21.

8. Lunz, E. J.
1938 Part I. Oyster culture with reference to dredging operations in South Carolina. Rept. to U. S. Engineers Office, Charleston, S. C., p. 1-135.

9. Lunz, E. J.
1942 Investigation of the effects of dredging an oyster leases in Duval County, Florida. In **Handbook of Oyster Survey, Intracoastal Waterway, Cumberland Sound to St. Johns River.** Spec. Rept. U. S. Army Corps of Engineers, Jacksonville, Florida.

10. May, E. B.
1973 Environmental effects of hydraulic dredging in estuaries. **Alabama Mar. Res. Bull.,** 9: 1-85.

11. Odum, H. T. and Wilson, R. F.
1962 Further studies on reaeration and metabolism of Texas bays. **Pub. Inst. Mar. Sci. (Texas),** 8: 23-55.

12. O'Neal, G. and Sceva, J.
1971 The effects of dredging on water quality. **World Dredging and Mar. Const.,** 7(14): 24-31.

13. Ryther, J. H. and Dunstan, W. M.
1971 Nitrogen, phosphorus and eutropication in the coastal marine environment. **Science,** 171: 1008-1013.

14. Sridharan, N. and Lee, G. F.
1972 Coprecipitation of organic compounds from lake water by iron salts. **Environ. Sci. Technol.,** 6: 1031-1033.

15. Verstraete, W. and Alexander, M.
1973 Heterotrophic nitrification in samples of natural ecosystems. **Environ. Sci. Technol.,** 7(1): 39-42.

16. Wilson, W.
1950 The effects of sedimentation due to dredging operations on oysters in Copano Bay, Texas. **Annual Rept., Mar. Lab. Texas,**

Game, Fish and Oyster Comm., 1948-1949.

17. Windom, H. L.
1972 Environmental aspects of dredging in estuaries. **Jour. Waterways, Harbors and Coastal Engineering Division, ASCE,** 98(WW4): 475-487.

18. Windom, H. L.
1973 Processes responsible for water quality changes during pipeline dredging in marine environments. **Proc. Fifth World Dredging Conf.,** Hamburg, Germany, June 1973.

MEIOBENTHOS ECOSYSTEMS AS INDICATORS OF THE EFFECTS OF DREDGING

by

Willis E. Pequegnat[1]

ABSTRACT

There is reasonable doubt that we possess the critical data for assessing the long-term effects of marine dredging activities on the benthos. Some subtle spatial and temperal effects of dredging are difficult to detect from long-cycle macrobenthos responses. Hence initial recolonization of spoil-disposal sites by macrobenthos may be an inadequate index of environmental recovery. Meiobenthos constituents respond definitively and for prolonged periods to mass sediment disruptions, as evidenced by changes in generation time, numbers, and diversity. Amelioration of these responses may prove to be a true indicator of habitat reconstruction, in part because of the close relationship of meiobenthos with sediment properties and their intrinsic short-cycle properties. Investigation of the significance of the difference in response between the macrobenthos and meiobenthos is needed.

INTRODUCTION

Most studies of the effects of pollution on the marine biota have been concerned with acute environmental changes of one type or another and their impacts upon organisms of commercial interest. This also has been the case with most studies of the effects of dredging upon marine ecosystems. There are

1. Oceanography Department, Texas A & M University, College Station, Texas 77843, and TerEco Corporation, 2508 Willow Bend Drive, Bryan, Texas 77801.

indications that biological studies of conventional format are reaching a point of sharply diminishing returns, which is to say that further studies of dredging effects that are field-oriented and deal principally with macrofauna components can yield few new insights into critical problems of marine pollution. Such a continuing trend would be unfortunate indeed in light of the fact that even the lay public appears to recognize that our ignorance of marine pollution is profound.

A feature article in a recent issue of *The Wall Street Journal* carried the title "The Sea: Pollution of the Oceans is an Enormous Threat, but Few People Care" (17). One disturbing point in the article is the contention that the ultimate cause of apparent governmental indifference and public apathy is a critical gap in scientific knowledge about pollution of the seas. Ignoring the matter of causes for the moment, it is apparent to most marine scientists that a gap certainly does exist and that two of its imponderables are (a) where particular pollutants go once they reach the sea, and (b) what effects they eventually will have on life in the oceans and on the land. The latter point emphasizes long time intervals and the need for predictive capability. These parameters are particularly relevant to the following discussion.

One can scarcely blame the lay public for appearing apathetic toward the welfare of the sea when they have been placed in the position of not knowing what or whom to believe. In the late 1960s elements within the scientific community predicted bio-catastrophies would result from the *Torrey Canyon* and Santa Barbara types of major oil spills. In response, the thinking public was roused to interested concern. Then in the early 1970s reports began to surface in public documents to the effect that the marine biota in these and other disturbed areas had made strong comebacks. Let me emphasize at this point that the problem of oil pollution has not been solved to the satisfaction of all scientists, and essentially the same is true of dredging and ocean-dumping of natural, and especially synthetic, organic compounds. Accordingly, we are in the process of defining environmental problems in terms of more penetrating parameters, and of devising study techniques that have higher degrees of reproducibility. It may yet prove to be that the above comebacks are, in fact, illusory when measured in longer time frames.

NATURE OF THE PROBLEM

A review of the literature seems to confirm the opinion that conventional field studies are inadequate for measuring the effects of dredging and spoil disposal on the marine biota, even when predredging studies are carried out. Those who believe this recommend that suitable laboratory techniques must be devised and used if we are to obtain critical information (6). Moreover, most data on the effects of pollutants have been measured on species of high trophic levels, leaving the effects on lower trophic levels largely untouched. Yet, as Gray and Ventilla (9) point out, "pollution in the sediment ecosystem could have far-reaching consequences at high trophic levels through alteration of the low trophic levels in food webs or even by elimination of organisms critical to the breakdown processes of carbon, nitrogen, and sulfur cycles."

My interest in marine pollution is focused on the long-term implications of synergistic interactions among low-levels of pollutants that elicit sublethal responses of organisms through impacts on basic life processes. If we are to understand and evaluate these phenomena within a decade, we must have appropriate bioassay organisms with reasonably short generation times.

Some meiofaunal species are considered to be promising subjects for this research, because they (a) have short life cycles, (b) produce three or more generations a year, and (c) are adaptable to laboratory culturing. Furthermore,

Use of Meiofauna as Environmental Monitors

Evidence is mounting that some of the consequences of disruptive environmental manipulations, of which dredging is one, may have more far-reaching effects in space and time than has been previously believed. One reason why possible subtle and long-term effects have not been recognized is certainly that attention has quite naturally been given to organisms that tend to have one or less generation, per year or less.

It is not now beyond feasibility to estimate the initial biological impacts of acute environmental degradations. But it is not logical to expect that we can measure in reasonable time the far-flung and long-term effects that dredging has on the benthos if we persist in using forms with long generation spans, or high mobility, or pelagic larvae. Some meiofaunal components appear to be excellent subjects for studies of the effects of low levels of pollutants on basic biological processes.

Numerous studies conclude that the submarine disposal of dredge spoil is only locally harmful and that benthic fauna recover rapidly from its effects, i.e., population densities are observed to return to normal one to two years after spoil has been dumped in the submarine environment (3). Rogers and Darnell (18), on the other hand, observed that meiofauna populations in shell-dredging cuts were not completely reconstituted a decade after dredging.

Recently, TerEco undertook a preliminary study of some dredging effects in coral-dominated bottoms in the U. S. Virgin Islands. Thus far we have observed that meiofauna populations in dredge cuts made six and ten years ago are today often monophyletic and of smaller numbers, as compared with controls 6-7 km upcurrent of the dredging sites. Actually, this observation is not beyond challenge because a certain amount of maintenance dredging has continued since channels were established. Even though the spoil has been deposited in holding reservoirs on shore a rather large amount of turbidity persists in the water adjacent to and downstream of the dredged channels. Our initial attempts to scan the distribution of a few heavy metals, pesticides, and PCBs in the sediments at the dredge sites and upcurrent and downcurrent are far from conslusive, but some trends are worthy of further investigation. The results in Table 1 indicate that the suspended material collected in jars supported 1 m above the bottom have substantially higher concentrations of those metals tested than the integrated sediment column. This clearly establishes a relationship between the fine sediments and concentration of metals.

TABLE 1.

Concentrations of metals and PCB in water-borne and Settled Sediment. St. Croix, August 1973. Pesticides below detectable levels. All water depths approximately 8 meters.

	ppm					Dry Wt. Sediment/
	Cd	Pb	As	Hg	PCB	m^2/day
Top 5 cm Sediments:						
Near dredging Site	0.1	1.3	0.6	—	—	—
6.6 km downstream	0.1	1.3	0.6	—	—	—
Top 1 cm Sediments:						
Near dredging Site	0.1	7	0.6	0.22	0.2	—
10.8 km downstream	0.05	3	0.4	0.17	0.07	—
Suspended Sediment:						
Particles:						
Near dredging Site	0.40	10	2.6	—	—	452
6.6 km downstream	0.25	4	0.1	—	—	186

In Table 2 we see some relationships between the top 5 cm of sediments and the meiofaunal characteristics at different dredge-related sites. The striking parallel between the amount of fine sediments and the number of meiofauna types and abundance of individuals is clearly worthy of further investigation. We see here also that fine sediments have apparently been carried considerable distances downcurrent. If they are the principal carriers of heavy metal ions, then dredging turbidity must have greater effects in space and time than previous studies indicate.

There is little doubt that the metal concentrations shown in Table 1 can have physiological effects upon meiofaunal organisms. Gray and Ventilla (9) found that the reproductive rate of a benthic ciliate (*Cristigera* sp.) was cut by a factor of 2 when exposed to 0.01 ppm of mercuric chloride and 0.1 ppm lead nitrate (Table 3). They found also (9) that the generation time of the archiannelid (*Dinophilus*) increased 50% (i.e. from 8 to 12 days) in 10 ppm and reproduction ceased in 20 ppm of sulfuric acid.

TABLE 2.

Some sediment characteristics and meiofaunal numbers. All water depths approximately 8 meters.

	Hand Cores: Top 5 cm Only			
Position (Depths, 5-8 m)	Percent Sand	Percent Silt And Clay	Meiofauna Types	Meiofauna No. $10cm^2$
Upcurrent of Dredge	95	1	12	2419
Dredge Area	50	50	2	46
Downcurrent of Dredge, 6.6 km	82	18	4	350
Downcurrent, 10.8 km	94	6	6	1260

TABLE 3.

Effect of heavy metals on the growth rate of the ciliate **Cristigera** spp. Time in hours for population to double in numbers (after Gray and Ventilla, 1971).

$HgCl_2$	$CuSO_4 \cdot 5H_2O$ 0 ppm		$CuSO_4 \cdot 5H_2O$ 2 ppm	
	$Pb(NO_3)_2$ 0 ppm	$Pb(NO_3)_2$ 0.1 ppm	$Pb(NO_3)_2$ 0 ppm	$Pb(NO_3)_2$ 0.1 ppm
0 ppm	2.5, 2.78	4.35, 4.55	2.78, 2.63	5.56, 4.35
0.01 ppm	5.0, 5.0	5.0, 5.26	8.33, 6.67	7.14, 9.09

Table 3 shows also some evidence for synergisms among low levels of chemical pollutants and the likelihood that such interactions are capable of reducing growth rates of meiofaunal organisms. It is doubtful that all bioassay techniques presently in use would be appropriate for detecting these effects.

Meiofauna Components as Bioassay Subjects

The above data suggest that some meiofaunal species will prove to be ideal for extended bioassay analysis. Accordingly, if subjects are selected with care, it seems likely that they will respond to synergistic effects of minimal concentrations of metal ions and synthetic organic compounds. By the same token, they might easily sort out harmless compounds and mark threshold levels for others.

Some of the other meiofaunal attributes that are advantageous for bioassay analysis are:

a. ease of collection, transportation, and establishment in the laboratory with small space requirement;

b. rather rapid growth;

c. favorable reproductive characteristics;

d. a total-culture growth rate that, in some cases, can be used as a criterion of toxicity rather than an estimate of the level of toxicity to individuals.

Meiofaunal Contribution to the Subtidal Ecosystem

The roles played by the meiofauna in the matter and energy cycles of the sea are not yet clearly defined. Their position on trophic levels ranges from nonselective detritus-feeders to carnivores. McIntyre et al. (15) suggest that the meiofauna through subsistence upon bacteria may be making use of the vast supply of dissolved organic matter. Marshall (16) and others believe that the importance of the food link from microbiota to meiofauna to macrofauna has been overstated. McIntyre and Murison (14) have suggested that the meiofauna as a whole occupies the top of a food chain and may actually compete with the macrofauna. Darnell (4), on the other hand, found that many estuarine macrofaunal species rely substantially on the interstitial fauna.

The principal role of the meiofauna may prove to occupy one step or another in nutrient recycling. Furthermore, although meiofauna biomass may amount to many of them exhibit a high level of sensitivity to sediment characteristics, and thus might be expected to reflect sediment ecosystem degradation and recovery.

NATURE OF THE MEIOFAUNA

Definition

McIntyre (11) has defined meiofaunal organisms as those metazoans that pass through a 0.5 mm sieve, but at the same time he points out that it is preferable to define the permanent meiofauna as a biological entity rather than as a statistical group (12). In so doing one takes cognizance of their ecological functions, short lifecycles, population size and seasonality, trophic attributes, and energy metabolism, which together set them apart within a continuum based on size alone. Intermediate in size between the microfauna and the macrofauna, meiofaunal species may range in length from 62 or so microns up to 1 or 2 mm. In marine ecosystems, where ciliate protozoans are generally quite large, it can be useful to include some species in the meiofaunal category. Nevertheless, this size reange includes that of sediment grain size extending from the silt-clay complex to coarse sand. This one-to-one relationship between whole animal and sediment-grain surfaces provides a unique responsiveness to chemical and physical changes attendant on many dredging operations. One is also inclined to speculate that some engineering properties of sediments may be modified by meiofaunal populations numbering a million or more per m^2.

Taxonomic Categories

A wide range of types of organisms has representatives that can be considered to be constituents of the meiofauna. In fact, some groups are found only in the meiofauna, but with few exceptions these groups seldom predominate either numerically or by weight. In general, sediment characteristics determine which groups occur in greatest numbers (1). The principal meiofaunal components are

a. Nematoda (Fig. 1A). Nematode worms are considered to be the most abundant metazoans in marine sediments. The degree of their predominance depends upon grain size, with clayey sediments favoring their development.

b. Copepoda. These crustaceans are represented to a large extent by the subgroup Harpacticoida (Fig. 1B) and are considered to be the second most abundant metazoan type of the meiofauna.

c. Turbellaria. Hulings and Gray (10) point out that, except for the polyclads, all of the marine turbellarian flatworms are meiofaunal.

d. Polychaeta. The segmented worms are frequently abundantly represented in the meiofauna. But all polychaetes are also represented in the macrofauna.

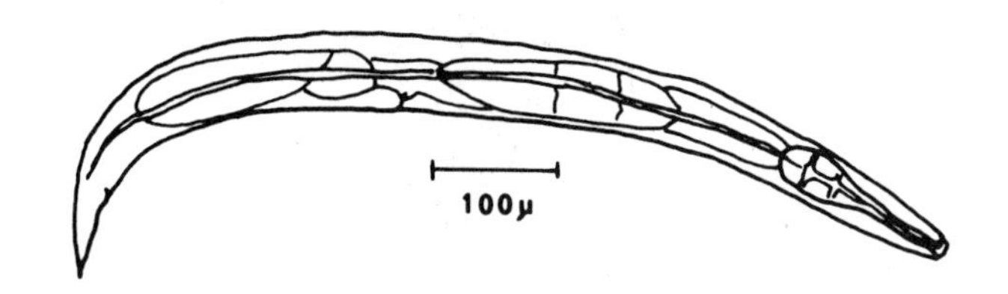

Figure 1A. Nematode

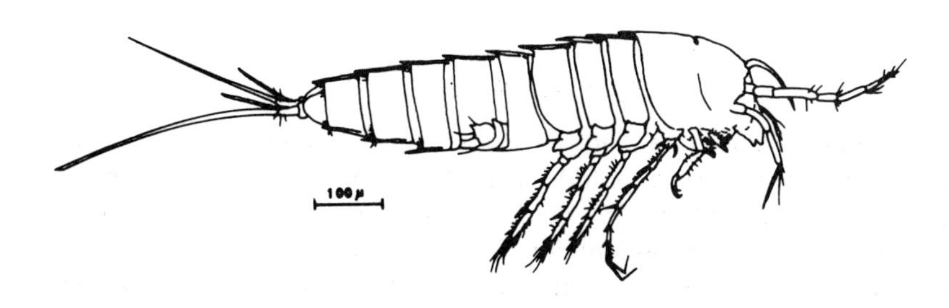

Figure 1B. Harpacticoida

In addition to the above four groups, there are a dozen or so other kinds of metazoan organisms that are less abundant but actually may be more characteristic of the meiofauna. Among these are the Kinorhyncha (Fig. 2A), all of which are of meiofaunal size and tend to be most abundant in muddy habitats; Tardigrada (Fig. 2B), the so-called water bears, the marine forms of which are exclusively meiofaunal and generally occur in sandy sediments; the interstitial Archiannelida, which are closely related to polychaetes; and finally the Gastrotricha are worthy of special mention as characteristic meiofaunal organisms.

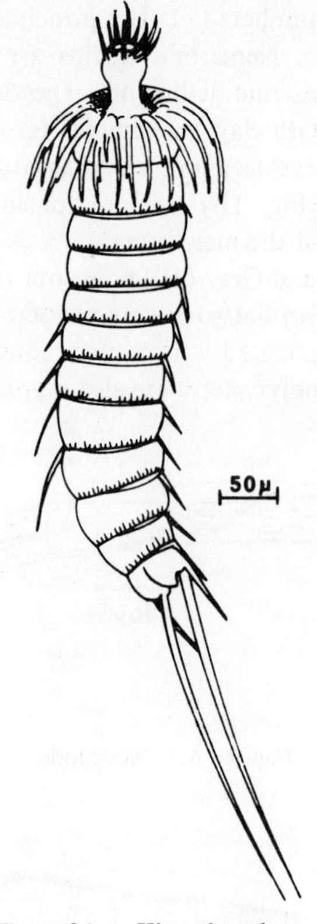

Figure 2A. Kinorthyncha

In addition to the above there are species among the Nemertea, Oligochaeta, Mollusca, Ostrocoda, Mystacocarida, and others that qualify for inclusion on the basis of size and habit. Among the Protozoa some benthic Foraminifera and Ciliata are eligible for inclusion with the meiofauna, but special care must be taken in sampling and extracting them. According to Fenchel (5), ciliate numbers and types are determined largely by sediment-grain characteristics. They tend to predominate in fine sands, in which they play an interstitial role,

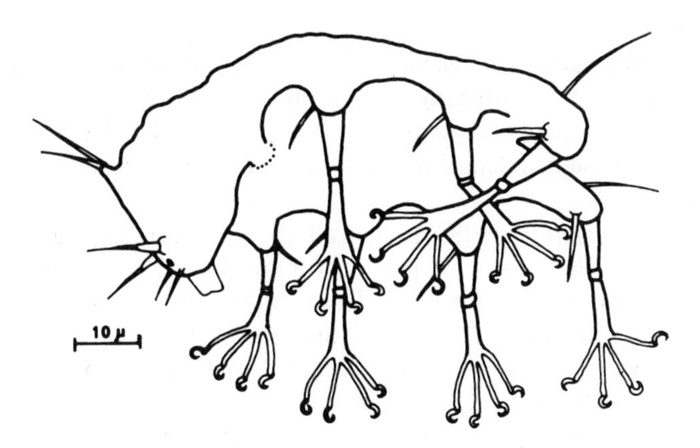

Figure 2B. Tardigrada

and reach lowest numbers in clays, where nematodes are abundant.

Sampling and Extraction Techniques

The meiofauna can be sampled with plastic core tubes, provided the tubes are equipped with some kind of external core retainer. To preclude turbulent disturbance of upper sediment levels with loss of meiofauna, coring is best carried out by divers. This is feasible down to depths of 50 m, which is ample for studies relating to dredging. The following additional factors must be considered if quantitative samples for comparative purposes are to be obtained:

a. The internal diameter of the cores should be adjusted to sediment type. Whereas a tube of 3.75 cm diameter (8) is satisfactory for sand, 10 cm is preferable for sampling in muds with a floc layer (11).

b. Core length must be tested in a given area to determine what percent of the meiofauna is collected with various lengths. Shorter cores are generally indicated in subtidal as compared with intertidal habitats. Craib (2) found that undisturbed cores 15 cm long were satisfactory for both sand and mud regions, in the subtidal, and our results are in agreement.

c. Meiofaunal distribution tends to be highly nonrandom; hence a sufficient number of cores must be taken in a relatively small area if one is to discern any between-area differences that might exist. In the subtidal I have obtained satisfactory results with quadruplicate samples in a 0.5 m^2 area.

d. Even in the subtidal, meiofaunal populations may exhibit some degree of seasonality. For instance, Rogers and Darnell (18) found that the nematode populations in bay muds increased 400 to 500 percent from spring lows to the populations highs of fall. Consequently some knowledge of peaks and troughs is necessary in order to reduce the numbers of samples required to obtain reliable population data.

Numerous techniques are available for extracting meiofauna from sediment samples. The manual prepared by Hulings and Gray (10) contains detailed information. It is generally advantageous to work with live samples. Fortunately, viable samples can be maintained in closed containers for several days, even in warm climates.

less than 5% of macrofauna biomass, Gerlach (7), estimates that in sandy sublittoral habitats the meiofauna may contribute about 15% as much organic matter to the food web as the macrofauna. McIntyre (11) and Gerlach (7) estimate that the annual production of biomass by the meiofauna may be approximately 10 times their standing stock, a factor that is about five times that of the benthic macrofauna.

This rapid cycling of organic matter by the meiofauna may well prove to have important implications regarding detoxification of waters and sediments charged with large burdens of metal ions.

Although the meiofauna are of no direct commercial value, it is concluded that because of their intimate dependence upon the integrity of the sediment ecosystem and because of their basic bioligical properties, they can respond to low-level environmental disturbances which if permitted to persist unabated would progressively degrade the marine environment, including the commercial macrofauna.

REFERENCES

1. Boaden, P. J. S.
   1962 Colonization of graded sand by an interstitial fauna. **Cahiers de Biol. Marine,** 3: 245-248.
2. Craib, J. S.
   1965 A sampler for taking short undisturbed marine cores. **Jour. Cons. perm. int. Explor. Mer.,** 30: 34-39.
3. Cronin, L. E.
   1970 Gross physical and biological effects of overboard disposal in Upper Chesapeake Bay. NR1 Special Rept. No. 3. U. S. Dept. of Interior, Washington, D. C.
4. Darnell, R. M.
   1961 Trophic spectrum of an estuarine community, based on studies of Lake Pontchartrain, Louisiana. **Ecology,** 42: 553-568.
5. Fenchel, T.
   1967 The ecology of marine microbenthos I. The quantitative importance of ciliates as compared with metazoans in various types of sediments. **Ophelia,** 4: 121-137.
6. Gannon, J. E. and Beeton, A. M.
   1971 Procedures for determining the effects of dredged sediments on biota-benthos viability and sediment selectivity tests. **Jour. Water Poll. Control Fed.,** 43: 392-398.

7. Gerlach, S. A.
1971 On the importance of marine meiofauna for benthos communities. **Oecologia,** 6: 176-190.

8. Gray, J. S.
1971 Sample size and sample frequency in relation to quantitative sampling of sand meiofauna. In **Proceedings of the First International Conference on Meiofauna,** p. 191-197 (ed. Hulings, N.C.). Smithsonian Institution Press, Washington, D. C.

9. Gray, J. S. and Ventilla, R. J.
1971 Pollution effects on micro- and meio-fauna of sand. **Mar. Pollution Bull.,** 2(3): 39-43.

10. Hulings, N. C. and Gray, J. S.
1971 A manual for the study of meiofauna. Smithsonian Contrib. to Zool., No. 78. Smithsonian Institution Press, Washington, D.C.

11. McIntyre, A. C.
1964 Meiobenthos of sub-littoral muds. **Jour. Mar. Biol. Ass. U.K.,** 44: 665-674.

12. McIntyre, A. C.
1969 Ecology of marine meiobenthos. **Biol. Rev.,** 44: 245-290.

13. McIntyre, A. D.
1971 Observations on the status of subtidal meiofauna research. In **Proceedings of the First International Conference on Meiofauna,** pp. 149-154 (ed. Hulings, N.C.), Smithsonian Institution Press, Washington, D.C.

14. McIntyre, A. D. and Murison, D. J.
1973 The meiofauna of a flatfish nursery ground. **Jour. Mar. Biol. Ass. U.K.,** 53: 93-118.

15. McIntyre, A. D., Munro, A. L. S., and Steele, J. H.
1970 Energy flow in a sand ecosystem. In **Marine food chains,** p. 19-31. (ed. Steele, J. H.) Univ. California Press, Berkeley and Los Angeles.

16. Marshall, N.
1970 Food transfer through the lower trophic levels of the benthic environment. In **Marine food chains,** p. 52-66. (ed. Steele, J. H.) Univ. California Press, Berkeley and Los Angeles.

17. Newman, B.
1973 The sea: pollution of the oceans is an enormous threat, but few people care. **The Wall Street Journal,** October 2.

18. Rogers, R. M. and Darnell, R. M.
1973 The effects of shell-dredging on the distribution of meiobenthic organisms in San Antonio Bay, Texas. Department of Oceanography, Texas A & M Univ. (Unpublished manuscript.)

Index

T

V

W